DATE DUE

DEC 0	

Compressor aerodynamics

Compressor
aerodynamics

N.A. Cumpsty

Department of Engineering
University of Cambridge

 LONGMAN

Addison Wesley Longman Limited,
Edinburgh Gate, Harlow,
Essex CM20 2JE, England
and Associated Companies throughout the world.

First published 1989
Reprinted 1996 and 1997

British Library Cataloguing in Publication Data
Cumpsty, N.A.
    Compressor aerodynamics
    1. Compressor. Aerodynamics
    I. Title
    621.5'1

ISBN 0-582-01364-X

Library of Congress Cataloging-in-Publication Data

Cumpsty, N.A.
    Compressor aerodynamics / N.A. Cumpsty
      p.   cm.
    Bibilography: p.
    Includes index.
    ISBN 0-470-21334-5 (Wiley):
    1. Compressors – Aerodynamics. I. Title.

TJ267.5.C5C86   1989
621.5'1—dc 19

Produced by Longman Singapore Publishers (Pte) Ltd.
Printed in Singapore

Contents

Preface

The idea for this book was given to me by Dr R C Dean early in 1973. His reason was that _Axial compressors_ by J H Horlock was out of date. That was 15 years after the publication of _Axial compressors_ and another 15 years have elapsed between then and now. I am sure that it is fortunate that I did not try to write a book at that time.

The title of this book may seem a little bald with so many books being entitled 'An introduction to' or 'Fundamentals of' but that is not a style which I like. I am however very aware that what I have written is only scratching the suface of a very big topic. It has also been a surprise to me how much I have had to leave out: in particular, discussion of ideas which are new (and may not be correct) but are currently interesting. Although I had set out to write a short book, I seem to have failed in this. In stressing that this is inevitably a brief account of a very big field, one occupying hundreds of people and vast expenditures of money, it must be realized that most of this work is being carried out by companies and the results are predominantly confidential. It is one of the main difficulties that much of what really matters is never released into the public domain, a trend that is more pronounced now than when _Axial compressors_ was being written, for example. One of the consolations for an author is that those people inside the large companies who have access to this knowledge are not likely to write a book about it.

Writing a book about compressors thirty or so years ago, it probably seemed reasonable to give a guide to design and this is evident in, for example, Eckert and Schnell's book _Axial- und Radialkompressoren_ as well as in Horlock's _Axial compressors_. Writing today this does not seem so appropriate. This is partly because of the greater concentration of knowledge and expertise in confidential form, referred to above, but also the greater use of numerical methods. The computer has come to dominate the design and assessment of compressors, so that it is not very likely that a major design would be carried out nowadays without the extensive use of numerical methods. Such methods are not necessarily described at all well in design manual form. In short this book does not set out to be a design guide but to provide ideas and clarification which will in turn lead to improvements in design. It is my hope that compressor aerodynamicists and designers will find it helpful in this way and that it will have a use in introducing newcomers to the topic. Some of the sections are relatively specialized, others are quite general and have application to all turbomachines and to wide areas of fluid dynamics.

On the whole I have not included much mathematics; someone wanting the details will be able to find them in one of the referenced works. Again the shift to numerical methods has somewhat diminished the attraction of the complete mathematical coverage of a topic and mathematical analysis has become relegated to certain specialist fields, mainly those such as stall or vibration where linear analysis is valid. As a rule equations are given when they are needed to follow the argument. In so far as I am able I have held my argument together by using results in graphical form and as a result there are a lot of figures in this book.

I would be giving a wrong impression if I suggested that to be modern is to relegate what was done in the past to obscurity. There is, I believe, a lot to be learned from the work which was done in the 1940s and 50s and I have included some of this. Naturally the areas where this is most apparent are those in which the early workers were most active, so in the chapters on blade-to-blade flows in axial machines and on centrifugal impellers there is quite extensive reference to much of the excellent early work.

This does not, however, set out to be an historical account attempting to put the record straight about who did what. Where it is appropriate I have referred to the originators of ideas, but this is not the primary purpose of the book or of the reference list. I have, of course, included any reference from which I have taken a diagram as well as those which I think will be useful, either because they are themselves interesting or because they in turn provide a very complete bibliography. Inevitably many first-class pieces of work are not referred to, though the person wishing to dig deeper will find them from the reference lists in other papers. In general I have tried to give references which are easily obtained and this necessarily distorts history to some extent. Thus, for example, I have referenced Day and Cumpsty (1978), which is easily available in libraries around the world, whereas the work is first described by Day in a doctoral dissertation held in the library of Cambridge University. To the majority of readers this distortion will probably seem a price worth paying for convenience.

As the author I feel more comfortable with some topics than others. In writing about blade vibration I was very conscious of having a fairly superficial grasp of the topic. I am very grateful to Dr D S Whitehead for his help with this and on the strength of this feel confident to see what I have written published. Similarly the computation of flows in turbomachines is a specialized field with which I felt unfamiliar. In Chapter 11 there is a relatively short section on numerical methods which is based on a lecture by Dr J D Denton who was kind enough to review what I wrote; again with this support I am prepared to see what I have written published, believing it to be free from the worst errors and a helpful introduction for the non-specialist.

A sense of realism tells me that there will be mistakes which slip through into the published text. In the hope that there will ultimately be a second edition I should be very pleased to have readers send me corrections, most particularly corrections of fact. If they have illustrations which they think would help a second edition to be clearer, perhaps even replacing existing examples, I should be very pleased to receive them.

Acknowledgements

I feel that in writing this book in the Whittle Laboratory I have been very fortunate; the combination there of resources, ability and a willingness to be helpful is all that one could ask for.

My contacts in the industry have served me very well and I can only list those who helped me most substantially. At the outset the encouragement of Dr L H Smith of General Electric was crucial and since then conversations with him and with Mr C C Koch and Dr D C Wisler of the same company have helped me a very great deal. Mr D P Kenny of Pratt and Whitney of Canada likewise helped by his encouragement and by discussion. My long-standing friend, Mr C Freeman of Rolls-Royce, has taught me a very great deal over the years and in particular whilst I have been working on this book. Dr M V Casey of Sulzer Escher Wyss has encouraged and helped me and given guidance in areas where I have had less experience. I also appreciate the help of Mr D Japikse of Concepts Inc. and Dr H G Weber of the Cummins Engine Company.

In academia no one could have done more to help me than Professor E M Greitzer of MIT, both by encouragement and by argument. Dr T P Hynes, with an office next to mine in the Whittle Laboratory, has been an invaluable colleague to test my ideas on and I owe him a great deal, both for his exceptional forbearance as well as his great abilities. (He also took a very active role in the section on matching of multistage compressors described in Chapter 2.) My debt to Drs Denton and Whitehead I have referred to in the preface, but to this must be added the use of computer programs they have written. Dr Dawes has helped me with calculations and advice. Others in the Whittle Laboratory have helped me substantially, in particular my students and former students who had the misfortune to be here when the book was being prepared; Drs S G Gallimore, Y Dong, and N McDougall and Mr Y S Li have helped both by their influence on my ideas and by carrying out calculations for me.

There is a special note of thanks I wish to record to those who have helped me with the book in its later stages. Foremost of these is Professor Greitzer who has done his best with savage injunctions to improve the book and has read most of the chapters for me. Dr Casey read three chapters and made invaluable contributions. Mr Freeman and Dr Gallimore took great pains over Chapter 3 when I was having a lot of trouble with it. Mr M Howard checked Chapters 4 and 5 for me. Dr McDougall has helped me in many ways, but I particularly appreciate his assistance with the later stages when he himself was under pressure. The contribution of the following in reading and checking the proofs is gratefully acknowledged: Dr D Andrews, Mr L He, Dr D Lambie, Mr Y S Li, Dr J Longley, Mr J A Storer and Dr M Zangeneh-Kazemi.

We are indebted to the following for permission to reproduced copyright material:

Academic Press Inc. Ltd. for figs. 10.21 from plate 3 (Cumpsty 1972) and 10.29 from fig. 5 (Mariano 1971); Advisory Group for Aerospace Research and Development, North Atlantic Treaty Organization (AGARD/NATO) and the respective authors for figs. 3.8 from p. 268 (Hirsch and Denton 1981), 4.47 from fig. 7.7 (Scholz 1977), 5.8 from figs. 6 & 7 (Wood *et al.* 1986), 8.5 from fig. 10 (Seyb 1972), 9.6 from fig. 11 (Riess & Blöcker 1987), 9.38 from figs. 9 & 10 (Mikolajczak & Pfeffer 1974), 9.45 & 9.46 from figs. 11 & 13 (Jansen *et al.* 1980) and 10.1 from fig. 3 (Sisto 1987a); American Institute of Aeronautics and Astronautics for figs. 3.14 from figs. 2 & 5 (Smith 1974), 4.49–4.51 from figs. 1, 6 & 8 (Schmidt *et al.* 1984), 5.2 from fig. 12 (Schreiber & Starken 1981), 8.6 from fig. 1 (Walker 1987), 10.4, 10.12 & 10.13 from figs. 7, 11 & 12 (Mikolajczak *et al.* 1975), 10.24 from fig. 2 (Lowrie 1975) and 10.28 from fig. 8 (Ginder & Newby 1977); American Society of Mechanical Engineers for material from *ASME Transactions*; Elsevier Science Publishers B.V., Physical Sciences & Engineering Division for figs. 3.15 & 9.10 from figs. 2 & 10 (Smith 1969) and 4.39–4.41 from figs. 10, 20 & 21 (Schlichting & Das 1969); Gas Turbine Society of Japan for fig. 2.4 from fig. 1 (Freeman & Dawson 1983); General Electric Company for figs. 2.1–2.3 (Wisler 1988) Copyright © 1988 by General Electric Co., U.S.A. All Rights Reserved; the Controller of Her Majesty's Stationery Office for figs. 4.5 from figs. 4, 6 & 14 (Andrews 1949), 4.34 from fig. 6 (Carter 1950), 4.37 & 4.38 from figs. 13, 25 & 36 (Rhoden 1952), 8.22 from fig. 13 (Lakshminarayana & Horlock 1967) and 10.9–10.11 from figs. 2, 10 & 12 (Halliwell 1975); Institution of Mechanical Engineers for figs. 2.10 & 4.10 from figs. 68 & 81 (Howell 1945), 6.11 from fig. 3 (Casey & Roth 1984), 8.8 from fig. 6 (Abu-Ghannam & Shaw 1980) and 9.32–9.36 from figs. 3, 8, 13, 14 & 20 (Day & Cumpsty 1978); Institute of Refrigeration for figs. 2.23, 2.24 & 6.18 from figs. 4, 5 & 7 (Casey & Marty 1986); the author, Prof W.C. Reynolds for table 1.1 and fig. 1.17 from pp. ix and 161 (Reynolds 1979); Rolls-Royce plc for figs. 4.25 (Carter 1961) and 4.44; Royal Aeronautical Society for fig. 10.2 from fig. 1 (Armstrong & Stevenson 1960); Society of Automotive Engineers Inc. for figs. 2.22, 2.13, 2.14, 6.20, 7.19 & 7.20 from figs. 16, 20, 25, 30, 32 & 34 (Kenny 1984) and 10.16 & 10.17 from figs. 13c & 15 (Tyler & Sofrin 1962) © 1984 & 1962 Society of Automotive Engineers Inc.; Von Karmen Institute, Belgium for figs. 7.17, 7.29 & 7.30 from figs. 18–39 (Stiefel 1972) and 8.16, 8.18, 8.21, 8.25 & 8.26 from figs. 1 and 33–44 (Freeman 1985).

Notation

There are very many notations in use for the consideration of turbomachines and it is just about impossible to evolve a system which has no duplication of symbols without recourse to excessive use of subscripts. It is hoped that the system adopted will represent a reasonable compromise which is fairly transparent and agrees with that generally used for the topic being discussed. The overlap that does exist here (for example m denotes mass flow rate and meridional distance) should not confuse the reader too much. There are other inconsistencies (as Emerson wrote, a foolish consistency is the hobgoblin of little minds) but this should not be too irritating. The list given is not an exhaustive one and various additional symbols are introduced throughout the book.

General points

Throughout the book all angles are measured from the meridional flow direction, which reverts to the axial direction for the blade-to-blade flow in axial machines and the radial direction towards the outlet of centrifugal compressors. (The terms radial or centrifugal compressor are used as alternatives without any implied difference; both are in common use).

The velocity magnitude and direction in the relative or rotating frame of reference are denoted by W and β whilst in the absolute or stationary frame of reference they are denoted by V and α.

As is common in British and American practice for compressors, a convention of positive and negative signs for flow or blade angles is *not* used; angles are taken as positive and the appropriate sense adopted.

The word stagnation is normally used, as for stagnation enthalpy $h_0 = h + V^2/2$, and not the word total. The usage total-to-static, as in total-to-static efficiency, is so widespread and the corresponding term based on stagnation so much harder to say that this is retained.

The outer diameter of axial machines is sometimes called the tip. This may be ambiguous, for the tip of the stators is at the hub. The word casing is therefore preferred for the outer diameter. The phrase hub—tip ratio is so common that this is occasionally used in place of hub—casing ratio for the ratio of the hub diameter divided by the casing diameter.

Variables commonly used

Geometric variables

b	passage width in spanwise direction, used for centrifugal compressors
c	blade chord
d, D	diameter
g	staggered gap, pitch resolved normal to the flow direction
h	blade height, used mainly for axial compressors
m	distance in meridional direction $dm = \sqrt{(dx^2 + dr^2)}$, $dx/V_x = dr/V_R$
r, R	distance in the radial direction
s	blade pitch
s	distance along streamline $ds = \sqrt{(dx^2 + dr^2 + r^2 d\theta^2)}$, $dx/V_x = dr/V_R = r d\theta/V_\theta$
t	blade thickness
t	tip clearance
x	distance in axial direction
y	distance in the pitchwise direction
z	distance normal to x and y
σ	solidity c/s

Angles Relating to Blading (see Fig. 4.1)

ϵ	angle between a blade filament and the radial direction in axial view (blade lean)
ξ	stagger (angle of chord line measured from the axial† direction)
θ	camber
θ	angle in the circumferential direction
χ_1	blade inlet angle (measured from the axial† direction)
χ_2	blade outlet angle (measured from the axial† direction)
λ	blade lean in radial compressors

Flow variables
Stationary frame of reference

α_1	flow inlet angle (measured from the axial† direction)
α_2	flow outlet angle (measured from the axial† direction)
V_1	inlet flow velocity
V_2	outlet flow velocity

Rotating frame of reference

β_1	flow inlet angle (measured from the axial† direction)
β_2	flow outlet angle (measured from the axial† direction)

---

† For radial and mixed flow machines, angles are measured from the meridional direction rather than the axial direction. For axial machines when the meridional streamlines are inclined at a substantial angle to the axial direction, the angles are also sometimes referred to the meridional.

W_1	inlet flow velocity
W_2	outlet flow velocity

Subscripted velocities

$V_{\theta 1}$	tangential component of velocity into blade row
V_{R1}	radial component of velocity into blade row
V_{x1}	axial component of velocity into blade row
	... likewise for other velocities, V_2, W etc
V_m	meridional component velocity, $V_m = \surd(V_x^2 + V_R^2)$

Special angles

i	incidence (angle between inlet flow direction and blade inlet direction, $i = \alpha_1 - \chi_1$ or $i = \beta_1 - \chi_1$ for stator or rotor respectively)
A	angle of attack (angle between inlet flow direction and the chord line, $A = \alpha_1 - \xi$ or $A = \beta_1 - \xi$)
δ	deviation (angle between outlet flow angle and blade outlet angle, $\delta = \alpha_2 - \chi_2$ or $\delta = \beta_2 - \chi_2$)
ϕ	inclination of meridional streamline to axial direction
γ	inclination of meridional streamline to axial direction (used for radial machines)

General variables

A	streamtube cross-sectional area
a	velocity of sound
$a*$	velocity of sound at condition when flow sonic (similarly $p*, \rho*$ etc.)
$AVDR$	axial velocity—density ratio $\rho_2 V_{x2}/\rho_1 V_{x1}$
b	streamtube depth measured normal to two-dimensional surface
B	blockage, 1-(mass flow ÷ mass flow across same section in ideal flow)
C	velocity of sound (Chapter 10)
C_D	dissipation coefficient or integral
c_f	skin friction coefficient, $\tau_w/(\frac{1}{2}\rho U^2)$
c_p	specific heat capacity at constant pressure
c_p	static pressure rise coefficient, $(p-p_1)/(p_{01}-p_1)$
DF	Lieblein's diffusion factor
F	flow function, $m(c_p T_0)^{1/2}/Ap_0$
h	specific enthalpy
h_0	specific stagnation enthalpy, $h + V^2/2$
I	specific rothalpy, $h + W^2/2 - U^2/2$
k	acoustic wavenumber
m	mass flow rate
M	Mach number

N	number of blades
N	angular velocity rev/min
p	static pressure
p_0	stagnation pressure, sometimes termed total pressure
Q	volume flow rate
R	gas constant
R	degree of reaction
s	specific entropy
T	static temperature
T_0	stagnation temperature, sometimes termed total temperature
U	blade speed
δ	boundary layer thickness
$\delta*$	boundary layer displacement thickness
δ_F	force deficit thickness
η	efficiency
θ	boundary layer momentum thickness
γ	ratio of specific heat capacities c_p/c_v
λ	swirl parameter V_θ/V_R, used for radial compressors
λ	acoustic wavelength
μ	dynamic viscosity
ν	kinematic viscosity, $\nu = \mu/\rho$
ρ	density
σ	slip factor, (absolute whirl velocity ÷ ideal absolute whirl velocity)
τ	shear stress
ϕ	flow coefficient; V_x/U for axial, different definitions for radial compressors
ϕ	velocity potential
ψ	stream function
ψ	loading, $\Delta h_0/U^2$
ω	loss coefficient, $\Delta p_0/(p_{01}-p_1)$
ω	angular velocity
Ω	vorticity

1 *Useful basic ideas*

1.1 Introduction

It is assumed in this book that the reader is familiar with the concepts and methods of fluid mechanics and engineering thermodynamics and no attempt will be made here to survey these. This chapter will look at topics which are of particular relevance to this book, either because they will be used or because they are basic ideas which are useful and have not been very satisfactorily treated elsewhere. Some of the sections are very basic and may serve some readers merely to establish the notation, others are much more challenging and presuppose a fairly thorough understanding of fluid mechanics and turbomachinery.

1.2 Blades and flow

The essential elements of turbomachines are the blades or vanes because it is these which impart the force and, more relevantly, the moment to the flow. Sometimes it is more convenient to think of the blades turning the flow and this is particularly the case with axial machines where the flow is very often at nearly constant radius. On other occasions it is the blade force per unit area (i.e. the pressure difference) which is more helpful.

Blades in axial compressors have some features in common with aircraft wings but the situation is more complicated. Thus attempts to take over successful concepts from aeronautics are by and large not very useful. The flow varies very strongly along the span, both because of endwall boundary layer effects and because the radius and the blade speed change markedly. Also a compressor blade is just one element surrounded by many other blades and blade rows; the whole flow is the result of all the blade rows and any one blade can itself have little effect. The design of blades or blade rows in multistage applications should be thought of as choosing a configuration which is compatible with a desired flow in which it is immersed. Fortunately the blades are, like wings, able to tolerate a range of inlet flow angles (i.e. a *range of incidence*). In addition to a first approximation the outlet flow angle remains constant because of the constraint of the bladed assembly.

Although the outlet flow angle is nearly constant it is not the same as the outlet flow direction of the blades themselves. For an axial turbomachine the difference between the outlet flow angle α_2 and the blade outlet angle χ_2 is referred to as the deviation defined by $\delta = \alpha_2 - \chi_2$. In the case of the radial impeller it is normal to define a slip factor σ by the ratio of the measured average absolute tangential velocity out of the impeller to the absolute tangential velocity if the outlet flow were uniform and in the direction of the blades at outlet. Both the deviation and slip factor are predominantly inviscid effects to which the boundary layer fluid makes only a small additional contribution, a point taken further in Chapters 4 and 6. In the idealized case with very thin boundary layers the streamline leaving the trailing edge would do so in the blade direction, a property associated with the Kutta–Joukowsky condition that the pressure difference should go to zero there. Out across the passage the flow is inclined to this direction, the sense of the inclination being that which reduces the force on the blades. Methods for estimating deviation and slip will be discussed in later chapters, but it is worthwhile in this introduction giving broad estimates: for a radial impeller with a typical number of vanes the slip factor is approximately equal to 0.9; for axial blades the deviation is given approximately by $\delta = 0.3\theta\sqrt{(s/c)}$ where θ is the camber, s is the blade pitch and c is the chord. The expression for deviation allows some simple generalization for axial blading. The outlet flow direction is given by

$$\alpha_2 = \xi - \theta/2 + \delta$$

where ξ is the blade stagger, the inclination to the axial direction of the chord, the line joining the leading and trailing edges. (This is valid only for circular arc camber lines but this is a very common choice.) Introducing the estimate for the deviation for a solidity $c/s \approx 1.0$ gives

$$\alpha_2 \approx \xi - \theta/2 + 0.3\theta$$
so that $\qquad \alpha_2 \approx \xi - 0.2\theta$

In other words the outlet flow direction depends to only a fairly small extent on the camber whereas it is the stagger angle ξ which really has a big effect. At low inlet Mach numbers most axial blades are able to tolerate quite a large incidence range, so again the camber is of secondary importance. At high inlet Mach numbers, even high subsonic ones, the blade performance is strongly affected by incidence and the overriding dependence on stagger to the relative exclusion of camber is no longer so true. It is nevertheless the principle behind the use of variable stagger stators in high-speed compressors. Stagger remains a very important variable at high Mach numbers because it has such a large effect on the blade passage areas and therefore on the mass flow capacity.

The flow in most turbomachinery blade passages is extraordinarily complicated and a major reason for this is the unsteadiness of the flow. It is essential that there should be stationary and rotating components and movement

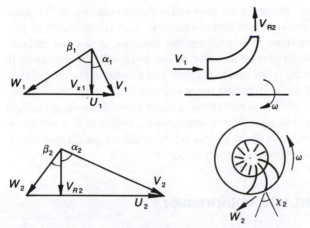

Fig. 1.1 The velocity triangles for flow entering and leaving a moving component, in this example an impeller with backsweep χ_2

of one past the other inevitably creates unsteadiness. For most purposes it is possible to ignore the unsteadiness by working in a frame of reference fixed to the component under consideration: for stator blades a coordinate system is used which is stationary (sometimes called the absolute frame) and for rotor blades or centrifugal impellers the frame of reference moves at the local blade speed (this is often referred to as the relative frame).

As a convenience in changing the frame of reference it is usual to describe the flow with vector triangles and Fig. 1.1 illustrates this for a backswept centrifugal impeller in which the flow at inlet has components in the axial and tangential directions only, while at outlet it has components only in the radial and tangential directions. Throughout this book the velocities in the stationary frame will be denoted by V and in the relative or moving frame by W. The corresponding flow angles will be denoted at inlet by α and β for the absolute and relative flows respectively.

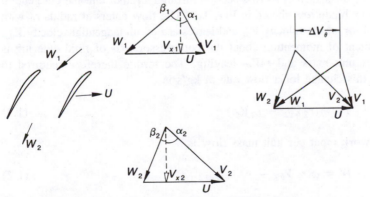

Fig. 1.2 The velocity triangles into and out of an axial rotor row

With axial machines, whenever the inlet and outlet radii are equal, it is quite common to overlay the inlet and outlet triangles. This is illustrated in Fig. 1.2. One of the attractions of overlaying the triangles is that the distance separating the two peaks gives a measure of the work done by the rotor. In general the axial velocity is not equal upstream and downstream of the blades, but simple analyses of performance are made so much easier if the axial velocity is constant that this is often assumed since in many cases this assumption still allows correct trends to be identified. Furthermore it is generally true that the variation in the axial velocity about a mean is kept small for good aerodynamic reasons for most blade rows in a compressor.

1.3 Work input into compressors

The pressure and shear stresses in the blades or impeller of a compressor produce a moment about the axis and these could in principle be evaluated and integrated. In reality this is an impossible task and instead it is usual to consider the moment of momentum for the flow entering and leaving. The advantage of this is that the precise nature of the flow processes can be ignored and the results are valid even for flow with large irreversibilities.

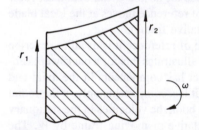

Fig. 1.3 An idealized rotor. Flow enters at radius r_1 and leaves at r_2

Consider the sketch of a cross-section through a rather unusual but general mixed-flow blade row shown in Fig. 1.3. The flow enters at radius r_1 with tangential (or whirl) velocity $V_{\theta 1}$ and leaves at r_2 with tangential velocity $V_{\theta 2}$. The moment of momentum about the compressor axis of fluid entering is $r_1 V_{\theta 1}$ per unit mass and $r_2 V_{\theta 2}$ leaving. The torque therefore required to produce this change for a flow rate m kg/s is

$$T = m(r_2 V_{\theta 2} - r_1 V_{\theta 1}) \tag{1.1}$$

and the work input per unit mass flow is

$$W = \omega(r_2 V_{\theta 2} - r_1 V_{\theta 1}). \tag{1.2}$$

This, or its derivatives below, is known as the Euler equation for turbomachinery. Sometimes it is written in terms of the local blade speed

$$W = U_2 V_{\theta 2} - U_1 V_{\theta 1} \tag{1.3}$$

and for the special case of axial machines where the flow enters and leaves at the same radius

$$W = U(V_{\theta 2} - V_{\theta 1}) = U\Delta V_\theta. \tag{1.4}$$

It is to be remembered that the Euler equation is valid no matter how the moment of momentum is produced, and even viscous drag on the blades can produce a positive work input. Because the tangential stresses at the casing and the hub are usually very small the total work of the blade row can usually be inferred from measurements of the velocity components upstream and downstream. In special cases, for example when there is casing treatment to delay stall, there may be large effective stresses at the walls and an expression such as equation 1.4 may give a misleading estimate for the blade work. The equation for the torque

$$T = m(r_2 V_{\theta 2} - r_1 V_{\theta 1})$$

can be applied to a small region of the machine, not necessarily across an entire blade row. The torque may then be replaced by the local blade force and radius

$$rF_\theta = m\delta(rV_\theta) \tag{1.5}$$

Writing $m = \rho VA$, where A is the streamtube cross-sectional area, and also writing $\delta(rV_\theta) = \partial(rV_\theta)/\partial s \; \delta s$, with s along the streamline direction, gives

$$rF_\theta = \rho VA\partial(rV_\theta)/\partial s \; \delta s$$

or
$$rf_\theta = V\partial(rV_\theta)/\partial s$$

$$= D(rV_\theta)/Dt \tag{1.6}$$

in steady flow, where f_θ is the tangential force per unit mass of fluid. The use of local forces such as f, often referred to as body forces, is very convenient in applying simplified mathematical formulations through the blade rows or through impellers.

In axial compressors it is often possible to consider flows in rotor passages by merely adopting a moving frame of reference. For radial or mixed flow machines it is necessary to be a little more careful because of the change in local blade speed with radius. It is convenient to begin with the familar energy equation

$$Q + W = h_{02} - h_{01}$$

where h_{02} and h_{01} are the stagnation enthalpies into and out of the region. Note that in this equation the work W is positive because in the Euler equation work input to the fluid was taken as positive. For adiabatic machines $Q=0$. Introducing the Euler equation and expanding the stagnation enthalpy gives after rearrangement

$$h_2 + V_2^2/2 - U_2 V_{\theta 2} = h_1 + V_1^2/2 - U_1 V_{\theta 1}. \tag{1.7}$$

Consider just the left-hand side, expanding V_2^2 as $V_{\theta 2}^2 + V_{x2}^2 + V_{R2}^2$ and then expressing the absolute tangential velocity in terms of that in the moving frame of reference $V_{\theta 2} = W_{\theta 2} + U_2$. After some manipulation to the left-hand side of the equation one obtains

$$h_2 + V_2^2/2 - U_2 V_{\theta 2} = h_2 + W_2^2/2 - U_2^2/2$$

Thus equation 1.7 can be written

$$h_2 + W_2^2/2 - U_2^2/2 = h_1 + W_1^2/2 - U_1^2/2 \tag{1.8}$$

or alternatively

$$(h_{02})_{\text{rel}} - U_2^2/2 = (h_{01})_{\text{rel}} - U_1^2/2 \tag{1.9}$$

where $(h_{02})_{\text{rel}}$ is the stagnation enthalpy in the moving or relative frame of reference. Following the suggestion of C H Wu it is common to define the quantity known as the *rothalpy* by

$$I = h + W^2/2 - U^2/2$$

so that equation 1.9 becomes simply

$$I_1 = I_2$$

In rotating blade rows rothalpy has properties analogous to stagnation enthalpy in stationary passages. In a moving passage the rothalpy is therefore constant provided:

(a) the flow is steady in the rotating frame;
(b) no work is done on the flow in the rotating frame (for example by friction from the casing);
(c) there is no heat flow to or from the flow.

Very clearly when the concept of rothalpy is applied to a stationary blade row the equation reverts to conservation of stagnation enthalpy.

   The objective of a compressor is usually to raise the *static* pressure and this means that there must be a rise in *static* enthalpy. Rearranging the equation for the conservation of rothalpy gives

$$h_2 - h_1 = \tfrac{1}{2}(U_2^2 - U_1^2) + \tfrac{1}{2}(W_1^2 - W_2^2). \qquad (1.10)$$

If W_1 and W_2 are relative to the blade row under scrutiny the equality is valid whether the blade row is moving or stationary. If $U_1 = U_2$, as it might in an axial machine, then it follows that

$$h_2 - h_1 = \tfrac{1}{2}(W_1^2 - W_2^2) \qquad (1.11)$$

as is usual for any stationary passage or diffuser. Note that the corresponding expression for the static pressure rise

$$p_2 - p_1 = \tfrac{1}{2}\rho(W_1^2 - W_2^2) \qquad (1.12)$$

is only valid for *loss-free* incompressible flow. If there are losses in incompressible flow the static pressure rise can be written

$$p_2 - p_1 = \tfrac{1}{2}\rho(W_1^2 - W_2^2) - \Delta p_{\text{loss}}. \qquad (1.13)$$

For compressible flow more care is necessary and it is convenient to begin with the thermodynamic relation $dh = dp/\rho + T ds$ so that on integrating

$$h_2 - h_1 = \int_1^2 \frac{dp}{\rho} + \int_1^2 T \, ds \qquad (1.14)$$

and the losses for adiabatic flow are contained in the entropy rise. The process can be shown graphically; Fig. 1.4 is for an axial blade row in which $(h_0)_{\text{rel}}$ is constant. With no losses the compression is isentropic and the outlet static pressure is p_{2s}; with losses the static pressure at outlet will be lower at p_2. The losses, by which is meant the rise in entropy, are produced by processes associated with the flow, mainly shear work (sometimes called viscous dissipation) and mixing of the flow.

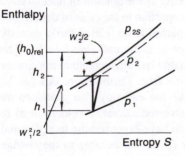

Fig. 1.4 The enthalpy diagram for an axial rotor with equal blade speed at inlet and outlet

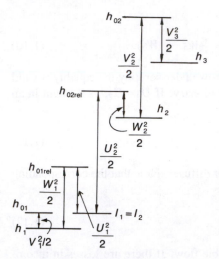

Fig. 1.5 The steps in the enthalpy change through a compressor, in this case with a large radius change in the moving element as, for example, in a centrifual compressor. $(h_{01})_{\text{rel}} = h_1 + W_1^2/2$ and rothalpy $I_1 = h_1 + W_1^2/2 - U_1^2/2$. Station 3 is after a stator or diffuser.

Consider now the case there is constant rothalpy but a change in the blade speed such that $U_2 > U_1$. The corresponding enthalpy–entropy diagram is shown in Fig. 1.5. In this case there is a marked change in the relative stagnation enthalpy such that

$$(h_{02})_{\text{rel}} > (h_{01})_{\text{rel}}.$$

With loss-free compression the static pressure would rise to p_{2s} but with loss this is reduced to p_2. As before the losses, represented here by the entropy Δs, arise only from the flow processes.

Some of the static enthalpy and pressure rise comes only from the term $\frac{1}{2}(U_2^2 - U_1^2)$ and since this is unconnected with the flow processes it does not have loss-making processes associated with it and gives rise to no losses and no entropy increase. The losses tend to increase as the amount of deceleration of the relative flow is increased and also in proportion to the cube of the relative velocity; in other words the loss may be expected to be related fairly directly to $\frac{1}{2}(W_1^2 - W_2^2)$ and not to the overall change in static enthalpy $h_2 - h_1$. The significance of this is that the part of the enthalpy rise attributable to the change in blade speed, $\frac{1}{2}(U_2^2 - U_1^2)$, is essentially loss free. Furthermore this can be increased without aerodynamic limit, unlike the enthalpy rise produced by decelerating the relative flow where excessive reductions in velocity lead to flow separation. It is these two factors which have favoured the use of radial compressors: if most of the static enthalpy rise is attributable to the change in blade speed between inlet and outlet the expected pressure rise will be obtained and the efficiency will be reasonably high even if the aerodynamics behaviour is poor with large regions of separated flow. In fact the principal limit on the maximum pressure rise from radial compressors is the strength of the material from which the impeller is made.

Radius changes may have important effects on axial machines as well. For a typical axial compressor stage the static enthalpy Δh might be approximately $0.4\ U^2$, where U is the local blade speed. Suppose that between inlet and outlet to the rotor the distance of the streamline from the compressor axis increases by 10 per cent. Then it follows that the quantity $\frac{1}{2}(U_2^2 - U_1^2)$ increases by about 10 per cent of U_1^2 too. In other words a small change in radius can produce 'free' changes in static enthalpy of the same order of magnitude as those produced by the deflection and deceleration of the flow in the blades. The changes are 'free' because they are without losses and do not contribute to the tendency of the boundary layer fluid to separate. This often has led to a pronounced effect at the hub of axial compressors; Figure 1.6a shows a compressor with a rising hub line where advantage is being taken of the change in radius, whilst Fig. 1.6b shows a constant hub radius and a falling casing where advantage is not being taken of this effect and difficulties of overloading near the hub are more likely to be encountered. In examining the flow in radial machines, or axial machines in which the distance of streamlines from the compressor axis changes significantly, it is proper to take account of the conservation of rothalpy by considering quantities such as $T - U^2/2c_p$, sometimes called the reduced temperature (and the pressure related to it) rather than the normal static temperature and pressure.

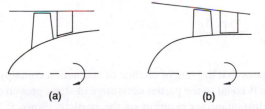

(a)   (b)

Fig. 1.6 Meridional sections through two hypothetical compressor stages: (a) has a rising hub, (b) a falling or contracting casing

There is a different way of looking at work input which is usually associated with Dean (1959). The one-dimensional equation of motion along an instantaneous streamtube can be written

$$\frac{\partial V}{\partial t} + V\frac{\partial V}{\partial s} = -\frac{1}{\rho}\frac{\partial p}{\partial s} \qquad (1.15)$$

and the stagnation enthalpy

$$\frac{\partial h_0}{\partial s} = \frac{\partial h}{\partial s} + \frac{1}{2}\frac{\partial V^2}{\partial s} \qquad (1.16)$$

Now introducing the thermodynamic relation $T\,ds = dh - dp/\rho$ and restricting consideration to reversible adiabatic processes, $T\,ds = 0$, gives

$$\frac{\partial h_0}{\partial s} = \frac{1}{\rho}\frac{\partial p}{\partial s} + \frac{1}{2}\frac{\partial V^2}{\partial s} \tag{1.17}$$

Combining equations 1.15 and 1.17 shows immediately that for reversible adiabatic flow

$$\frac{\partial h_0}{\partial s} = -\frac{\partial V}{\partial t} \tag{1.18}$$

Thus, with the restriction to isentropic flow, the rate of change of stagnation enthalpy in the flow direction is equal to minus the rate of change of velocity with respect to time at that point. If the flow were steady, $\partial V/\partial t = 0$, there could be no work input. Equation 1.18 may be modified by multiplying by V and adding $\partial h_0/\partial t$ to each side so that

$$\frac{Dh_0}{Dt} = \frac{\partial h_0}{\partial t} + V\left\{ -\frac{\partial V}{\partial t} \right\} \tag{1.19}$$

The term $\partial h_0/\partial t$ may be replaced by a process analogous to equation 1.16 to give

$$\frac{Dh_0}{Dt} = \frac{1}{\rho}\frac{\partial p}{\partial t} \tag{1.20}$$

In words this states that the rate of change of stagnation enthalpy of a particle along a pathline is equal to the partial derivative of static pressure with respect to time at the instantaneous position of the particle. Since it is the partial derivative with respect to time it is apparent that stagnation enthalpy change is impossible without unsteadiness.

For a stationary observer in a turbomachine the pressure distribution from a rotor is unsteady and the function of time may be idealized by a linear sawtooth distribution. As the number of blades in increased the amplitude of the sawtooth can be reduced for the ideal blade row but the slope $\partial p/\partial t$ remains constant so that the rise in stagnation enthalpy of fluid passing through the machine remains the same.

The unsteady view of work input has found little practical application although it dispels any illusion that conditions inside a turbomachine can be rendered steady. The convenience of the Euler equation for turbomachines lies with its simplicity and with the fact that by changing the frame of reference for rotors and stators it is possible to carry out most calculations by treating the flow as steady. The real complexity of unsteady flow is beyond the present capabilities of calculation methods.

1.4   Dynamic scaling

This section would often be called 'dimensional analysis' but the present title has been chosen because it is more appropriate. The cancelling of dimensions is rarely used to analyse real problems and in addition the scaling groups which are finally used are often left with dimensions. So long as the correct scaling is taking place it does not matter whether the groups have dimensions or not even though there are obvious advantages of generality in having groups with the same value in any system of units.

Geometric scaling

The simplest scalings to consider are the simple geometric groups, most of which are obvious such as the solidity (the inverse of pitch−chord ratio), aspect ratio (the ratio of mean blade length to mean blade chord), hub−casing ratio (the ratio of the diameter of the hub to that of the casing; often this is referred to as the hub−tip ratio but tip is ambiguous because the stator tip may be on the hub). The geometric scaling could also be taken to include angles but these are usually viewed differently.

Whilst the specification of the geometric ratios is simple it is not always clear what is the most appropriate choice. For example the tip clearance is known frequently to have a crucial effect on the blade performance but the most appropriate ratio in which to express it is not at all evident. In design, particularly mechanical design, it is often convenient to scale tip clearance with the blade height but this does not take proper account of the relevant fluid mechanics for long blades (i.e. high aspect ratio blades) because it is only conditions near to the blade tip which will be directly significant. More suitable lengths with which to scale tip clearance, all of which are used, are the blade chord, the pitch, the staggered gap (essentially the distance between adjacent blades at the trailing edge measured normal to the flow direction) and the maximum thickness of the blade at the tip. All the possibilities have some merit and the choice depends on the process or model being used. It is here that the judgement of the aerodynamicist must be exercised. With the exception of the last of the lengths all the other three are very similar in magnitude and it may be difficult to identify which is most appropriate from the measured trend.

Overall aerodynamic scaling

Very often the choice of aerodynamic scaling depends on the use to which the information is to be put. The customer for a compressor will be most interested in the mass flow, the pressure ratio, the rotational speed and the efficiency. Sometimes instead of mass flow it will be volume flow which is required and as an alternative to efficiency, the power input could be used. The pressure ratio is non-dimensional but it needs to be decided what pressures are most suitable for the application or investigation. Most compressors are able to use the inlet dynamic pressure, so the inlet stagnation pressure p_{01} is a suitable denominator. It is not always the case that the downstream device can utilize the dynamic pressure at outlet and if this is the case it is then more sensible to use the downstream *static* pressure p_2 in the numerator of the pressure ratio. When the gas leaving one compressor enters another or when the air is to be used directly for propulsion (as for the bypass stream of a jet engine) then the outlet stagnation pressure p_{02} is appropriate. Clearly p_2/p_{01} (known as the total-to-static pressure ratio) is less than p_{02}/p_{01} (known as the total-to-total pressure ratio); in an idealized case with uniform outlet flow the amount can easily be quantified in terms of outlet Mach number

$$p_{02} = p_2 \left\{ 1 + \frac{\gamma - 1}{2} M_2^2 \right\}^{\gamma/\gamma - 1}$$

Uniform outlet flow also corresponds to the highest value of p_2 relative to p_{02} and the amount by which these differ is a measure of how good is the outlet flow from a compressor.

The pressure ratio (or pressure rise) and the mass flow through a compressor depend on the compressor speed. The compressor speed is most naturally converted into a Mach number based on the blade tip speed U, i.e. $U/\sqrt{(\gamma R T)}$, where T is the local static temperature which is an unknown quantity since the static temperature depends on the local velocity of the gas. It is therefore normal to use a speed of sound based on the inlet stagnation temperature, $\sqrt{(\gamma R T_0)}$. For compressors passing only one type of gas γR is constant so that the ratio $U/\sqrt{T_0}$ is often used. If only one machine is of concern the tip radius is fixed and then $N/\sqrt{T_0}$ suffices, where N is the rotational speed in rev/min, radian/s, Herz etc. A convenient method for expressing the speed is to use the corrected variable so that

$$N_{\text{corr}} = \frac{N}{\sqrt{\theta}} \quad \text{where here } \theta = \frac{T_0}{T_{0\text{ref}}}$$

T_0 is the stagnation inlet temperature and $T_{0\text{ref}}$ is a reference stagnation temperature, usually the sea-level temperature for the standard atmosphere, 15°C. The corrected speed has the advantage that with the value of θ usually not very far from unity the corrected speed is fairly similar in magnitude to the actual speed and as a variable it is less abstract than, for example, $N/\sqrt{T_0}$.

An important variable for the customer and for determining the overall performance of a compressor is the mass flow rate. In considering the operation of blades and what sets the level of incidence it is apparent that it is the volume flow rather than mass flow which is crucial, so the density of the incoming mass flow is also relevant. As the speeds become large in relation to the speed of sound then the density in the inlet flow is no longer constant. A convenient non-dimensional group can therefore be formed as the ratio of the actual mass flow rate m to the choking mass flow rate m/m_{choke} where m_{choke} is the mass flow of a uniform sonic velocity flow through the same cross-sectional area A. For the sonic flow the density must be evaluated at the sonic conditions (with properties denoted as p^*, ρ^*, T^* and a^*) and the choking mass flow rate becomes

$$
\begin{aligned}
m_{choke} &= A\,\rho^*\,a^* \\
&= A(p^*/RT^*)\sqrt{(\gamma RT^*)} \\
&= A[p^*/\sqrt{T^*}]\sqrt{(\gamma/R)}
\end{aligned}
$$

Now it is easily shown that for the flow of a perfect gas

$$
T^* = T_0/\{1+(\gamma-1)/2\}
$$

and for isentropic flow

$$
p^*/p_0 = (T^*/T_0)^{\gamma/(\gamma-1)}
$$

so that p^* and T^* are both proportional to their corresponding stagnation properties. It is then easy to show that the group

$$
\frac{m\sqrt{(RT_0)}}{Ap_0}
$$

is a satisfactory non-dimensional combination of parameters being proportional to m/m_{choke}. The name flow function is sometimes given to the group or those related to it. In some work R is replaced by the specific heat c_p with no particular advantage. Clearly if the same gas is being used the same effective scaling is produced with R (or c_p) dropped so that the group has dimensions and this is common industrial practice. Furthermore if a single machine is being considered then the area term A is redundant. What is then left is frequently reframed as the corrected mass flow rate, which can be written

$$
m_{corr} = m\,\frac{\sqrt{\theta}}{\delta} \quad \text{where here } \theta = \frac{T_0}{T_{0ref}} \text{ and } \delta = \frac{p_0}{p_{0ref}}
$$

The reference stagnation pressure is usually taken as sea-level static, 101 kN/m^2. Again the use of corrected mass flow has the advantage that the numerical magnitude, and with it the physical significance this has, is not obscured as it might be with truly non-dimensional scaling.

The use of a group to relate mass flow to the choking mass flow of a uniform flow along a streamtube of the same cross-sectional area has wide application in compressible flow. In the present work, for example in Chapters 4 and 5, it will be used in considering axial blade-to-blade flows. There it will be used in the form

$$\frac{m\sqrt{(c_p T_0)}}{A p_0},$$

denoted by F and referred to as the flow function. An equally valid and perhaps neater formulation compares the streamtube area A to the streamtube area for a uniform sonic flow with the same mass flow rate; this ratio is usually written A/A^* and can be easily related to F. Sometimes the interest is directed to the conditions downstream of a compressor, for example when a second compressor is to be connected, and then the flow function is expressed in terms of the outlet variables p_{02} and T_{02}. If the dynamic pressure at exit is to be lost it may be more appropriate to use the static outlet pressure p_2.

For a compressor with no variable geometry (or variables held fixed or to a schedule itself fixed to corrected mass flow and corrected speed) for a fixed gas and where the dependence on Reynolds number can be neglected, the overall functional dependence can be written

$$\frac{p_{02}}{p_{01}} = f\left\{ \frac{m\sqrt{T_0}}{p_0} , \frac{N}{\sqrt{T_0}} \right\} = f'\left\{ m_{\text{corr}} , N_{\text{corr}} \right\}$$

The same functional dependence would exist if p_{02} were replaced by the static pressure p_2 or if the compressor efficiency η were being sought.

Aerodynamic scaling

The dynamic scaling so far considered has been of an overall nature such as might be most relevant to a customer or user for obtaining an overall measure of performance. Often different machines will be compared on the basis of these variables and then it is important to bear in mind the geometric difference between the machine — again knowledge and insight are called for in recognizing when it is appropriate to compare dissimilar machines. When it comes to finding appropriate scaling to assess the aerodynamics a rather different formulation is appropriate; for example pressure ratio gives no measure of the severity of blade loading and the flow function or corrected mass flow does not relate directly to the incidence on the blades.

Stage and blade loading

The enthalpy rise of a stage is related to the square of the rotational speed $\Delta h/U^2$ where U can either be taken as the blade tip speed or the speed at mid-radius, the latter being quite common for axial machines. The enthalpy change can be the static or stagnation enthalpy, depending on the context, though stagnation is more common. For compressors in which the pressure rise is small compared to the absolute pressure, such as low-speed machines, the density may be reasonably approximated as constant and it is convenient to define the pressure rise coefficient $\Delta p/\rho U^2$, with a 1/2 sometimes put into the denominator. As before the use to which the group is put will determine whether static or stagnation quatities should be used. It is a common convention to define $\psi = \Delta h/U^2$ although sometimes ψ is used for $\Delta p/\rho U^2$: it follows of course that for incompressible operation $\Delta p/\rho U^2 = \eta \Delta h/U^2$, where η is the efficiency, and both non-dimensional groups are very similar in magnitude and often in their trends. The ratios $\Delta h/U^2$ or $\Delta p/\rho U^2$ provide a measure of the actual input to the potential work available, i.e. to U^2. Clearly the demand on the stage is more taxing if a large enthalpy input is required from a low blade speed machine than a high-speed one and the magnitude of ψ gives a measure of this. For radial machines it is quite common to define the loading by

$$\psi = \Delta p/(\rho N^2 D_2^2)$$

and sometimes to replace $\Delta p/\rho$ by the isentropic (sometimes called the polytropic) head rise

$$H_{\mathrm{p}} = \int \mathrm{d}p/\rho$$

Sometimes pressure ratio is used to present low-speed compressor results and values like 1.005 are quoted. This is not a helpful presentation because it shows the pressure rise to be orders of magnitude below the ambient pressure and gives no indication of the severity of loading of the blades or stages. If Δp denotes the pressure rise which is small in relation to the inlet pressure p_1 then it is easy to show that

$$\frac{p_2}{p_1} = \frac{p_1 + \Delta p}{p_1} = 1 + \frac{\gamma \Delta p}{(\gamma p_1/\rho_1) \cdot \rho_1} = 1 + \frac{\gamma}{a^2/U^2} \frac{\Delta p}{\rho_1 U^2}$$

$$= 1 + \gamma M_{\mathrm{u}}^2 \frac{\Delta p}{\rho_1 U^2}$$

where second order terms in Δp have been neglected. M_{u} denotes the Mach number based on the blade speed. For low-speed machines, when M_{u} is much less than unity, it is better to use the pressure rise coefficient $\Delta p/\rho U^2$ than the pressure ratio since the latter is essentially providing less information. For values of M_{u} significantly less than one the non-dimensional performance of the machine is essentially independent of M_{u}. However when M_{u} becomes large the pressure ratio *is* a very useful parameter because the rise in pressure is comparable to the absolute inlet pressure. Furthermore, since there is an upper limit to the possible tip speed, the experienced engineer gets an implicit assessment of the loading from the value of pressure ratio.

Flow coefficient

The work input to a stage depends on the flow through it and for an axial stage it is easy to show that ψ is related to V_x/U, the ratio of axial velocity to the blade speed often called the flow coefficient and denoted by ϕ. Similarly the flow coefficient determines the incidence into the first rotor and then in turn into the blade row downstream. For a particular stage the incidence is one of the crucial flow quantities in determining the performance of a blade row. The flow coefficient therefore effectively determines the performance of the stage. The axial velocity might be the local value or the mean value across the annulus, depending on the use to which it is being put. As ϕ is reduced the incidence rises, bringing changes in the blade operation. Most axial compressors are designed to have ϕ in the range $0.3-0.9$, with the lower end of this range being more common nowadays.

For radial machines a somewhat different convention for the flow coefficient is sometimes used. The symbol ϕ is still used but is defined differently sometimes in terms of the impeller outlet speed U_2, mass flow rate m and a density, typically the inlet density ρ_1; $\phi = m/(\rho_1 U_2 D_2^2)$. The flow coefficients defined in this way have rather low values, typically of order $0.01-0.1$. Another not dissimilar expression which is sometimes used is $\phi = m/(\rho N D_2^3)$.

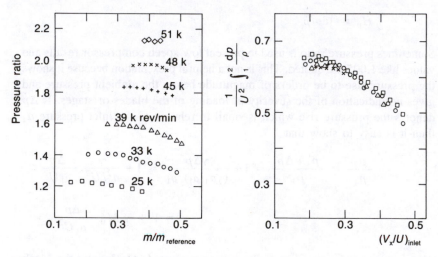

Fig. 1.7 The measured performance of a turbocharger compressor presented in two different ways; the symbols on the right are for the same rotational speeds as those on the left. (From Fink, 1988)

Figure 1.7 shows measurements of a turbocharger compressor made by Fink (1988) and presented here in two different ways. The left-hand plot is the overall performance that might be passed to a hypothetical customer and it shows the pressure ratio plotted versus corrected mass flow rate for a range of rotational speeds (the corrected mass flow rate is actually non-dimensionalized by a reference value). The right-hand diagram shows the same data plotted with the aerodynamic scaling: the ordinate is the compressible version of $\Delta p/\rho U^2$ and the abcissa is the inlet flow coefficient V_x/U_1. Expressed in these aerodynamic variables it can be seen that the impeller performance is almost independent of the rotational speed; Mach number is not an important variable in the range at which this machine was operated. (It should be noted that the constant speed lines in the left-hand diagram are short at high speeds only because the facility had insufficient power to drive the compressor at higher mass flow rates.)

The flow coefficient V_x/U can be related to the overall flow function $m\sqrt{(c_pT_0)}/Ap_0$ introduced above. The axial velocity is really the volume flow per unit area, or $m/\rho A$, and the density is, of course, given by $\rho = p/RT$. For a given flow rate the static and stagnation variables are related by functions only of flow Mach number, so the mean axial velocity is given by $V_x = kmRT_0/Ap_0$ where k is a constant. A compressor operated at constant corrected speed, i.e. $N/\sqrt{T_0} = $ constant, is actually one for which the Mach number based on blade speed is constant, so the mean blade speed is given by $U = \kappa\sqrt{T_0}$ where κ is another constant. The flow coefficient V_x/U can then be written

$$\frac{V_x}{U} = \frac{k(m/A)\ R(T_0/p_0)}{\kappa\sqrt{T_0}} = K\frac{m\sqrt{(c_pT_0)}}{Ap_0}$$

so that the flow coefficient is directly proportional to the flow function for a compressor at constant corrected speed.

Reynolds number

Real compressors suffer losses attributable to shear stresses and mixing and the amount of this is strongly affected by the extent to which the flow is laminar or turbulent. If the flow remains laminar the shear stresses are low and the loss directly attributable will be correspondingly small, but on the other hand there will probably be extensive regions of separated flow leading to a lower pressure rise and a lower efficiency than that which might be achieved with a more turbulent flow. The transition to turbulence is ultimately an indication that the inertial stresses have overwhelmed the viscous dissipation and the non-dimensional group which describes this is the Reynolds number, $R = Vl/\nu$ where V is a representative velocity and l a representative length. The values of Reynolds numbers used in turbomachines are usually large, often 10^5 or more based on characteristic lengths of the machine. Normally the correct non-dimensional group has a magnitude near to one if it is relevant to the flow because it represents the ratio of two related or competing effects; a good example of this is Mach number which is relevant only when it is of order unity. In the case of Reynolds number very large values are relevant because macroscopic quantities such as the dimensions of the compressor or of the blade are used whereas the process which is being assessed is a comparatively microscopic one taking place with dimensions comparable to or smaller than the boundary layer. This use of the 'wrong' scale is possible because the boundary layer thickness itself depends on the overall or macroscopic Reynolds number.

In a compressor it is not obvious what are the relevant velocity, length and viscosity. If machines of a particular geometry are being compared the sense of the Reynolds number can be retained with any relevant specification: ND^2/ν would be quite suitable with N being rotational speed, D the inlet or outlet diameter and ν the viscosity at a reference condition such as inlet stagnation. This could be added to corrected mass flow and corrected speed in the specification of pressure ratio given above. The choice becomes more difficult when machines are not identical and the comparison requires the salient variables to be chosen. For axial compressors a crucial effect at low Reynolds numbers is separation of the flow on the blades, a topic considered further in Chapter 4. For the axial machines the natural choice is therefore based on blade chord and the inlet velocity relative to the blade. In most cases the changes across an axial blade row are small enough that the viscosity could be defined anywhere upstream or downstream but by convention upstream conditions are used as the reference. Decisions need to be made about which blade row to use and at which radial station in the evaluation of Reynolds number. Figure 1.8 shows normalized stage pressure rise coefficient at stall for multistage axial compressors as a function of blade chord Reynolds number, taken from the paper by Koch (1981).

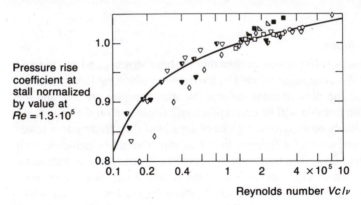

Pressure rise coefficient at stall normalized by value at $Re = 1.3 \cdot 10^5$

Reynolds number Vc/ν

Fig. 1.8 The variation in static pressure rise at stall for axial stages with Reynolds number. (From Koch, 1981)

The choice of the appropriate Reynolds number for radial machines requires there to be an idea of where the most serious viscous loss source is likely to be. This is believed to be at or near the impeller outlet, largely because the passage is narrowest here, see Casey (1985). An appropriate choice is therefore $U_2 b_2/\nu$, where U_2 is the impeller tip speed and b_2 is the impeller tip width at exit. With the viscosity set at the inlet value this group is now the basis of a recommendation reported by Strub *et al.* (1987) of an international group who have devised correlation procedures for assessing Reynolds number effects in radial compressors. Rather than correlate efficiency it is the non-dimensional loss, $1 - \eta$, which is used. The dependence of loss on Reynolds number is assumed to follow the same form as the flow in roughened pipes and the prediction of the effect of Reynolds number change on performance needs the effective surface roughness to be known as a fraction of the impeller outlet width b_2.

The effect of Reynolds number on the performance of axial compressors was discussed by Wassell (1968) with the Reynolds number based on the blade chord and the inlet velocity to the blade. The effect on efficiency, flow rate and maximum pressure rise was considered. Loss rather than efficiency was presented in the form

$$1 - \eta = kR^{-n}$$

with n varying significantly from compressor to compressor. A rapid rise in loss with decrease in Reynolds number occurred below $0.3 \cdot 10^5$ for compressors, compared with about $1 \cdot 10^5$ in the case of cascade tests. Schäffler (1980) found Wassells correlation for efficiency to work well for hydrodynamically smooth blades but not adequately for rough blades at high Reynolds numbers. The loss decreased quite rapidly in the turbulent region, proportional to R^{-n} with n in the range 0.10 to 0.13, until a value of Reynolds number of about $5 \cdot 10^5$ was reached at which loss remained more or less independent of Reynolds number. This is the condition at which roughness comes to dominate. The Reynolds number at which roughness becomes dominant is a function of the blade pressure distribution and, of course, the roughness height. The behaviour with roughness can be described in a manner analogous to the flow in rough pipes and a critical Reynolds number for roughness height was found to be about 90. In the discussion on Schäffler's paper the evidence on roughness was supported by Koch and Smith of the General Electric Company.

Specific speed

For radial machines it is common to define a group known confusingly as specific speed. It is the ratio of two non-dimensional groups, a loading denoted conventionally by ψ and given by $\Delta p / (\rho N^2 D_2^2)$ and a flow coefficient $\phi = m / (\rho N D_2^3)$. They are combined in such a way as to remove the dimension of size, in this case the impeller outlet diameter D_2. What results is the specific speed

$$N_s = \frac{\phi^{1/2}}{\psi^{3/4}} = \frac{m^{1/2} \rho^{1/4} N}{\Delta p^{3/4}}$$

and a plot of the efficiency against specific speed is shown in Fig. 1.9, taken from Rodgers (1980) for a range of different impellers of the type with axial inducers. In this particular figure the average density has been used to calculate

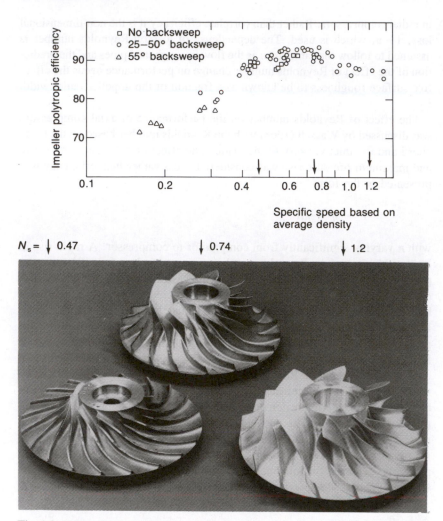

Fig. 1.9 The variation in polytropic efficiency with specific speed. The impellers shown are reproduced by permission of Sundstrand Turbomach and were used in the test program (N_s = 0.47, 0.74 and 1.2). (From Rodgers, 1980)

the specific speed. Very often inconsistent units are used to calculate specific speed so that a dimensional quantity results but the form which is given here is the true non-dimensional one. (The true non-dimensional specific speed is equal to the common value in Imperial units divided by 129.) Peak efficiency occurs in Fig. 1.9 at a value of N_s equal to about 0.7. This figure also shows three of the impellers used in deriving this data with specific speeds, from left to right, of 0.47, 0.74 and 1.2. The similarity in the overall geometry should be noted because for the specific speed to provide useful guidance it is necessary that it be used on a single family of designs and it cannot be relied upon to give the correct trend for machines with radially different geometrical layout.

The specific speed can be thought of as a comparison of flow rate to pressure rise. Now the pressure rise depends *primarily* on the outlet diameter and rotational speed. The mass flow rate depends, however, *primarily* on the inlet diameter and the rotational speed. The specific speed then contains or reflects the ratio of the inlet diameter to the outlet diameter and the magnitude of N_s for a particular family of designs is determined almost entirely by the ratio of these diameters. Low specific speed compressors pass a relatively small flow and have small inlets in relation to the outlet diameter, conversely high specific speed impellers have large inlets. For the impellers with the highest efficiency at around $N_s=0.7$ in Fig. 1.9 the inducer tip diameter is between about 75 per cent and 80 per cent of the impeller outlet diameter. (More recent designs suggest that still higher efficiency can be obtained at rather lower specific speed with a diameter ratio of about 0.6, see Chapter 6.) In some organizations it is recognized that specific speed gives geometric information which can be handled more easily and transparently by using simply ratios such as the ratio of inlet to outlet diameter. Given the additional effects at high Mach number which are omitted from the derivation of N_s this would seem to be a very satisfactory decision.

1.5 Losses

It is often assumed that everyone knows how the losses arise but discussion often reveals that there is great confusion surrounding this. Agreement exists on the end result, a rise in entropy and a reduction in stagnation pressure compared to the inlet value or to an ideal value. (Actually the equivalence of entropy rise and fall in stagnation pressure is only true for adiabatic flow, but this is usually an acceptable assumption for compressors.) In considering loss it is usually most helpful to consider the mass weighted average of the stagnation pressure,

$$\bar{p_0} = \frac{\int p_0 \, \rho \, V dy}{\int \rho \, V dy}$$

At the microscopic level the losses may be thought to have a single cause, viscous shearing leading to a rise in internal energy, but to the aerodynamicist working at a macroscopic level the sources of loss seem very different. A possible list of loss sources, which other people might draw up rather differently, is:

1. drag at solid surfaces;
2. mixing;
3. shock losses;
4. shear work.

Rather than try to unify the loss generation it is probably more helpful to keep the different categories as separate entities and to illustrate and amplify them.

1. *Drag at solid surfaces*

The effect of drag alone may be seen by considering incompressible flow through a cascade of uncambered and unstaggered blades, Fig. 1.10. Upstream of the cascade conditions are uniform so that the static pressure and velocity can be denoted by p_1 and V_1. At the trailing edge of the cascade the velocity v_2 is not uniform, although to a very good approximation the static pressure p_2 is. Well downstream the conditions will revert to uniform with velocity V_3 and pressure p_3. Because the flow is incompressible it follows immediately that $V_1 = V_3$ and also

$$V_1 s = V_3 s = \int_0^s v_2 dy = V_2(s - \delta^*)$$

$$\text{where } \delta^* = \int_0^s \left(1 - \frac{v}{V}\right) dy$$

is the total displacement thickness on both sides of each passage. Applying conservation of momentum between planes 1 and 3 leads to the very simple result, since $V_1 = V_3$, that

$$F/s = p_1 - p_3,$$

where F is the force on each blade. Because $V_1 = V_3$ it also follows that

$$F/s = p_{01} - p_{03},$$

the loss in stagnation pressure. As the flow is uniform at stations 1 and 3 it is clear that the stagnation pressures are also equal to the mass weighted values. In this simple case then the overall loss can be obtained entirely by knowing the drag force on the blades and the argument could also be extended to compressible flow. The simplicity of the result gives ground for caution and it can be seen that the same result would have been predicted for a grid of rods on each of which the drag force F would have been produced by the pressure distribution consequent upon separation of the flow around the point of maximum thickness. In other words this simple analysis has provided no information about the mechanism of loss but has merely given an expression for evaluating it.

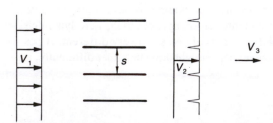

Fig. 1.10 Loss production by a cascade of uncambered unstaggered plates. Conditions uniform at stations 1 and 3

Suppose that the loss were assessed at the trailing edge. In this case the mass weighted stagnation pressure is not equal to the value in the middle of the passage and different values would be obtained depending on the shape of the velocity profile at the trailing edge. It is for this reason that it is usual to choose a reference condition for the losses when all the nonuniformities have mixed out. This condition is obtained a long way downstream. However, it is usual only to make measurements sufficiently far downstream that the largest changes have taken place when a reasonable approximation to the mixed out loss can be obtained.

This identification of loss with blade force has had seriously misleading effects. First it is only true in rather special cases. Second it caused people to identify loss with momentum changes and the trailing edge momentum thickness θ as *the* crucial variable. This was presumably because for an isolated aerofoil the drag force is proportional to the total momentum thickness in the wake. As will be seen below momentum thickness is not the most relevant variable, although by using correlations the true loss can be inferred from the momentum thickness, and more insight can be obtained by using the correct variable which is the so-called energy displacement thickness usually denoted by δ_3.

2. *Mixing loss*
Mixing loss has been referred to above in connection with two-dimensional cascade flows but it is also a familiar process which takes place in many other instances in turbomachinery and elsewhere. The mixing taking place after the sudden enlargement in a pipe is a standard example of a thermodynamically irreversible process; the conservation of mass and momentum allows the rise in static pressure and the fall in stagnation pressure to be calculated. Mixing usually occurs in much more complicated forms in turbomachines and the flow is generally three-dimensional. An example of the pattern downstream of a rotor blade with tip clearance is given in Chapter 8, Figs 8.19 and 8.20. Here a jet is formed by the tip clearance flow which has ultimately to be mixed out to the uniform condition and it is in achieving this that the major losses are to be expected. The crucial point here is that the high kinetic energy of the jet cannot be properly utilized (at least no one knows how) and is therefore entirely lost. The only sure reference condition is the fully mixed out one.

In considering mixing one recognizes that the essential feature is non-uniformity in the flow. Care is needed in distinguishing between those non-uniformities which lead to mixing and those which do not. A loaded blade or an isolated wing produces a nonuniform flow as a necessary part of producing lift. This does not necessarily produce loss because it may be possible to decelerate the high velocity flow by a process which is virtually thermo-dynamically reversible without requiring mixing. This is the way in which one hopes that the flow will behave on a well-designed aerofoil. If, on the other hand, the flow separates and produces a wide wake there is then no mechanism but mixing to return the flow to uniform and the losses will be high at the completion of the mixing process. Flow separation does not necessarily lead to large mixing loss because the flow may reattach to form a bubble. The increment in loss attributable to the presence of the bubble is very small.

It is known that many blades operate with a separated region near the trailing edge and that the corner regions often contain separated flow. One can make estimates for the loss that this incurs in returning to the uniform condition and the easiest case to consider is a two-dimensional cascade. The model is shown sketched on Fig. 1.11 where the flow separates in the blades so as to leave at the trailing edge as a core or jet with uniform velocity V_2 and a wake with zero velocity in it. For simplicity the flow is treated as incompressible. The flow enters at α_1 to the axial direction and for simplicity is assumed to leave the blade row in the axial direction (even after separation). The flow is assumed uniform across the core region with velocity V_2 and stagnation pressure p_{02} and with no flow across the wake. The static pressure at the trailing edge plane is taken to be uniform and equal to p_2 across the entire passage. Since it is assumed that there are no losses to the core flow outside the separated region it is clear that at the trailing edge $p_{02} = p_{01}$ in the core.

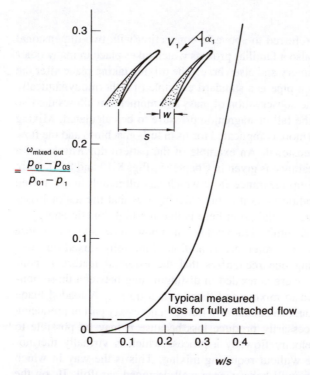

Fig. 1.11 Calculated loss for fully mixed out flow with idealized wake of width w. No flow in wake. Incompressible flow. Inlet flow angle $\alpha_1 = 35°$, outlet flow angle $\alpha_2 = 0°$

Because there is no flow at all in the separated region the mass-weighted stagnation pressure across the trailing edge plane would also be equal to p_{01}. Clearly a loss-making process has occurred in the passage but the usual mass flow weighted book-keeping process does not record it; it might be better to say that a loss-initiating process has taken place which could hypothetically be undone at least in part by some device which would decelerate the core flow reversibly.

Some distance downstream the flow mixes out to a uniform condition and here the velocity is V_3, the static pressure p_3 and stagnation pressure p_{03}. Applying continuity of mass and axial momentum (only axial momentum is needed here because of the simplification to axial outlet flow) between station 2, the trailing edge, and station 3, the station for fully mixed out flow, results in

$$\omega_{\text{mixed out}} = \frac{p_{01} - p_{03}}{\frac{1}{2}\rho V^2} = \cos^2\alpha_1 \left(\frac{w}{s-w}\right)^2 \tag{1.21}$$

The curve plotted in Fig. 1.11 is for $\alpha_1 = 35°$, a reasonable value for inlet flow angle with an outlet flow which is axial. Also shown is the level of measured loss in an unstalled cascade with this inlet angle and a solidity of one. Rather surprisingly the wake needs to extend over about one-eighth of the passage before the mixing loss attributable to it equals that of other regular losses in the cascade. At higher widths of wake the loss rises very rapidly. Very similar conclusions would be reached for other inlet and outlet flow angles and for more general three-dimensional separations: small areas of wake can mix out with surprisingly small loss.

The treatment in this section is predominantly from the point of view of incompressible flow because the ideas and mechanisms can be understood much more clearly with this simpler flow. In many practical cases the flow Mach numbers will be quite high and it is reasonable to ask how valid is the assumption of incompressible flow. Stewart (1955) looked into this from the point of view of flow downstream of a row of axial blades. Calculations were performed varying parameters to explore the effects and one of the most revealing is shown here as Fig. 1.12 in which the ratio of the fully mixed out loss

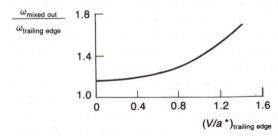

Fig. 1.12 The effect of compressibility in the mixing out of non-uniform flow at the trailing edge $\omega = (p_{01} - p_0)/(p_{01} - p_1)$, a* is the velocity of sound when flow is sonic. (From Stewart, 1955)

ω_3 is compared with loss just downstream of the blades ω_2 for a range of Mach numbers. The ratio ω_3/ω_2 rises quite markedly with Mach number and this trend for mixing loss to rise with Mach number is a general one which is not restricted to two-dimensional flow downstream of cascades.

3. Shock losses
These are another of the standard thermodynamic irreversibilities. The largest loss in total pressure occurs when a flow is decelerated by a single normal shock and the deceleration can be more nearly reversible if it takes place in a series of weaker oblique shocks. Inlet Mach numbers to compressor rotors exceeding about 1.4 are unusual and at this condition the loss in stagnation pressure across a normal shock is only 4.2 per cent of inlet stagnation pressure. Such a normal shock produces a static pressure rise of 2.1 and this combination therefore represents quite an efficient compression system. One of the aims of the design of supersonic blades is to keep the Mach number from rising too high downstream of the leading edge; in some cases it may get as high as 1.6. The losses rise as $(M-1)^3$ so a small increase in M may give a substantial increase in loss which is somewhat redeemed by the corresponding rise in static pressure ratio.

The most serious aspect of shocks is their tendency to cause boundary layers to separate. A shock incident on a flat plate will cause a turbulent boundary layer to separate if the static pressure ratio exceeds about 1.8 and for a normal shock wave this corresponds to an incoming Mach number of about 1.3. For a laminar boundary layer separation occurs with shocks at very much lower Mach numbers. Furthermore a turbulent boundary layer which has overcome an adverse pressure gradient prior to the shock (and has an elevated value of form parameter H) will separate with much weaker shock waves. If after separating the flow can reattach the loss may not be that great, but if a wide wake is created, for example downstream of a blade row, then the contribution to loss may be very high.

4. Shear work dissipation
Shear work is taking place wherever there are velocity gradients but it is only in regions where the gradients are very steep that its magnitude is of concern. These regions are shear layers including boundary layers, wakes and the regions dividing fast and slow regions of separated flow. At the fundamental level all the loss mechanisms can be related to viscous dissipation by shear work but it is not always profitable to think of it in this way. A complete treatment could be given in three dimensions but the salient features can be found from a very simple one-dimensional analysis which suffices for the present purpose. The flow will be assumed to be in a boundary layer as sketched in Fig 1.13. The normal viscous stresses are expected to be very much smaller than the shear stresses so the momentum equation in the streamwise direction can be written

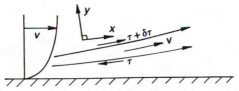

Fig. 1.13 The velocity profile and shear stresses operating near to a solid surface with
x and y parallel and normal to local streamlines respectively

$$\rho V \frac{\partial V}{\partial x} = \frac{\partial \tau}{\partial y} - \frac{\partial p}{\partial x} \tag{1.22}$$

The balance of energy can be written, including the heat transfer q across the
streamlines but not in the direction along them, as

$$\rho V \frac{\partial h_0}{\partial x} = -\frac{\partial q}{\partial y} + \frac{\partial (\tau V)}{\partial y}$$

$$= \frac{\mu}{Pr} \frac{\partial^2}{\partial y^2} \left(h_0 - \frac{V^2}{2} \right) + \mu \frac{\partial^2}{\partial y^2} \left(\frac{V^2}{2} \right) \tag{1.23}$$

where h_0 is the stagnation enthalpy, $Pr = \mu c_p / k$ is the Prandtl number and the
bulk properties have been taken to be uniform. Equation 1.23 may be simplified
to

$$\rho V \frac{\partial h_0}{\partial x} = \frac{\mu}{Pr} \frac{\partial^2}{\partial y^2} \left\{ h_0 + (Pr - 1) \frac{V^2}{2} \right\} \tag{1.24}$$

which is the form given by Denton (1986). For a Prandtl number of one and
inlet conditions with uniform stagnation enthalpy a solution to equation 1.24
is clearly that $h_0 =$ constant. In most gases the Prandtl number is of order one
(for air at standard conditions it has a value of 0.73) and it is common experience that the stagnation enthalpy is nearly constant through a boundary layer
on an adiabatic wall. This is more generally true in turbulent flow when the
effective Prandtl number is very nearly one for all gases. Constant stagnation
enthalpy indicates that there is heat transfer through the gas of an amount to
balance the work input by viscous shear. Using the thermodynamic relation
$T ds = dh - dp/\rho$ with equations 1.22 and 1.24 gives

$$T \frac{\partial s}{\partial x} = -\frac{1}{\rho} \frac{\partial \tau}{\partial y} \tag{1.25}$$

This can be written in terms of the substantive derivative, $D/Dt = V \partial/\partial x$ (i.e.
the rate at which the entropy of a particle changes as it is carried by the flow)

by multiplying both sides by V to give

$$T\frac{Ds}{Dt} = -\frac{V\partial\tau}{\rho\partial y} \tag{1.26}$$

Although equation 1.26 has been derived for a Prandtl number of one it can be used in a wide range of gases since the processes which dominate the flow in compressors are turbulent. The generalization of equation 1.26 to full three-dimensional flow is straightforward if tedious and leads to

$$T\frac{Ds}{Dt} = -\frac{V_i}{\rho}\frac{\partial\tau_{ij}}{\partial x_j} \tag{1.27}$$

It would be wrong to leave the impression that shear work is invariably a loss and invariably detrimental. If one considers the upper and lower faces of the streamtube in Fig. 1.13 work is done on the lower surface at the rate of $V\tau$ and on the upper surface of $V\tau + \partial(V\tau)/\partial y \cdot \delta y$. If the flow is laminar, or if the shear stress can be taken as locally proportional to the velocity gradient, it is clear that the net shear work, the difference in the work in at the bottom and out on the top, is given by

$$\frac{\partial}{\partial y}(V\tau)\delta y = \frac{\partial}{\partial y}\left(\mu V\frac{\partial V}{\partial y}\right)\delta y$$

It follows immediately that the net shear work can be expanded into two terms

$$\frac{\partial}{\partial y}\left(\mu V\frac{\partial V}{\partial y}\right) = \mu\left(\frac{\partial V}{\partial y}\right)^2 + \mu V\frac{\partial^2 V}{\partial y^2} \tag{1.28}$$

The first term on the right-hand side of equation 1.28 is positive definite and is the irreversible component, which translates directly into internal energy and entropy increase of the gas. The second term is not a loss but is the 'useful' work done by the shear; it is this component which, for example, provides the momentum transfer to keep the low velocity fluid in a boundary layer moving against an adverse pressure gradient.

Macroscopic loss

In considering loss it is easier to address attention to a particular geometry than to talk in more abstract terms. The geometry chosen is the familiar two-dimensional cascade, Fig. 1.14, but the principle to be discussed is much more general. The inlet stagnation pressure is assumed uniform at p_{01} and downstream of the passage the velocity distribution is sketched, v_2 is the resultant velocity but the direction is assumed to be uniform. The static pressure

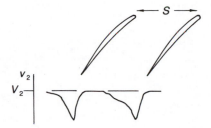

Fig. 1.14 The velocity profile downstream of a blade row

p_2 is assumed uniform across the exit plane. Near the middle of the passage the velocity is V_2 and the stagnation pressure is equal to the inlet value, $P_{02}=p_{01}$, where the capital letters will be used to denote the 'freestream' values for the outlet flow. For simplicity the flow is treated as incompressible.

Integrating over the whole flow the loss in total pressure is given by

$$\int_{\text{area}} \rho v_2(p_{01}-p_{02})\mathrm{d}A \tag{1.29}$$

The integration here may be with respect to y over a pitch and advantage may be taken of the fact that there is a core of fluid for which $P_{02}=p_{01}$. Using the incompressible relation for stagnation pressure, $p_0=p+\frac{1}{2}\rho v^2$, the integral 1.29 can be rewritten as

$$\int_0^s \rho v_2(V_2^2 - v_2^2)\mathrm{d}y \tag{1.30}$$

and the integral 1.30 is familiar in boundary layer theory as the energy displacement thickness where it is usual to write it as

$$\delta_3 = \int_0^s \frac{v_2}{V_2}\left\{1-\left(\frac{v_2}{V_2}\right)^2\right\}\mathrm{d}y \tag{1.31}$$

(In this case the same notation as the rest of the section has been retained rather than switching to that more usual for boundary layers.) The energy displacement thickness has found extensive use in boundary layer calculation methods, particularly those from Germany of which the best known is by Truckenbrodt (1952).

Lieblein and Roudebush (1956) had found the form of equation 1.30 for the loss from a two-dimensional cascade. Such, however, was the status of momentum thickness,

$$\theta = \int v/V(1 - v/V)\mathrm{d}y,$$

derived from its importance in aeronautics where the drag of an aircraft is

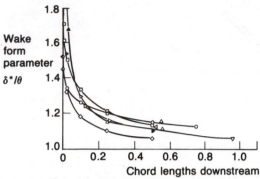

Fig. 1.15 The form parameter measured in the wake of blades and aerofoils. (From Lieblein and Roudebush, 1956)

proportional to θ, that they found it worth recouching the loss in terms of momentum thickness. It is an important part of turbulent boundary layer calculation methods to assume that the various integral thicknesses are in a fixed relation to one another. When the form parameter $H = \delta^*/\theta$ (with $\delta^* = \int (1 - v/V)\,dy$ being the displacement thickness) tends to one it is easy to show that the energy displacement thickness δ_3 tends to 2θ. Figure 1.15 shows the form parameter $H = \delta^*/\theta$ measured downstream of cascades and isolated aerofoils from Lieblein and Roudebush (1956) and from this it can be seen that within one third of a chord downstream of a cascade the form parameter is less than 1.2. At this condition the energy displacement thickness would be in the range $1.9\theta < \delta_3 < 2.0\theta$. Measurements of loss would normally be made at least this far downstream from a blade in cascade but inside compressors, measurements often have to be made very close to the trailing edges. Lieblein and Roudebush used a power law profile to represent the wake and were then able to derive the loss in terms of the momentum thickness and form parameter. Because the form parameter was so close to unity, Lieblein (1956) was able to adopt the approximation $\delta_3 = 2\theta$ when he wrote Chapter VI in the report by the staff of the NACA Lewis Research Center 'The aerodynamic design of axial-flow compressors'. He recommended the approximate expression for the mass averaged stagnation pressure loss

$$\bar{\omega}_1 = 2 \frac{\theta}{c} \frac{\sigma}{\cos\alpha_2} \left[\frac{\cos\alpha_1}{\cos\alpha_2} \right]^2 \tag{1.32}$$

as adequate for unstalled cascades. (This also required the restriction that the momentum thickness was much less than the pitch, again reasonable for unstalled cascades.) The loss is here nondimensionalized with respect to the inlet dynamic pressure.

Lieblein and Roudebush also considered the mixing loss from the measurement station to the fully mixed out condition downstream where the flow has returned to uniform. In doing this it is necessary to consider the conservation of mass and momentum in the axial and tangential directions so that there are changes in the downstream static and stagnation pressures and flow direction.

The mixing loss is a function of the form parameter and the momentum thickness but in going from a typical measuring station to the mixed out condition the loss will not increase by more than about 5 per cent for a value of the form parameter H up to 1.2 and by not more than 12 per cent for H up to 1.4. If the mixing were being considered from the trailing edge then the additional loss in going to the mixed out, uniform condition would make a very much larger contribution, even as high as 100 per cent for the idealized case considered in the section above on mixing loss.

There is, of course, nothing wrong with expressing the loss in terms of the momentum thickness provided the momentum thickness is not thought of as indicating the mechanism by which loss is generated or can be controlled. However, the use of momentum thickness as the usual correlating parameter has had the unfortunate effect of directing minds towards the skin friction drag on the blades or hub and casing walls as the origin of the loss.

The form for the mass-weighted loss given in equation 1.30 leads, as already noted, to the identification of the loss with the energy displacement thickness δ_3 defined in equation 1.31 which has formed an important part of turbulent boundary layer calculation methods. Boundary layers will be discussed in more detail in a later chapter but it may be noted here that the equation for flow in a two-dimensional boundary layer, with the assumptions which are usually known as the boundary layer assumptions, is

$$u\frac{\partial u}{\partial x} + v\frac{\partial u}{\partial y} = -\frac{1}{\rho}\frac{dp}{dx} + \frac{1}{\rho}\frac{\partial \tau}{\partial y} \tag{1.33}$$

where u and v are here the local mean velocities in the x and y directions, x being in the direction of the flow outside the boundary layer. If equation 1.33 is integrated with respect to y from the surface $y=0$ to a distance out from the surface $h>\delta$, where δ is the boundary layer thickness, what results is the so-called momentum integral equation

$$\frac{d\theta}{dx} + \frac{\theta}{U}\frac{dU}{dx}(H+2) = \frac{1}{2}c_f \tag{1.34}$$

where U is here the free-stream velocity, $c_f = \tau_w/(\frac{1}{2}\rho U^2)$ is the skin-friction coefficient and τ_w the wall shear stress. This equation forms the basis of most integral calculation methods. If, however, equation 1.33 is multiplied through by u before integrating with respect to y a different equation is formed

$$\frac{1}{U^3}\frac{d}{dx}(U^3\delta_3) = \frac{2}{\rho U^2}\int_0^\delta \frac{u}{U}\frac{\partial \tau}{\partial y}dy$$

$$= 2\int_0^\delta \frac{\tau}{\rho U^2}\frac{\partial}{\partial y}\left(\frac{u}{U}\right)dy = 2C_D \tag{1.35}$$

The integral on the right-hand side, denoted by C_D, is known as the dissipation coefficient or integral. The integrand, it will be noted, is, apart from the terms to make it non-dimensional, the same quantity derived in the section on shear work. The mainly turbulent processes making up C_D were discussed by Fernholz (1964), the main contributor being the Reynolds stress multiplied by the mean flow velocity gradient. Fernholz derived the values of C_D for a number of sets of measurements and C_D tends to increase as the form parameter H gets larger. Nevertheless the spread is quite small, most values lying within a factor of two (by way of comparison, the skin friction coefficient varies from 0 to a finite value).

As part of a calculation strategy, simplified correlations were devised for dissipation coefficient C_D as well as for the skin friction coefficient c_f. Schlichting (1979) gave an approximation for the dissipation coefficient of the form

$$C_D = \frac{\beta}{R^b} \tag{1.36}$$

The Reynolds numbers R in this correlation is based on the energy displacement thickness and the local free-stream velocity. From Figure 22.7 of Schlichting (1979) b may be taken as 0.15 for a wide range of pressure gradients, rising to 0.17 in strong favourable pressure gradients and down to 0.1 in regions close to separation. The parameter β varies little over the same range and may reasonably be approximated by 0.005.

Equations 1.30 and 1.35 are equally valid for laminar as for turbulent flow. Truckenbrodt (1952) gave a relation for the dissipation coefficient in laminar flow and this is of the simple form

$$C_D = \beta(H)/R_\theta \tag{1.37}$$

where here H is the usual form parameter and the Reynolds number here is based on momentum thickness. The first power of Reynolds number is predictable from similarity conditions. β was shown to vary very little with the form parameter H, which is perhaps surprising.

Figure 1.16 seeks to give some more concrete comparison of the dissipation coefficient, and therefore the losses, for laminar and turbulent flow. It has been assumed that for the turbulent case a profile exposed to a modest adverse pressure gradient would be appropriate and therefore $\beta=0.005$ and $b=0.15$ have been chosen as representative. A form parameter $H=\delta^*/\theta=1.6$ has been taken from which it can be found that $\delta_3/\theta=1.65$ as a means to giving the results in terms of the usual Reynolds number based on momentum thickness. For the laminar flow a slightly favourable pressure gradient has been assumed, since it is under such conditions that laminar flow is usually found in compressors, and for this $\beta=0.19$ was taken. The corresponding values of C_D have been plotted against Reynolds number based on momentum thickness in Fig. 1.16. The turbulent loss is very much greater than the laminar,

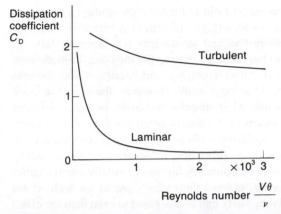

Fig. 1.16 Dissipation coefficients for turbulent and laminar flow useful for estimating loss production

an order of magnitude greater at all but the lowest Reynolds numbers. If C_D can be approximated by a constant value, a better approximation for turbulent than laminar flow, equation 1.35 can be rewritten as

$$U^3 \delta_3 = \int_0^X 2C_D U^3 \, \mathrm{d}x = 2C_D \int_0^X U^3 \, \mathrm{d}x \qquad (1.38)$$

in which the left-hand side term will be recognized as the flow-weighted loss in stagnation pressure. In other words the principal factor determining the loss production for a laminar or turbulent boundary layer is the *cube* of the freestream velocity.

The dissipation coefficient for turbulent flow varies only slowly with both Reynolds number and form parameter (and hence pressure gradient). Laminar flow offers huge reductions in loss production, but the laminar flow is unable to resist substantial adverse pressure gradients and in some instances the reduction in turning by the blades would offset any gains due to reduction in loss. Very often the pressure surfaces of axial compressor blades have largely laminar boundary layers and it is clear from Fig. 1.16 why the pressure surface loss is usually assumed negligible. There is a benefit to be had by delaying transition to as late as possible on any surface and this strategy is adopted on the suction surface of low-loss blades such as the new so-called supercritical ones. The means by which transition can be delayed is to continue accelerating the flow, which in turn leads to peak high freestream velocities.

The dependence of the loss production on the cube of the free stream velocity gives a strong incentive to keeping the peak velocity as low as possible and decelerating as rapidly as possible once the flow is turbulent. Since the loss production is an integral along the surface in the flow direction it also follows that for low loss one wants as few surfaces as possible, giving an incentive to have the lowest solidity and the highest aspect ratio possible. Loss is only one factor to be considered in the design of compressors and there are other strong reasons for having low aspect ratio and fairly high solidity; a successful compressor is one in which low loss (high efficiency) is balanced with wide operating range, large pressure rise and satisfactory resistance to stall.

The analysis in this section has taken as an example the blade-to-blade flow in an axial machine. This was for convenience and because it was the area which had in the past received the most study. However, the ideas are much more general. The two-dimensional assumption would not be exact, for even on the blades of axial compressors the boundary layers are three-dimensional. Close to corners, on the hub and casing walls and generally in centrifugal compressors the flow is markedly three-dimensional. Nevertheless the crossflow velocities in the three-dimensional boundary layers are usually much smaller than the streamwise component, at least until very close to the wall. If for this purpose a turbulent eddy viscosity can be imagined to exist then the effect of the crossflow velocity on the three-dimensional energy dissipation integral is likely to be second order. As a means of assessing loss production the treatment here is probably an adequate guide, pointing to the low levels for laminar flow, the dependence to the third power on free-stream velocity and the insensitivity to the shape of the velocity profile (and therefore to the pressure gradient). Furthermore the ideas are probably transferable to more general flows where the boundary layer approximation is not valid and a full turbulent Navier–Stokes type solution is required.

1.6 Efficiency

The term efficiency finds very wide application in turbomachinery. Sometimes it is used for cascades or blade rows to relate the static pressure rise to the drop in stagnation pressure, but this is much less important than its use applied to complete stages or machines. For all machines or stages efficiency is defined as

$$\eta = \frac{\text{work into ideal compressor}}{\text{work into actual compressor}}$$

where the pressure rise from the ideal compressor is equal to that from the actual. There are several different ways of evaluating efficiency and these reveal different information. The ideal compressor will be reversible in the thermodynamic sense and depending on the duty it may be isothermal (constant temperature) or adiabatic (no heat flow to the gas). For duties where the gas is to be stored for a long time before use or where it is to be cooled further, for example in refrigeration plant, the isothermal compressor is the appropriate ideal. When the gas is to be used directly for propulsion or to be heated or burned then the adiabatic and reversible ideal is the sensible choice. The adiabatic application is more common and will be considered first.

Adiabatic compression

An ideal compressor which is both adiabatic and reversible cannot alter the entropy of the gas flowing through it; such compressors are usually referred to as isentropic with the corresponding efficiency described as the isentropic efficiency,

$$\eta_{\text{isen}} = \frac{W_s}{W_{\text{actual}}}$$

where the subscript s denotes entropy held constant. The restriction to adiabatic means that the work input is equal to the rise in stagnation enthalpy, $W = h_{02} - h_{01}$, but for brevity at this stage the distinction between stagnation and static enthalpy will be overlooked and so $W = h_2 - h_1$. The efficiency is then given by

$$\eta_{\text{isen}} = \frac{h_{2s} - h_1}{h_2 - h_1} \tag{1.39}$$

Ideal gases

Equation 1.39 is valid without restriction to the nature of the gas being compressed but great simplification is possible if the gas can be assumed to be ideal. In this case $h = c_p T$, with c_p, the specific heat at constant pressure, either constant or a function of temperature alone. If c_p is constant (i.e. independent of temperature as well as pressure) the gas is often referred to as perfect. For all ideal gases it follows that $p/\rho = RT$, with ρ the density and R the gas constant. For gases at low pressures (low in relation to their critical pressure) the ideal gas assumption is very good but in many engineering applications of compressors there are significant departures from the ideal gas relations. A convenient measure of the closeness to an ideal gas is the compressibility factor, defined by $Z = pv/RT$, which is identically equal to unity for an ideal gas. Most substances show very similar properties when expressed in terms of the reduced variables, 'the law of corresponding states', p/p_c and T/T_c, p_c and T_c being the critical pressure and temperature. Figure 1.17, taken from Reynolds (1979), plots values of Z against reduced pressure and Table 1.1 shows approximate values of reduced pressure and temperature for a range of common gases taken from the same source.

Fortunately most air compressors work within a range of reduced temperatures between about 2 and 5 so that the excursion of the compressibility factor from unity is very small even when the pressure is well above the critical value. The specific heat capacity c_p and the ratio of specific heats γ vary a little over this temperature range: in the case of air, which is very thoroughly documented, c_p ranges from 1.003 to 1.066 kJ/(kg K) and γ from 1.401 to

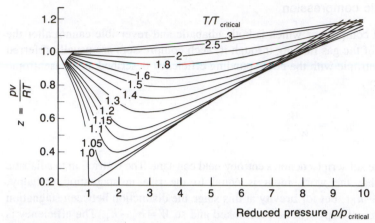

Fig. 1.17 Generalized compressibility chart, valid for most gases and vapours. (From Reynolds, 1979)

1.368. For the newer gas turbines with overall pressure ratios of around 40:1 it is necessary to include the variations in c_p and γ, as discussed below. For many gases, including refrigerants and hydrocarbons, the operating temperatures are not significantly higher than the critical temperature and extensive deviations from ideal gas behaviour are normal. Aerodynamic design, however, is always on the basis of an ideal gas and the recognition of the nonideal nature is relegated to the overall aspects of the machine, including assessment of efficiency, as discussed below.

For an ideal gas it is possible to write the isentropic temperature ratio in terms of the pressure ratio

$$\frac{T_{2s}}{T_1} = \left(\frac{p_2}{p_1} \right)^{(\gamma-1)/\gamma}$$

This yields the simple expression for the isentropic efficiency

$$\eta_{\text{isen}} = \frac{(p_2/p_1)^{(\gamma-1)/\gamma} - 1}{(T_2/T_1) - 1} \tag{1.40}$$

The irreversible compression and its reversible counterpart can be shown on the $T-s$ diagram, Fig. 1.18, and if the ordinate were replaced by enthalpy it would apply to nonideal gases as well. The constant pressure lines on the $T-s$ diagram have a slope proportional to the temperature so that they diverge as the temperature and the entropy increase. In other words the minimum temperature rise to produce a given pressure rise increases as either the initial

Table 1.1 Critical pressures and temperatures for common engineering substances

Values as given by the computational equations in Section 3 of Reynolds (1979); these may differ slightly from the true values.

Substance	Chemical formula	Critical temperature (K)	Critical pressure (MPa)
Helium-4	He^4	5.2	0.228
Hydrogen (para)	$H_2(p)$	32.9	1.28
Neon	Ne	44.4	2.65
Nitrogen	N_2	126	3.4
Air	—	133	3.77
Argon	A	151	4.86
Oxygen	O_2	155	5.04
Carbon dioxide	CO_2	304	7.38
Ammonia	NH_3	407	11.6
Water	H_2O	647	22.1
Methane	CH_4	191	4.60
Ethane	C_2H_6	306	5.01
Propane	C_3H_8	370	4.24
Isobutane	C_4H_{10}	409	3.68
Butane	C_4H_{10}	424	3.72
Pentane	C_5H_{12}	467	3.24
Hexane	C_6H_{14}	506	2.93
Heptane	C_7H_{16}	538	2.62
Octane	C_8H_{18}	568	2.40
Ethylene	C_2H_4	283	5.08
Propylene	C_3H_6	365	4.61
Refrigerant 14	CF_4	228	3.75
Refrigerant 503	Mixture	293	4.33
Refrigerant 23	CHF_3	299	4.84
Refrigerant 13	$CClF_3$	302	3.87
Refrigerant 502	Mixture	355	4.07
Refrigerant 22	$CHClF_2$	369	4.98
Refrigerant 500	Mixture	379	4.43
Refrigerant 12	CCl_2F_2	385	4.12
Refrigerant C-318	C_4F_8	389	2.78
Refrigerant 114	C_2Cl_2F	419	3.27
Refrigerant 11	CCl_3F	471	4.41

temperature or the entropy increases. In a multistage compressor this means that the work input required to produce a given pressure rise is greater for the later stages because the temperature is higher and also that the work input required by the later stages is raised because of the losses in the early stages.

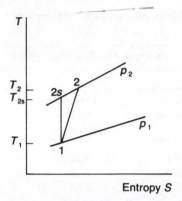

Fig. 1.18 Schematic temperature entropy diagram showing reversible adiabatic compression 1–2s and irreversible adiabatic compression 1–2

The isentropic efficiency of compressors of identical aerodynamic quality therefore gets lower as the overall pressure ratio is increased and this trend may be confusing or misleading. It can be avoided by using another definition for efficiency for adiabatic compressors, the so-called polytropic or small-stage efficiency, η_p. The polytropic efficiency removes the penalty for higher pressure ratios so that compressors of equal aerodynamic quality but significantly different pressure ratio would have the same polytropic efficiency though a different isentropic efficiency.

For all processes $T\mathrm{d}s = \mathrm{d}h - \mathrm{d}p/\rho$ and for an adiabatic and reversible one $\mathrm{d}h_s = \mathrm{d}p/\rho$. In a real compression the enthalpy rise $\mathrm{d}h$ will be larger than $\mathrm{d}h_s$ and it can be written

$$\mathrm{d}h = \frac{1}{\eta_p} \mathrm{d}h_s = \frac{1}{\eta_p} \frac{\mathrm{d}p}{\rho}$$

If η_p is assumed constant over a finite change in pressure it is easy to show for a perfect gas that

$$\frac{T_2}{T_1} = \left(\frac{p_2}{p_1} \right)^{(\gamma-1)/\eta_p\gamma}$$

and this may be rearranged to give the polytropic efficiency as

$$\eta_p = \frac{\gamma-1}{\gamma} \frac{\ln(p_2/p_1)}{\ln(T_2/T_1)} \tag{1.41}$$

Hence if T_2/T_1 and p_2/p_1 are known the overall or average polytropic efficiency for the compression process can be calculated. In a real multistage compressor if each stage had an equal value of η_p the polytropic efficiency for the whole machine would be equal to that of each stage. Using the temperature ratio derived above for a given polytropic efficiency one can then calculate the corresponding isentropic efficiency

$$\eta_{\text{isen}} = \frac{(p_2/p_1)^{(\gamma-1)/\gamma} - 1}{(p_2/p_1)^{(\gamma-1)/\eta_p\gamma} - 1} \qquad (1.42)$$

and the results obtained are displayed in Fig. 1.19 for a perfect gas with the ratio of specific heats equal to 1.4. The difference between the polytropic and isentropic efficiencies increases with raising of the pressure ratio and lowering of the efficiency, the two being equal for $p_2/p_1 = 1.0$ or for $\eta_p = 1.0$.

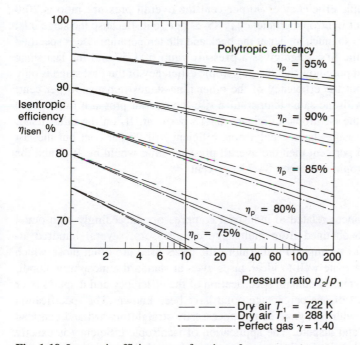

Fig. 1.19 Isentropic efficiency as a function of pressure ratio and polytropic efficiency for a perfect gas ($\gamma = 1.40$) and for dry air. (From Koch, 1964)

Even though air at pressures and temperatures around the standard atmosphere is quite a good approximation to a perfect gas, the pressure ratios now considered for gas turbines are sufficiently high and the level of precision required sufficienty fine that the deviations of air from the properties of a perfect gas must be allowed for. Figure 1.19 also shows a comparison of the predicted relation between isentropic efficiency, pressure ratio and polytropic efficiency for air obtained by Koch (1964) using the real properties of air tabulated by Keenan and Kay (1948). With the real gas the temperature has an effect and lines are drawn for two different inlet temperatures. As Fig. 1.19 shows, the isentropic efficiency can be significantly underestimated if air is treated as a perfect gas, the discrepancy becoming larger as the pressure ratio is increased, the inlet temperature increased or the polytropic efficiency reduced. The calculations giving rise to the lines in Fig. 1.19 were for dry air, and water vapour also affects the properties of air to an extent which needs to be considered at high pressure ratios.

As has been remarked above, the various stages of a multistage compressor contribute differently both to the overall pressure ratio and to the overall efficiency. Algebraic expressions can be derived to show this but it is perhaps more instructive to give a numerical example. Consider a ten-stage compressor with inlet air temperature 300 K and equal temperature rises per stage of 47.6 K. For this purpose air will be treated as a perfect gas. If each stage has a polytropic efficiency of 90 per cent the overall pressure ratio is 20.0 and the overall isentropic efficiency is 85.22 per cent. Because the stage inlet temperature is so much higher at the back and the temperature rise is specified as constant, the first stage gives a pressure ratio of 1.59 and the last stage only 1.22. Suppose now that the polytropic efficiency of the first stage is only 85 per cent but the efficiency of the other nine stages remains 90 per cent: in this case with the same temperature rise the overall pressure ratio would be 19.5 and the isentropic efficiency 84.15 per cent. If, on the other hand, the first nine stages have a polytropic efficiency of 90 per cent but the last stage only 85 per cent then the overall pressure ratio would be 19.8 and the overall isentropic efficiency 84.73 per cent.

Real gases

In many instances related to industrial compressors exceedingly high outlet pressures are obtained from multistage compressors, several hundred atmospheres, for example. Furthermore the gases used are often those which do not approximate well to ideal gases even at standard atmospheric conditions. Contracts may contain specification of the efficiency and disputes over as little as half of one percentage point have been known. The specification of efficiency with real gases is not necessarily straightforward and can lead to confusion and dispute. The application of isentropic efficiency is exactly the same as that for ideal gases except that the enthalpy rise corresponding to the isentropic pressure rise must be obtained from an appropriate empirical data correlation. It is normal, however, to use a polytropic efficiency as the measure of performance to take out the bias which would otherwise be introduced in favour of low pressure ratio machines. Given a satisfactory equation of state the differential expression defining the polytropic efficiency, $dh = (dp/\rho)/\eta_p$, can be integrated numerically to deduce an overall value for η_p from the measured values of T_2/T_1 and p_2/p_1. This process can be as accurate as the equation of state allows and is consistent in the sense that no additional assumptions have been introduced: within the limits of the assumption of equal η_p for each stage and a given equation of state the method is exact provided the numerical scheme is accurate. Examples of this procedure have been given by Nathoo and Gottenberg (1983) and independently by Huntington (1985).

The aim is normally to derive the overall value of η_p, so this is a constant in the integration, but if the intention were to investigate the thermodynamics of the compressor the numerical integration could be performed with η_p allowed to vary from stage to stage. The actual machine will not have the same efficiency for each stage but one which varies; which stage is more efficient depends on the designer's choice. Fozi (1985) gives an example of a six-stage compressor with ethylene entering at 25 bar and 37°C and leaving at 224 bar and 262°C. The polytropic efficiency across the entire machine is about 75 per cent, but representative polytropic efficiencies for the individual stages were estimated to be, from front to back 81.0, 75.9, 71.6, 73.1, 64.0 and 63.7 per cent (the lower efficiencies towards the rear were because the flow passage area was reduced as the density was increased). Taking a single average efficiency is a perfectly satisfactory procedure for establishing an overall assessment of the performance of a machine but it is possible to become confused in thinking about the real machine. A different outlet temperature T_2 would be produced for the same pressure ratio if the schedule of efficiencies for the different stages were altered: the overall performance of the machine depends on the distribution of efficiency in relation to the temperature and pressure through the machine, remembering that pressure and temperature are not related in this case by a simple ideal gas relation.

Nowadays numerical methods are so common and computers so readily available that the use of numerical integration to derive the polytropic efficiency seems natural. Twenty fives years ago this did not seem to be the case and Schulz (1962) devised a simple approximate method for calculating η_p. The method essentially treats the fluid as an ideal gas, so that the compression is assumed to follow a polytropic compression $p/\rho^n = $ constant, with the value of n determined from the end points of the compression. The Schulz method has become the industry standard as part of the ASME Power Test Code 10-65. It is very much an approximation which is difficult to assess precisely because the errors depend on the exact form of the equation of state for the substance being compressed and on the route in property space (i.e. how p varies with ρ) which the compression takes. An accurate equation of state is clearly essential. Huntington (1985) compared his numerical integration with Schulz's method and another approximate method by Mallen and Saville (1977) for four cases of compression with a range of conditions, one with Refrigerant-12, two with ethylene and one with carbon dioxide. Errors between Huntington's numerical scheme, which is essentially exact, and Schulz's method of 0.2, 1.7, 1.6, and 1.3 per cent were found. Errors of roughly similar magnitude but sometimes different sign were found with the method of Mallen and Saville. Huntington was able to derive an approximate method which agreed very closely with his numerical method (which is essentially exact if the equation of state is known). It was pointed out in the discussion of Huntington's paper that ultimately what matters most is that the supplier and customer should agree in advance on what method of calculation should be used since the differences are too small to be interesting from the point of view of fluid mechanics and either will give an adequate relative assessment of quality. In comparing tests with different substances the manufacturer will want a consistent, reliable and realistic method of calculating efficiency and for this a numerical scheme or the approximate method devised by Huntington would seem superior to earlier methods.

Isothermal compression

The ideal isothermal compression is shown in Fig. 1.20 as the path $1-2T$ on the temperature entropy diagram. The ideal process will, of course, be reversible. The corresponding process for an irreversible adiabatic compressor is shown by the path $1-2$ and roughly identifying the enthalpy rise with temperature rise it can be seen that a great deal of unnecessary work input has taken place which must then be 'undone' by cooling from T_2 down to T_{2T}. An alternative irreversible process is shown with the path $1-2'$, where at the intermediate pressure p_{int} the gas is cooled before a second stage of compression to the final pressure p_2. Intermediate cooling brings marked savings in energy input and frequently cooling may be introduced at several stages through a high pressure ratio compressor.

To calculate efficiency for an isothermal process more careful reasoning must be introduced than in the case of the adiabatic process. If the gas can be adequately approximated as ideal the enthalpy is a function only of the temperature and will be equal at stations 1 and $2T$ in Fig. 1.20. For real gases this is no longer the case and the enthalpy will be lower at $2T$ than at 1 even though the temperatures are equal. Now it can be shown that the smallest work input to produce a given change in property is equal to the change in the availability function, Haywood (1980), so for unit mass flow the minimum work input to produce the compression from $1-2T$ is

$$W_{min} = (h_{2T} - T_0 s_{2T}) - (h_1 - T_0 s_1)$$

Here T_0 is the temperature of the surrounding environment to which heat can be rejected. This expression is valid for real gases as well as ideal ones but if the gas is ideal then it can easily be shown that the entropy change is $s_{2T} - s_1 = -R\ln(p_2/p_1)$, where R is the gas constant, and $h_{2T} = h_1$. A formal statement of the isothermal efficiency is therefore

$$\eta_{isothermal} = \frac{(h_{2T} - T_0 s_{2T}) - (h_1 - T_0 s_1)}{W_{actual}} \tag{1.43}$$

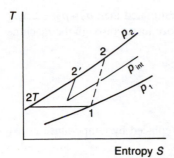

Entropy *S*

Fig. 1.20 A schematic temperature entropy diagram showing an irreversible adiabatic process 1—2 (broken line), an adiabatic process with intercooling at pressure p_{int} and an isothermal compression 1—2T

which for an ideal gas reduces to

$$\eta_{\text{isothermal}} = \frac{R\ln(p_2/p_1)}{W_{\text{actual}}}$$

Efficiency in terms of pressure loss

It is sometimes convenient to think of efficiency as the ratio of the actual pressure rise to that which would be produced with the same work input if the process were ideal. Restricting attention to ideal gases under adiabatic conditions the same work input corresponds to equal temperature rise. The comparison is illustrated in Fig. 1.21 where p_2 is the pressure actually produced whereas p_{2i} is the ideal pressure rise for the same work input. The efficiency may then be written

$$\eta = \frac{\Delta p}{\Delta p_{\text{ideal}}} = \frac{p_2 - p_1}{p_{2i} - p_1} \tag{1.44}$$

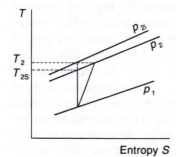

Entropy *S*

Fig. 1.21 A schematic temperature entropy diagram for a compression process with pressure loss $p_{2i} - p_2 = \Delta p_L$

If the pressure losses in the machine can be estimated then $p_2 = p_{2i} - \Sigma \Delta p_L$ where $\Sigma \Delta p_L$ is the sum of the stagnation pressure losses through the machine then the efficiency can be rewritten as

$$\eta = 1 - \frac{\Sigma \Delta p_L}{p_{2i} - p_1}$$

For a perfect gas the pressure ratios may be replaced by temperature ratios so that

$$\eta = \frac{(T_{2s}/T_1)^{(\gamma-1)/\gamma} - 1}{(T_2/T_1)^{(\gamma-1)/\gamma} - 1}$$

If the temperature rise is much less than the inlet or outlet temperature, which it will be for most axial stages and for many entire low-speed compressors, then this equation may be expanded by the binomial theorem to give

$$\eta = \frac{T_{2s} - T_1}{T_2 - T_1}$$

which is the form for the isentropic efficiency, equation 1.39, in the case of an ideal gas. Giving efficiency in terms of the pressure loss is convenient when the temperature changes are very small, for example in low-speed machines used for research, or when the correlation for loss is given in terms of pressure loss. In the next chapter the loss correlation will be used in this way to estimate hypothetical stage efficiency.

The loss is given for blade rows in terms of the relative flow velocities in the appropriate frame of reference fixed to the blade row. Care is needed because the stagnation pressures (and the loss in stagnation pressure) change when the frame of reference is changed. This difficulty can be avoided by working with the entropy, since entropy is equal by definition for stagnation and static conditions and is therefore independent of the frame of reference adopted. Use is made of $T_0 \Delta s = \Delta h_0 - \Delta p_0 / \rho_0$, where Δ denotes the changes because of loss. By definition the loss is the drop of stagnation pressure at a fixed value of stagnation enthalpy so that $\Delta h_0 = 0$ and the entropy can be related directly to the loss in stagnation pressure. The rise in entropy depends on the local value of stagnation temperature T_0 which is varying throughout the compression process, and for complete consistency the entropy change should be obtained from an integration. Since the object of converting to entropy rise from stagnation pressure loss is to allow a proper summation to be made this does not matter and if the value of Δp_0 is given at the blade row or impeller trailing edge it can be converted to entropy loss using the local value of temperature. A more fundamental problem arises if the temperature is very non-uniform at any section, but for compressors this is usually small enough that it is not important.

What can be very important in determining efficiency is that the stagnation temperature is often significantly non-uniform downstream of blade rows. For reasons which will be addressed elsewhere in the book the stagnation temperature is often higher in rotor wakes than in the free-stream, but the wake fluid has a lower axial velocity so that it tends to collect towards the pressure surface of stators, see Kerrebrock and Mikolajczak (1970). Positioning temperature measuring devices towards the pressure side of stator passages will give an overestimate of the stagnation temperature whereas measuring the temperature towards the suction side will give an underestimate. If efficiency is being deduced from temperature measurements, as it often is, the measurement of temperature near the pressure and suction surfaces will give under- and over-estimates of efficiency respectively.

Total-to-total and total-to-static efficiency

The work input to an adiabatic compressor is equal to the rise in stagnation enthalpy. The ideal work input to which this is compared in the efficiency can be the difference in either stagnation or static enthalpies. Usually the inlet condition is taken as the stagnation enthalpy but at outlet either stagnation or static enthalpy can be used. The former is the so-called total-to-total efficiency η_{tt}, the latter the total-to-static efficiency η_{ts}. The decision as to which is more appropriate rests on the same issues as the choice of pressure ratio discussed earlier in this chapter. If the outlet dynamic pressure is usable then the stagnation outlet conditions may be more relevant but if the dynamic pressure is to be wasted then the static outlet condition is far more appropriate. If the outlet flow is uniform and the Mach number known, the static and stagnation conditions can be interchanged very easily for the reversible process, assuming that the gas is ideal. If conditions are not uniform there is no simple way of converting between total-to-total and total-to-static efficiencies. The total-to-total efficiency is always higher than the total-to-static and there is an incentive to use it when it is the aim to stress the quality of a compressor. For machines with very high pressure ratios the difference between the outlet stagnation and static pressures becomes a small fraction of the overall pressure rise and the distinction between the two efficiencies ceases to be significant.

2 *General design considerations*

2.1 Introduction

In the design of any compressor the initial decisions on the layout and duty determine to a large extent problems to be encountered and the level of efficiency to be achieved. It needs to be recognized that the single most important design decision is the choice of stage loading, usually meaning the pressure rise in relation to the number of stages and the rotational speed. If, for example, unduly high loading is required of one or more components it is probable that no subtlety of design will render the overall performance satisfactory. Great skill and extensive commercial databases may be involved in making the initial decisions, steering the choice between ambitious goals and safely realizable ones. Occasionally the preliminary design is not given the serious attention it deserves and the results may be catastrophic.

The decision to have an axial or a radial compressor (radial compressors are very often termed centrifugal) is one of the basic preliminary decisions of this section and this excludes a potentially wide class of mixed flow machines. The mixed flow compressor is rarely used, probably because of the limited experience and data existing for it, although it would seem to have a very natural niche. Amongst the problems of the mixed flow compressor compared to the axial or radial is weight, with the mixed flow machine coming out longer than the radial but of similar massive construction. The diffuser downstream of the mixed flow impeller has also been found to be a problem, with performance well down on what was expected of the radial machine. With sufficient effort and appropriate design there is reason to expect that the diffuser performance could be greatly improved.

The decision to choose either an axial or a radial compressor rests on many factors, not least the experience of the company building the machine. For aircraft propulsion the high flow rate per unit area of the axial is a big advantage but when the blade height becomes very small the advantage swings to the radial: helicopter engines usually employ radial compressors and they have even been proposed for the later stages of large jet engines with very high pressure ratios. Highly loaded radial compressors seem to have generally lower efficiencies than axial machines, but this is not altogether clear. In cases where the impeller can be precision cast, such as those in turbochargers for automotive use, the simplicity of the radial compressor means that it has a huge cost advantage over the axial. In the remainder of this chapter it will be assumed that the decision of axial versus radial has been made.

It is a common factor with all compressors that when several stages are used together in series there is a serious problem of matching the stages so that the outlet flow from one stage is acceptable to the next. This becomes more acute as the overall pressure ratio across the machine increases because of the large density changes that result. Because the pressure ratio, and therefore the density ratio, is roughly proportional to the square of the rotational speed it is a common difficulty to match multistage machines at both the full design speed and at reduced speed, leading to many problems not least that of starting the compressor or engine. This aspect of compressors is considered in this chapter and is illustrated with reference to the particular problems of axial compressors.

What follows in this chapter are some fairly simple ideas relating to the overall performance of compressors with the treatment being essentially one-dimensional. This begins with the axial and then moves on to the radial compressor. The final section of the chapter is the elementary consideration of stage matching for axial compressors.

2.2 The axial compressor

In the preliminary design calculations are usually performed at a mean radius, called the pitchline in some work. Refinements may be introduced to assess the blade loadings at hub and casing, particularly if the ratio of the hub and casing radii is low. (NB: This is sometimes referred to as the hub–tip ratio. Here the word tip will not be used because of its possible ambiguity; for a stator cantilevered inwards, does tip refer to the hub or the casing end of the blade? The word casing will be preferred instead.) Criteria have to be chosen for satisfactory blade loading, pressure rise at the walls and maximum Mach number.

The blade loading is now usually assessed by diffusion factor or alternatively equivalent diffusion ratio, both derived by Lieblein and described in Chapter 4. Here diffusion factor will be used. Essentially this relates empirically the peak velocity on the suction surface of the blade to the velocity at the trailing edge, with one component due to the one-dimensional deceleration of the flow and the second due to the turning of the flow. The term related to the turning introduces the blade solidity. For a simple two-dimensional geometry diffusion factor reduces to

$$DF = 1 - \frac{V_2}{V_1} + \frac{\Delta V_\theta}{2\sigma V_1} \qquad (2.1)$$

where V_1 and V_2 are the average velocities into and out of a blade row in a frame of reference fixed to the blade, ΔV_θ is the change in whirl velocity in the row and σ is the solidity, equal to blade chord/blade pitch. Values of DF in excess of 0.6 are thought to indicate blade stall and a value of 0.45 might be taken as a typical design choice. Over the last few years attention has been focussed more on the endwall region as the limit for loading and the weight given to the diffusion factor has decreased.

The criterion to be adopted for endwall loading or pressure rise is less clear, mainly because the fluid mechanics is still not understood. Methods analogous to that produced by de Haller (1953) are still current and this will be discussed more in later chapters. de Haller deduced that the velocity out of a blade should not be less than about 0.75 times the inlet velocity if the performance is to be satisfactory. This is equivalent to requiring that the static pressure rise at the wall should not exceed about 0.44 times the dynamic pressure into a blade row. The de Haller criterion has not been found to be entirely satisfactory. More recently Koch (1981) has published a method which relates stage pressure rise capability to the mean height (i.e. mid-span) solidity averaged over the stage; it is based on a large number of measurements in multistage compressors and will be discussed more fully in Chapter 9. The most common method of assessing what is acceptable loading at the wall is probably by reference back to previous designs by the same manufacturer, it is now very rare for an organization to be designing an axial compressor for the first time! The general view seems to be that a stage pressure rise not exceeding about $0.4U^2$ is reliable.

In looking at the trends in multistage compressor design it is very helpful to take advantage of the results given by Wisler (1988) in a comprehensive set of lecture notes, relating mainly the work of his company, General Electric. These will be referred to many times in this chapter.

The limit on maximum Mach number is flexible and depends to a large extent on the balance between high efficiency and high pressure ratio per stage being sought. The loss in efficiency with Mach number is nowhere near as serious as was once thought. Inlet relative Mach numbers of 1.4 are now common at the tips of first-stage rotors in multistage compressors for aircraft and the flow may even be slightly supersonic into the third stage. As the Mach number is increased the operating range reduces, i.e. the difference between the mass

Table 2.1 Compressor developments by General Electric

Year	Designation	Design pressure ratio	Number of stages	Corrected tip speed (m/s)
late 50s	CJ805/J79	12.5	17	291
1969	CF6-50	13.0	14	360
1974	CFM56	12	9	396
1982	E^3 engine	23	10	456

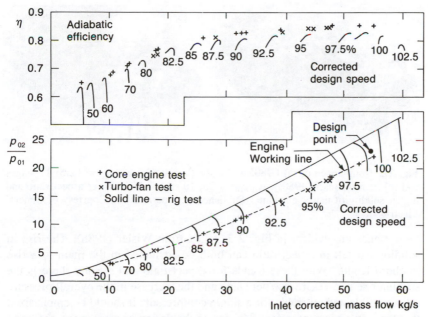

Fig. 2.1 The pressure ratio and efficiency characteristics of the General Electric E³ compressor. (Published with permission, Courtesy of General Electric Co.)

flow for choke and surge is reduced. An important reason for keeping the speeds of industrial compressors down is to maintain the widest possible operating range. The numbers given in Table 2.1 are taken from Wisler (1988) and show the trend for much higher tip speeds from one manufacturer of jet engines, General Electric, but similar trends would be found for other companies as well as for land-based machines.

The pressure rise—mass flow and efficiency—mass flow characteristics for the compressor of the E³ engine are shown as Fig. 2.1, the solid lines being from tests of a compressor rig and the crosses from engine tests. Just prior to surge at 102.4 per cent speed the very high pressure ratio of 29:1 was achieved. It is interesting that the engine performance is better than the rig, partly because the Reynolds number was higher but mainly because the tip clearances were smaller for the engine. The compressor was designed with six variable stagger stator rows but only four were used for the performance map shown. The peak adiabatic efficiency corresponds to a polytropic efficiency of 90.4 per cent, a high value, and evidently the high pressure ratio per stage does not have to be bought at the expense of low efficiency. A photograph comparing the rotors for the E³ compressor with that of the much earlier CJ805/J79 is shown in Fig. 2.2 from which the very much higher solidity and lower aspect ratio of the more recent compressor is very obvious.

Decisions have to be taken regarding the blade chord and the number of blades. Increasing the chord reduces the aspect ratio (height/chord) and increases solidity (chord/pitch) for the same annulus and number of blades. Both

(a)   (b)

Fig. 2.2 Comparison of (a) CJ805/J79 rotor (late 1950s) $p_{02}/p_{01} = 12.5$; 17 stages and (b) E^3 rotor (early 1980s), $p_{02}/p_{01} = 23$; 10 stages. Note lower aspect ratio and higher solidity of newer machine. (Published with permission, Courtesy of General Electric Co.)

these trends are evident in Fig. 2.3, taken from Wisler (1988). The rise in solidity and fall in aspect ratio can both be attributed in the main to a rise in chord length. With these trends for aspect ratio and solidity there is the striking rise in pressure rise per stage and the increase in the overall pressure ratio possible and utilizable for a single compressor. It should be emphasized that the single most important decision in the design process is the choice of a realistic stage loading. An over-ambitious choice may lead to untold problems later with little possibility of actually achieving the combination of efficiency, pressure ratio, mass flow and range originally intended.

Back in the 1950s it was believed that the trend would be towards high aspect ratio blades to give a short compressor, mainly, it seems, because the blade behaviour well away from the endwalls was comparatively well understood and this was the direction of development which consideration of the blades seems to indicate. The trend was reversed mainly because large chord blades are more effective in the endwall regions and it is these regions which are crucial in determining both the efficiency and the stall point. High aspect ratio blades were long and thin and had atrocious vibration problems. The change towards low aspect ratios was not the result of an understanding of the processes involved but consideration of the trends for performance of different designs. Wennerstrom (1986) has described the catastrophic effect of adopting high aspect ratio blading.

There are several performance goals to be compared, in particular pressure rise, efficiency and operating range (operating range might be defined as the ratio of the difference between maximum and minimum mass flow to the design value). The evidence suggests that for a good compressor near the design point efficiency tends to be slightly lower if the solidity is on the high side (and the aspect ratio low) but the pressure rise and operating range are greater. The major trend over the last 30 years has shown a rise in efficiency but a more marked rise in overall pressure rise as Fig. 2.4, from Freeman and Dawson (1983), shows for Rolls-Royce compressors.

There are special problems that arise from combining stages to form

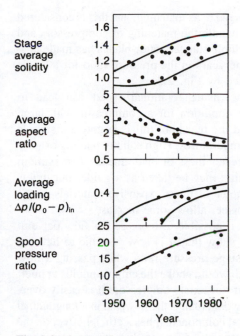

Fig. 2.3 The trend in compressor geometry (solidity and aspect ratio) and in performance (stage loading and spool pressure ratio) with time. (From Wisler, 1988)

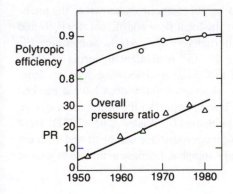

Fig. 2.4 The variation in overall pressure ratio and in polytropic efficiency for gas turbine compressors. (From Freeman and Dawson, 1983)

multistage compressors, usually referred to as matching, and this is considered later in this chapter. The ability to handle the matching of compressors and the operation of several rows of variable stagger stator blades has made possible the very large increase over the years in the pressure ratio for a single compressor spool which is illustrated by Fig. 2.3.

Increasing Mach number by increasing rotational speed can lead to mechanical problems. The limiting condition for a compressor with large pressure ratio is normally reached at the rear hub; this is largely a materials problems connected with the high temperatures. High solidity blading exacerbates the problem because of its greater mass of blade metal. A maximum hub rotational speed of about 380 m/s may be taken as a guide, but this is not a firm boundary because the choice of more expensive materials or the use of a heavier disc would, at a price, allow some increase.

Increased rotational speed makes it possible to increase the flow per unit area; Freeman and Dawson (1983) show that it is now possible to have an efficient compressor giving a high stage pressure ratio while passing a flow approaching 90 per cent of that which would choke the empty annulus at inlet.

With the emphasis on blade design for axial compressors it is easily overlooked that the overall meridional flowpath (that is the flowpath in a longitudinal cross-section showing axial and radial components) has a crucial effect on the design and the performance of a compressor. The aerodynamic problems are, for example, greatly relieved if the hub radius can increase from front to back, whilst they are made worse if the annulus area is too large towards exit. Decisions taken at the preliminary stage in laying down the annulus shape and choosing the inlet and outlet radii can effectively determine whether a compressor will be satisfactory or not and may be far more influential than subsequent decisions regarding the blade shape.

Fundamental to all of the aerodynamic design are the basic decisions of an aerodynamic nature. At the blade mid-height (sometimes known as the pitch-line radius) a choice must be made for the local flow coefficient $\phi = V_x/U$ and the stage loading $\psi = \Delta h_0/U^2$ (or alternatively $\Delta p_0/\rho U^2$). Sometimes the degree of reaction $R = \Delta h_{rotor}/\Delta h_{stage}$ (or the equivalent in terms of static pressure rise) is treated as important too. Such decisions are separate from choice of solidity, blade section etc., although solidity does have a marked effect on the choice of loading. A fascinating report of the preliminary design of a multistage compressor has been published by Wisler *et al.* (1977). Here the interactions between different design decisions are demonstrated with the advantage of the realistic estimates and extensive database available to a large company.

Parametric study for a repeating axial stage

The decisions on aerodynamic design take into account amongst many other things the compressibility of the flow, and this will be considered in later chapters. However, many of the trends in most stages of a multistage compressor are not related to compressibility and the flow can be understood adequately by treating it as incompressible. In this section some parametric studies will be used to demonstrate the effect of various choices of design variables. The blading is two-dimensional, so that all endwall effects are ignored. A simple axial stage will be considered with identical velocities in and out; such a stage is often called a repeating stage.

In the parametric study the diffusion factor will be taken as the measure of blade loading and comparisons will be made varying flow coefficient ϕ, blade loading ψ and degree of reaction R as independent variables. Some estimate of efficiency is made utilizing data measured in two-dimensional cascades by Lieblein (discussed in Chapter 4). This shows that as a reasonably good approximation the blade profile loss is given by

$$\omega = \frac{p_{01} - p_{02}}{p_{01} - p_1} = \frac{0.007 \cdot 2\sigma}{\cos\alpha_2} \tag{2.2}$$

where $\sigma = c/s$ is solidity and α_2 is the flow angle out of the blade row. For incompressible or low Mach number flow it is possible to write the dynamic pressure in the form

$$p_{01} - p_1 = 1/2 \, \rho V_1^2.$$

In calculating the stage loss the individual contributions of rotor and stator are added so that the efficiency η can be written in terms of the loss in stagnation pressure in the stator and rotor row as

$$\eta = \frac{\text{useful work}}{\text{work input}} = \frac{\text{work input} - \text{losses}}{\text{work input}} = 1 - \frac{(\Delta p_{0\text{rotor}} + \Delta p_{0\text{stator}})}{\rho \Delta h_0} \tag{2.3}$$

where Δh_0 is the stage enthaply rise and Δp_0 are the losses.

Because the efficiency η is often nearly equal to unity it is more helpful to consider

$$1 - \eta = \frac{\Delta p_{0\text{stator}} + \Delta p_{0\text{rotor}}}{\rho \Delta h_0}$$

The velocity triangles for the examples considered are shown in Fig. 2.5. The

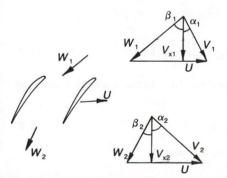

Fig. 2.5 The velocity triangles for an axial rotor row

restrictions adopted mean that

$$V_{x1} = V_{x2} = V_{x3},$$

where station 3 is at stator outlet, and a number of expressions suited to the parametric study are readily found. For example the relative velocity into the rotor is given by

$$W_1^2 = V_x^2 + W_{\theta 1}^2$$

so that

$$(W_1/U)^2 = \phi^2 + (W_{\theta 1}/U)^2 \qquad (2.4)$$

The degree of reaction can be written

$$R = \frac{W_1^2 - W_2^2}{2\Delta h_0} = \frac{W_1^2 - W_2^2}{2U(W_{\theta 1} - W_{\theta 2})} \qquad (2.5)$$

and since $V_{x1} = V_{x2}$ it follows that

$$W_1^2 - W_2^2 = W_{\theta 1}^2 - W_{\theta 2}^2$$

and the degree of reaction is given by

$$R = (W_{\theta 1} + W_{\theta 2})/2U \qquad (2.6)$$

Rearranging equation 2.6 gives

$$2RU = W_{\theta 1} + W_{\theta 2}$$

whilst from the definition for the stage loading ψ

$$\psi U = W_{\theta 1} - W_{\theta 2}.$$

So that on rearranging

$$W_{\theta 1} = (2R + \psi)U/2 \text{ and } W_{\theta 2} = (2R - \psi)U/2. \qquad (2.7)$$

The relative velocity into the rotor can then be written, using this equation and 2.4, as

$$(W_1/U)^2 = \phi^2 + (\psi/2 + R)^2. \qquad (2.8)$$

Similarly for the absolute velocity into the stator

$$V_2^2 = V_{x2}^2 + V_{\theta 2}^2$$

leading to $(V_2/U)^2 = \phi^2 + (\psi/2 + 1 - R)^2.$ (2.9)

The expressions for V_2 and W_1 are needed in the calculation of loss. The flow angles are needed in order to find loss and diffusion factor. With the repeating stage condition $V_x = $ constant, it is easy to show that

$$\cos\beta_2 = \phi\{\phi^2 + (R - \psi/2)^2\}^{-1/2}$$

and $$\cos\beta_1 = \phi\{\phi^2 + (R + \psi/2)^2\}^{-1/2}.$$ (2.10)

The traditional view has been that 50 per cent reaction, i.e. an equal split of pressure rise between rotor and stator, is a natural and superior choice for an axial compressor. Figure 2.6 shows the variation of various parameters with stage loading ψ for a 50 per cent reaction stage. In this comparison the flow coefficient ϕ is constant and equal to 0.6. The efficiency in Fig. 2.6a has been calculated here and elsewhere in this section taking $\omega=0.007$ $(2\sigma/\cos\alpha_{out})$ for each blade row. In Figs 2.6a and c the solidity σ has been chosen to give a constant value of diffusion factor set here at the typical design value of $DF=0.45$. The rise in solidity with loading ψ so as to keep DF constant, shown in Fig. 2.6c is the principal cause of the fall in efficiency. The corresponding variation in static pressure rise is given in Fig. 2.6b, showing that with this combination of parameters the de Haller criterion is exceeded for $\psi>0.37$, about equal to a typical stage loading at design.

The choice of diffusion factor is arbitrary and the manner in which this varies with loading and solidity is shown in Fig. 2.6d. By raising solidity it is possible to operate at higher loading ψ for the same diffusion factor or at a lower

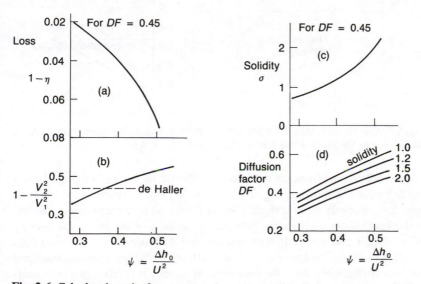

Fig. 2.6 Calculated results for a parametric study of a two-dimensional stage. Degree of reaction 50 per cent. Flow coefficient $\phi = V_x/U = 0.6$

diffusion factor for the same loading. Within the current assumptions the highest efficiency is produced by having the highest diffusion factor, a conclusion which would not have been materially altered if the degree of reaction had been somewhat greater or less than 50 per cent.

A common alternative choice of compressor to that with a given reaction is one for which the absolute flow into the rotor is axial (such a compressor would have no inlet guide vanes). The variation of various parameters with loading is shown in Fig. 2.7. As Fig. 2.7b shows, the degree of reaction for such machines is variable and high (for the repeating stage it is given by $R = 1 - \psi/2$), so that most of the pressure rise (and most of the loss) is produced in the rotors. For an equal diffusion factor in both rotor and stator the solidity is different in each, Fig. 2.7c — solidities in excess of 2 are normally unacceptable, not only because the losses are high but because the passages are inclined to choke. Because the static pressure rise is greatest across the rotor it is here that the de Haller number is first exceeded at $\psi = 0.36$, Fig. 2.7d. For a diffusion factor of 0.45 the efficiency of the 50 per cent reaction compressor is slightly higher for levels of loading ψ which do not exceed 0.42.

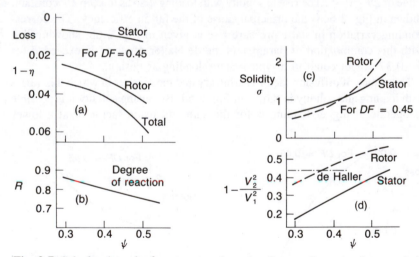

Fig. 2.7 Calculated results for a parametric study of a two-dimensional stage. Axial inlet flows to the rotor. Flow coefficient $\phi = V_x/U = 0.6$

The variation in efficiency with flow coefficient for a fixed loading, $\psi = 0.4$ and $DF = 0.45$, is shown in Fig. 2.8 both for 50 per cent reaction and for axial inlet flow. The efficiency is slightly lower for the 50 per cent reaction machine at the lower flow coefficients, principally because of the increased solidity in both rotor and stator. If both solidity and diffusion factor are each held constant (with the choice of $\sigma = 1.2$, $DF = 0.45$ and 50 per cent reaction) and the flow coefficient varied, the maximum efficiency is reached at a flow coefficient, ϕ, just above 0.6.

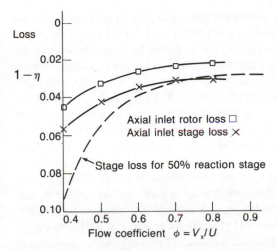

Fig. 2.8 The calculated variation in efficiency for a two-dimensional stage, considering *only* profile loss, as a function of flow coefficient. Diffusion factor held constant at 0.45, stage loading $\psi = \Delta h_0 / U^2$ constant at 0.4. Profile loss calculated from $(\omega \cos \alpha_{out})/2\sigma = 0.007$

In academic accounts of compressors the degree of reaction has attracted considerable attention in the past. The efficiency calculated from the blade profile losses is, however, only a very weak function of reaction. For a particular stage loading $\psi=0.4$, flow coefficient $\phi=0.6$ and blade loading $DF=0.45$ in each blade row the efficiency is shown as a function of reaction in Fig. 2.9. The variation, which is symmetric about $R=0.5$, shows a range of only about 1/2 per cent in efficiency from $R=0.1$ to $R=0.9$. A parametric study by Casey (1987) including the endwall losses has shown a similar insensitivity to the degree of reaction or, equivalently, to the shape of the velocity triangles.

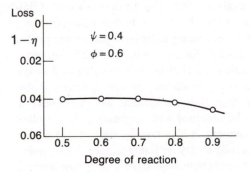

Fig. 2.9 The calculated variation in efficiency for a two-dimensional stage, considering *only* profile loss as a function of degree of reaction. Diffusion factor = 0.45. Profile loss calculated from $(\omega \cos \alpha_{out})/2\sigma = 0.007$

The efficiencies which have been shown above have been generally in excess of 95 per cent; for the combination $\psi=0.4$ and $\phi=0.6$ they are about 96 per cent. Real compressors of good aerodynamic design have an efficiency in the range 86 to 92 per cent so it is clear that the loss incorporated in this analysis fails to include the major contributors. This is not at all a new discovery and had been recognized by Constant (1939). Even if slightly different values had been assumed for the profile loss, or if a variation in loss with blade shape or diffusion factor had been included, it would still have been true that the efficiencies calculated by the above approach were much higher than those likely to be realized in practice.

The primary source of losses in an axial compressor is in the end-wall regions but the rules for estimating them, analogous to those for the profile losses, are less soundly based. It is still common to quote the estimate for the different loss sources given by Howell (1945), shown here as Fig. 2.10. Nowadays one would probably lump annulus and secondary loss together. Also modern designs tend to have a design flow coefficient a lot smaller than that shown in Fig. 2.10 and the entire diagram should be shifted to lower flow coefficients.

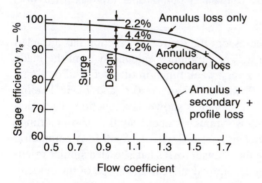

Fig. 2.10 Howell's breakdown of loss for an axial stage. (From Howell, 1945)

Efficiency does not seem to be strongly affected by the design value of flow coefficient over a reasonably wide range (this is true both for the efficiency measured for the whole stage and for efficiency calculated just on the profile loss). On the other hand the work input ψ, for a given level of diffusion factor and solidity, does rise rapidly with flow coefficient. A high value of ϕ would tend to raise the relative Mach numbers, which may be very deleterious, but the higher values of ψ possible at high flow coefficients would allow comparable pressure rises per stage to be obtained with operation at lower blade speeds and therefore lower Mach numbers. A high flow coefficient also implies a high mass flow per unit area, generally a very desirable feature. These points raise the question why most compressors nowadays have a fairly low average flow coefficient, ϕ near to 0.5 might be a typical value. In fact ϕ often varies at design through the length of the compressor, perhaps being higher at the front, to get a large mass flow, and lower at the rear, to reduce the dynamic pressure at outlet where it would be wasted. The wish to have low dynamic pressure at outlet is a considerable incentive to keep the flow coefficient low. The inlet flow to many compressors is already high in relation to the choking mass flow, and in some machines cannot be increased much further without risk of choking; given the high levels of blade speed this in turn tends to keep the flow coefficient at inlet fairly low.

There are other reasons for preferring flow coefficients nearer to 0.5 than one. At the simplest level it can be said that the low ϕ compressors have been found to operate better and this has encouraged designs to follow that route. Two explanations are offered; the first, a very simple one, relating to the stability of the whole machine, the other, based on an analysis by Smith (1956), relating to stability of the elements. In both cases the analysis given here is incompressible because the underlying processes are weakly affected by compressibility for most of the stages of the multistage compressor and the ideas are so much more transparent in incompressible flow.

First consider an axial stage as sketched in Fig. 2.5. The work input per unit mass is $U(V_{\theta 2} - V_{\theta 1})$ which can be written nondimensionally as

$$\psi = \Delta h_0 / U^2 = (V_{\theta 2} - V_{\theta 1})/U.$$

It is useful to express this is terms of the outlet angles relative to blade rows because, as a reasonable approximation, these remain constant as the flow rate is varied. The axial velocity V_x is here assumed equal into and out of the rotor so

$$V_{\theta 1} = V_x \tan\alpha_1$$

where α_1 is the flow direction into the stage, possibly the leaving angle from an inlet guide vane or an upstream stator.

For the flow leaving the rotor with relative flow angle β_2

$$V_{\theta 2} = U - W_{\theta 2}$$
$$= U\{1 - (V_x/U)\tan\beta_2\}$$

It follows that

$$\psi \quad = 1 - \phi(\tan\beta_2 + \tan\alpha_1) \tag{2.11}$$

where here the $\phi = V_x/U$ is considered a parameter which can be varied (as by a throttle). The result can be plotted on Fig. 2.11; the larger the values of α_1 and β_2 the steeper the slope of the line. Consideration of Fig. 2.5 shows that large values of $(\tan\beta_2 + \tan\alpha_1)$ correspond to blades chosen for low values of ϕ at the design point. The angles are fixed primarily by the design of the machine; they are not strictly constant, for the flow does not exactly follow the blade outlet angle but deviates by a few degrees and the deviation

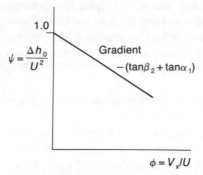

Fig. 2.11 The variation in work input with flow rate for a stage with equal inlet and outlet velocity

is a function of incidence, though a weak function until close to stall. The expected non-dimensional pressure rise characteristic would lie below ψ by an amount fixed by the losses.

Now stability considerations, discussed in Chapter 9, make it clear that steep characteristics, that is for large values of $(\tan\beta_2 + \tan\alpha_1)$, are advantageous in avoiding stall or surge. Even away from the stall point it is clear with a steep characteristic that a small reduction in ϕ for any reason brings about a large rise in work input and pressure rise which tends to correct it: such changes can take place over part as well as the whole of the annulus. It means that if a non-uniformity develops in the radial or the circumferential direction a compressor with a steep $\psi-\phi$ characteristic will work more strongly to smooth out the flow. An extreme example would be a stage in which the flow from both the rotor and stator rows leaves in the axial direction. For this case $(\tan\beta_2 + \tan\alpha_1)=0$ giving $\Delta h_0/U^2=1$ so there would be no attenuation of any non-uniformity. The ability to smooth out non-uniform flow entering the stage seems to be one reason for the preference for compressors designed to have a low flow coefficient ϕ at design.

An alternative and more searching way of considering this was given by Smith (1958). His analysis was compressible and also included a correction for the variation in deviation with local incidence. Exactly the same conclusions can be arrived at, with considerable simplification, by neglecting compressibility and assuming deviation is constant. The analysis can be simplified without loss of relevance by assuming an efficiency of unity so that across the rotor

$$p_{02} - p_{01} = \rho\Delta h_0 = \rho U(V_{\theta2} - V_{\theta1})$$
$$= \rho U\{U - V_{x1}\tan\alpha_1 - V_{x2}\tan\beta_2\} \qquad (2.12)$$

Here V_{x1} and V_{x2} are *not* assumed equal since the section considered is a small part of a stage of substantial circumferential and spanwise extent wherein there can be flow movements allowing the axial velocity to vary. Considering perturbations to equation 2.12 gives

$$\delta p_{02} = \delta p_{01} - \rho U[\delta V_{x2}\tan\beta_2 + \delta V_{x1}\tan\alpha_1] \tag{2.13}$$

where β_2 and α_1 are assumed to vary by a negligible amount as the local flow rate changes (as remarked above this is normally quite a good assumption over a wide operating range).

At entry to the rotor

$$p_{01} = p_1 + \tfrac{1}{2}\rho V_1^2$$

and if p_1 is assumed constant

$$\delta p_{01} = \rho V_1 \delta V_1 = \rho V_{x1} \delta V_{x1}/\cos^2\alpha_1$$

or $\qquad \delta p_{01}/U^2 = \rho\phi_1\delta\phi_1/\cos^2\alpha_1 \tag{2.14}$

Similarly at outlet from the rotor

$$p_{02} = p_2 + \tfrac{1}{2}\rho V_{x2}^2 + \tfrac{1}{2}\rho V_{\theta 2}^2$$

so that $\qquad \delta p_{02}/U^2 = \delta p_2/U^2 + \rho\phi_2\delta\phi_2 + \rho(1-\phi_2\tan\beta_2)(-\delta\phi_2\tan\beta_2)$
$$\tag{2.15}$$

It is plausible to assume that there will be no spatial variation in static pressure downstream of the blade row and therefore δp_2 is set to zero. Equations 2.13 and 2.15 for δp_{02} may then be equated to give

$$\rho(1+\tan^2\beta_2)\phi_2\delta\phi_2 = \delta p_{01}/U^2 - \rho\delta\phi_1\tan\alpha_1$$

Using 2.14 to remove the upstream perturbation in stagnation pressure

$$(1+\tan^2\beta_2)\phi_2\delta\phi_2 = \{\phi_1/\cos^2\alpha_1 - \tan\alpha_1\}\delta\phi_1 \tag{2.16}$$

so there is now an equation linking the axial velocity perturbations upstream and downstream of the blade row.

Now Smith defined a recovery ratio

$$R_R = 1 - \delta p_{02}/\delta p_{01}$$

and for a blade row which causes any disturbance in the upstream stagnation pressure to be completely eliminated downstream the recovery ratio is unity. It follows, using equations 2.16 and 2.13, that

$$R_R = (\cos^2\beta_2\tan\beta_2)/\phi_2 - (\cos^2\alpha_1\cos^2\beta_2\tan\alpha_1\tan\beta_2)/\phi_1\phi_2$$
$$+ (\cos^2\alpha_1\tan\alpha_1)/\phi_1 \tag{2.17}$$

which is the form arrived at by Smith (1958) when the flow is treated as incompressible and the expression intended to allow for the variation in deviation is dropped.

In a discussion of this paper Ashby remarked that the NACA had been working along similar lines and that they had arrived at the expression

$$R_R = 1 - \left[\cos(\beta_1 + \alpha_1) \cos(\beta_2 + \alpha_2) \frac{\cos\alpha_1 \cos\beta_2}{\cos\beta_1 \cos\alpha_2} \right] \quad (2.18)$$

which, it can be shown, is exactly equivalent to equation 2.17. This expression brought out a more easily grasped physical picture because it shows that complete elimination of the upstream perturbation is obtained, i.e. $R_R = 1.0$, when either $(\beta_1 + \alpha_1)$ or $(\beta_2 + \alpha_2)$ is equal to 90°. By drawing out the velocity triangles (Fig. 2.12) it can be seen that these conditions will be satisfied when the flow coefficient is small. There is, of course, no need for the recovery ratio R_R to be exactly equal to one, rather it is that values which do not reduce the disturbance significantly (so that $R_R \ll 1$) have obvious disadvantages. If either $(\beta_1 + \alpha_1)$ or $(\beta_2 + \alpha_2)$ is greater than 90° then R_R is greater than one and regions of high loss in one frame of reference become regions of excess stagnation enthalpy and pressure in the frame of reference downstream.

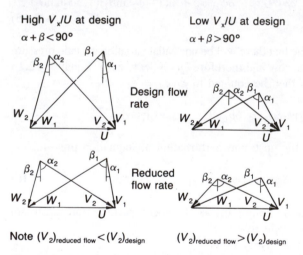

Fig. 2.12 Velocity triangles for axial stages for different design values of V_x/U. At reduced flow α_1 and β_2 (exit flow angles) equal to those at design

2.3 The radial compressor

The centrifugal or radial compressor finds the most widespread use of any type. At one extreme there are machines for aviation producing pressure ratios of more than 8:1 from one stage. These have impellers machined from solid

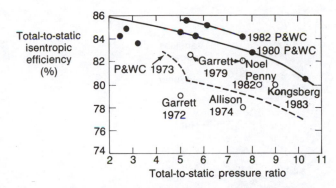

Note: All efficiencies are at 11% surge margin

Fig. 2.13 The advances in centrifugal compressors for aircraft gas turbines up to 1983. (From Kenny, 1984)

billets of titanium alloy. At the other end of the range very low cost devices in domestic appliances are fabricated out of sheet metal where the overriding concern is to keep the cost low. Research and development activity has concentrated on those areas where the performance is high, in particular those related to aviation. Figures 2.13 and 2.14 are taken from a fairly recent paper by Kenny (1984) and give some idea of what can now be achieved and how this has varied over the years. (Most of the measured points in these figures were obtained by Pratt and Whitney of Canada who have been active in the development of centrifugal compressors for aircraft application. Most are for small machines passing on the order of 1 or 2 kg/s.) Kenny also notes that over the past 20 years efficiency for a given pressure ratio has risen steadily at about $\frac{2}{3}$ per cent per year.

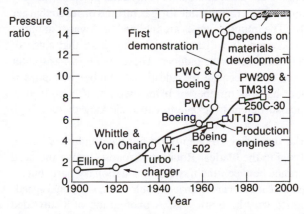

Fig. 2.14 The evolution of centrifugal compressors reflected in the pressure ratio per stage. (From Kenny, 1984)

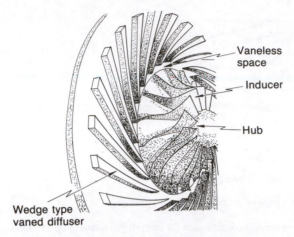

Vaneless
space

Inducer

Hub

Wedge type
vaned diffuser

Fig. 2.15 The geometry of the centrifugal compressor: the example shown has an inducer, an unshrouded impeller and a wedge-type vaned diffuser

Radial machines have a nomenclature of their own and some of this is illustrated in Fig. 2.15. The rotating part is usually referred to as the impeller and the blades as vanes. The stationary part is known as the diffuser and it may have vanes or be vaneless, depending on the duty; that shown in Fig. 2.15 is vaned. Very often the diffuser is surrounded by a volute or scroll to duct the flow away and, it is hoped, to decelerate it further. It is one of the special features of radial turbomachinery that the moving and stationary components are very different, whereas in the axial machine they are relatively similar.

Radial compressors for use in gas turbines or turbocharges form a special class and have features in common. The impeller is unshrouded, that is to say the moving blades move past the stationary shroud with a small clearance, and the impeller is predominantly axial at the inlet. This axial portion is usually referred to as the inducer. A modern impeller for gas turbine or turbocharger use is shown in Fig. 2.16. The impeller vanes are inclined backwards at outlet and this is known as backsweep. It will be noted in Fig. 2.16 that some of the blades do not extend from outlet to the inducer. These are known as splitter blades or vanes. They are used because the additional number of blades is needed towards the outlet but this number of blades would, if continued to the leading edge, produce sufficient blockage to cause choking there at high speeds and flows.

Machines for industrial duty very frequently do not have inducers but do have rotating shrouds fixed to the blades. Rotating shrouds remove the need to maintain a small clearance between the moving and stationary parts but do add significant extra mass and are therefore unacceptable when rotational speeds are high enough to produce very high stresses. A photograph of a shrouded impeller is shown in Fig. 2.17. There is no inducer and the impeller vanes are highly backswept, the inclination of the vanes to the radial being given by $\chi_2 = 42°$.

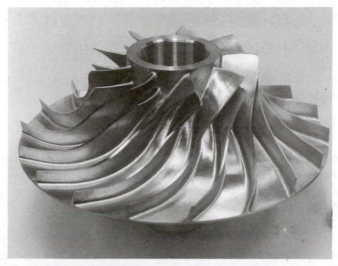

Fig. 2.16 An unshrouded backswept impeller with inducer. (Reproduced by permission of Sundstrand Turbomach)

Fig. 2.17 A shrouded impeller. (Reproduced by permission of Sulzer Escher Wyss)

Whether it is a low cost device or a high performance high pressure ratio compressor the attraction of the radial machine is the same in each case; much or even most of the static enthalpy and static pressure rise for the complete stage comes about from a centrifugal process unrelated to the fluid mechanics of the machine and depending *only* upon the impeller blade speed at inlet and outlet. Furthermore the work input is almost independent of the nature of the flow and, unlike the axial machine, is not liable to drop discontinuously if the blades are overloaded at or near to compressor stall.

The specific work input to the gas flowing through an impeller is given by

$$W = U_2 V_{\theta 2} - U_1 V_{\theta 1}$$

and, for the present, restricting attention to impellers with no inlet swirl, this simplifies to

$$W = U_2 V_{\theta 2}.$$

The simplest case is a radial impeller in which the blades are radial at outlet, $\chi_2 = 0$. For this a hypothetical ideal flow would have the relative flow leaving in a purely radial direction and the absolute whirl velocity would be

$$V_{\theta 2 \text{Ideal}} = U_2.$$

In fact the flow from a radial outlet compressor would be inclined backwards and the actual absolute whirl would be $V_{\theta 2} = \sigma V_{\theta 2 \text{Ideal}}$ where σ is known as the slip factor.

Many radial compressors have vanes or blades which are sloped backwards at outlet, backsweep, and for these the ideal flow is also inclined backwards. For a backsweep angle χ_2 the corresponding ideal whirl velocity

$$V_{\theta 2 \text{Ideal}} < U_2$$
$$= U_2 - V_{R2} \tan \chi_2$$

The slip factor can still be invoked to write the actual outlet whirl as

$$V_{\theta 2} = \sigma V_{\theta 2 \text{Ideal}}$$

and the work input is

$$W = \sigma U_2 (U_2 - V_{R2} \tan \chi_2). \tag{2.19}$$

V_{R2} is proportional to mass flow rate so the work–mass flow characteristic takes the form sketched in Fig. 2.18.

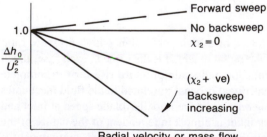

Fig. 2.18 The effect of backsweep on the work input in a centrifugal compressor, slip faster assumed constant

The slip factor will be discussed in greater detail in Chapter 6 but some explanation for the phenomenon can be given briefly here. To make the flow follow the direction of the blades it is necessary that a force should be produced on the flow by the impeller and this is accomplished by the pressure difference between the leading (pressure) and trailing (suction) surfaces of the vanes. In the outer part of the impeller where the relative flow is predominantly radial this produces the Coriolis force that the flow needs. At the blade trailing edge the pressure difference and the force go to zero and since this is not an abrupt change but a gradual one the blade force is reduced some way upstream. With the blade force reduced the flow is no longer able to follow the blade direction and it therefore becomes inclined backwards, giving what is referred to as slip. In fact the streamlines adjacent to the blades are required to follow the blade direction and this in turn requires the flow towards the middle of each passage to deviate more. The magnitude of the slip factor depends primarily on the number of blades but also to some extent on the geometry, including backsweep. It appears to be a quite weak function of the flow rate. The most used correlation is that given by Wiesner (1967). For a typical impeller the slip factor is about 0.9. If the slip factor is regarded as more or less constant the work input and pressure rise are therefore proportional to the square of the impeller tip speed.

The production of pressure rise by the centrifugal effect has already been referred to. By equating the rothalpy at impeller inlet and outlet the rise in static enthalpy across the impeller is given by

$$h_2 - h_1 = \tfrac{1}{2}(U_2^2 - U_1^2) - \tfrac{1}{2}(W_2^2 - W_1^2). \tag{2.20}$$

The contribution to the enthalpy rise from the first term, the difference in the squares of blade speed, is the centrifugal effect, whilst the second term is the rise produced by the deceleration of the relative flow.

The relative importance of the two contributions to the static enthalpy rise in equation 2.20 can be assessed by a simple example taking an impeller with an inducer. (Impellers with inducers generally have a better internal aerodynamic performance than those without so that the contribution from the deceleration term is likely to be relatively larger.) Suppose that the inducer tip diameter is equal to 0.7 times the outlet diameter so that $U_{1tip}=0.7U_2$. As a reasonable approximation assume that $V_{x1}=0.5U_{1tip}$. The smallest centrifugal contribution and the largest deceleration contribution will occur for the streamline passing close to the inducer tip where it is immediately apparent that the centrifugal term is

$$\tfrac{1}{2}(U_2^2 - U_1^2) = 0.255\,U_2^2.$$

Some assessment needs to be made for the likely deceleration of the relative flow: it will be shown in Chapter 6 that a reasonable estimate for a good impeller would be $W_{1tip}/W_2=1.6$ and with the other approximations adopted $(W_{1tip}/U_2)^2 = 0.61$. It then follows that

$$\tfrac{1}{2}(W_2^2 - W_1^2) = 0.186\,U_2^2.$$

In words this means that even for a rather good impeller with an inducer, the static enthalpy rise from the centrifugal effect amounts to 58 per cent of the total with only 42 per cent from the relative flow deceleration.

A rise in enthalpy from the deceleration term depends on the flow, just as in a conventional stationary diffuser. Likewise there are losses associated with the deceleration of the flow and the actual pressure rise will be less than the isentropic one corresponding to the enthalpy rise. There are no losses associated with the centrifugal rise in enthalpy $\frac{1}{2}(U_2^2 - U_1^2)$, so that

$$T ds = dh - dp/\rho = 0$$

and therefore

$$p_2 - p_1 = \int_1^2 \rho dh$$

for the contribution from the centrifugal effect with the integration from states 1 to 2. In other words the enthalpy rise from the centrifugal effect is entirely converted to its equivalent reversible pressure rise.

To continue the example of the impeller used above, suppose that the efficiency measured for the static pressure rise in the impeller was 90 per cent. Some 58 per cent of the pressure rise was produced by a loss-free process so the 10 per cent of lost efficiency is produced by a process giving only 42 per cent of the pressure rise; the decleration process would evidently be a very inefficient one in this example.

The existence of this 'free' pressure rise from the centrifugal effect is fortunate because the flow path of most radial compressors is not naturally conducive to low loss. The flow path is long and the aspect ratio (the ratio of blade height to blade chord or length in the flow direction) may be very low. The aspect ratio is very much related to the specific speed of the compressor, but even for the optimum value of this an average aspect ratio for the impeller would turn out to be about 0.2 or 0.3, a very low value by the standards of axial machinery. The aspect ratio of the diffuser may be even lower, with a large ratio of 'wetted area' to cross-sectional area. Such geometries are known for axial compressors and ordinary diffusers to lead to high levels of loss.

In Chapter 1 the specific speed was defined and explained; specification of pressure rise and flow rate leads to an optimum configuration and thence the optimum size and speed. In effect this is equivalent to an optimum geometry, characterized by a ratio of inlet to outlet radius. Very often there is not the freedom to choose the optimum configuration according to such a criterion because the rotational speed may be fixed or because, for example, there must be a large diameter shaft passing through the compressor. As Fig. 1.9 shows the loss in efficiency is small provided the excursion from the optimum value of N_s is not large.

So far attention has been directed entirely at the impeller but there must also be a stationary diffuser. Very often this will be vaneless, consisting of an annular region in which the flow spirals outwards and is decelerated, but there may be vanes consisting of aerofoil sections, wedges or cambered plates to make the deceleration more rapid.

The diffuser is an important component because the absolute velocity leaving the impeller is usually high. For a radial vaned (no backsweep) impeller the outlet whirl is equal to about 90 per cent of impeller tip speed U_2, i.e. the slip factor is about 0.9. A typical design choice is $V_{\theta 2} \approx 3V_{R2}$. With this combination $V_2 \approx U_2$ and $h_{02} - h_{01} = W = 0.9 U_2^2$. Using these values as an example it is easy to write for the static enthalpy at impeller outlet

$$h_2 - h_1 = (h_{02} - h_{01}) - V_2^2/2 + V_1^2/2$$
$$= W - V_2^2/2 + V_1^2/2$$

or
$$h_2 - h_1 \approx 0.9 U_2^2 - U_2^2/2 + V_1^2/2$$
$$\approx 0.4 U_2^2 + V_1^2/2$$

and at diffuser outlet, noting that $h_{02} - h_{01} = h_{03} - h_{01}$

$$h_3 - h_1 = (h_{03} - h_{01}) - V_3^2/2 + V_1^2/2$$

or
$$h_3 - h_1 \approx 0.9 U_2^2 - V_3^2/2 + V_1^2/2.$$

A good diffuser may halve the mean velocity so that

$$V_3^2/2 \approx 0.25 V_2^2/2.$$

For the representative impeller used above the absolute flow velocity entering the impeller, V_{x1}, is 0.5 times the inducer inlet tip speed and a representative radius ratio of the inducer inlet to impeller outlet is 0.7. This gives $V_1 = 0.35 U_2$. Thus $h_2 - h_1 \approx 0.46 U_2^2$ and $h_3 - h_1 \approx 0.84 U_2^2$, so there is a very significant rise in static enthalpy (and static pressure) in the diffuser. The degree of reaction is

$$R = (h_2 - h_1)/(h_3 - h_1) \approx 0.46/0.84 \approx 0.55$$

(The enthalpy rise $h_2 - h_1$ by this method of estimation is slightly higher than that obtained on the previous page, $0.44 U_2^2$: this discrepancy merely reflects inconsistency in the round numbers used as input parameters.)

With backsweep the contribution from the diffuser is smaller, because the absolute velocity out of the impeller is reduced, and the degree of reaction is corespondingly higher: with similar numbers to those used above, but with 30° of backsweep, the degree of reaction would be about 0.67. The flow in the diffusers is frequently not very satisfactory and the pressure rise there may be less than expected, with a large loss in stagnation pressure. This will be considered in greater depth in Chapter 7. One incentive for choosing backsweep is that the impeller is a more efficient device and the stage efficiency can be raised by having a larger proportion of the pressure rise occur in the impeller than in the diffuser.

Most axial compressors look very much the same: industrial compressors of axial design are actually fairly similar to those for high performance jet engines. With the radial machine there is a wide variation in type and some of these differences have already been touched upon. One reason for this range of types is the fully three-dimensional geometry of centrifugal machines, unlike axial compressors which are still mainly thought of in two systems of two dimensions. Crucial factors in specifying the geometry are:

1. Is the impeller shrouded or unshrouded, in other words is there a clearance gap between the ends of the impeller vanes and a stationary shroud or is there a rotating shroud fixed to the impeller vanes.
2. Does the impeller have an inducer section so that the vanes are continued around into the axial direction and the compressor resembles an axial compressor at inlet.
3. Are the impeller vanes radial at outlet or are they inclined backwards (the word backswept is used to describe this).
4. Does the compressor have inlet guide vanes and if so are they at a fixed angle or can they be rotated to vary the inlet flow angle.
5. Is the diffuser vaned or vaneless.

Since each of these is crucial to the design of the machine they will be considered individually.

1. Shrouded or unshrouded impeller
There is known to be serious degradation in performance caused by large clearance between moving impeller vanes and a stationary shroud. Small clearance is possible if axial movement of the impeller with respect to the shroud is very small and under such circumstances the loss from the clearance may not be serious enough to require a moving shroud. In multistage compressors the axial movement of the shaft is sufficiently large that the tip clearances would be too great for use with an unshrouded impeller. Shrouded impellers are therefore normal with multistage configurations, one of which is shown in Fig. 2.19. Even in a multistage compressor an unshrouded impeller may be used for the stage close to the axial thrust bearing. A meridional view of the flow path for a typical unshrouded impeller with an inducer and for a shrouded impeller without an inducer is shown in Fig. 2.20.

A shroud fixed to the impeller also reduces the magnitude of the axial load on the thrust bearing and this may be very important in high pressure applications. Lastly the shrouded impeller is a very rugged device from which blade vibration problems are mainly eliminated. However, the shroud complicates the design of the impeller and the greatly increased mass means that lower

Fig. 2.19 A multistage centrifugal compressor for industrial use. Note the complicated flow path and sharp radii of curvature which are needed. (Reproduced by permission of Sulzer Escher Wyss)

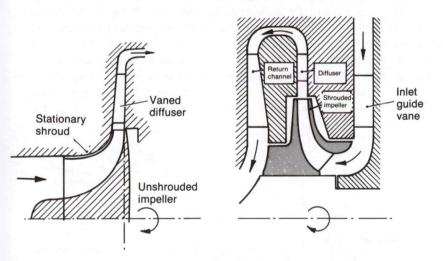

Fig. 2.20 Somewhat simplified meridional sections through two types of compressor. One has an unshrouded impeller with an inducer, the other has an impeller with no inducer but a rotating shroud

peripheral speeds are required for stress reasons. With a steel impeller for an industrial compressor, for example, a maximum tip speed of 380−430 m/s would be acceptable with an unshrouded wheel, but only 300−360 m/s with a shroud.

2. *Inducer on the impeller*

The aerodynamic performance of an impeller will generally be improved by an inducer of good design. For impellers without an inducer there is frequently a fairly tight radius of curvature on the shroud side. The velocity at the shroud side can then be a great deal larger than that at the hub. The existence of such large gradients is undersirable for two reasons. The first is that the high velocity at the shroud implies low static pressure and the flow must be decelerated again to bring the static pressure back up. This deceleration is a source of loss but more seriously there is a maximum pressure rise which can reasonably be expected of the boundary layers without separation and separation is very likely to occur near the inlet on the shroud side. The second is that the high velocities near the shroud will lead to a reduction in density and possible supersonic velocity even though the mean throughflow velocity is well below this. The high non-uniform velocity across the passage greatly complicates the design of the impeller blades.

An inducer is able to raise the average static pressure before the full effect of the shroud curvature is realised. Furthermore the swirl velocity V_θ which the inducer imparts to the flow produces a static pressure gradient, with higher pressure near the shroud, and therefore tends to counter the low static pressure from the meridional flow around the shroud curvature itself.

One of the most important features of the inducer is that it gives a larger throat area than that which is produced by the build without it. Figure 2.21 shows the meridional view through two impellers with several dimensions in common. The tip and inlet radii are identical, so too is the axial tip width at outlet and each compressor was run with a vaneless diffuser. Compressor A has the vanes swept back over the outer 20 per cent of the radius to give a backsweep of 30°. Compressor B is an industrial type with no inducer and

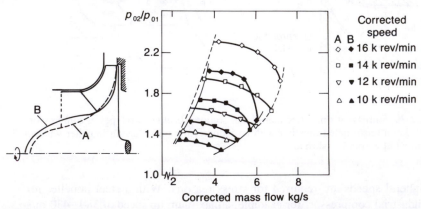

Fig. 2.21 A comparison of the performance of two impellers with the same outer diameter and width. Impeller A has an inducer and 30° of backsweep, impeller B has no inducer and 40° of backsweep. (From Eckardt, 1977)

with 40° of backsweep. The vane height of compressor B at the leading edge is the same as the meridional passage height of compressor A at the same position in the annulus. The throat area, that is the minimum cross-sectional area in the flow passage, was significantly larger for impeller A. The pressure ratio–mass flow map for the two impellers from a report on the measurements by Eckardt (1977) is also given in Fig. 2.21. Of particular note here is the reduced maximum mass flow for the compressor without an inducer, so that at 16 000 rev/min the flow appears choked at 6 kg/s, whereas the same sized machine with an inducer is able to pass more than 7 kg/s with no sign of choking.

With industrial machines the speeds are generally lower so that choking is a less serious problem than for aircraft or turbocharger applications. Industrial machines are very often designed to be run as multistage compressors with many stages on the same shaft. The overall length of the shaft then becomes of enormous significance. As the length increases so must the shaft diameter and this in turn affects the compressor aerodynamics by increasing the inner hub diameter. It is therefore important to reduce the axial length as much as possible. Not only does the inducer itself occupy axial length but producing an axial flow to enter the inducer would make this still longer. By locally reducing the shaft diameter in the entry region it is possible to increase the inlet flow area for impellers without an inducer in a way which would be impossible if there were an inducer. Very high levels of efficiency and unchoked flow may be obtained in impellers without inducers by restricting the speed and pressure rise to lower levels than those normally needed for gas turbines or turbochargers.

3. *Radial or backswept impeller vanes at outlet*
The aerodynamic repercussions of backsweep will be discussed more fully in Chapter 6. Only two aspects will be addressed here and these are the tendency for the backswept impeller to be more stable but to produce a smaller pressure rise for the same tip speed. This latter effect is easily understood. If the flow leaves the impeller inclined by the vanes in a relative direction opposite to that of the rotation then the absolute whirl velocity will be lower than for a similar wheel discharging its flow radially, equation 2.19. Very high pressure ratio impellers generally have no more than small amounts of backsweep because they are operating near the limit of acceptable tip speed and backsweep reduces the enthalpy rise for a given tip speed. Moreover backsweep introduces non-radial filaments into the impeller and this has the effect of creating bending stresses in the blades. Figure 2.22, taken from Kenny (1984), shows the effect of backsweep for a class of designs related to aircraft gas turbines. The low cycle fatigue limit and the creep limit refer to just one titanium alloy, Ti-6-2-4-6β, but similar trends would be found for other materials. The temperature of this impeller backface is clearly very important. For automotive turbocharger applications, where the full speed range may be cycled at every gear change, fatigue of the compressor bore is a particularly serious constraint.

74 *General design considerations*

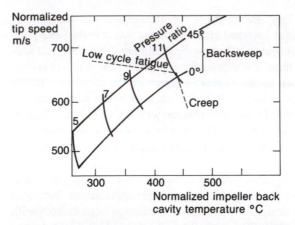

Fig. 2.22 A correlation between impeller tip speed and materials in an aircraft gas turbine application. Material for the correlation is titanium alloy Ti-6-2-4-6β. (From Kenny, 1984)

The bore stress is increased by non-radial (i.e. backswept) vanes; using cast aluminium impellers, a rule of thumb for one manufacturer is a maximum tip speed of 550 m/s with 20° of backsweep and 580 m/s with no backsweep.

The effect of backsweep on stability comes from the way in which the pressure rise increases with decrease in flow for backswept impellers. The change in pressure rise is solely a result of the rising enthalpy change,

$$\Delta h_0 = U_2 V_{\theta 2} - U_1 V_{\theta 1}$$

and
$$V_{\theta 2} = U_2 - V_s - V_{R2}\tan\chi_2$$

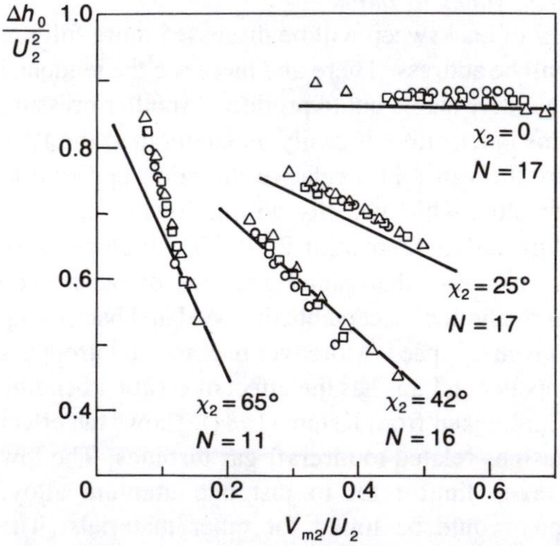

Fig. 2.23 The measured and predicted variation in work input as a function of flow rate for impellers of different backsweep χ_2. The number of blades is denoted by N. (From Casey and Marty, 1986)

where V_s is the slip velocity which is assumed constant. As the radial velocity rises the absolute outlet whirl velocity falls and therefore so too does the work input per unit mass flow. This is shown very clearly for a range of industrial compressors in Fig. 2.23 taken from Casey and Marty (1986), in which the points show measurements at a range of impeller Mach numbers and the solid lines represent predictions. Stability of compressors will be discussed in a later chapter but here it suffices to state that for stability it is highly desirable that the pressure rise should drop sharply with increase in mass flow rate.

4. Inlet guide vanes

Variable guide vanes find application in industrial compressors, particularly multistage compressors, where it is necessary to accommodate large variations of flow rate. Most compressors for gas turbine or turbocharging applications do not have inlet guide vanes, primarily because at the high inlet flow Mach numbers these would be likely to choke; for industrial machines the inlet Mach numbers are generally low enough that this is not a serious problem. Inevitably inlet guide vanes introduce some extra loss.

The inlet guide vane imparts an inlet swirl velocity $V_{\theta 1}$ and this reduces the work input if it is positive (i.e. in the direction of rotation of the impeller) and increases the work if negative. Figure 2.24, taken from Casey and Marty (1986) shows what can be achieved by varying the angle setting of inlet guide vanes and of diffuser vanes. The ordinate in this figure is the so-called polytropic head, $H_p = \int (1/\rho)\mathrm{d}p$, a measure of the reversible part of the work input. The effect of varying the inlet guide vane direction is to alter the pressure rise *and* flow rate. The diffuser vanes, however, are unable to influence the work input so varying these allows them to be rematched so that they are

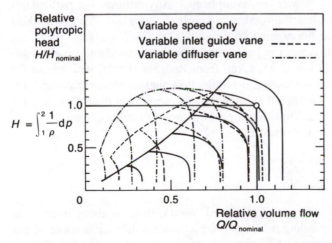

Fig. 2.24 Comparison of the operating characteristics of a single-stage centrifugal compressor with variable speed, variable inlet guide vanes and variable diffuser vanes. (From Casey and Marty, 1985)

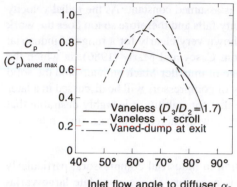

Fig. 2.25 The variation in diffuser pressure coefficient with inlet flow angle for three types of diffuser. The pressure coefficient is normalized by the maximum value for the vaned diffuser. (From Rodgers, 1982)

operating efficiently for the combination of radial and tangential velocity leaving the impeller. The diffuser vanes therefore alter the efficiency without altering the head rise and for the same rotational speed allow high efficiency operation at about the same pressure ratio over an extended flow range. A very complete investigation of varying both inlet guide vane and diffuser vane settings is described by Simon *et al.* (1987).

5. Vaned or vaneless diffuser

The two types of diffuser will be discussed in much greater detail in Chapter 7. The flow in the diffuser of either type may look superficially simple but this is wrong — the flow is very complicated. Any diffuser, but particularly a vaneless one, is often followed by a scroll collector and this gives added pressure rise but at the expense of range.

The principal differences in performance between vaned and vaneless diffusers are shown in Fig. 2.25, taken from Rodgers (1982), which is in the form of pressure rise versus average outlet flow angle from the impeller. This flow angle may be thought of as showing the variation in mass flow rate for a given impeller speed. The diffuser performance is usually expressed in terms of pressure coefficient defined by

$$C_p = \frac{p_3 - p_2}{p_{02} - p_2}$$

A typical peak value for a well-designed vaned diffuser is about 70 per cent whereas the corresponding peak value for a vaneless diffuser is about 50 per cent. For applications where maximum range is required the vaneless diffuser is the preferred choice but for maximum pressure recovery the vaned diffuser is better. In some very low specific speed compressors the axial width of the impeller at outlet is relatively large and in consequence the radial velocity is very small compared with the tangential velocity. The flow entering the diffuser is therefore more nearly tangential than is optimum for the vaneless diffuser and for this type of compressor wider operating range can be obtained with a vaned diffuser.

The vaneless diffuser decelerates the circumferential velocity as the radius increases so that for inviscid flow angular momentum is conserved. In a multistage compressor the flow after the diffuser from one stage must then follow a path of decreasing radius to take it back to the inlet of the next stage impeller. If there are no vanes present the tangential velocity would increase again as the radius decreased on the inward path. In multistage compressors it is therefore essential to have some vanes after each stage to take out the swirl.

The vaneless diffuser also decelerates the radial or meridional velocity but in many applications, certainly those for high pressure ratio stages, the radial velocity is so much less than the whirl velocity (typically about a third) that the dynamic pressure attributable to the radial component is comparatively unimportant. For this reason irregularities in shape as shown in meridional views are often much less serious than they appear because they only affect the relatively small dynamic pressure from the radial component of velocity.

With the wide variation in type of component it is convenient to consider the impeller flow separately from that downstream of it. There is no *a priori* reason why the pressure field of the vaned diffuser should not affect the impeller, and likewise the field from the impeller affect the diffuser, but in fact it does seem to be a quite good approximation to neglect this interaction.

Of greatest concern is the mass flow swallowing capacity of the diffuser in relation to that of the impeller. If the maximum mass flow rate through the diffuser were less than the flow at which the impeller stalls the stage would never operate satisfactorily; likewise if the impeller chokes at a mass flow smaller than that needed to unstall the diffuser the performance would be disastrous. The matching of the capacities of the impeller and diffuser is particularly acute for vaned diffusers because of their narrower operating range. The problem gets worse as the speed increases because of the large density variations produced by the work input (similar considerations apply to axial stages but for these the rotor and stator density rises are much smaller and a marked difference in swallowing capacity for the two is less likely to occur). The variable diffuser vanes referred to above extend the region over which the impeller and diffuser are both matched by changing their stagger to alter the diffuser throat area.

At this preliminary stage some description should be given of what constitutes a good centrifugal compressor; the reasons for these choices will be discussed in Chapters 6 and 7. A good impeller will have a sufficient number of blades so that the slip will not be excessive, though slip is not in itself detrimental if it can be predicted. The impeller may well be backswept and the blade height at outlet (which is in the axial direction) will not be too small in relation to diameter. If the impeller is unshrouded the clearance between the blade ends and the shroud will be fairly small. Frequently the blade height at outlet is larger than the optimum on the basis of flow deceleration so as to keep the ratio of the tip clearance to the blade height reasonably small. The impeller inlet (or the inducer, if there is one) will be curved in such a way that the flow will enter with very small incidence at design conditions over the entire height of the blade. Furthermore the shape of the impeller will be such that curvature will vary smoothly and will be as gentle as possible given the overall constraints; a particularly important feature is to keep the shroud curvature as gentle as possible. The diffuser, whether vaned or vaneless, will have the correct flow area and the selection of this is one of the most sensitive parts of the design and development. If the diffuser is vaned then the vane leading edges will be inclined so that at design point they have only a small incidence to the average inlet flow.

There are also quantitative measures of a one-dimensional type that can be applied, particularly for the impeller. Normally a ratio of a relative velocity or relative Mach number at inlet to a corresponding quantity at outlet is specified. The maximum relative velocity at inlet will be at the impeller inlet close to the shroud. The flow in the impeller is often assessed loss free for the purpose of setting the overall velocity ratio. Most impellers have extensive regions of flow separation in the passage on the trailing (or suction) blade surface or the shroud (sometimes on both): the above assumptions are very sweeping and are a means to design rules rather than a description of the flow. A typical value of diffusion ratio based on relative Mach number is 1.6 (see for example Japikse, 1987) and this seems to represent about the highest value which the designer can reasonably expect to obtain. If the impeller is designed for a higher value, which implies a larger cross-sectional area at outlet, the amount of deceleration achieved is unlikely to be increased because the separated region can expand to limit the overall degree of deceleration.

2.4   The matching of multistage compressors

All compressors are designed with a particular condition in mind, usually referred to as the design point or matching point. The design point is normally chosen where the compressor will operate for much of the time or where its performance is particularly critical, generally near the maximum speed and pressure rise that the machine can deliver. It is usually essential that an adequate pressure ratio and efficiency are produced at other conditions and for gas turbines, for example, an important condition is low speed operation during starting. The problem of matching the inlet flow requirements of the one stage to the outlet flow of those upstream is general to all multistage compressors. There is one combination of nondimensional speed and mass flow, the design or match point, at which the inlet flow at each stage corresponds to its optimum. Difficulties arise when the compressor is operated at different corrected speeds or at the same speed but different mass flow, i.e. on a different working line.

Although flow is usually expressed in terms of mass flow, which remains constant from stage to stage unless there is some bleed, the requirement of a stage is best thought of in terms of the volume flow rate. This is because it is the flow angles into each blade row or impeller which really determine whether a stage is operating correctly and the angles depend on the ratio of flow velocity to rotational speed. It is the volume flow which fixes the flow velocity. A suitable dimensionless group would thus be Q/AU, where Q is the volume flow rate, A is the cross-sectional area and U the mean blade speed. For axial machines, or radial machines with an inducer, Q/AU can be simplified to V_x/U, the usual variable for displaying the stage pressure rise characteristic V_x being a mean axial velocity at inlet. As shown in Chapter 1 $m\sqrt{(c_p/T_0)}/Ap_0$ is proportional to V_x/U *for a constant corrected blade speed* $U/\sqrt{(c_pT_0)}$, often written as $N/\sqrt{T_0}$.

The problem of matching stages is similar for both axial and centrifugal compressors. The axial compressor, however, generally has a narrower operating range (a smaller range of mass flow between stall and choke) and because all the temperature and pressure rise of the axial comes from the deceleration of the flow, it is much more sensitive to the correct matching of the stages. In this section a brief introduction to the matching of stages at design will be given, followed by an account of the difficulties encountered when operating a multistage compressor at off-design conditions.

Design point operation

Once the overall pressure ratio, blade speed and general layout of the compressor are fixed it is the correct matching of the stages which is probably most critical in deciding if it will be successful. A small error can put a stage or blade row so far from its design inlet flow or incidence that any subtlety of blade or impeller design is entirely lost. It therefore follows that a large amount of work has gone into the development of methods for calculating the correct way to combine the stages, often referred to as stage stacking. Most of these are proprietary.

For most compressors the design point will occur at or near the maximum speed for the machine. The blading and the flow cross-sectional area will have been chosen for each stage to be appropriate for the design mass flow and rotational speed. Under the high speed conditions envisaged the density rise through each stage is likely to be significant and to compensate for this, and to prevent the axial or throughflow velocity falling too much, the cross-sectional area of the compressor must decrease in the downstream direction. A satisfactory match of the later stages of a compressor therefore requires that the density at entry to each stage be accurately known, which in turn requires that the performance of the upstream stages should be accurately predicted.

The specification of the stages can take place initially with a one-dimensional analysis at mid-span of the blades, sometimes referred to as a pitchline analysis. More recently the analysis has been carried out at additional spanwise positions as well. The approach used for representing the blades is usually fairly simple, with work input found from velocity triangles and pressure rise found from the work with empirical rules for the losses. If the annulus shape is specified the pressure and temperature rises fix the axial velocity into each stage and the calculation can then be refined to get improved estimates for the stage and the overall pressure rise and efficiency. If a blade row is predicted to be operating so that the inlet flow direction is very different from the minimum-loss incidence angle, the blading can be changed or the annulus area altered and the calculation repeated until a satisfactory result is obtained.

Problems arise in matching the stages even at the design point, mainly because of the growth of the annulus wall boundary layers. (As discussed in Chapter 8 the viscous regions of the annulus walls are not very accurately described as boundary layers but the term is used here as a convenient shorthand.) The boundary layer introduces losses near the hub and casing but, much more seriously, introduces blockage as a result of the displacement thickness on the endwalls. The blockage is equivalent to a reduction in the flow area. Small changes in flow area have a very large effect on the stage performance both because it affects the mass flow at choke and because it affects the amount of work done by the rotor. The blockage needs to be accurately specified if the stage performance is to be correctly predicted. Unfortuantely there is no generally accurate method for predicting blockage and errors in its estimation are probably the greatest single cause of inaccuracy in predicting multistage compressor performance.

The blockage reduces the work done by the rotor outside the boundary layers by raising the throughflow velocity. Inside the boundary layers there is additional work input but also additional losses; in fact most of the losses for the entire stage occur in the endwall region.

To estimate performance of axial stages it is now normal to use estimates for the blockage and loss, with the blockage used to estimate the reduced work done by each stage as a result of the higher throughflow velocity over most of the span. In the past it was common to define a work-done factor, the ratio of the actual mass average work into a stage to the theoretical value assuming uniform axial flow, and this still finds occasional application. The work done factor allows for the lower work in the middle of the passage which result from the higher velocity caused endwall blockage and it also allows for the compensatory higher work inside the boundary layers themselves. Howell (1945) recommended a value of 0.86 for the work-done factor well inside a multistage compressor, the value being 1.00 for the first stage. The work-done factor rolls too many effects into one empirical term and is not the best method to allow for the boundary layers on the hub and casing. However, there is no wholly satisfactory method for the prediction of blockage or loss due to the endwall regions, though there are correlations, and this will be discussed in Chapter 8. A rule-of-thumb for blockage is that 0.5 per cent is added per blade row until 4 per cent is reached when it is assumed to remain constant.

Off-design operation

Most compressors will operate at least some of the time at pressure ratios or

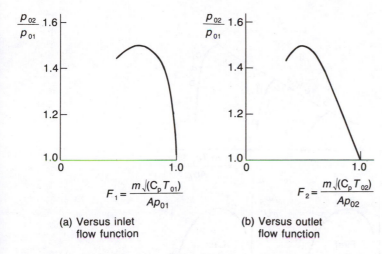

(a) Versus inlet
flow function

(b) Versus outlet
flow function

Fig. 2.26 A comparison of a stage characteristic with flow function based on (a) inlet conditions; (b) outlet conditions

speeds other than design. The fundamental difficulty in operation at these conditions stems from the fact that an excursion in non-dimensional mass flow (i.e. the flow function $F = m\sqrt{(c_p T_0)}/A p_0$) at inlet to a stage generally leads to a larger excursion at outlet. The same is true of the volume flow and V_x/U. The amplification of changes in nondimensional flow can be understood from Fig. 2.26 which shows the stage pressure rise plotted against the inlet flow function $F_1 = m\sqrt{(c_p T_{01})}/A p_{01}$ in the left-hand sketch and against outlet flow function $F_2 = m\sqrt{(c_p T_{02})}/A p_{02}$ in the right-hand one. The pressure ratio $PR = p_{02}/p_{01}$ depends strongly on the flow function and for unstalled blades increases rapidly as the flow is reduced. The temperature ratio T_{02}/T_{01} also rises with reduction of flow function and can be expressed by PR to the power $(\gamma - 1)/\gamma\eta_p$, with η_p being the stage polytropic efficiency; for reasonably high efficiency machines this index is typically equal to about 0.3, and so T_2/T_1 is much smaller than the pressure ratio. The ratio of inlet and outlet flow functions can then be approximated

$$F_2/F_1 \approx PR^{-0.85} \approx PR^{-1}$$

for stages of reasonably high efficiency. Consider, for example, a 1 per cent reduction of flow function into such a stage: the reduction in flow rate will increase the pressure ratio PR and the flow function at outlet will decrease by more than 1 per cent. Likewise an increase in flow function into the stage will produce a larger increase in the outlet flow function. The volume flow rate, proportional to mT_0/p_0 is also amplified through the stages, the magnitude of the amplification rising with the pressure ratio.

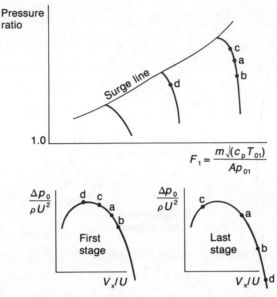

Fig. 2.27 Overall pressure ratio–mass flow characteristic for a multistage compressor and identical stage characteristics for the first and last stages

The effect of mismatching is so very large in multistage compressors because of their multiplicative nature: the overall pressure ratio $PR_{overall}$ can be written

$$PR_{overall} = PR_1 PR_2 \dots PR_n$$

where PR_1 is the pressure ratio of the first stage and so on. The effect is shown diagrammatically in Fig. 2.27 where the overall pressure ratio–mass flow characterstic is shown along with the stage characteristics for the first and last stage. The characteristics of all the stages are assumed to be the same. Point a denotes the design point, point b is a reduced pressure rise at design speed and point c an increased pressure rise at the same speed. For point a the position is the same on the first and last stage characteristics. For the increased flow, point b, stage 1 produces a smaller pressure rise and therefore smaller density rise than that assumed in the design. The excursion in flow rate from the design value is therefore larger into stage 2 than stage 1. The situation gets progressively worse through the compressors until at the last stage the operating point is a long way from design. For a high-speed compressor at a mass flow above that for design it is possible for the pressure ratios of the rear stages to decrease to such an extent that they are less than unity. If this occurs the last stage is then acting as a throttle and is often referred to as choked. It should be noted that the term choked is sometimes an incorrect usage; a very rapid drop in pressure with rise in mass flow, so rapid that when drawn on the pressure rise–mass flow characteristic it seems infinitely steep, can be observed when sonic or supersonic velocity is absent. The apparent choking is severe negative incidence stall, caused by the stage being so far off-design that the losses and deviation have very high values and are in the range where they increase so rapidly that precipitous changes in stage performance reminiscent of choking are seen.

Just as a reduction in the stage pressure rise from the design value leads to a choking of the rear stages, an increase in the pressure rise at the design speed leads to their stalling. Point c on Fig. 2.27 shows this case. The choking or stalling of the rear stage (or stages) puts a limit on the range of flow function into the whole machine.

To understand the effect of change in rotational speed of a compressor is somewhat more involved. Point d on Fig. 2.27 is for a case where the rotational speed is reduced, leaving the effective throttle downstream of the compressor unchanged. The consequences can best be understood by considering a sequence of events which, in reality, occur together. A reduction in speed means that the density rise from each stage is reduced below that for which they were designed. With the reduced density the mass flow is limited by the rear stage, i.e. it is choked, so that the flow function there reaches a limiting value. This in turn means that the nondimensional flow function and volume flow into the front stages will be much lower than that for which they are designed.

At speeds well below design it is normal for the front stage or stages to be heavily stalled, operating to the left of the peak on the stage characteristic and producing a lower pressure rise and efficiency than at the design point, though the temperature rise is relatively high. Stall in the front stages then reduces the density of the flow and, since the flow function is fixed for the choked rear stage, a further reduction in the mass flow is required as a result, exacerbating conditions in the front stages. It is quite common to find the front stages operating stalled at low speed with the stall taking the form of part-span rotating stall or an axisymmetric nonrotating stall. Stable operation for the whole compressor, but with one or more stages stalled, is only possible because of the stabilizing behaviour of the later stages, an aspect discussed in Chapter 9. Although the consequences of mismatching are felt throughout the machine, particularly in the front stages, it should be recognized that the problems are established by the inability of the rear stages to pass the desired flow.

Numerical examples for a multistage axial compressor

A feature of the operation at off-design conditions is that the behaviour is highly non-linear and depends on the performance of the stages in terms of both pressure and temperature rise. Generalizations are difficult to make and to illustrate the effects it is more helpful to consider a numerical example. Most of the methods for predicting the behaviour of compressors off-design are proprietary, relying on the in-house data bases. Some of the few methods in the open literature are by Novak (1976), Serovy (1976), Wall (1976) and Howell and Calvert (1978). Some methods evaluate performance at an average condition on the meanline or pitchline radius, others use blade-to-blade analysis at several radii. In any method the performance of the stages needs to be found in terms of the blade speed and non-dimensional flow rate into it. This can be achieved by using empirical characteristics (for example temperature rise and efficiency versus V_x/U) or the temperature rise can be calculated after specifying the blade outlet angle using a correlation for deviation, with the pressure rise then being found from a correlation for loss.

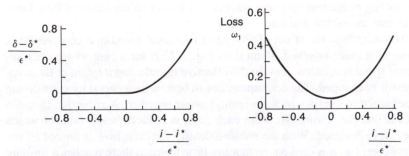

Fig. 2.28 Deviation and loss variation with incidence assumed in calculations of results displayed in Figs 2.29–32. $i*$, $\epsilon*$ and $\delta*$ are the incidence, flow deflection and deviation for the optimum condition

Up to this point the description has been fairly general but it is more informative to show some calculated results† performed at the blade mid-height. The method used to give the stage characteristics is one in which blade outlet angle and correlations for deviation and loss are specified. For this purpose the procedure has been kept as transparent as possible, with the fewest and simplest empirical inputs. The correlation for deviation comes from Howell (1942), where it is specified in terms of the deflection as $\epsilon/\epsilon* = f\{(i-i*)/\epsilon*\}$ with $\epsilon*$ and $i*$ being the deflection and incidence at the 'nominal' condition, see Chapter 4. The loss is assumed to be a minimum at the 'nominal' condition and parabolic with respect to $(i-i*)/\epsilon*$: the minimum loss is taken as 0.066. These are shown graphically in Fig. 2.28. Although better empirical inputs might improve the prediction for a specific machine the correct trends are evident with the present simple model. At each of the corrected speeds for which the performance is calculated the method first searches for the choke and stall conditions, the latter being taken to be when the rate of change of the overall total-to-static pressure rise with mass flow is zero. The performance is then evaluated at a number of intermediate points to give performance parameters along a constant speed line.

The test case is intended to be representative of a wide class of compressors. It is a six-stage axial compressor, similar in type to the high pressure compressor of a modern three-spool aircraft engine. The pressure ratio at design is approxi-

---

† These calculations were performed by Dr T P Hynes using a method he had written for analysing off-design conditions in multistage axial compressors.

mately 3.7:1 which shows the effects of mismatching very clearly, but the effects of mismatching become markedly greater at higher pressure ratios. At design the work input is chosen to be the same for each stage and the mean radius is constant from front to back. There are no inlet guide vanes so that at design the flow leaving each stator is axial — it is therefore a high reaction design and the nominal deflections for the rotor and stator rows are about 10° and 30° respectively. Like many high-pressure compressors the axial velocity is designed to decrease from front to back to give a lower dynamic pressure at outlet. For the present example the axial velocity is assumed to decrease in equal increments across each stage at the design condition, all the decrease in axial velocity occurring across the rotor row, with $V_x/U = 0.61$ into the first stage and 0.47 into the sixth.

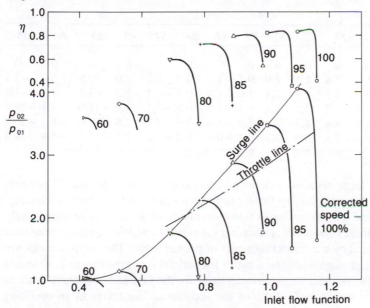

Fig. 2.29 Predicted pressure ratio and efficiency characteristics for a hypothetical compressor with all stages matched at optimum incidence at the design point, which is on the throttle line at 100 per cent corrected speed

Figure 2.29 shows the overall pressure ratio and efficiency lines for a range of rotational speeds in the case of a compressor in which the blades are chosen so that at the compressor design point the inlet flow angles correspond to the minimum loss condition. For the present purpose these are taken to be the zero incidence angles. A working line is drawn on Fig. 2.29: the working line here is taken to be a line of constant outlet flow function and is equivalent to a constant throttle setting. (For a gas turbine a working line defined by $m\sqrt{[c_p(T_{02} - T_{01})]}/Ap_{02} = $ constant gives better approximation to the turbine than the ordinary outlet flow function.)

The large drop in pressure rise as the speed is reduced is hard to assess in terms of what is to be expected from good machines but the enormous drop in efficiency is evidence that the conditions inside the compressor are very far from optimal at low speeds. The working line crosses the surge line somewhere between 80 and 85 per cent speed, which is clearly unacceptable for most applications. A good measure of the changes which have occurred when the speed is reduced can be gained by examining the blade row incidence (the angle between the flow direction at inlet and that at minimum loss) for conditions of peak efficiency at each speed. They are set out in the table below with R1, R2 etc. denoting entry to the rotors; S1, S2 etc. denoting entry to the stators.

Table 2.2 Incidence in degrees into blade rows for various corrected speeds

Speed	R1	S1	R2	S2	R3	S3	R4	S4	R5	S5	R6	S6
100%	0	0	0	0	0	0	0	0	0	0	0	0
95%	1.7	2.8	1.1	2.0	0.7	1.2	0.4	0.2	0.0	−0.8	−0.3	−2.1
90%	3.7	4.2	1.9	2.8	1.0	1.0	0.3	−1.1	−0.5	−3.7	−1.3	−7.3
80%	7.9	1.7	3.2	2.8	1.2	−0.5	−0.3	−5.1	−1.9	−11.9	−4.1	−24.2
70%	11.2	−4.8	4.3	2.9	1.5	−1.2	−0.6	−7.7	−2.9	−17.6	−6.1	−36.9
60%	12.9	−7.8	6.0	3.8	2.4	−0.1	−0.2	−7.2	−2.7	−17.6	−5.9	−34.8

The very large negative incidences into the last stator at 60 and 70 per cent speed are the most striking feature of these results; it is this which is 'choking' the flow and it is interesting to note that conditions do not change significantly between these two speeds, indicating the extent to which the compressor flow is dominated by the restricting effect of the last stage. The positive incidence into the first stage would be a cause for serious concern. Figure 2.30 shows the non-dimensional stage characteristics for the first and last stages in terms of $\Delta h_0 / U^2$ versus V_x / U. Each of the sections of line refers to an operating speed and indicates the range experienced by the stage at that speed when the whole compressor goes from 'choke' to stall. The sections do not overlay completely because of compressibility effects. The very large negative incidences into the last stage at the low speeds mean that this stage is operating either like a turbine or a throttle, while the high positive incidence into the first stage means that it is operating heavily stalled, well to the left of the peak pressure rise on the stage characteristic. Even when the flow rate is at its largest (i.e. the throttle is wide open) the first stage is operating to the left of its peak at all speeds below 80 per cent of design. At and near the design speed it can be seen that the variation in $\Delta h_0 / U^2$ and V_x / U for the front stage is quite small, whereas for the last stage it is many times larger, a result of the effect considered in the discussion of Fig. 2.26.

The choice of minimum-loss incidence into all stages at the design point is not necessarily the most satisfactory choice and large benefits can be ob-

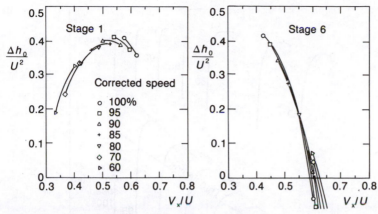

Fig. 2.30 The stage characteristics for a hypothetical compressor with all stages matched at optimum incidence at the design point

tained by anticipating the type of incidence encountered at low speed. This means choosing negative incidence in the front stages at design speed, to compensate for their tendency to move towards stall, and positive incidence in the rear stages to compensate for their tendency to move to choke or negative incidence stall. This was applied to the hypothetical compressor discussed earlier and the table of the incidences, the difference between the flow angle and minimum loss flow angle, is set out below in Table 2.3.

Table 2.3 Incidences in degrees into blade rows for various corrected speeds

	R1	S1	R2	S2	R3	S3	R4	S4	R5	S5	R6	S6
100%	−2.0	−4.0	−1.0	−2.0	0.0	0.0	0.0	0.0	1.0	2.0	2.0	4.0
95%	−1.0	−1.8	−0.1	−0.2	0.7	0.9	0.3	−0.1	0.9	0.8	1.6	2.1
90%	0.0	0.5	0.7	1.1	1.1	1.0	0.3	−1.0	0.6	−1.3	0.8	−0.9
80%	2.8	4.6	2.3	3.1	1.8	1.1	0.3	−2.9	−0.2	−5.7	−0.6	−8.4
70%	5.7	7.1	3.6	4.4	2.4	1.0	0.3	−4.3	−0.7	−9.3	−1.9	−15.2
60%	8.0	8.6	4.9	5.6	3.1	1.4	0.5	−4.8	−0.9	−11.2	−2.5	−18.9

The deliberate choice of incidences different from the minimum loss values at 100 per cent speed has led to a big improvement at part speed. The effect of the choice is more than additive; reducing the losses in the early stages decreasing the extent to which the compressor becomes mismatched and the later stages pushed to negative incidence. For this configuration it is worth noting that the 60 and 70 per cent speed cases are distinctly different, indicating that, unlike the previous case shown in Table 2.2, the last stage is not fully 'choked' at 70 per cent speed.

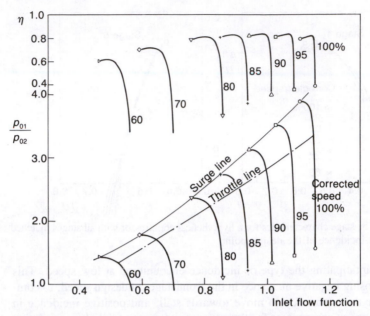

Fig. 2.31 Predicted pressure ratio and efficiency characteristics for a hypothetical compressor with front stages having negative incidence and rear stages positive incidence at the design point, which is on the throttle line at 100 per cent of corrected speed

The pressure ratio flow rate characteristics for the compressor with the negative incidence in the front stages and positive incidence in the rear stages at design point are shown in Fig. 2.31. Compared with earlier results in Fig. 2.29, the notable feature is the much higher efficiency and pressure ratio at the low-speed end. With this buildup the throttle line, corresponding to a constant outlet flow function, remains well away from the surge line. With the blades set for incidence other than that for minimum loss at the design point there is, as one would expect, a small reduction in pressure rise and efficiency at 100 per cent speed. At this speed there is, in addition, a reduction in the mass flow at choking and an increase in the mass flow at stall, both leading to a reduction in the range there. These sacrifices at full speed may be worthwhile for the much improved low speed performance and it is very common to find the highest efficiency occurring at speeds in the middle of the range.

As the design pressure ratio gets larger it gets more difficult to arrive at an acceptable compromise between full-speed and part-speed performance. For example, for a compressor producing about 20:1 pressure ratio the following values were obtained for the mean V_x/U into the first stage on the working line:

$$N_{\text{corrected}} = 100\% \qquad V_x/U = 0.55$$
$$90\% \qquad\qquad 0.37$$
$$80\% \qquad\qquad 0.22$$

The variation in V_x/U is so large here that it cannot be accommodated by staggering the blades to give incidence other than the minimum loss value at the design point because such large incidences would be required at design that this would bring unacceptable penalties in pressure rise and efficiency. Furthermore as the relative inlet Mach number to the front stages increases, the incidence range of the blades for acceptable performance reduces: when the inlet flow is supersonic the range can be very small, no more than about say 4° at a Mach number of 1.4. It is then impossible to design in an increment of negative incidence to allow for low speed operation.

The solution is to change the compressor configuration at corrected rotational speeds substantially lower than design. Basically two different approaches are adopted: bleed and variable stators. Quite often both are used.

The cause of difficulties at low speed can be traced to the rear stages being unable to pass the large volume flow required to prevent the front stages from stalling. One can therefore fit bleed ports somewhere about the middle of the compressor. These can be arranged to open at reduced speed, which increases the mass flow through the front of compressor, reducing the tendency to stall, and at the same time reduces the mass flow and pressure loss in the rear stages. Both these changes lead to a rise in the overall pressure rise at low corrected rotational speed. Because the flow bled off is normally dumped, bleed is a relatively inefficient method of rematching, more suited to starting than to continuous running.

The most widely used method of adapting a multistage axial compressor to run at low speeds is to use stators with variable stagger (the same is also true of radial compressors but only the axial case will be examined here). In some of the compressors now in service in aircraft engines the inlet guide vanes and the stators for the first six or so stages, out of a total of about 12, are arranged to be variable. To demonstrate the strong effect of a variable stator the prediction method used above has been run with a row of variable inlet guide vanes ahead of the compressor which was described above having all the blades matched to minimum loss incidence at the design speed. Figure 2.32 shows the overall compressor performance at 60 per cent speed for a range of inlet guide vane settings, expressed here in terms of the flow swirl angle into the first rotor. Very clearly the performance with 0° inlet swirl case, which was shown in Fig. 2.29, is extremely poor and very great improvements occur for the first few degrees of swirl. Swirl angles above about 40° lead to a marked change in the characteristic, with much reduced mass flow rate for about the same maximum pressure rise and efficiency, and 40° would seem like a good design choice. The effect of introducing inlet swirl is to reduce the incidence onto the first rotor but at the cost of raising it on the first stator and second rotor. Further improvements can then be achieved by restaggering the first stator, then the second stator and so on. The amount of restaggering required decreases as one moves back in the compressor, as seen in Tables 2.2 and 2.3

With several rows of variable stators the practical problem of optimizing the scheduling, that is the variation in setting of the individual rows with rotational speed, is formidable. It is simplified if the optimization is begun from

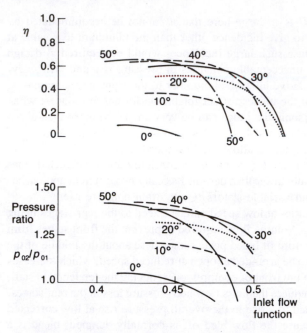

Fig. 2.32 Pressure ratio and efficiency characteristics for the hypothetical compressor with all stages matched at optimum incidence at the design point. Results shown for operation at 60 per cent of design corrected speed. Angles refer to swirl into first stage produced by variable inlet guide vanes

the last (i.e. most downstream) set of variable stators, adjusting this to give the best performance before adjusting the next row upstream. The rationale for this seems to be that it is the rear stages which are the root of the problem and it is expedient to obtain the best performance for the combination of stages downstream of the set of variables being adjusted before moving on to the next row upstream. The process is iterative and the rear variables would be readjusted after the upstream ones were initially optimized.

The treatment here of matching at off-design conditions has been highly simplified and is only valid in the trends it shows, in particular its indication of the sensitivity to changes in speed and to small changes in blade stagger. In carrying this out for a real compressor it would clearly be important to use the best available information about the stage performance; the limit of what can usefully be achieved is related to the accuracy of the information available. Even at the design condition, when the blade rows should all be operating unstalled, the accuracy of the estimates for blade row peformance and annulus blockage is not usually adequate and significant uncertainties arise. At conditions well away from the design point it is generally necessary to provide performance information for one or more of the front stages operating in their stalled mode and this is likely to be even less accurate. Thus, despite its obvious importance, the matching of compressors at design and even more off design is an uncertain business. Inadequacies are usually remedied by expensive development programmes. One of the merits of the compressor with many rows of variable stators is that there is more scope for correcting in development any errors incurred early on in the design; to be set against this is the much increased complexity of the machine and the number of variable parameters to be considered.

Overview

The ideas set out in this chapter give only a crude view of the constraints on the design of axial or radial compressors. They are essentially one-dimensional in their treatment of the flow, occasionally nodding towards the existence of two and three dimensionality, but retaining the simplicity of one dimension. Until the two-dimensional calculation methods became widely available the simple one-dimensional methods played a very big part in the whole design process; it is now possible to do much better than this though the simple methods still have their application. The preliminary design of compressors even today is normally carried out on a one-dimensional basis at the mean radius albeit augmented by extensive experience and correlation.

The following chapters consider the flow inside machines in more detail, first for axial compressors and then for radials. The treatment follows the way in which the subject has developed, which consists of treating the flow as two intersecting flows, each of two dimensions, rather than the much more complicated three-dimensional flow which really exists. One of the two-dimensional flows is normally the meridional flow, sometimes also known as the through-flow, which is the flow in the plane including the axial and radial coordinates. The intersecting two-dimensional surface is usually a surface of revolution and is very often referred to as the blade-to-blade flow. Sometimes this approach is inadequate, and such instances will be introduced as special cases, but the body of knowledge is so related to the two-by-two approach that this is the only realistic basis. Of relevance to axial and radial compressors the next chapter, Chapter 3, therefore considers mainly the meridional throughflow.

The following two chapters then consider blade-to-blade flows in axial compressors, beginning with the blade rows with subsonic inlet flows in Chapter 4 and then supersonic inlet flow in Chapter 5. The treatment of radial compressors is similar but the meridional and blade-to-blade flows need to be treated more closely; Chapter 6 looks at the flows in radial impellers and Chapter 7 at the flows in the downstream diffusers.

All compressors attempt to decelerate the flow and thereby increase the static pressure; the axial compressor depends entirely on this for the pressure rise, the radial compressor gets part of its static pressure rise in this way. The viscous (or turbulent) dominated regions close to the solid surfaces, often referred to as the boundary layer regions, tend to become thick and even to separate. The behaviour of these regions therefore provides the effective limit on what can be achieved and it is appropriate to devote Chapter 8 to aspects of boundary layer and viscous theory related specifically to the flow in compressors.

For all compressors stall or surge is one of the major barriers to performance and Chapter 9 considers this. The topics of noise and blade vibration are addressed in Chapter 10, both being undesirable unsteady phenomena. Finally in Chapter 11 the direction of compressor design is looked at with particular reference to the developments in experimental and numerical methods for computing the flows.

3 Throughflow on the hub–casing surface and some aspects of flow in three dimensions

3.1 Introduction

The flow in compressors is inherently three-dimensional. The form that this takes and the importance it has depends on the configuration: an axial fan is different in this respect from the rear stages of an axial compressor and both are different in the character of their three-dimensionality from a centrifugal compressor.

Three-dimensional calculation methods are relatively new and furthermore most people find it difficult to think, let alone design, in three dimensions. For this reason it is still normal to consider the flow in two separate but inter-related two-dimensional surfaces. This chapter is mainly concerned with the flow in the surface from hub to casing, usually a meridional plane. It is conventional to refer to this as throughflow. For the axial compressor the throughflow calculation essentially connects the blade-to-blade flow in the spanwise direction, requiring the flow at the various sections to be compatible and to satisfy the momentum equation in the radial direction. In essence this means that the performance of the blades at a given spanwise position is not determined by that section alone but by the same blade row at other spanwise positions, by the annulus shape and by the compressor as a whole. It is important to design the blades so that at every spanwise position the geometry is compatible with what the flow is constrained to do; the constraint being imposed in most designs by the throughflow analysis. This is very important for the design of compressors or fans of low hub–casing ratio (i.e. the hub radius much smaller than the casing radius, leading to blades of large span in relation to the mean radius). When the hub–casing ratio becomes nearly equal to one, as it does to the rear of many multistage compressors, the flow is still three-dimensional in the sense that there are strong gradients in the spanwise direction, but the variation in the meridional direction is then dominated by viscous effects.

For the radial compressor the main significance of the throughflow calculation on the hub–casing surface is not so much in connecting the flow at different blade sections along the span but in taking account of the strong curvature of the hub and shroud walls. The shroud curvature is of most concern because, being convex to the flow, it sets up high meridional velocities, $V_m = \surd(V_x^2 + V_R^2)$, in the regions where the radius of curvature in the meridional plane is smallest. As the meridional flow is straightened towards the impeller outlet it must therefore be decelerated and separation in this area is very common. This aspect of radial compressors is considered in Chapter 6.

Throughflow calculation methods are probably the most used of all the calculation procedures, being applied in several different ways. One application is the prediction of flow given the annulus and blade shapes and after introducing some information about the blade (or impeller) performance. This is probably the least frequent use to which throughflow calculations would be put. A more common use is to predict the geometry having prescribed the pressure and temperature rise and some estimates for blade performance. A third very common and important application is to use available measurements, which are normally only stagnation temperature and pressure, and to predict velocity and temperature everywhere inside the compressor so as to fill in the gaps in knowledge. This then allows the conditions at spanwise blade sections to be found, for example diffusion factor, axial velocity–density ratio† (*AVDR*), deviation and loss. Finding the local performance of blades is an important step for axial compressors in producing compatibility along the entire span. When the calculation reveals that a change in the blading is desirable the throughflow method is used again in verifying that proposed blade changes will produce the correct trend in performance.

The throughflow calculations are invariably approximate to some degree: because they may be inviscid, or carried out on a plane and not a streamsurface (a point addressed below) or because terms in the equations may be neglected. In many instances this does not matter too much because the methods are incorporated in a scheme which allows for their deficiencies. Much of the input data will have been derived from measurements with the aid of a throughflow method and when this data is then used as an input for design purposes some of the inaccuracies and approximations can be expected to cancel. It also needs to be remembered that not only the calculation method but also much of the input data is approximate. The blockage of the endwall boundary layer, for example, is known only approximately, yet for multistage axial compressors or for radial compressors this has a large effect on the calculation of the throughflow, far outweighing many of the approximations in the calculation procedure.

In recent years it has become possible to calculate three-dimensional flows numerically. The most complete methods solve the compressible Navier–Stokes equations in Reynolds averaged form so that the turbulence appears only as stress applied to the mean flow. The other major three-dimensional method treats the flow as inviscid and uses the Euler equations, again in compressible form. It is possible to get very useful results from the inviscid calcula-

---

† Axial velocity–density ratio is discussed in Section 4.1.

tions in regions where the boundary layers are thin, for example in an axial fan, the first blade rows of an axial compressor or the front of a radial compressor. The solution of the Navier—Stokes equations has wider applicability: as well as transonic fan blades, where it is the three-dimensional nature of the transonic flow which is of most interest, the solutions can encompass regions such as the rear of multistage axial compressors or the rear of centrifugal compressors where viscous effects are paramount. The three-dimensional methods are described in a little more detail in Chapter 11 but for the present it can be taken that with different levels of accuracy they can predict all aspects of the flow, although at the present time most are restricted to steady flow. They are therefore applied to one blade row at a time, and one of the problems of applying them, particularly those that solve the Navier—Stokes equations, is the specification of the inlet flow with adequate precision.

Until recently there was not the computing power to handle three-dimensional flows and a different approach was adopted. This approach has proved so successful, and highly satisfactory design systems have been developed around it, that it remains the principal method of describing, analysing and calculating the flow in compressors to the present. As mentioned above the three-dimensional flow is analysed as two intersecting and interrelated two-dimensional flows. Figure 3.1 shows the system originally proposed by Wu (1952) in which the flow is considered on streamsurfaces S1 and S2: the S1 surfaces start upstream as surfaces of revolution but twist and warp as they pass through the blade row. The S2 surfaces form meridional (i.e. $r-x$) planes upstream of the blades extending from hub to casing but these too warp as they pass through the blades. This approach lends itself to a possible exact solution of the inviscid three-dimensional flow but this has rarely been attempted because of the need to iterate with the streamsurface shape changing with each iteration.

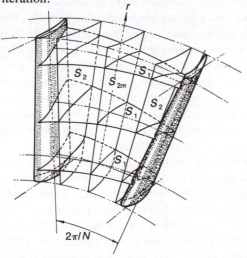

Fig. 3.1 Intersecting S1 and S2 streamsurfaces in a blade row

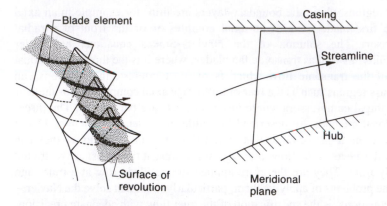

Fig. 3.2 Conventional description of flow in compressors on surface of revolution (the blade-to-blade surface) and on meridional plane

The more common method of analysing the flow in two intersecting two-dimensional surfaces is shown in Fig. 3.2. The flow is analysed on cylindrical blade-to-blade surfaces at several spanwise positions and on meridional planes, usually on only one average or mid-pitch meridional plane. These surfaces are not in general streamsurfaces and in most implementations no allowance is made for this, so that an extra level of approximation is introduced. Because the methods are inviscid much greater inaccuracies arise from other effects, so that errors arising out of the approximate formulation on these simple surfaces are overwhelmed. Principal sources of error are the estimates for the boundary layers on the annulus walls, particularly the blockage, and the specification of such quantities as the flow deviation in axial blade rows and the slip factor in radial compressors. The centrifugal impeller usually has strong curvature in both the meridional and blade-to-blade surfaces, sometimes in the same part of the impeller. For this reason the impeller is a component where the streamsurfaces twist and warp most noticeably and their deviation from blade-to-blade surfaces of revolution and the meridional plane is likely to be greatest. However, the passages in a centrifugal compressor are long compared to their cross-sectional area, more specifically their hydrualic diameter, and the flow experiences strong viscous effects. Because of this discrepancies arising in the formulation of any inviscid method of solution for the radial compressor are also likely to be overwhelmed by viscous effects.

Performing full calculations for the inviscid flow on the meridional surface was at one time too laborious for frequent use and further approximations had to be used. The general equations describing the flow in axial and radial compressors are, of course, identical but the different geometries mean that dominant effects and appropriate simplifications are not the same. The next section of this chapter therefore looks at the particular approximations which were developed for the axial compressor. For most axial compressors the dominant effect in the radial equation of momentum is the centripetal acceleration V_θ^2/r set up by the swirl velocity. Only when the flow path is highly curved in the meridional plane, as it might be near the hub of a fan or the front stages of a multistage compressor, are there other accelerations of comparable importance to V_θ^2/r. The approximation in which only this term is considered in the radial direction is usually known as simple radial equilibrium. It has more than just historical interest because it can be used to demonstrate rather clearly the manner in which the flow in most axial stages is connected in the spanwise direction. Simple radial equilibrium also allows the prediction of trends with changes in blade geometry or flow coefficient from the original design point, a topic addressed in the last section of this chapter.

After the section on simple radial equilibrium there is a brief account of the historical background, partly for the useful insights that can be obtained and partly because many of the ideas are still quite often referred to and occasionally used. Following this the practical implementation of methods for calculating the flow in the meridional plane is discussed, followed by some examples calculated for axial compressors.

One of the basic assumptions of the calculation methods used for the meridional flow is that they are inviscid and non-conducting. These assumptions can lead to serious errors which are often compensated for in an approximate manner. Unfortunately the compensation to correct one variable leads to inconsistencies in another. The penultimate section therefore describes how the inclusion of spanwise mixing to simulate the turbulent shear stress and heat transfer processes can be used to greatly improve the accuracy and validity of meridional calculation methods.

3.2 Approximations applicable to axial compressors: simple radial equilibrium

In the past the practical design needs of axial machines made it necessary to have approximate methods convenient for hand calculation. These approximations can still be used to show many of the effects predicted by the more complete inviscid methods. There are two quite separate issues to be addressed in this way. One is the specification of work input along the span of the blades from hub to casing so as to produce the required outlet conditions. What is ultimately of most concern is the spanwise variation in stagnation pressure which is related to the work input by either empirical loss coefficients or local efficiency. The other main issue is the way in which the flow adjusts itself in the radial direction in response to the constraints imposed by the blades. (Both of these topics are more of a problem for the axial compressor because the blade span in a centrifugal compressor is small in relation to the blade length and because the outlet of a centrifugal impeller is usually at constant radius.) As a first approximation the blades may be thought of as devices which cause the flow to leave in a prescribed direction; this is an oversimplification, as will be discussed in Chapter 4, but is reasonably accurate.

The simplest case to consider is an axial stage in a parallel annulus with the rotor and stator well separated. Across the rotor there will be a work input for steady flow along a mean streamtube given by

$$W = h_{02} - h_{01}$$

$$= U_2 V_{\theta 2} - U_1 V_{\theta 1} = \Omega(r_2 V_{\theta 2} - r_1 V_{\theta 1})$$

and clearly the work input is affected by the local value of the blade speed, U. For many axial compressors, particularly those in a parallel annulus, the radial distance of the streamtube from the rotational axis will be almost equal at inlet and outlet to the rotor. With this restriction the work input is

$$W = U(V_{\theta 2} - V_{\theta 1}).$$

If the work input is to be uniform in the radial direction then, since the blade speed is proportional to radius, it is essential that the change in V_θ is inversely proportional to radius. Now $V_\theta = \text{const}/r$ is the description of a free vortex and one way of achieving the uniform work input is for both $V_{\theta 1}$ and $V_{\theta 2}$ to be of the free vortex type. (It was common in the past to refer to *any* specification of the whirl velocity as the vortex distribution because for certain flows it might actually have this form.) To produce radially uniform work input it is only necessary that the difference $V_{\theta 2} - V_{\theta 1}$ is inversely proportional to radius; for example

$$V_{\theta 1} = ar - b/r \text{ and } V_{\theta 2} = ar + b/r$$

would be an acceptable solution, giving

$$W = 2Ub/r = 2\Omega b,$$

where Ω is the rotor angular velocity. Such a choice was advocated quite widely in the past. What is actually required is normally not radially uniform work input but uniform stagnation pressure rise and nowadays it is usual to allow the work input to vary in such a way that it compensates for the predicted radial distribution in loss.

The choice of the free vortex distribution for the whirl velocity $V_\theta = \text{const}/r$, was often supported by another reason. This relates to the second issue referred to above, the effect of the radial distribution on the flow field in which the blade is situated. The radial momentum equation can be written as

$$V_R \frac{\partial V_R}{\partial r} + V_x \frac{\partial V_R}{\partial x} - \frac{V_\theta^2}{r} = -\frac{1}{\rho}\frac{\partial p}{\partial r} + F_R \qquad (3.1)$$

where the assumption that there is no variation in the circumferential θ direction is already included and the flow is treated as inviscid. In equation 3.1 F_R is the radial force exerted by the blades on the gas and for most axial compressors this will be small enough to be ignored. In a parallel annulus and with the blades operating near design the radial velocities will be very much smaller than the other two components so that the first term in equation 3.1 can be dropped. Similarly the rate of change in the axial direction, $\partial/\partial x$, is small if there are only small radial velocities, and therefore small radial shifts in the streamtube, across a blade row. This means that the second term in equation 3.1 can be neglected to leave the very simple equation

$$\partial p/\partial r = \rho V_\theta^2/r \qquad (3.2)$$

showing that the whirl velocity generates a radial pressure gradient. This is often known as the equation of simple radial equilibrium and could be deduced immediately by considering the equilibrium of radial force and acceleration. Equation 3.2 can be integrated to yield

$$p_{casing} - p_{hub} = \int \{\rho V_\theta^2/r\}\, dr \qquad (3.3)$$

showing that static pressure p can never be lower at the casing than the hub for this simplified case of the parallel annulus. If the stagnation enthalpy h_0 is uniform then upstream or downstream of a blade row one can write

$$\partial h_0/\partial r = 0$$
$$= \partial h/\partial r + \tfrac{1}{2}\partial/\partial r(V_\theta^2 + V_x^2 + V_R^2). \qquad (3.4)$$

Because V_R is much smaller than either the axial velocity V_x or tangential velocity V_θ, it is neglected. If the entropy is uniform in the radial direction, it also follows from $\partial h_0/\partial r = 0$ that the stagnation pressure will also be independent of radius and one can write

$$\partial p_0/\partial r = 0 = \partial p/\partial r + \tfrac{1}{2}\rho\partial/\partial r(V_\theta^2 + V_x^2). \qquad (3.5)$$

The expression for the static pressure gradient equation 3.2 can be introduced and it then follows that

$$-V_\theta^2/r = [V_\theta \partial V_\theta/\partial r + V_x \partial V_x/\partial r].$$

Simple rearrangement yields

$$V_x \frac{\partial V_x}{\partial r} + \frac{V_\theta}{r} \frac{\partial}{\partial r} (rV_\theta) = 0 \qquad (3.6)$$

which immediately shows that if the term rV_θ is independent of r then the axial velocity is also uniform. The term rV_θ is the moment of momentum per unit mass and is a crucial quantity in determining the behaviour of the flow. Clearly if the whirl velocity is inversely proportional to radius, as it is in a free vortex, then the moment of momentum is independent of radius and, by equation 3.6, the axial velocity is also uniform in the radial direction. Because there is a preference for keeping the axial velocity more or less uniform in the radial direction this was a further reason for choosing the free vortex type of whirl velocity distribution.

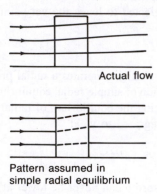

Actual flow

Pattern assumed in
simple radial equilibrium

Fig. 3.3 Schematic meridional streamline pattern for actual flow and for model of simple radial equilibrium through an axial blade row

The implications of the approach adopted here are illustrated in Fig. 3.3 in which a sketch shows the streamlines in the meridional plane for a parallel annulus. In the actual flow the streamlines are not straight and change radius as they pass through the blade row. The change is gradual, beginning some way upstream of the leading edge and continuing some way downstream. The equivalent streamline pattern for the model of simple radial equilibrium is also shown. In this case behaviour inside the blade row is not considered but all the changes in streamline radius and gas properties are assumed to occur inside the blade passage; therefore outside the blade row the streamlines are straight, have no radial velocity and the only acceleration is that due to the centripetal effect V_θ^2/r. In the actual flow the radial velocity is likely to become negligible only a short distance upstream or downstream of the blade row provided the annulus has parallel inner and outer walls. For some compressors the annulus is nowhere near parallel but nevertheless, as will be shown later, the ideas of simple radial equilibrium usually give the correct trends and the more detailed methods often provide fairly small corrections.

To achieve uniform stagnation pressure in equation 3.5 it was necessary to assume uniform entropy, which in turn implies radially uniform losses. In practice the losses are greater near the inner and outer walls of the annulus and losses also get higher when the relative Mach number is significantly greater than unity. To achieve uniform stagnation pressure it is necessary to put more work into these regions and uniform rV_θ is no longer adequate. In fact the blades may do more or do less turning in the endwall regions and the rV_θ distribution obtained there will rarely be that predicted by the simple deviation rules which are reasonably adequate over most of the span.

The simple radial equilibrium approach lends itself to solving algebraic equations to determine the axial velocity distribution for prescribed distributions of whirl velocity or blade outlet flow direction. From these the possible blade angles can be specified. Some examples of the solutions obtained to the simple radial equilibrium equation can be found in Horlock (1958). Until the 1960s the simple radial equilibrium approach was a major tool in the design of axial compressors. This is not any longer the way compressors are designed because using numerically based methods it is possible to relax the assumptions of simple radial equilibrium and provide something which is more generally accurate.

Although simple radial equilibrium is no longer a major tool for the design of compressors, the idea does however warrant treatment here for it explains some characteristics of compressors. First it shows for all axial compressors with the hub−casing ratio low ($r_{hub}/r_{casing} \ll 1$) that the change in V_θ across the rotor must be larger towards the hub than near the casing if very large variations in stagnation pressure and temperature are to be avoided. This in turn requires that there will be much larger deflections of the flow near the hub and therefore much greater camber. Second, because the rotor puts work into the flow, the whirl velocity is higher downstream of the rotor than upstream. The static pressure variation

$$\partial p / \partial r = \rho V_\theta^2 / r$$

will therefore be greater downstream of the rotor than upstream. This means that the rise in static pressure across the rotor will be smaller at the hub than at the casing. Put another way the degree of reaction (being the ratio of static pressure rise across the rotor to the static pressure rise across the whole stage) will be lower near the hub and higher near the tip.

An extension of this idea demonstrates a limit of operation for all compressors. The static pressure will be low downstream of the rotor hub compared with the casing. Stator blade rows reduce the absolute whirl velocity, so downstream of the following stator the radial gradient in static pressure will be much smaller than upstream of it. The static pressure rise across the stator hub is therefore necessarily higher than that at the casing and may be unacceptably high if the hub−casing ratio is low. If the design is such that the stagnation pressure is locally lower near the hub, because of smaller work input, this can lead to a marked drop in axial velocity across the stator hub, which exacerbates the loading on the stator blades in that region.

The ideas of simple radial equilibrium are also used in the last section of the chapter to examine effects in axial compressors operating away from the design point.

3.3 Early developments

Theories for the meridional flow in turbines and pumps were being produced from the early part of the twentieth century, see Wu (1952) for references. Simple radial equilibrium was an idea which was introduced very early and, as Marble (1964) states, the description was so widespread by 1942 that it is impossible to cite accurately any priority in the matter. However, the work of Traupel (1942) is one of the earliest complete accounts in which it is employed. Ruden (1937) had taken the ideas further towards the actuator-disc treatment which was to receive much greater attention in the 1950s.

The chapter by Marble (1964) was prepared many years before the work appeared and provides a broad treatment of the axisymmetric flow through turbomachines with axial machines very much in mind. The treatment is mathematical with the aim of taking analysis as far as possible, a very natural tendency before the widespread availability of computers. To make progress the blades are approximated by one or more surfaces of discontinuity across the annulus at a constant axial position. These surfaces are referred to as actuator discs and across them the axial and radial velocities are continuous but the tangential velocity changes discontinuously. They are equivalent to blade rows with an infinite number of blades of zero chord. In the simplest case the real blades are replaced by a single actuator disc. The mathematical attraction of using the actuator disc is that the solution can then be performed in a blade-free region for which the differential equations are homogeneous.

With a single actuator disc across the annulus certain simplifications are possible. Consider, for example, the actuator disc at $x=0$. If at a particular radius the axial velocity a long way upstream is $V_{x-\infty}$ and that a long way downstream is $V_{x+\infty}$, then at the actuator disc the axial velocity on either side is given by $V_{x0}=[V_{x+\infty}+V_{x-\infty}]/2$. In other words half the change between upstream and downstream is accomplished at the plane of the actuator disc. Marble was able to show that as a very good approximation the axial velocity at stations upstream and downstream of the actuator disc changes exponentially so that

$$V_x(x,r) = V_x(x=0,r) \pm \tfrac{1}{2}\{V_x(-\infty,r) - V_x(+\infty,r)\}\{1-\exp(\pm \pi x/h)\},$$

where x is the axial distance from the actuator disc and h is the height of the annulus. This remains a most useful way of looking at the overall interference or interaction of blades in a semiqualitative way. Some explanation for the exponential decay in the axial direction can be obtained by considering an idealized loss-free incompressible flow for which Laplace's equation must be satisfied. Writing Laplace's equation in terms of velocity potential and treating the annulus as if it were a two-dimensional strip with x and r as Cartesian coordinates gives

$$\frac{\partial^2 \phi}{\partial x^2} + \frac{\partial^2 \phi}{\partial r^2} = 0$$

It is a natural consequence of the properties of Laplace's equation in two dimensions that the decay in the axial direction will be exponential with the rate set by the length scale in the other direction, that is $e^{\pm x/h}$. For the axisymmetric geometry the treatment is more complicated and the radial variation needs to be in terms of Bessel functions leading to the decay rate being well approximated by $e^{\pm \pi x/h}$.

The actuator disc approach to the prediction of the axisymmetric meridional flow was later adopted by several other workers, prominent among them being Horlock who has given a very thorough treatment of the topic in his book, Horlock (1978). It became possible to include effects of compressibility and the effect of variations in the annulus cross-section. The interaction between several blade rows could be treated and various levels of approximation were explored. The theory was shown to be capable of predicting the radial variation in axial velocity for the cases on which it was tried. It could not, of course, handle the variation produced by the boundary layer near the annulus walls, which had the effect of making the predictions look unsatisfactory in some regions of the flow. (The endwall region remains the major problem with the methods in use today so that the overall accuracy achieved for predictions of the meridional flow is compromised more by the effects near the wall than by inaccuracy in modelling over the bulk of the flow.) As far as one can tell the actuator disc method has had very little effect on compressor design; at first designs were based on simple radial equilibrium theory and then later on numerical schemes, almost always a type of streamline-curvature solution.

The most quoted development was a general theory produced by Wu (1952) which gave rise to the stipulation of S1 and S2 streamsurfaces shown in Fig. 3.1. (In fact this is just one paper of a series which appeared in quick succession at about that time.) The S1 surfaces will be recognized as similar to the conventional blade-to-blade surfaces. Although however the S1 surface is a streamsheet at constant radius upstream of the blades, it distorts as it passes through the passage and at outset there is no way of knowing what shape the surface will have. If the circulation along the blades is uniform, so that the change in rV_θ is uniform along the span, no circulation is shed at the trailing edge and the S1 surface will return to being a surface of constant radius downstream. Although blades are not always designed to have uniform rV_θ they do normally have a distribution at the design point which is not too different from this because of the wish to have a uniform increase of stagnation pressure. (This may well be why the neglect of the deformation of the streamsheet implicit in the usual blade-to-blade treatment on surfaces of revolution does not seem to lead to serious error.) The S2 surfaces are also streamsheets; upstream of the blades these are formed by generators which are purely radial but the surfaces twist and warp as they pass through the blade passage. The full calculation therefore needs an iteration in which the S1 and S2 surfaces would first be guessed and calculations performed in each surface, the shape of each then being refined iteratively on the basis of calculations in the other surface.

It is widely recognized that Wu's work was truly pioneering in its description of the three-dimensional flow and in the way it identified problems to be tackled. Wu's general theory was not properly implementable at the time it was created because computers were in their infancy, but even now it is hardly used. One reason for this is because the overwhelming sources of inaccuracy in throughflow calculations come from the uncertainties associated with the endwall boundary layer and the prediction of such quantities as deviation. The errors introduced by simpler schemes (for example using a blade-to-blade surface which is a surface of revolution and treating the throughflow as being axisymmetric) are not serious in comparison with these. Another reason Wu's general theory has found little application now that powerful computers are available is that it is an awkward procedure to implement and there are better ways of carrying out full three-dimensional calculations.

Because it was an important feature of Wu's general theory that S1 and S2 refer to warped streamsurfaces it is unfortunate that these designations are used in the more normal formulations where the blade-to-blade surface is a surface of revolution and the surface in the hub—casing direction is the meridional plane. It is sometimes written, or at least implied, that the widely used throughflow calculation methods based on a single meridional surface are the outcome of Wu's general theory. This does not seem to be tenable, for some such methods were already in existence for calculating the meridional flow before the general theory was published, including one by Wu and Wolfenstein (1949).

Novak and Hearsey (1977) described the first steps towards the iterative solution of the flow on two surfaces, in the manner envisaged by Wu, in the main restricting attention to a blade-to-blade surface which was a surface of revolution. Krimerman and Adler (1978) carried out one of the few iterative applications of a method based on Wu's (1952) general theory when they calculated the flow in a radial impeller. Step 1 was an axisymmetric throughflow calculation, step 2 was a blade-to-blade calculation using the results of the first step, step 3 was a throughflow calculation (not axisymmetric) using the results of the second step. The results for the Mach number distribution near the shroud calculated in steps 1,2 and 3 are shown in Fig. 3.4, together with the results of a full three-dimensional calculation. The results for steps 2 and 3 agree quite well with the full three-dimensional calculations but since the results of steps 2 and 3 differ significantly from one another these still seem some way from convergence. Figure 3.5 shows the computed streamsurface shapes on passage cross-sections at two positions along the impeller passage where there is curvature in two directions and the three-dimensional effects would be expected to be largest. In these views the lengths have been scaled to make the cross-section square in each case. The broken lines show the simple approximation

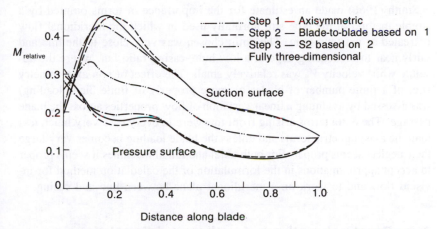

Fig. 3.4 Predicted relative Mach numbers along blade at shroud of centrifugal impeller. (From Krimerman & Adler, 1978)

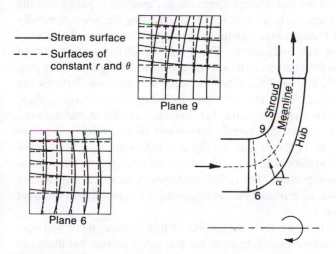

Fig. 3.5 Streamsurface distortion on planes in the inducer of a centrifugal impeller. (From Krimerman & Adler, 1978)

with the blade-to-blade flow on surfaces of revolution and the other surface as a meridional plane. Generally the three-dimensional streamsurfaces lie fairly close to those of the simplified model. However all the predictions are so different to the measurements in the radial impeller, as discussed in Chapter 6, that this refinement is academic, because of viscous effects. It is essentially because the benefits from the refinement given by the iteration between surfaces are so much less than the errors inherently present from the neglect of major effects, like those associated with viscosity, that the method of iterating between surfaces has attracted so little application.

Smith (1966) made an estimate for the importance of terms omitted by a simple method of the type still normally used in which the meridional flow is treated as axisymmetric. The point chosen was very close to the hub and fairly near to the leading edge of a low hub—casing ratio fan, where the absolute whirl velocity V_θ was relatively small. The effect of non-axisymmetry (i.e. of a finite number of blades or more properly of finite blade loading) was assessed by assuming a linear variation of flow properties across the blade passage. The extra terms arising from this were in every case very much less than the axisymmetric terms and unless the blade loading becomes very large their neglect seems proper. For both axial and radial machines it seems proper to accept approximations in the formulation of the calculation method for inviscid flow and to work on a meridional plane between hub and casing.

3.4   Practical methods for the meridional flow

In the form in which the full, though approximate, analysis is performed the flow is considered separately on two sets of intersecting surfaces. Normally one set will be the blade-to-blade surfaces and several stations will generally be chosen between hub and casing. These will be surfaces of revolution (although it is known that the true streamsheets will twist or warp as they pass through the blades). For axial machines the behaviour on these surfaces can be related to two-dimensional cascade flow, including a correction to blade angles to allow for the change in radius between inlet and outlet and the axial velocity—density ratio, see for example Jansen and Moffatt (1967). A recent report by Cetin *et al.* (1987) is specifically directed to the consideration of loss and deviation correlations for use with meridional throughflow calculations in transonic axial compressors. For radial compressors empirical information for the blade-to-blade flow is used but, as discussed in Chapter 6, in a different manner to the axial.

In the other direction, which is the subject of this chapter, there will normally be a single surface extending from the hub to the casing, but there can be more. The calculation of the flow along this hub—casing surface is usually referred to as the throughflow calculation and a detailed review and application of a range of such methods for axial turbomachines is given by Hirsch and Denton (1981). With the formally correct approach of Wu (1952) the hub—casing surfaces are streamsurfaces. If only one surface is used it can be a streamsurface so that, for example, in a given blade passage half the flow is on one side of it and half on the other. This was the choice made by Novak and Hearsey (1977). Horlock and Marsh (1971) modelled the flow as a circumferential average and concluded that though the model can never be strictly correct the errors involved in using it are acceptably small. (It is more relevant to say that the errors and inaccuracies attributable to viscous effects and other uncertainties like correlations for flow angle are much greater.) Outside the blades, loss-free flow would rapidly return to being axisymmetric and the different methods for averaging the flow all then become equivalent.

The model which is normally adopted, and which seems to be generally adequate, is that the flow is uniform in the circumferential direction, in other words $\partial/\partial\theta=0$. Since blade forces in the tangential direction must be allowed, and these require tangential pressure gradients, putting $\partial p/\partial\theta=0$ is equivalent to stipulating an infinite number of blades so that over the infinitesimal thickness of the hub–casing surface the variation with θ across the blade pitch is averaged out. What one now has is a mean streamsurface from hub to casing on which, at a given axial and radial position, the velocity components are all equal to their circumferential average. The calculation is normally carried out after projecting the velocities onto a different surface, the meridional plane, which includes the axial and radial (x and r) coordinates and is normal to the θ-direction.

There are several methods for calculating the flow on the meridional surface of which the most widely used is the so-called streamline curvature method and this will be discussed first.

Streamline curvature

What is required is an equation for the pressure (or equivalent property) gradients in the radial or the spanwise directions. This may be obtained from the equations of motion, for example Smith (1966). The description and approach that will be adopted here, however, is that given by Denton (1986) because it emphasizes the physical basis, with applicability to a wide range of geometries. Figure 3.6 gives two views of the machine, one showing the meridional plane, the other a view along the axis. For the view along the axis, Fig. 3.6b, the vector **e** lies in the mean hub–casing streamsurface and the vector **q** lies in the meridional plane. The streamsurface is inclined at an angle ϵ to the radial; with the flow taken as uniform in the θ direction this will be the local inclination of the blade camber line to the radial. ϵ will not normally be constant with radius although generally it will be small.

In the view of the meridional plane, Fig. 3.6a, the streamline ABC can be seen projected onto it. The local tangent to this is given by the unit vector **m** and the normal to the streamline in the meridional surface is given by **n**. However attempts to work with the true normal to the streamlines get into difficulties because the directions are not known in advance and they change

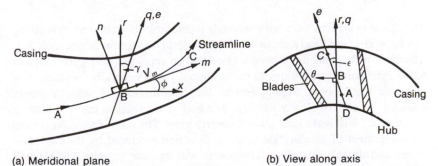

(a) Meridional plane   (b) View along axis

Fig. 3.6 Coordinate system for streamline curvature calculation of meridional flow

as the calculation progresses. It is therefore common to choose in advance directions which are roughly perpendicular to the streamlines and which do not change during the calculation. These directions are called quasi-orthogonals and are shown on the mean streamsurface of Fig. 3.6 denoted by **e**. The projection of **e** onto the meridional plane gives the unit vector **q**; both **q** and **e** are inclined at an angle γ to the r-θ plane so that for axial machines γ will be nearly zero whereas near the exit of radial compressors it will approach $-90°$. The streamline direction and the true normal to the streamline in the meridional plane, **m** and **n**, are each inclined at an angle ϕ to the axial and radial directions respectively.

The acceleration of a fluid particle at B can be built up from the following components:

1. $V_m \partial V_m / \partial m$ in the m direction, being the familiar substantive acceleration in the direction of the flow;
2. $-V_\theta^2 / r$ in the radial direction, this being the centripetal acceleration produced by the absolute swirl velocity which was included in simple radial equilibrium;
3. V_m^2 / r_m in the n direction, where $r_m = \partial m / \partial \phi$ is the radius of curvature of the streamlines projected onto the meridional plane, this being the centripetal acceleration produced by the flow following a path with radius of curvature r_m in this plane;
4. $a_\theta = (V_m / r) \partial (r V_\theta) / \partial m$ in the circumferential or θ direction, being the local blade loading.

The first and fourth terms contain substantive derivatives with respect to m whereas the true acceleration of particles is with respect to distance along the streamline direction in the curved hub–casing surface. With the flow taken to be axisymmetric, so that $\partial / \partial \theta = 0$, the differentials in the meridional plane are equal to those in the hub–casing surface, but in more general calculation schemes this approximation could not be applied.

The first three acceleration components may be combined to give the total acceleration a_q in the direction of the unit vector **q**,

$$a_q = V_m \frac{\partial V_m}{\partial m} \cos[90 - (\phi + \gamma)] + \frac{V_m^2}{r_m} \cos(\phi + \gamma) - \frac{V_\theta^2}{r} \cos\gamma \qquad (3.7)$$

The cosine in the first term is immediately replaceable by $\sin(\phi + \gamma)$. Something can be said immediately about the magnitude of the terms in equation 3.7. The quasi-orthogonal direction q is chosen to be approximately perpendicular to the meridional streamlines so the angle $\phi + \gamma$ will generally be small. It is also true that designs usually aim to keep the meridional velocity nearly constant, or at least slowly varying. For both these reasons the first term in equation 3.7 will make only a small contribution. The second term represents the component of acceleration in the q-direction produced by the curvature of the meridional streamline. The cosine will be near to unity and V_m will be of the same order as V_θ. In axial machines the radius of the meridional streamlines, r_m, will generally be large, and the second term of equation 3.7 small, except for the front stages of low hub–casing ratio machines. The second term is, however, very important for radial compressors. The last term will be small when γ is large, for example, in the radial part of a centrifugal compressor. For most axial compressors (as well as in the inlet and inducer regions of centrifugal machines) γ is very small and the last term can give a significant acceleration in the quasi-orthogonal direction. The magnitude of V_θ will depend on the loading and on the position in the machine, being always larger downstream of rotors than stators. In axial machines this third term in equation 3.7 is usually dominant and is the basis of the simple radial equilibrium approach discussed near the beginning of this chapter.

Equation 3.7 gives the acceleration a_q in the meridional (x,r) plane. Now the component perpendicular to the q and θ directions is also perpendicular to the e direction, that is normal to the streamline in the hub–casing mean streamsurface. The acceleration in the e direction can then be written

$$a_e = a_q \cos\epsilon + a_\theta \sin\epsilon \qquad (3.8)$$

where ϵ is the local inclination of the mean streamsurface to the meridional plane. The momentum equation applied in the streamsurface in the direction of e is then

$$-1/\rho\, \partial p/\partial e = a_q \cos\epsilon + a_\theta \sin\epsilon \qquad (3.9)$$

The distance $dq = de\cos\epsilon$ so that

$$-1/\rho\, \partial p/\partial q = a_q + a_\theta \tan\epsilon \qquad (3.10)$$

It is customary to remove the static pressure gradient and replace it with gradients of enthalpy and entropy,

$$-1/\rho\, \partial p/\partial q = T\partial s/\partial q - \partial h/\partial q$$
$$= T\partial s/\partial q - \partial h_0/\partial q + \tfrac{1}{2}\partial/\partial q(V_m^2 + V_\theta^2) \qquad (3.11)$$

where h_0 is the stagnation enthalpy. The equation in its conventional form for gradients in the direction of the quasi-orthogonal in the meridional surface is therefore

$$\frac{1}{2}\frac{\partial}{\partial q}V_m^2 = \frac{\partial h_0}{\partial q} - T\frac{\partial s}{\partial q} + V_m\frac{\partial V_m}{\partial m}\sin(\phi+\gamma) + \frac{V_m^2}{r_m}\cos(\phi+\gamma)$$

$$- \frac{1}{2r^2}\frac{\partial}{\partial q}(r^2 V_\theta^2) + \frac{V_m}{r}\frac{\partial}{\partial m}(rV_\theta)\tan\epsilon \qquad (3.12)$$

This equation is sometimes called the radial equilibrium equation and is the basis of all streamline curvature throughflow calculation methods, which take their name from the fourth term on the right hand side containing V_m^2/r_m.

The radial equilibrium equation contains the term $V_m \partial V_m / \partial m$ and in many methods this is removed by using the equation of continuity along a stream-tube. The procedure is algebraically quite involved but more seriously has the undesirable effect of introducing into the denominator a term $(1 - M^2)$, where M is the local relative Mach number. The equations therefore become singular at $M = 1$. Denton (1978) has shown that the derivative $V_m \partial V_m / \partial m$ can be quite adequately determined numerically from the previous iteration and this computational problem disappears.

Equation 3.12 gives the gradient of meridional velocity and a further constraint is required to fix the level. This is provided here by requiring that the meridional velocity integrated across the annulus gives the total mass flow across the quasi-orthogonal.

$$\int_{\text{hub}}^{\text{casing}} \rho V_m \cos(\phi + \gamma) w \, \mathrm{d}q = \frac{m}{N} \tag{3.13}$$

Here m is the total mass flow rate, N the number of blades and w the stream-surface thickness given by $w = 2\pi r B / N$, where B is a measure of the blockage and would be equal to unity in an ideal flow. B is strongly affected by the boundary layer displacement on the annulus walls and blades. B is also reduced, probably to a much smaller extent, by the fact that the flow is not genuinely uniform in the circumferential direction.

The radial equilibrium equation and the equation for mass flow are solved together. The most common way of doing this is to begin with a guess or previous iteration for the streamline shape in the meridional plane and the distribution of V_m along the quasi-orthogonals. From this V_θ can be determined from the blade or impeller geometry using correlations or blade-to-blade calculations. The entropy is assumed to be conserved along streamlines in calculations where losses are ignored and allowed to increase by the amount of the losses more generally. In stationary blade passages or outside blade rows the stagnation enthalpy h_0 is conserved along streamlines, in moving blades it is the rothalpy $I = h + W^2/2 - U^2/2$ which is conserved. From this the terms in the radial equilibrium equation, equation 3.12, can be calculated. The result is the rate of change of V_m across the annulus. The integral of V_m is compared with the overall mass flow and on the basis of this the overall level of V_m is adjusted up or down. The mass flow between streamlines is used to fix the position of the intersections of the streamlines with the quasi-orthogonals to be used for the next iteration. The method has been widely used in various forms for many years. Katsanis (1965), Smith (1966), Novak (1967) and Denton (1978) give a range of approaches which have been widely adopted and a fairly complete list of references is given by Hirsch and Denton (1981).

There are stability constraints and the streamline shape must only be adapted by a small fraction of its predicted change for each iteration if the process is to remain stable. This is expressed by the relaxation factor which gets smaller as the distance between quasi-orthogonals is reduced in relation to the distance between hub and casing. In axial machines this has meant that quasi-orthogonals are very often only put at the leading and trailing edge of blades. The relaxation factor must also be reduced as Mach number rises in order to maintain stability. A thorough treatment of this has been given by Wilkinson (1970) who has showed that the optimum relaxation factor is given by

$$R = 1/\{1 + kA^2(1 - M^2)\},$$

k being in practice likely to be of order 0.2 and A here denoting the grid aspect ratio, the distance between hub and casing divided by the distance between quasi-orthogonals.

The method of solving the streamline curvature equation is rather specialized and many approaches exist for doing this: one is illustrated in the flow chart at the end of this chapter.

Alternative methods for calculating the meridional flow

Another approach to analysing the meridional flow uses the streamfunction. In conjunction with Fig. 3.7, showing the meridional plane, a streamfunction is defined by

$$r\rho BV_q = -\partial\psi/\partial p$$

and $\qquad r\rho BV_p = \partial\psi/\partial q$ $\hfill$ (3.14)

where B is the blockage factor, as above.

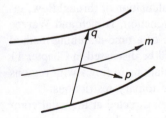

Fig. 3.7 Coordinate system for matrix method of calculating meridional flow

The pressure gradient in the q direction was given by equation 3.10 and it is now convenient to write the acceleration in the direction of the quasi-orthogonal

$$a_q = V_q\frac{\partial V_q}{\partial q} + V_p\frac{\partial V_q}{\partial p} - \frac{V_\theta^2}{r}\cos\gamma \hfill (3.15)$$

On replacing the static pressure with the stagnation enthalpy and entropy and using the streamfunction defined by equation 3.14 one obtains

$$\frac{\partial^2 \psi}{\partial p^2} + \frac{\partial^2 \psi}{\partial q^2} = \rho \frac{rB}{V_p} \left\{ \frac{\partial h_0}{\partial q} - T\frac{\partial s}{\partial q} - \frac{V_\theta}{r} \frac{\partial (rV_\theta)}{\partial q} + a_\theta \tan\epsilon \right\}$$

$$\frac{\partial \psi}{\partial p} \frac{\partial}{\partial p} \ln(\rho rB) + \frac{\partial \psi}{\partial q} \frac{\partial}{\partial q} \ln(\rho rB) \qquad\qquad (3.16)$$

The acceleration in the θ direction is given by $a_\theta = (V_m/r)\partial(rV_\theta)/\partial m$. Equation 3.16 may be solved iteratively with the right-hand side evaluated from the initial guess or the previous estimate. Marsh (1968) solved the equation by a matrix inversion and this has caused the method to be given the name 'matrix throughflow'.

There appears to be little relative advantage (and indeed no fluid mechanical difference) between the streamline curvature and matrix throughflow methods. The streamline curvature method does appear to be overwhelmingly the more popular and some possible reasons may be advanced for this. One is that the meaning of the terms in the equations is more obvious. Another is that the streamfunction is inherently less satisfactory when the resultant Mach number is close to unity because the Mach number is two-valued, one value less than unity and the other greater, for a particular value of streamfunction. The stream-function method therefore fails when the Mach number becomes even locally supersonic. In duct regions, that is outside the blade rows, it is the meridional Mach number which must remain below unity, and this is normally satisfied, but inside blade passages it is the local relative Mach number which is relevant and for compressors this often exceeds unity. The streamline curvature method considers continuity across the whole annulus and small patches of supersonic flow can be accommodated fairly easily.

Other methods have been developed for the calculation of throughflow, including a finite element solution to the streamfunction, Hirsch and Warzee (1979). A more significant departure is the use of a time-marching method Spurr (1982). Time-marching as a technique will be discussed in Chapter 11 but it can be noted here that it does allow calculations in flows with regions of subsonic and supersonic flow without the customary restrictions.

Most of the discussion in this section has been directed at the axial compressor. For radial compressors the same approaches are adopted and streamline curvature is the most common. Normally several quasi-orthogonals are placed inside the impeller and, with few exceptions, the flow is treated as axisymmetric. (It will be recalled from Fig. 3.5 that the refinement Krimerman and Adler (1978) obtained by allowing for the non-axisymmetric nature of the flow was very small compared with the discrepancy between measurement and inviscid calculation.) The axisymmetric throughflow calculation gives the velocity distribution along the hub and shroud; the velocity always reaches high values along the shroud because the streamline curvature term in equation 3.12 is dominant and the shroud forms the convex surface in the meridional plane. As will be discussed in Chapter 6 the velocity on the shroud has a crucial importance for the performance of centrifugal compressors. This is because the boundary layer flow (using this term for the complicated viscous region) tends to separate in the region where the high velocity is decelerated as the shroud curvature decreases in the more downstream part of the impeller. Such effects cannot be predicted by any inviscid method although such methods can indicate where separation is likely to occur.

Once the velocity distribution has been determined along the hub and shroud of the radial compressor the velocity distribution in the blade-to-blade surface can be estimated. This is sometimes performed by a very simple method of assuming a linear pressure gradient in the θ-direction just sufficient to give the necessary Coriolis acceleration. A wide range of methods have been brought to bear to calculate the blade-to-blade distribution but only those able to give a satisfactory prediction of the complicated viscous flow are likely to lead to very significant improvements over the very simplest methods. It seems most useful then to treat the axisymmetric throughflow method as a useful tool but one with major approximations. If the approximations are unacceptable there are now more precise ways of analysing the flow using three-dimensional viscous methods.

Throughflow calculations have been routine tools for the design and analysis of flows in compressors for quite some time and there is now a very large number of them available. The assumptions and methods overlap a good deal and it is not always clear how one differs from another. One numerical analyst remarked with droll resignation that in most papers describing methods it is so difficult to understand what exactly has been done that it is generally easier to write a new method; this may well have contributed to the large number of them. Streamline curvature methods and their combination with blade-to-blade methods continue to receive refinement, for example Jennions and Stow (1985, 1986). This is partly because they are quick and inexpensive to run compared to the full three-dimensional methods but also because they have become familiar methods to the users and large computing systems have been built around them. The refinements offer some significant corrections for components such as low hub–casing ratio fans where there is strong curvature in the meridional plane and a large deflection near the hub while at the same time the annulus boundary layer has a sufficiently small effect that it is not the principal source of the inaccuracy of prediction. In most cases the evidence is clear that the largest inaccuracies in describing the flow with a surface of revolution for the blade-to-blade flow and a meridional plane for the throughflow do not arise from errors in the method of averaging the flow to take account of variations in the circumferential direction or from the neglect of the streamsurface distortion. It is errors in predicting the flow turning by the blades, spanwise mixing and the presence of regions of high loss and low velocity near to the solid surfaces (boundary layers, in the usual terminology) which are much more serious.

Some of the concern with the details of throughflow methods, such as the averaging, was connected with the wish to use the methods to get information inside blade passages of axial machines. If details of the flow are required inside a blade passage it now seems that it is far more satisfactory to use a full three-dimensional method, either inviscid or better still viscous, than try to adapt the rather cumbersome method utilizing different planes and surfaces. The relatively simple methods of meridional throughflow are likely to continue to be very useful for a long time to come, but as methods recognized to be approximate in which the inconsistencies in averaging method or the neglect of terms relating to the non-uniformity in the circumferential direction are consciously overlooked.

The discussion so far has concentrated on the throughflow method as a part of design. A use which is at least as important is in the processing of measured data. In this case the input is generally stagnation temperature and pressure, a few casing static pressures, mass flow and rotational speed; the output includes the radial distributions of velocity components.

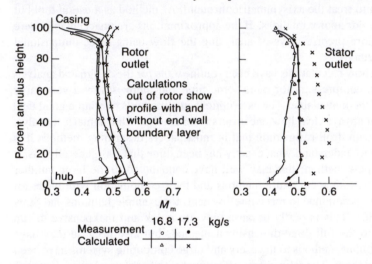

Fig. 3.8 Profiles of meridional Mach number for a single stage fan. Inlet hub–casing ratio 0.5. (From Hirsch & Denton, 1981)

3.5   Applications of streamline curvature methods in Axial compressors

The streamline curvature method can be thought of as a tool for the analysis and design of compressors and for the analysis of experimental results. In this section the results of some calculations are considered.

Figure 3.8 shows measured and predicted meridional Mach number for a transonic single-stage compressor with inlet hub–casing ratio of 0.5. The figure is taken from Part II of Hirsch and Denton (1981). The compressor was designed and tested by DFVLR and the calculation was performed by Hirsch. The comparison is at the design speed with two different mass flow rates. The discrepancies revealed are not the minor ones attributable to averaging procedures but quite major ones indicating serious deficiencies. Downstream of the rotor two sets of computed results are shown, one set with the endwall boundary layer included and the other with it omitted. The inclusion does bring some local improvement in the prediction close to the casing wall but does little for the discrepancies over the majority of the annulus. The majority of the discrepancy comes from errors in the input to the throughflow calculation of the flow outlet angle from the blades, i.e. errors in the correlation for deviation.

After the first or second stages of a multistage axial compressor the endwall boundary layer tends to dominate the uncertainties, mainly through the end-wall blockage. (The same is true of radial impellers too.) The blockage is usually defined by $B = m / \int \rho V_x \mathrm{d}A$, where m is the total mass flow, A the annulus cross-sectional area, ρ and V_x the density and axial velocity neglecting the presence of annulus boundary layers and assuming that the 'free-stream' values can be extrapolated to the annulus walls. To some extent the errors in endwall blockage and flow deviation can be set against one another so that overestimates for deviation can compensate for underestimates of blockage. There is no unambiguous method for estimating either of these quantities from the measurements normally made in multistage compressors. Generally there is believed to be a greater proportional uncertainty in the blockage than in the deviation.

The problems associated with matching were briefly discussed in Chapter 2 where it was shown that for multistage axial compressors very large variations in stage performance can be produced by quite small changes in the axial velocity. It can be appreciated that relatively small errors in the estimate of blockage can have large effects on the matching of the stages and therefore lead to highly inaccurate predictions of the throughflow. No refinement of the throughflow calculation method can compensate for this fundamental weakness in the input data.

It is useful to present some results from a throughflow calculation for a three-stage transonic compressor. In this case no comparison with measurements is possible although the input data do correspond to a real compressor. The compressor has a design total-to-total pressure ratio of 2.5 with an inlet relative Mach number to the first rotor of 1.45 at the tip and no inlet guide vanes. The hub–casing radius ratio at inlet is only 0.367 increasing to 0.60 at outlet from the third stator. The streamline curvature method was written by Denton (1978). The calculations have been performed for the artificially simple case of no losses and at 90 per cent of design speed to avoid problems of choking. The design will have included estimates for losses and the neglect of these must lead to there being a progressively larger mismatch for the blade rows after the first rotor. Two geometries have been investigated, one with an inlet bullet (such as the front of a jet engine might have) and the other with an upstream annulus. A meridional view is given in Fig. 3.9 for the two geometries together with the quasi-othogonals and the computed streamline shapes. (For the sake of clarity in the pictures only half the streamlines used in the

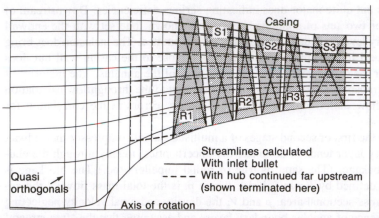

Fig. 3.9 Computed meridional streamlines for a three-stage compressor of low hub–casing ratio, with and without inlet bullet

calculations are shown in the meridional views of this compressor.) There is clearly much greater streamline curvature near the hub of the first rotor with the bullet and the effect of this is evident in the corresponding meridional velocities shown in Fig. 3.10 along quasi-orthogonals at entry to the first rotor, first stator and second rotor. The velocity is nondimensionalized by the mean blade speed at inlet to the first rotor. Only at entry to the first rotor are the differences significant; the more pronounced curvature with the inlet bullet leads to markedly higher meridional velocities near to the hub and, since the mass flow is the same for each, lower velocities elsewhere. Because the curvature effects are larger for the case with the bullet it is this which will be used in the other comparison described below.

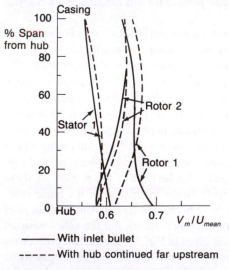

Fig. 3.10 Profiles of meridional velocity V_m for a three-stage compressor with hub altered from first rotor forwards. Calculations using streamline curvature method

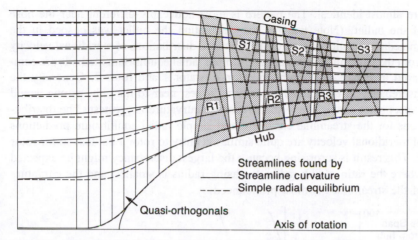

Fig. 3.11 Computed meridional streamlines for a three-stage compressor of low hub—casing ratio

To assess the effects of streamline curvature the calculations have also been performed using the method of simple radial equilibrium (so that the only pressure gradient in the radial direction is that necessary to balance the centrifugal acceleration, V_θ^2/r). Figure 3.11 shows the streamlines in the meridional plane predicted with the full streamline curvature method (solid lines) and by simple radial equilibrium (broken lines). For the simple radial equilibrium case the streamlines are made up of a series of straight lines from one quasi-orthogonal to the next. The streamline layout from the two calculations is generally fairly similar and around the first rotor the streamline patterns

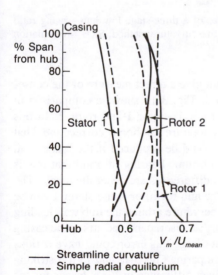

Fig. 3.12 Profiles of meridional velocity for a three-stage low hub—casing ratio compressor calculated by streamline curvature method and simple radial equilibrium

are almost identical. The pattern is most noticeably different near the nose of the bullet. (Note that for the calculation the bullet was cusped in the immediate vicinity of the axis.) Calculated distributions of meridional velocity from the streamline curvature and simple radial equilibrium methods are compared in Fig. 3.12 at entry to rotor 1, stator 1 and rotor 2. Generally the discrepancy in meridional velocity is largest near to the hub, as one would expect since this is where the streamline curvature is greatest. The distributions for the streamline curvature and simple radial equilibrium predictions of meridional velocity are quite similar at entry to rotor 1 but less so for rotor 2. This result is surprising because the largest discrepancy might be expected where the ratio of hub radius to casing radius is smallest and the curvature of the streamlines is most pronounced.

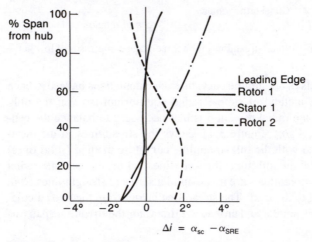

Fig. 3.13 Increment in incidence for blades of a three-stage low hub–casing ratio compressor. SC = calculation using streamline curvature method; SRE = calculation assuming simple radial equilibrium

The plots of meridional velocity do not give a direct measure of the consequences for the blades and to remedy this Fig. 3.13 shows a comparison in terms of an increment in incidence to rotor 1, stator 1 and rotor 2. In this it has been assumed that the streamline curvature solution is correct and what is shown is the extra incidence which would be produced if the blades had been designed using simple radial equilibrium. Bearing in mind that this is for a low hub–casing ratio machine the differences are generally small. The negative incidence increment towards the hub for rotor 1 and stator 1 can be understood as a result of the streamline curvature around the hub wall leading to increased meridional velocity. The largest discrepancy occurs at the casing for stator 1 and is about 3°; although not large this error could have serious consequences at high Mach numbers. Comparing Figs 3.12 and 13 it is striking that the increment in incidence for a given difference in V_m is larger for the stator than the rotor, a consequence of the fact that this is a high reaction compressor with no inlet swirl to the first rotor. The difference in incidence attributable to streamline curvature is very small near the rotor tips for both rotor 1 and 2; this is significant because blades with supersonic tips have a very limited capacity to tolerate incidence changes. This small discrepancy in incidence goes some way to explain the success which was achieved before the advent of proper throughflow calculation methods in designing high-speed rotors using only simple radial equilibrium.

In Figs 3.11, 12 and 13 the predictions by the streamline curvature and the simple radial equilibrium methods were generally similar, but particularly so for rotor 1 which has a very small hub radius. This was investigated by examining a bare annulus with the same hub shape but with no blades present. Calculations with simple radial equilibrium and the streamline curvature methods showed remarkable agreement in the region of the annulus which is occupied by rotor 1, implying that locally the effects of the additional terms in the full radial equilibrium equation, equation 3.12, cancel. To some extent then the similarity between the predictions of the simple radial equilibrium and streamline curvature methods for rotor 1 must be fortuitous. Nevertheless the discrepancies elsewhere in this low hub−casing ratio compressor of quite high loading are not large on a proportional basis.

The effect of the extra terms included in the streamline curvature method for a simple geometry such as this is to provide small corrections, probably smaller in some instances than the errors admitted with the empirical inputs. This will be progressively more true for later stages in an axial compressor where the streamline curvature becomes very small but the flow is highly non-uniform because of the viscous effects from endwalls. As noted earlier, the greatest uncertainties in predicting the performance of axial compressors of high hub−casing ratio arise from estimates for the endwall blockage and for deviation. To some extent errors in one can compensate for the other and organizations develop a consistent set of inputs rather than correct values for each. It is also true that the blockage is crucial in the prediction of the throughflow in radial impellers and large errors can arise from errors in the estimates for it.

For simple geometries of axial compressor, even of low hub−casing ratio, the dominant term in fixing the radial pressure gradient is that considered in simple radial equilibrium, $\rho V_\theta^2/r$, and the streamline curvature terms give comparatively small corrections, though these are not necessarily negligible. An example where the streamline curvature produces a major part of the span-wise or radial acceleration in an axial compressor is shown in Fig. 3.14. This is taken from Smith (1974) which describes the evolution of a design for a high bypass-ratio fan. The lines shown are streamlines calculated using a streamline curvature method and here several calculating stations were used inside the main rotor row and the outlet guide vane row. Although not evident in the view shown, the blade row labelled 'inner outlet guide vanes' is significantly leaned so that the blade force is partly directed radially inwards and contributes significantly to the spanwise pressure gradients. Many aspects

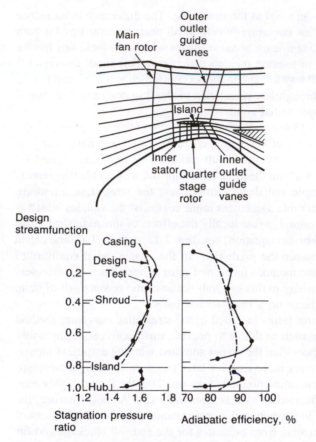

Fig. 3.14 The fan and booster stage of a high bypass-ratio engine. (From Smith, 1974)

of fan design are touched on in this paper with explanations for the design choices. Of particular interest is the work input which is shown in terms of pressure ratio and efficiency in Fig. 3.14. Here the ordinate is the fraction of streamfunction to give a better impression of the extent of the core. In the design the work input was reduced towards the hub and the tip: towards the hub, because otherwise excessive blade loading would be produced; towards the tip, mainly because stall is found to be initiated there and by lowering the loading this could be delayed. Near the hub the inner stage was able to make up the extra work so that downstream of the island there was not a gross mismatch of pressure which, if there had been, could have led to serious flow separation. It is very clear from the measurements that the design pressure ratio distribution has been well achieved with satisfactory estimates for the efficiency.

The strongly non-uniform work input means that vorticity is shed into the flow downstream of the rotor and the stage which sets up a secondary flow field. The effect of the secondary flow field was found to be small enough to neglect.

3.6 Mixing in multistage axial flow compressors

Throughflow calculation methods are almost invariably inherently inviscid, but to get satisfactory predictions in multistage compressors it is necessary to include loss estimates. A fairly arbitrary spanwise profile must be used for the loss because very little is known of the loss mechanisms in three dimensions. In other words an estimate has to be made for the endwall losses, the profile losses and the corner losses so that the total loss adds up to a value to give reasonable agreement with the overall measured performance. It has been known for a long time that the endwall and corner losses, which contain many different mechanisms, make up most of the loss. If, however, the loss were distributed radially in a way which took realistic account of this the inviscid throughflow method would fail numerically if several stages were calculated. To avoid this the methods are used with the loss spread out radially, pretending that the profile loss near mid-span is much higher than is realistic and correspondingly that the loss is lower near the endwalls. If this expedient is adopted the stagnation pressure distribution may be reasonably well predicted but with a quite erroneous trend in stagnation temperature. The spreading of loss across the span contributes to the emphasis put on minimizing profile loss, even though this is not a large proportion of the total loss.

Quite separately it was found that when the stagnation temperature and pressure measurements in multistage compressors were analysed using throughflow methods the predicted loss near mid-span seemed unreasonably high, sometimes exceeding the loss near the endwalls. Near the endwalls the loss was occasionally even inferred to be negative from the measurements. Quite clearly the loss creation was not being properly handled by the inviscid throughflow methods.

Howell (1945) showed measured axial velocity profiles at various stages in a six-stage compressor. The velocity profile became less and less uniform the further downstream the measurements were made, suggesting that the loss could be collected in the regions in which it was produced. For a long time this was accepted as the normal behaviour until Smith (1969) showed results in a much better 12-stage compressor. These results, reproduced here as Fig. 3.15, showed that the profiles of velocity and temperature settled down to give a more or less repeating pattern after only about three or four stages. If a repeating condition can be produced it means that some process must be redistributing the loss in the spanwise direction.

Adkins and Smith (1982) addressed the essential issue, which is to find a mixing process in the spanwise direction which needs to be included in any method for predicting the flow and in interpreting measurements to deduce the loss. They constructed a method for estimating the spanwise mixing which was assumed to be caused by the spanwise velocities set up by the secondary flow. The secondary flow was taken here to include not only the secondary flow due to non-uniform blade circulation (i.e. non-uniform work) and flow from the pressure to suction surface in the endwall region but also the tip-clearance flow and the spanwise flow in the blade boundary layers. The mixing

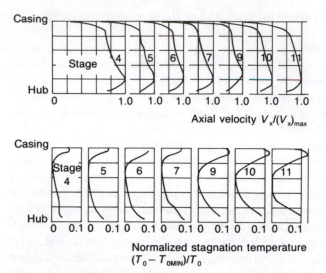

Fig. 3.15 Axial velocity and stagnation temperature profiles in a 12-stage compressor. (From Smith, 1969)

coefficient β was assumed to relate properties in the axial and spanwise directions by an equation of the form

$$\frac{\partial P}{\partial x} = \beta \, \frac{\partial^2 P}{\partial y^2} \tag{3.17}$$

where P is a circumferential average of a scalar property and x and y are the axial and spanwise coordinates respectively. In the Adkins and Smith model the mixing coefficient is assumed to be given, in simplified form, by

$$\beta = \frac{x}{a} \int_0^a \frac{V_y^2}{V_x^2} \, da \tag{3.18}$$

with a slight change in definition of coordinates, so that a is the passage width at blade exit and da is taken as normal to the flow and parallel to the axi-symmetric surface on which the mean flow is supposed to lie. The crucial point is that the radial mixing is supposed to be proportional to the square of the predicted spanwise secondary flow, V_y, and therefore to be deterministic.

A number of empirical factors were included by Adkins and Smith in obtaining the overall secondary flow and the prediction of flow direction in the axial/tangential direction was shown to be very good. In the spanwise direction there is very good reason to believe that the magnitude of the secondary flow is significantly overestimated, see Gallimore and Cumpsty (1986). Using these spanwise secondary flows in equation 3.18 Adkins and Smith were able

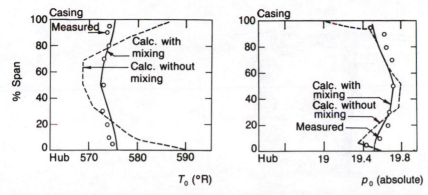

Fig. 3.16 Exit stagnation temperature and pressure from a low aspect-ratio three-stage compressor. (From Adkins and Smith, 1982)

to produce very much improved predictions of the radial distribution in properties. Figure 3.16 shows the measured and predicted spanwise distributions of stagnation temperature and pressure for a three-stage compressor of fairly low Mach number. The predictions were obtained by Adkins and Smith using a streamline curvature throughflow method both with no mixing and with their predicted mixing. The improvement with the mixing included is very striking. A similar improvement was shown for the prediction of temperature profile from a nine-stage compressor and this is shown in Fig. 3.17.

Gallimore and Cumpsty (1986) began from a different hypothesis for the cause of the mixing, that it was turbulent in origin. Here turbulence was used rather loosely to include any unsteadiness which was non-deterministic and the definition might require refinement when the understanding of the unsteady flow in compressors is better. By injecting ethylene gas into the compressor and measuring the spread it was possible to deduce the mixing level directly in two four-stage compressors. A plot of the contours measured downstream

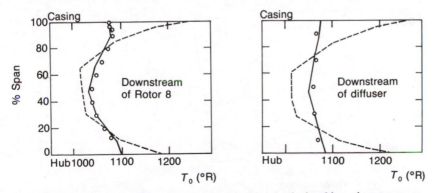

Fig. 3.17 Stagnation temperature profiles measured and calculated in a nine-stage compressor. Solid line shows calculation with mixing, broken line calculation without mixing. (From Adkins and Smith, 1982)

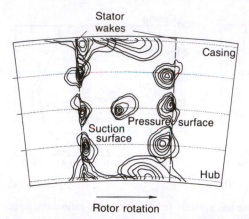

Fig. 3.18 Ethylene contours measured downstream of third stator with injection upstream of same blades. Injection at radii shown by dotted lines. (From Gallimore and Cumpsty, 1986)

of the third stator row with injection upstream of the same stator is shown in Fig. 3.18. Some evidence of secondary flow in the spanwise direction can be seen near mid-span near the blade trailing edge but the predominant cause of the mixing is the turbulent spread to give more-or-less round contours. The mixing is somewhat higher near the endwalls but as a first approximation can be taken as uniform. Contours measured downstream of rotor 4 after injecting ethylene at mid-span upstream of rotor 1 are shown in Fig. 3.19. What is particularly striking about this is that there is no evidence of a core flow with a boundary layer, but instead mixing takes place right across the annulus. The magnitude of the mixing was found from the measured contours to be similar to that predicted by Adkins and Smith in the one case where a direct comparison was possible. However the mixing predicted by Adkins and Smith's model, equation 3.18, using the radial secondary velocities inferred from the

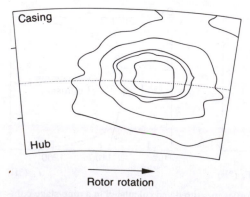

Fig. 3.19 Ethylene concentration contours measured downstream of stator 4 for injection upstream of rotor 1 at mid-span. (From Gallimore and Cumpsty, 1986)

movement and distortion of the measured ethylene contours, was only about one-third or one-quarter of the true value. The level of mixing in the compressor was higher than would be expected in a flat-plate boundary layer but similar to that in fully developed flow in a channel or pipe with dimensions approximately equal to the annulus height. A very simple analysis was given so that the mixing could be expressed in terms of the loss

$$\frac{\epsilon}{V_x L_s} = \frac{0.4t}{L_s} \left[\frac{2\omega^*(t/L_s)}{3\phi^2} \right]^{1/3}$$

(3.19)

In this equation L_s is the axial stage length, t is the blade maximum thickness, $\phi = V_x/U_{mean}$ is the flow coefficient and ω^* is the loss non-dimensionalized with the mean blade speed. (The Adkins and Smith mixing coefficient β is defined differently so that $\epsilon = V_x\beta$.) Equation 3.10 matched the values deduced from measurements in the two compressors described by Gallimore and Cumpsty and has since been found to give an accurate prediction of the mixing near mid-span of the very efficient compressor tested by Wisler *et al.* (1987).

Wisler *et al.* carried out very detailed measurements with the aim of establishing definitively the mechanism of mixing generation. They found clear evidence of the turbulent type of mixing of nearly uniform magnitude over a large part of the span. Near the endwalls the mixing was higher and the contours were highly distorted. Wisler *et al.* attribute the higher mixing in this region to the secondary flow mechanism originating with Adkins and Smith, but this was challenged in the discussion of the paper.

Gallimore (1986) described the inclusion of the mixing modelled as a turbulent diffusion-type process in a simple streamline curvature method and its application to the calculation of the meridional flow in axial flow compressors. The mixing coefficient ϵ was taken to be radially uniform. The mixing produces stresses and heat transfer and brings changes in the momentum and the stagnation enthalpy (or rothalpy for rotors) between quasi-orthogonals. For axial machines in which the meridional streamlines are nearly parallel to the axis, i.e. the radial velocity is very small, the inclusion of mixing has no direct effect on the streamline curvature equation, equation 3.12, since the quasi-orthogonals are radial and there is no stress in that direction. Gallimore achieved a comparable fit with the measurements to that shown in Fig. 3.16 or 3.17. He was also able to show that the improvement produced by including mixing was remarkably insensitive to the exact level of mixing, so that the mixing could be reduced by a factor of two or three without making the agreement with measurements significantly worse (it was for this reason that Gallimore and Cumpsty chose to ignore the radial variation in the level in mixing). Gallimore also showed rather graphically the effect of mixing on the radial distribution of loss needed to give a match with the data. Figure 3.20 shows the loss distributions found by trial and error to give a good match for the axial velocity distribution in a small four-stage compressor. In one case the calculation included mixing, whilst in the other mixing was neglected, just as is normal in most compressor designs. The neglect of mixing requires too

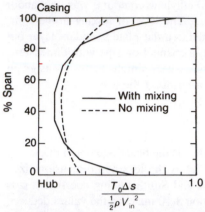

Fig. 3.20 Input loss distributions to give reasonable velocity profile predictions in a four-stage compressor using streamline curvature method. (From Gallimore and Cumpsty, 1986)

much loss near mid-span and too little near the endwalls. This in turn leads to too much attention being given to reducing profile losses when the major losses are being generated elsewhere. Although the incorrect loss distribution can allow a correct prediction of one property, in this case axial velocity, the errors are felt in other quantities; with high-speed compressors the error usually appears in the stagnation temperature if the loss has been distributed to give a reasonable prediction of stagnation pressure, as in Figure 3.16.

The prediction methods of Adkins and Smith or of Gallimore represent corrections to the inviscid streamline curvature method to include a real effect. The term mixing is unnatural, implying that a special or unusual process is taking place. In fact all that is being included is the radial effects which have been neglected in the formulation of the inviscid throughflow methods: the turbulent shear stress and heat transfer arising from the radial gradients in velocity and temperature. The implementation is still not altogether straightforward because of the difficulty of including the high level of losses in the region near the endwall. The usual throughflow methods adopt assumptions compatible with inviscid flow so that 'slip' occurs at the endwall, i.e. there is a discontinuity in the velocity from a free-stream value to zero on the surface without the physical region of large but finite gradients. Full consistency requires that the correct loss, mixing level and velocity profile are all used together and at this point what is being demanded is a solution of the Navier–Stokes equations in the meridional direction with appropriate turbulent terms.

3.7   Axial compressor off-design trends

It has already been stated that one of the special features of compressor blading is the way in which the flow along the whole blade span is connected together and that each spanwise section cannot be treated independently. For most axial compressors it is the constraint of radial equilibrium which is the main link between the different spanwise sections. At the design point the blade sections at each spanwise position will, if the design is accurate, be compatible with the flow in which they are placed. If the flow rate is changed the conditions stipulated in the design will no longer be achieved and the blades can be forced to operate locally at very different conditions.

Two examples may serve to show this. In each case the method of simple radial equilibrium will be used. (It was shown earlier in this chapter that simple radial equilibrium adequately predicts the main trends in axial machines and has the advantage of being simple to the point of transparency.) In the present application the flow is assumed to have uniform entropy and effects related to compressibility are omitted. It is further assumed that the relative flow direction out of blade rows, i.e. the deviation, is unchanged as the flow

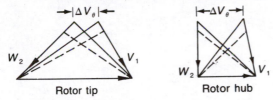

Fig. 3.21 Variation in velocity triangles at rotor hub and tip with flow rate. Solid lines show design flow, broken lines reduced flow

rate changes. Figure 3.21 shows the velocity triangles assumed to exist at hub and tip of a rotor, with the design condition shown by solid lines. At the design point the design is such that the stagnation pressure rise and the axial velocity are radially uniform, $dh_0/dr=0$ and $dV_x/dr=0$. Suppose that the mass flow rate is now reduced — the broken lines in Fig. 3.21 indicate this. Near the tip the reduction in mass flow brings about a large increase in work input, since the change ΔV_θ is large, but near the hub the shape of the velocity triangles means that the increase in work input is smaller. Whereas at design $dh_0/dr=0$, at the reduced mass flow dh_0/dr is positive. Assuming that the entropy remains radially uniform and the flow effectively incompressible

$$\frac{1}{\rho}\frac{dp}{dr}=\frac{dh_0}{dr}-\frac{1}{2}\frac{dV^2}{dr}$$

It is then easy to show that the radial equilibrium equation is given by

$$\frac{1}{2}\frac{dV_x^2}{dr}=\frac{dh_0}{dr}-\frac{V_\theta^2}{r}-\frac{1}{2}\frac{dV_\theta^2}{dr}=\left(\Omega-\frac{V_\theta}{r}\right)\frac{d(rV_\theta)}{dr} \qquad (3.20)$$

To derive the final expression in equation 3.20 it has been necessary to assume $d(rV_\theta)/dr$ is negligible upstream of the rotor compared with values downstream of it. (It has also been assumed here that off-design as well as on-design the assumption of negligible radial velocity is valid.) For a compressor it will be the case, with very few exceptions, such as near the root of low hub—casing ratio fans, that the rotor angular velocity Ω is greater than V_θ/r. For mass flow less than design then it follows from equation 3.20 that dV_x/dr is also positive. In other words the constraint of radial equilibrium requires that the axial velocity falls towards the hub relative to that at the casing.

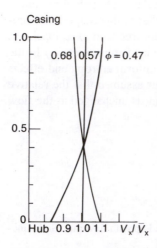

Fig. 3.22 Variation in axial velocity profile predicted by streamline curvature method downstream of stage 3 in four-stage compressor, hub—casing ratio 0.75, low speed. No losses included. Design point: $\phi = V_x/U_{mean} = 0.57$, $\psi = \Delta h_0/U_{mean}^2 = 0.4$

Such trends can be confirmed by streamline curvature calculations and Figure 3.22 shows results of calculations using an axisymmetric streamline curvature method for a multistage axial compressor of hub—casing ratio equal to 0.75. The rotational speed was low enough for compressibility effects to be neglected and for the present calculation the losses were omitted. At the design condition the stage loading, and flow coefficient were uniform across the span and the degree of reaction was 50 per cent at mid-span. Figure 3.22 shows the axial velocity after the third stage stator, but essentially similar plots would have been obtained at other stations. Three flow coefficients are shown, one of which is the design value. Just as predicted by the argument based on simple radial equilibrium the reduction of the overall flow from the design value has the effect of further lowering the axial velocity near the hub and increasing it near the casing, exacerbating the already more severe conditions in the hub region. Similar calculations were also performed for a different design, having axial inlet flow to each rotor row and a very high degree of reaction, but the same stage loading and flow coefficient. The off-design trends were virtually identical.

The second example concerns the procedure common during development of compressors, the twisting of blades. With perfect designs such a procedure would be unnecessary and indeed detrimental since it would destroy the perfect balance achieved. In fact it is still common practice for a number of reasons, including the correction of aerodynamic design errors, the correction for the effect of blade untwist because of the high centrifugal loads and the removal of certain types of blade flutter. The complete effect, provided that no radical change in blade behaviour like massive flow separation occurs, can be predicted by a streamline curvature method. However, the trend can be seen from simple radial equilibrium. Suppose a rotor blade row is twisted so that the tips are opened up, i.e. the tip stagger is reduced. Near the tip the blades will now tend to pass more mass flow, that is V_x will be locally higher. The lower stagger means a rather smaller static pressure rise across the rotor tip but a greater absolute whirl velocity V_θ downstream of it. By the consideration of simple radial equilibrium the whirl velocity gives rise to a radial pressure gradient.

$$\partial p / \partial r = \rho V_\theta^2 / r$$

and an increased V_θ means a larger radial pressure gradient. Since the opening up of the rotor tip reduced the static pressure downstream of the tip, the increased radial pressure gradient means a still greater lowering of the static pressure at the rotor hub. The lower static pressure will, if the work input is more or less unchanged near the hub, mean a higher axial velocity, further unloading the hub. In other words a simple gross change to the tip section has an effect along the whole span; the same would also be true for more subtle changes, for example to the rotor tip profile.

3.8 Flow chart — Use of a streamline curvature method in analysis mode for an axial compressor

The method in analysis mode is applied to the meridional plane with information given about the blade angles. The starting point is a view of the meridional plane. It should be recognized that there are very many variations of the method and this indicates only one. In many schemes the blade-to-blade step, step 4, would not be repeated each iteration and might only be performed once using the guessed input at the start of the calculation.

The sequence is as follows:

1

Choose positions for quasi-orthogonals (q-os), for example at blade leading and trailing edges

2

Guess positions of streamlines in meridional plane and then evaluate the streamline curvature and streamtube contraction at intersections with q-os.

3

Guess meridional velocity V_m at each intersection of q-o and streamline, guess T_0, entropy etc. along first q-o.

4

Use blade-to-blade calculation or correlation in conjunction with specified geometry and estimates for V_m, T etc. to calculate flow outlet direction and loss. (Streamline convergence gives important contribution to *AVDR* which strongly affects blade performance.) Hence V_θ and p_0 along the q-o. Note that stagnation enthalpy (or rothalpy) is conserved along meridional streamlines and losses are convected along meridional streamlines.

5

Evaluate terms on RHS of equation 3.12 along q-o, beginning with first, using current estimate for shape of meridional streamlines.

6

Integrate dV_m^2/dq along q-o to get V_m with an arbitrary or guessed constant

7

Calculate overall mass flow rate from equation 3.13 and adjust constant in predicted V_m distribution to get prescribed overall mass flow. Return to 6 unless no adjustment needed, in which case go to 8

8

Integrate V_m to find new locations of meridional streamlines along q-o for correct mass flow between them and store this information

9

Move to next q-o and repeat steps 4 to 8; after last q-o go to 10

10

Allow intersection of streamlines with q-os to move towards new position stored in step 8 but use relaxation factor (so that movement is a small fraction of difference between current and stored positions) to ensure stability. This gives new streamline shape and thence curvature

11

Go to 5 unless movement required of streamlines is less than a convergence threshold, i.e. meridional solution is converged, in which case go to 12

12

Print out results

As an alternative to going from step 11 to 12 the calculation could return to step 4 and recalculate the blade-to-blade flow in the light of the improved estimate for the meridional flow. The meridional flow could then be recalculated in steps 5 to 11. This refinement seems to be rarely performed.

The streamline curvature method is extremely sensitive to the shape of the hub and casing. Great care needs to the taken to make the surfaces used in the calculation smoothly curved in the meridional plane, even if the actual compressor has significant discontinuities of radius or curvature. (The real discontinuities are smoothed out by the boundary layer.) Because this requires some smoothing, for example the fitting of a low-order polynomial to describe the endwall, there can be problems when the meridional curvature of the endwalls is large.

4

Blade-to-blade flow for axial compressors with subsonic inlet flow

4.1 Introduction

The specification of the blades for an axial compressor is at the very heart of compressor design. It has over the years attracted the greatest part of the attention of designers and researchers and because there are so many aspects to be addressed this is a long chapter. Compressor blades operate in regions ranging from incompressible to supersonic; in this chapter attention will be given to blades with subsonic inlet flow, although inside the blade passage the flow may become locally supersonic. The chapter following this restricts itself to blades where the inlet flow is supersonic.

The blade-to-blade surface description of the blades of an axial compressor is a familiar and seemingly natural way to describe the flow. Figure 4.1 shows the geometry and defines the appropriate notation. The chord line is a straight line through the leading and trailing edges of the blade, while the camber line is curved and runs down the middle of the profile. The camber line and chord line meet at the leading and trailing edges. (For a rotor blade the angle and magnitude of flow velocities relative to the blades would be denoted by β and W respectively but in this chapter the notation appropriate to the stator, α and V, will be used for both.)

The blade-to-blade surface suggests modelling as a linear (or two-dimensional) cascade and this has been widely adopted both experimentally and theoretically. The decision to describe the flow in this way is only a model and is approximate, since the real flow is three-dimensional and the neglect of three-dimensionality may have serious consequences. There is a fundamental difference between a cascade flow and the blade-to-blade flow in a compressor. In the cascade the outlet flow direction and the static pressure rise are determined by the cascade geometry and by the inlet flow, principally the inlet flow direction. For the blade-to-blade surface at a particular spanwise position the local blade shape does *not* fix the outlet flow direction or pressure rise; these are determined by the whole blade and not just the spanwise section being considered. To some extent the performance is also determined by the adjacent blade rows. The art of design is therefore to select blade shapes compatible with the flow produced by the blade section surrounding it. The real use of

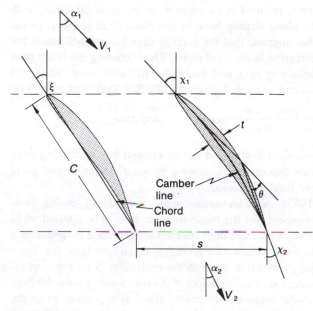

Fig. 4.1 Blade-to-blade geometry and notation

two-dimensional information such as that derived from cascade tests is to provide the information on what turning, static pressure rise and stagnation pressure loss are realistic design inputs and what blade shapes are needed to achieve these.

After the introduction the first topic is the effect of blade shape and then attention is given to the most important issue, what maximum loading can blades realistically and safely be expected to stand. Then what is the optimum way to achieve this and what is the optimum incidence. The next two sections look at the consequences of the design, how well the flow follows the blade outlet direction (or how much deviation there is) and how much loss is produced. There are particular effects associated with Reynolds number and Mach number and a section is devoted to each of these. The remainder of the introduction is used to discuss general topics relating to blade-to-blade flows.

Non-constant radius blade-to-blade surfaces

In simple geometries, like compressors with constant hub and casing radii, the blade-to-blade surface will be approximated by a cylinder which is a surface of revolution of constant radius. For more complicated geometries, like the front stages of multistage axial compressors, where the hub radius changes rapidly in the axial direction, it is usual to take the blade-to-blade surface to be a surface of revolution which projects as a straight line in the meridional plane (i.e. the $r-x$ plane) meeting the mean streamlines at the leading and trailing edges. Suppose that at the leading edge of a blade the mean streamline in the meridional plane is inclined at an angle ϕ_1 to the axial direction, with the leading edge of the blade sloping back in this plane at an angle γ_1 to the radial direction. Further suppose that the leading edge has an inclination (or lean) ϵ_1 to the radial direction in the $r - \theta$ plane. Then denoting the blade inlet angle on the streamsurface by $(\chi_1)_s$ and the blade inlet angle on a cylindrical surface by $(\chi_1)_c$ the two can be related geometrically (see Tysl *et al.*, 1955) by

$$\tan(\chi_1)_s = \frac{\cos(\gamma_1 + \phi_1)}{\cos\gamma_1} \tan(\chi_1)_c - \sin\phi_1 \tan\epsilon_1$$

Clearly an exactly equivalent expression can be created for the trailing edge or for flow angles. This expression is solely a geometric relation and gives no consideration to the fluid mechanics.

Smith and Yeh (1963) gave an approximate analysis to enable two-dimensional data and concepts for the blade-to-blade flow to be applied when the mean axisymmetric streamsurface does not intersect the blade at right angles to the spanwise direction (the spanwise direction is taken here to be the direction of the stacking line, very often through the centroids). Two types of inclination, sweep and dihedral, were considered. In the case of sweep the flow is not perpendicular to the spanwise direction; dihedral is present when the blade is not normal to the endwalls or to the axisymmetric streamsurfaces of the meridional flow. A radial blade in the front of a compressor where there is a significant cone angle on the hub will have quite large sweep; if the blade is staggered it will also have dihedral where it intersects the hub. Smith and Yeh showed that away from the endwall the two-dimensional cascade information should be used by taking a blade section formed by the intersection with the axisymmetric streamsurface but then considering the section projected onto a plane normal to the stacking line of the blade. The approach is not valid close to endwalls and approximate corrections had to be given for this region.

The three-dimensionality of the flow also includes, at the simplest level, substantial changes in the radial distance between blade-to-blade streamsurfaces, i.e. changes in streamsurface thickness, bringing additional acceleration or deceleration of the flow. When changes in streamsurface radius and thickness are included the methods are sometimes referred to as quasi-three-dimensional.

The theoretical and numerical background

Until a few years ago calculating two-dimensional inviscid flow in cascades was a major undertaking. A number of methods were devised for this purpose and a summary of these is provided by Gostelow (1984). The majority of the methods were for incompressible flow and before the advent of computers even this typically required several weeks work per cascade. In Germany approximate methods were developed to take isolated aerofoil performance and transform this to give the performance of a cascade. The nature of this procedure directed the design towards low camber blades of high stagger because this is the area when the approximations hold most accurately. A detailed summary of the extensive German work stretching right back to the early days, as well as much else, is given by Scholz (1977). A summary stressing the US contribution to the calculation of flow in cascades is given by Roudebush (1956).

The advent of quick, accurate computer-based calculation methods for two-dimensional cascade flows has transformed the consideration of blading. General methods are now well established to calculate the inviscid flow when it is incompressible, subsonic, transonic and supersonic without restriction to cascade geometry (see Chapter 11). The uncertainties and difficulties now arise primarily from limitations which are inherently viscous (including turbulent) in nature. The effect of this is to relegate the previous methods with all their restrictions to history and they will not therefore be described further here, but the interested reader will find abundant references in Gostelow (1984) or in Scholz (1977). In what follows, modern calculation methods will be used to illustrate effects and attention will be concentrated on the fluid mechanics. The improvements in computing now mean that the entire blade shape may be specified arbitrarily to achieve the desired velocity distribution (or, equivalently, the Mach number distribution or the pressure distribution). Such shapes have sometimes been referred to as prescribed velocity distribution (PVD) blades and some aspects of them are discussed below. There is then no longer any need to use the profile families, with the profile thickness wrapped around a camber line chosen to give the desired turning. Nevertheless the profile families are still widely used — they provide a good starting point for designs and sometimes appear to work as well as the specially designed blades. In some sense describing them is looking backwards but most of the knowledge about compressor blading has been built on such families and there is much to be learnt from them.

Geometric parameters

In the blade-to-blade surface the following may be considered the geometric variables for aerofoils specified with profile families:

stagger ξ
solidity $\sigma = c/s$
camber angle θ
camber line shape (usually a circular arc, sometimes a parabola)
thickness—chord ratio t/c
thickness distribution (e.g. C4, NACA-65, DCA).

The succinctness of the term solidity, more common in American usage, makes it preferable to the equivalent pitch—chord ratio, $s/c = \sigma^{-1}$. The most important variables are the first two, followed by the camber angle. The last three only become very important for the overall performance as the inlet Mach number rises and the flow starts to have supersonic patches. An additional geometric variable which is sometimes important is the surface roughness, characterized by the roughness height k_s. The leading edge radius may be thought of as part of the definition of profile shape but sometimes it is considered as a separate variable. It has a large effect at high Mach numbers.

Although the direction of the blade at the trailing edge is denoted by χ_2 the flow does not leave in this direction. The deflection $\epsilon = \alpha_1 - \alpha_2$ is then smaller than that which would be implied by the blade exit angle. The loss in deflection is usually referred to as the deviation which is defined by

$$\delta = \alpha_2 - \chi_2,$$

and this quantity has the advantage for correlation purposes of being fairly small. Defined in the same way, but with very different significance, is the incidence

$$i = \alpha_1 - \chi_1$$

being the angle between the mean flow direction into the blade and the projection of the camber line at the leading edge. This is not the same as the definition of incidence used in aeronautics, referred to here as the angle of attack A, which is the angle between the inlet flow direction and the chord line. The chord line is inclined to the axial direction by the stagger angle ξ so it follows that

$$A = \alpha_1 - \xi.$$

Unfortunately a large body of data was obtained by NACA in terms of the angle of attack. If the camber angle of a blade θ is known or can be estimated and if the camber line can be approximated by a circular arc, then it is easy to see that the incidence can be found from $i = A - \theta/2$.

Aerodynamic parameters

The principal aerodynamic inputs may be summarized as the inlet flow direction α_1, the inlet Mach number M_1 and the Reynolds number based on blade chord and inlet velocity, $\rho_1 V_1 c / \mu$. There is another very important aerodynamic input which is the ratio of the axial velocity out of and into the blade V_{x2}/V_{x1}. In the early days of compressor blade investigations this was often not given the attention it deserved, particularly in Britain, but it is immediately apparent that changes in V_x across the blade row have a direct effect on the blade boundary layers; $V_{x2}/V_{x1} > 1$ implies that the conditions are relieved, $V_{x2}/V_{x1} < 1$ implies that they are worsened. For flows where density changes are significant it is the axial velocity—density ratio,

$$AVDR = \rho_2 V_{x2}/\rho_1 V_{x1},$$

which matters. This abbreviation has now become widespread and will be used even when the density change is negligible.

Given these aerodynamic inputs or constraints it is possible to decide what is an acceptable or realistic value of outlet flow angle, α_2, and the deflection, $\epsilon = \alpha_1 - \alpha_2$, an issue addressed in the section of this chapter on loading. From this, it is possible to make decisions about solidity, stagger and camber, as well as thickness and the distribution of thickness and camber.

The object of the blades is to produce a rise in static pressure or a deflection of the flow and normally one effect is necessary for the other. The rise in static pressure may be conveniently non-dimensionalized by the inlet dynamic pressure to the blade row,

$$c_p = (p_2 - p_1)/(p_{01} - p_1)$$

or, for incompressible or low Mach number flow,

$$c_p = (p_2 - p_1)/(\tfrac{1}{2}\rho V_1^2).$$

Neither the deflection nor the pressure rise coefficient is itself a sufficient description of blade loading: blades of high stagger can produce little deflection but large pressure rise whereas for blades of very low stagger a high deflection can be accompanied by almost no pressure rise.

For all blades there is a loss in stagnation pressure and it is again convenient to non-dimensionalize the loss by the inlet dynamic pressure so that

$$\omega_1 = (p_{01} - p_{02})/(p_{01} - p_1)$$

where p_{02} denotes the average stagnation pressure measured a short distance downstream from the blades (usually the mass-average value of p_{02} is used). For a typical blade row operating near its optimum condition at an inlet Mach number sufficiently low for no strong shocks to occur in the passage the loss in total pressure will be small, perhaps $\omega_1 < 0.02$. This loss, usually referred to as profile loss, makes up only a fairly small fraction of the total loss in the stage. There is not very much to choose between the losses of the various families of profiles, again until the Mach number becomes high enough for supersonic patches and strong shocks to occur. Modern designs of PVD type blading, the so-called supercritical or controlled diffusion blades, have been shown in cascade tests to reduce the losses by perhaps 20–30 per cent, but for most purposes the blade losses are not of first importance. What matters far more is that the blades should produce the deflection specified so that the flow leaves the blades inclined at the desired outlet angle α_2.

The universal use of lift and drag coefficients in aeronautics led to their employment in compressor cascades and Fig. 4.2 shows an isolated blade row with the axial and tangential forces on the blade shown as X and Y respectively.

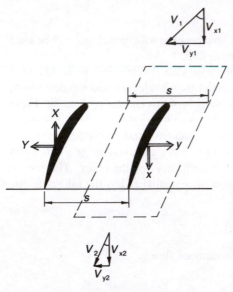

Fig. 4.2 The momentum balance about a blade in cascade

It is assumed that the fluid can be treated as incompressible and the axial velocity in and out taken as equal. Conservation of momentum then shows that the axial force applied by the blade to the gas is

$$X = (p_2 - p_1)s$$

and application of Bernoulli's theorem gives

$$X/s = \tfrac{1}{2}\rho(V_1^2 - V_2^2) - \Delta p_0 \tag{4.1}$$

where Δp_0 is the loss in stagnation pressure.

The tangential force Y can also be found from conservation of momentum as

$$Y = \rho V_x s(V_{y1} - V_{y2})$$
$$Y/s = \rho V_x^2(\tan\alpha_1 - \tan\alpha_2) \tag{4.2}$$

With constant axial velocity it makes sense to define mean direction and velocity by

$$V_m = V_x \sec \alpha_m, \tag{4.3}$$

$$\tan\alpha_m = \tfrac{1}{2}\left(\frac{V_{y1} + V_{y2}}{V_x}\right) = \tfrac{1}{2}(\tan\alpha_1 + \tan\alpha_2) \tag{4.4}$$

By analogy with isolated aerofoil theory the lift force L can then be defined as the resultant force perpendicular to the mean velocity and the drag force D as that parallel to the mean velocity. Thus

$$L = X \sin\alpha_m + Y\cos\alpha_m.$$

Introducing equations 4.1 and 4.2 and rearranging one obtains

$$L = \rho s V_x^2 \sec \alpha_m [\tan\alpha_1 - \tan\alpha_2] - s \Delta p_0 \sin\alpha_m \tag{4.5}$$

and for the lift coefficient

$$C_L = \frac{L}{\frac{1}{2}\rho V_m^2 c} = \frac{2}{\sigma}[\tan\alpha_1 - \tan\alpha_2] \cos\alpha_m - \frac{\Delta p_0}{\frac{1}{2}\rho V_m^2} \frac{\sin\alpha_m}{\sigma} \tag{4.6}$$

where $\sigma = c/s$ is the cascade solidity.

The drag force

$$D = Y\sin\alpha_m - X\cos\alpha_m$$

and using equations 4.1 and 4.2 this reduces to

$$D = s \Delta p_0 \cos\alpha_m \tag{4.7}$$

The drag coefficient is then given by

$$C_D = \frac{D}{\frac{1}{2}\rho V_m^2 c} = \frac{\Delta p_0}{\frac{1}{2}\rho V_m^2} \frac{\cos\alpha_m}{\sigma} \tag{4.8}$$

However it is conventional to express the loss in terms of the inlet velocity

$$V_1 = V_m \cos\alpha_m/\cos\alpha_1.$$

Hence

$$C_D = \frac{\Delta p_0}{\frac{1}{2}\rho V_1^2} \frac{1}{\sigma} \frac{\cos^3\alpha_m}{\cos^2\alpha_1} = \frac{\omega}{\sigma} \frac{\cos^3\alpha_m}{\cos^2\alpha_1} \tag{4.9}$$

The lift coefficient may be rearranged with Δp_0 replaced by C_D so that

$$C_L = \frac{2}{\sigma}[\tan\alpha_1 - \tan\alpha_2]\cos\alpha_m - C_D\tan\alpha_m \tag{4.10}$$

Since C_L/C_D is typically greater than 40, and $\tan\alpha_m$ is not very different to unity, it often suffices to neglect the last term in equation 4.10. With the loss small, i.e. $C_D \ll C_L$, it is also easy to show that

$$C_L = (2/\sigma)(\psi/\phi)\cos\alpha_m \qquad (4.11)$$

where $\psi = \Delta h_0/U^2$ and $\phi = V_x/U$ are the stage loading and flow coefficients.

By considering a closed contour around a blade it is easy to see that the circulation around an individual blade in the cascade is found to be

$$\Gamma = \oint V.dl = (V_{y1} - V_{y2})s$$
$$= V_x(\tan\alpha_1 - \tan\alpha_2)s \qquad (4.12)$$

With the restriction of constant axial velocity and the neglect of the total pressure loss term equation 4.5 may be immediately rewritten as

$$L = \rho V_m \Gamma,$$

the Kutta Joukowski formula derived in isolated aerofoil theory.

The expressions for lift and drag have been given for completeness and because some of the work referred to is couched in terms of them. Theoretical work on cascades around the 1940s, much of it German, used lift and drag (see for example Scholz, 1977), and the NACA cascade tests in the 1940s and 1950s made extensive use of these parameters. Compressibility and allowance for changes of axial velocity, however, so complicates matters that they are not normally included. It is now generally recognized to be more convenient to use flow deflection and non-dimensional loss directly and the use of C_L and C_D has almost disappeared.

The aim in the specification or design of a blade row is to achieve the turning with the least loss and the largest tolerance to incidence changes, but there are often varying strategies to be used. For example the flow may be turned by a cascade of aerofoils with low camber operating at moderate positive incidence, or with aerofoils of higher camber and zero or negative incidence. Alternatively one could use a high solidity cascade showing small deviation or with lower solidity cascade with higher camber blades to offset the higher deviation. In designing for the blade row it is important to bear in mind that each blade section should be chosen to be compatible with the whole blade since the actual performance is the composite of the whole.

4.2   The effect of blade shape

Profile families

It is possible to specify the blades of an axial compressor in many different ways. The traditional way is to base the design on families of profile such as the C-series or the NACA-65 series and information about these is given in the Appendix. A third very frequently used section is the double-circular-arc (DCA) or biconvex blade which, although not looking like a traditional

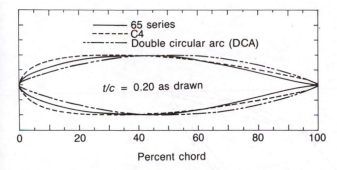

Fig. 4.3 A comparison of the thickness distribution for three common profiles

wing aerofoil section, has great advantages at transonic Mach numbers. The various thickness distributions are shown superimposed in Fig. 4.3, drawn at a thickness—chord ratio of 20 per cent for clarity although such thick sections would rarely if ever be used. The early investigations were carried out with sections of 10 per cent thickness—chord ratio but nowadays most applications would call for much thinner blades, typically around 5 per cent for subsonic inlet flow, because the high Mach number performance is so much better. The C4 section has its maximum thickness at 30 per cent chord, the NACA-65 at 40 per cent and the double circular arc (DCA) at 50 per cent. The C4 has a rather stubby nose and the DCA a rather sharp one and this difference has a large effect on the velocity distribution around the blades.

The thickness distribution then has to be laid along a camber line. Normally this is a simple curve, usually a circular arc and occasionally a parabolic arc; once the camber line shape is chosen it is only necessary to specify the magnitude of the camber angle. Strictly speaking the camber line for the NACA-65 series is uniquely defined, with infinite slope at the leading and trailing edges, but away from these regions the camber line is very well approximated by a circular arc and this has become the normal camber line used with the NACA-65. The nomenclature for the shape and camber of the 65-series thickness distribution is in terms of the lift of an isolated aerofoil in incompressible flow and this is explained in the Appendix. (A curve for getting the equivalent circular arc camber angle from the NACA specification in terms of the lift coefficient of the blade acting as an isolated aerofoil is also given in the Appendix.)

The effect of blade thickness distribution is demonstrated in Fig. 4.4 where calculated pressure distributions are presented for the C4, NACA-65 and DCA, all on a circular arc camber line with the same solidity, camber, stagger and maximum thickness—chord ratio. The choice of stagger and camber, which is the same for each, is representative of blades in a multistage compressor and the loading has been chosen to be consistent with current practice (the diffusion factor, a measure of blade loading discussed later in this chapter, is approximately 0.48). The choice of zero incidence is arbitrary but this is close to the normally recommended values, as will be discussed below. In Fig. 4.4

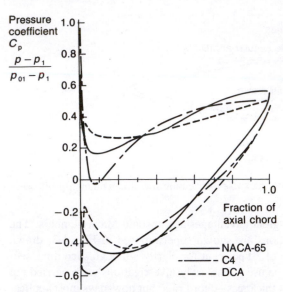

Fig. 4.4 The predicted pressure distribution about three common blade profiles in cascade. Stagger 33.6°, camber 27.3°, solidity 1.25, thickness/chord ratio 0.10, $M_1 = 0.6$, incidence 0°

it can be seen that the NACA-65 blade gives a higher pressure rise and this is attributable to the lower deviation for this section. The NACA-65 and even more the C4 profile, show an undesirable feature on the pressure surface with a low pressure region near the leading edge; this not only reduces the blade force but also leads to unnecessary deceleration of the pressure surface boundary layer. If these sections were at slight positive incidence this pressure surface velocity peak would decrease but with a corresponding worsening of conditions on the suction surface. Nearly always attention is concentrated on the suction surface because loss generation is greatest when the velocities are high and in decelerating the high velocity on the suction surface towards the trailing edge there is the risk of separation. The DCA blade has a much more satisfactory pressure distribution on both surfaces with little unnecessary acceleration except for the spike near the leading edge. These spikes are often missed by calculation methods unable to resolve conditions in sufficient detail but are common whenever the leading edge radius is small, as it is for the DCA here, or for other sections of small thickness–chord ratio. What is strange is that these spikes appear to have no deleterious effect on the loss, a point which will be developed further below. The DCA shows an increasingly strong adverse pressure gradient from the suction peak to the trailing edge. This tends to cause flow separation before the trailing edge, though the contribution of this to the loss seems small.

At low Mach numbers the effect of blade shape seems to be very small. Felix and Emery (1953) tested C4 and NACA-65 blades of the same thickness–

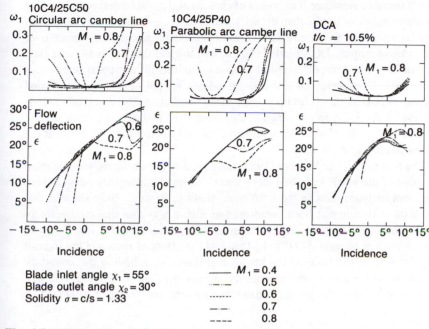

Fig. 4.5 Measured flow deflection and loss coefficient, $\omega_1 = \Delta p_0/(p_{01} - p_1)$, from Andrews (1949) for C4 aerofoil of circular and parabolic camber and for double circular arc. $Re = 2.10^5$ at $M_1 = 0.5$

chord ratio and camber at low Mach numbers and found them to behave very similarly with virtually identical losses but slightly wider operating range for the C4. Andrews (1949) tested a series of blades including the C4 on circular arc and on parabolic arc camber lines and the double circular arc (DCA). The DCA was of 10.5 per cent thickness–chord ratio with leading and trailing edge radii of 5 per cent of maximum thickness. The C4 blades were both of 10 per cent thickness–chord ratio. Measured losses and deflections are shown for all three cases in Fig. 4.5. At low Mach numbers the loss is similar for all three profiles but the range of low loss operation is notably higher for the C4 on a circular arc camber line. As the Mach number into the cascade is raised the minimum loss tends to rise slightly, but more noticeably the range narrows very sharply for the C4 section on either camber line. Even worse for the C4 on the parabolic arc the condition for minimum loss moves to significant negative incidence so that the opportunity for flow turning falls. It was the design intention of the parabolic camber that the low speed performance would be better than the circular arc equivalent but with an earlier rise in loss with increase in Mach number. In the event the blades with the parabolic camber seem worse at all conditions.

It is at the high Mach numbers that the superiority of the DCA blade becomes apparent, as Figure 4.5 shows, with very much wider operating range and no rise in minimum loss or reduction in flow turning at $M_1 = 0.8$. The loss rise at negative incidence is the result of choking and could be delayed somewhat by use of a thinner profile than the thick one used by Andrews, a point discussed in a later section. Despite this evidence there has been some reluctance over the years to employ DCA blades and one can suspect that this is because they do not look like 'proper' aerofoil shapes — indeed in his report one senses that Andrews experienced just this prejudice. Wright (1970) recommended the use of NACA-65 blades up to inlet Mach number of 0.78 and recommends the use of the DCA in the range of M_1 from 0.70 to 1.20, but there is little to suggest that the NACA-65 would be noticeably better than the DCA at the lower Mach numbers.

The data of Felix and Emery (1953) have already been referred to. The main purpose of the work described there was to reconcile differences which were apparent between tests in the USA and others in Britain. They showed that the tests in Britain had been performed in such a way that the axial velocity was significantly higher at outlet than at inlet to the cascade, i.e. the axial velocity−density ratio $AVDR>1$. This had the effect of reducing the overall deceleration of the flow on the blades and therefore relieving the boundary layers on the blades. In consequence the British tests had shown larger deflections at stall and lower deviation than those for tests performed at $AVDR=1.0$.

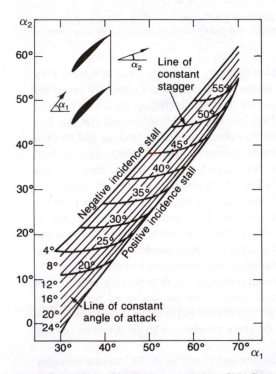

Fig. 4.6 A sample of Mellor's presentation of NACA cascade data for 65-(12)10 cascade with solidity 1.0 at low Mach number. (Equivalent circular arc camber 30°)

In the USA a large data base had been built up by the low speed testing of NACA-65 series profiles, for example Emery *et al.* (1958), using sidewall suction to maintain $AVDR=1$. This information was presented in a generally rather inconvenient manner with tests performed at a constant air inlet angle for a range of blade staggers. Mellor (1956) replotted this data in a more helpful form as illustrated by Fig. 4.6. (This work was never published but the complete set of diagrams is given in Horlock, 1958.) In Mellor's diagrams stall is defined as the condition at which the loss is 1.5 times the minimum; others such as Lieblein and Howell took the stall to be when the loss was twice the minimum.

Prescribed velocity distribution blades

The improvements in computing now mean that it is no longer necessary to be constrained to use blades of a profile family but the camber and thickness may both be specified arbitrarily to achieve the desired velocity distribution (or equivalently the Mach number distribution or the pressure distribution). Such shapes have sometimes been referred to as prescribed velocity distribution (PVD) blades. In the case of compressors PVD was held back by the difficulty of satisfactorily including the effect of the boundary layer on the suction surface; the boundary layer becomes thick enough to alter the pressure distribution but also tends to separate near the trailing edge so that the existing calculation methods become inaccurate. Nevertheless Goldstein and Mager (1950) studied the prescribed velocity distributions with a view to obtaining the maximum possible blade circulation, i.e. the maximum possible loading. The boundary layer methods were necessarily approximate and the effect of the boundary layer on the potential flow was not considered, but Goldstein and Mager made the essential step of choosing velocity distributions which would not cause the boundary layer to separate and therefore allowed the boundary layer to be calculated reasonably accurately. The velocity distributions look remarkably similar to those chosen more recently as the optimum type, with very rapid acceleration near the leading edge, a portion of nearly constant velocity along the suction surface and then a deceleration region, beginning very steep but relaxing so as to keep the boundary layer form parameter, $H=\delta^*/\theta$, approximately constant.

More recently Papailiou (1971) adapted an inverse boundary layer method by Le Foll (itself based on Truckenbrodt's method) to design the optimum blade. Papailiou had a rapid acceleration to a plateau, at the end of which the rapid deceleration was assumed to produce boundary layer transition. The adverse pressure gradient weakened continuously to keep the form parameter of the turbulent boundary layer approximately constant. As the start of the adverse pressure gradient was moved further back along the blade suction surface (and, consequently, the transition to turbulent boundary layer flow was delayed) the predicted losses were reduced, but if it was put too far back the overall pressure rise also decreased. Transition at about 30 per cent chord was found to be the optimum. A separate study by Fottner (see Scholz, 1977) derived similar velocity distributions and came to the similar conclusion: maximum lift compatible with the prescribed type of velocity distribution was obtained with the distribution of free-stream velocity set to start the deceleration and to give transition on the suction surface at about 30 per cent chord.

More recently, driven by the wish to design blades for operation with supersonic patches of flow but no shock waves (so-called supercritical blades), it has again been found that pressure distributions can be found which will avoid boundary layer separation at or near design conditions. The current design philosophy which has evolved for supercritical blades has been summarized by Hobbs and Weingold (1984) and is illustrated in Fig. 4.7. From the leading edge the flow on the suction surface is allowed to accelerate continuously, to retain the boundary layer laminar, until the point where the strong adverse pressure gradient begins where a rapid transition is likely to take place. On the pressure surface the velocity is kept nearly constant, the aim again being to retain laminar flow. The blade shape is selected so that no shock wave is predicted in the deceleration of the supersonic flow and what this turns out to give is an adverse pressure gradient which begins severe and gradually weakens. The type of pressure or velocity gradient on the suction surface has been referred to rather graphically as a 'ski-jump' distribution. The pressure gradient on the suction surface is chosen to avoid separation of the turbulent boundary layer. Expressed in terms of the free-stream velocity gradient and momentum thickness it is necessary to keep $\theta/U \, dU/dx$ constant if the form parameter of the boundary layer profile is to remain nearly constant and the deceleration is not to cause separation. Very often a boundary layer approaching

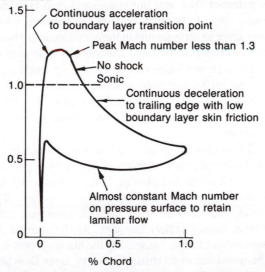

Fig. 4.7 A schematic representation of the design Mach number distribution for a supercritical (controlled diffusion) blade. (From Hobbs and Weingold, 1984)

the Stratford (1959) zero skin friction case is aimed for because this is expected to minimize the loss.

From what was said in the section on losses in Chapter 1 it will be clear that it is an oversimplification to assume that a boundary layer close to separation will give low losses. The loss production of a turbulent boundary layer is more or less independent of the profile form and skin friction but depends overwhelmingly on the cube of the local free-stream velocity. The loss generation of laminar boundary layers is very much lower for the same free-stream velocity except at very low Reynolds numbers. The supercritical type of velocity distribution does mean that the free-stream velocity becomes very high on the suction surface and loss production will be high in the early stages of the turbulent boundary layer. The choice of a boundary layer with the maximum possible deceleration rate may, however, lead indirectly to the low losses for a given loading because this can be made to produce the longest possible region of laminar flow. Very clearly the precise condition for transition becomes very important both in determining the loss and in ensuring that the chosen pressure gradient is accurately matched to the thickness of the boundary layer.

The flow does not always turn out to be that which is calculated in the design process, particularly at transonic conditions, and this is discussed in the final section of this chapter. Early designs had a plateau of nearly constant high velocity on the suction surface but this did not seem to be realizable at transonic conditions and the design shown by Hobbs and Weingold avoids having a plateau on the suction surface. Originally inverse design methods such as those of Garabedian and Korn (1976) were required to arrive at shock-free distributions; the output of such inverse method is the coordinates of the blade. Since then the iterative use of other easier and less restrictive direct methods has taken over: a shape is specified, the pressure distribution calculated and if it is not satisfactory a modified shape is tried. The shape of the supercritical blade is fairly characteristic, with a flat straight region towards the trailing edge and most of the camber in the forward part, so that from a reasonable first guess it may be possible to converge rapidly on the final shape. A

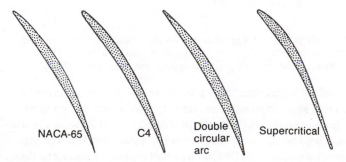

Fig. 4.8 Four blades for nominally identical turning, inlet flow angle 43.6°, outlet flow angle 23.5°, for NACA-65, C4 and DCA

comparison of a supercritical blade with the NACA-65, the C4 and the double-circular-arc, all chosen for the same duty, is given in Fig. 4.8. The super-critical cascade in this case has a solidity of only 0.83, compared with 1.00 for the others, and an important reason for the lower losses of the supercritical blade is probably the lower solidity at which it is possible to operate.

What came as a surprise about the supercritical cascades, designed with particular inlet Mach numbers in mind, was that they also operated very well at all inlet Mach numbers below that for which they were designed. Some measurements reported by Hobbs and Weingold for a blade operating at Mach numbers lower than that for which it was designed are shown in Fig. 4.9; the loss at low Mach numbers is very similar to that from conventional profiles but the useful operating range of incidence has been significantly increased. The pressure distribution at low inlet Mach numbers on a blade designed to be shock-free at transonic conditions turns out to be very similar to an 'optimum' blade designed to have a turbulent boundary layer fairly close to separation over the entire deceleration region on the suction surface.

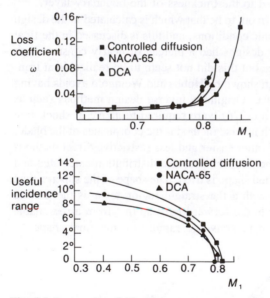

Fig. 4.9 Loss and useful incidence range measured for three cascades, one a controlled diffusion (supercritical) type. (Solidity = 0.933 for all cascades; at design $M_1 = 0.7$, $\alpha_1 = 60°$, $\alpha_2 = 46.4$, $AVDR = 1.07$.) (From Hobbs and Weingold, 1984).

The PVD blade of the supercritical type is the extreme opposite to the double-circular-arc: the DCA shape minimizes the velocity peak on the suction surface (neglecting any leading edge spike) but the pressure gradient increases in the flow direction and separation is expected to occur before the trailing edge. As noted in the section on losses in Chapter 1, small regions of separated flow mix out with only small losses.

One of the greatest benefits of the PVD blade is that the pressure gradient is chosen to eliminate separation, so that calculations including the boundary layer become more reliable. Furthermore the shape of the blades puts very little camber near the trailing edge and this has the effect of producing very little deviation. With a small value of deviation, perhaps less than 4° for a high camber blade, the design of the compressor can be more precise. The combined effect of this is to produce a cascade for which all aspects of performance (deviation, loss, incidence range and choking characteristic) can be better predicted than the profile family sections; this rather than the lower losses is probably their greatest advantage.

The high Mach number aspects of the supercritical type of blade will be discussed later in the section of this chapter devoted to Mach number effects in general. Certain worries remain, however, even at low Mach numbers. The performance will be seriously affected if the prediction of the boundary layer is incorrect. Boundary layer prediction is not very good for blades in cascade, in particular the transition; it is not at all accurate inside multistage compressors where there is a high level of unsteadiness and turbulence. The shape of the supercritical blade, with the camber near to the front, is not unlike the parabolic arc blades tested in the past (e.g. Andrews, 1949). The potential advantages of the 'ski-slope' type of pressure distribution in controlling the boundary layer development were understood (Carter, 1961), but so too was a potential problem. With a conventional profile on a circular arc camber line the pressure gradient on the suction surface tends to increase towards the rear, leading to the boundary layer getting closer to separation as the trailing edge is approached. For such a blade a modest increase in incidence beyond the optimum is likely to move the separation a small way upstream with only small deleterious consequences. With a pressure gradient chosen so that the turbulent boundary layer is near to separation along its entire development, a similar increase in incidence can lead to a separation well forward on the blade with catastrophic effects on turning and loss.

To summarize, it appears that blade shape has a quite small effect on turning, pressure rise and loss for Mach numbers such that the flow remains subsonic over the whole chord. The supercritical type may offer a small improvement in loss and range at low Mach numbers, but the effect is small; the largest benefit with this type of blade is that the flow is more accurately predictable. At high inlet Mach numbers, such that the flow is locally supersonic on the suction surface, the blade shape is important and with conventional profiles requires the blades to be thin, lightly cambered and with small leading edge thicknesses.

4.3 Loading limits for blade rows

The most useful outcome of the consideration of blade-to-blade flows and of cascade testing is to allow realistic loading to be chosen for the blade sections. This was recognized early on in the development of the axial compressor and some of the earliest work which still has an obvious influence on the way people think is described by Howell (1942 and 1945). Some workers proposed a maximum lift coefficient as the limit on loading but following the work of Howell the prevailing approach has been to think of the blades as forming passages rather than isolated aerofoils. What Howell recognized is that for low Mach number flows the turning due to incidence and to camber are additive and what matters most is the overall deflection because it is this which brings about the deceleration of the flow and possible separation of the boundary layers. Howell defined a nominal condition which occurred when the deflection produced by the cascade was 80 per cent of the maximum possible. (If the deflection did not show a clear peak then nominal was defined as 80 per cent of the deflection at which the loss was twice the minimum value.) The choice of 80 per cent was arbitrary but was intended to give some margin for error and some operating range. The correlation he obtained is shown in Fig. 4.10 taken from the later paper in which some of the detail concerning Reynolds number was omitted.

The interchangeability of turning by camber and incidence breaks down badly as the Mach number rises and Fig. 4.11 shows results of an inviscid calculation for two cascades of C4 blades with an inlet Mach number of 0.7. The overall deflection is the same in each case but in one there is 5° less camber and a positive incidence of 2.5°. With the positive incidence the flow accelerates to a local Mach number of 1.2 and this patch would be terminated with a shock, probably sufficient to lead to a boundary layer separation. The same two cascades operating at a low inlet Mach number (e.g $M_1 < 0.3$) have very similar distributions of Mach number, differing by no more than 10 per cent, along the blade surfaces.

The most important development in assessing the loading of cascades is attributable to Lieblein and is the diffusion factor, abbreviated here to DF. This was derived and reported in a confidential NACA report (Lieblein *et al*. 1953) but was not released to the public until Lieblein (1956). The ideas and arguments had changed considerably during that time and it is more straightforward to base the description here on the later account. First it must be said that Lieblein considered only the case envisaged for design, the so-called reference incidence, which he chose to be the minimum loss incidence angle. The definition of this is shown in Fig. 4.12. Because attention was directed to blades at design and unstalled the total pressure loss was adequately related to the momentum thickness of the wake by

$$\omega = 2\left(\frac{\theta}{c}\right)\frac{\sigma}{\cos\alpha_2}\left[\frac{\cos\alpha_1}{\cos\alpha_2}\right]^2$$

The boundary layer on the suction surface was recognized as limiting the pressure rise and turning of the blades and since the condition of this would be affected most by the deceleration Lieblein defined a term which he called 'local diffusion factor' and which gives a simple measure of this,

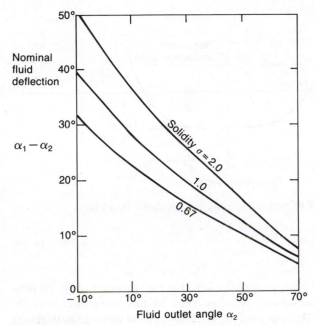

Fig. 4.10 Cascade loading correlation of Howell. Nominal deflection is defined as 80% of deflection at stall. (From Howell, 1945)

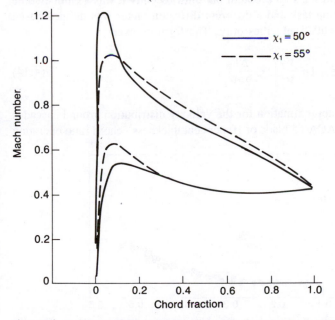

Fig. 4.11 Computed Mach number distributions for two C4 cascades of different *camber* so that one is at 2.5° positive incidence and the other at 2.5° negative incidence. Both cascades have outlet angle $\chi_2 = 25°$. Inlet Mach number 0.7, solidity 1.0, thickness–chord ratio 0.1 for both

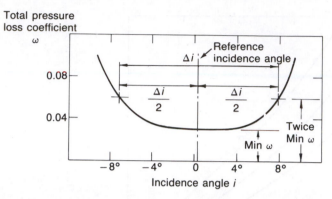

Fig. 4.12 Definition of reference minimum loss incidence by Lieblein

$$D_{\text{loc}} = \frac{V_{\text{max}} - V_2}{V_{\text{max}}} \qquad (4.13)$$

where V_2 is the mean velocity at outlet from the cascade and V_{max} is the peak velocity on the suction surface. A better name for this would have been the true diffusion factor. This was used to correlate the wake momentum thickness of NACA-65 blades at minimum-loss incidence, Fig. 4.13. Nowadays the computation of D_{loc} presents no problem but until recently it was a cumbersome and time-consuming task and a different diffusion factor was defined based on the inlet and outlet velocities only. The form evolved

$$DF = \left[1 - \frac{V_2}{V_1} \right] + \frac{\Delta V_\theta}{2\sigma V_1} \qquad (4.14)$$

was based on an approximation for the velocity distribution around cascades made up of the NACA-65 blade of 10 per cent thickness–chord ratio operating

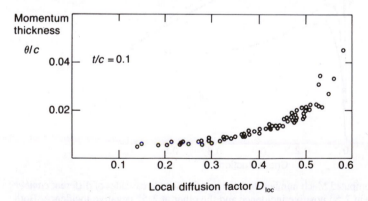

Fig. 4.13 Wake momentum thickness versus *local* diffusion factor for NACA-65 aerofoils at minimum loss incidence. (From Lieblein, 1965)

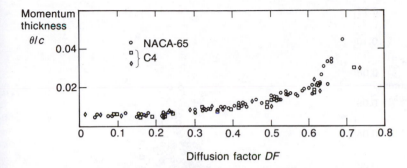

Fig. 4.14 Wake momentum thickness versus overall diffusion factor for NACA-65 and C4 aerofoils at minimum loss incidence. (From Lieblein, 1965)

at their minimum loss incidence. The solidity appears in the second term because the strength of the cross-passage pressure gradient depends on the centripetal acceleration of the flow being deflected and, as the solidity is reduced, the pressure difference across the blades must rise. For incompressible flow with equal axial velocity into and out of the cascade ($AVDR=1$) this can be simplified to

$$DF = \left[1 - \frac{\cos\alpha_1}{\cos\alpha_2} \right] + \frac{\cos\alpha_1}{2\sigma} (\tan\alpha_1 - \tan\alpha_2) \qquad (4.15)$$

A copy of the original plot given by Lieblein of wake momentum thicknesses against the diffusion factor for NACA-65 and C4 blades at minimum loss incidence is shown in Fig. 4.14. For diffusion factors above about 0.6 there is a steep rise in the momentum thickness and therefore, it was thought, in the loss: this is conventionally taken to be the beginning of blade stall and the limit for confident operation of blades.

A few years after the public release of the work on diffusion factor Lieblein (1960) published a paper in which he described another way of looking at the correlation of blade performance. In this he defined the diffusion ratio, which will be denoted here by DR, as

$$DR = V_{max}/V_2,$$

with V_{max} and V_2 being the peak velocity and mean outlet velocity as before. The choice of name was unfortunate because of its similarity to diffusion factor. Indeed the concept itself is very similar because it follows immediately that

$$D_{loc} = 1 - 1/DR.$$

At minimum loss incidence the correlation of wake momentum thickness with diffusion ratio was similar to that obtained earlier with diffusion factor, a result

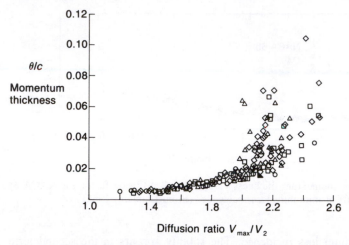

Fig. 4.15 Wake momentum thickness for NACA-65 aerofoils versus diffusion *ratio DR* at angles of incidence greater than that for minimum loss. (From Lieblein, 1959)

which is understandable because the diffusion ratio and diffusion factor are themselves well correlated. The rise in momentum thickness became rapid for diffusion ratios in excess of about 1.9.

Lieblein then went on to correlate the momentum thickness for blades at incidence greater than the minimum loss and this result is shown in Fig. 4.15. The results in this case correlate much less well but it is still true that above a diffusion ratio of 2.0 the momentum thickness may have risen sharply and almost certainly will have done so by 2.5. Because the calculation of V_{max} was a tedious quantity Lieblein derived an equivalent diffusion ratio D_{eq} for the case of blades at minimum-loss incidence and at greater positive incidence based on the inlet and outlet flow directions and the blade solidity,

$$D_{eq} = \frac{\cos\alpha_2}{\cos\alpha_1}\left[1.12 + a(\Delta i)^{1.43} + 0.61\frac{\cos^2\alpha_1}{\sigma}(\tan\alpha_1 - \tan\alpha_2)\right]. \quad (4.16)$$

Δi is the amount by which the incidence exceeds that for minimum loss. In this equation a is to be taken as equal to 0.0117 for the NACA-65 series blades and 0.007 for the C4 on a circular arc camber line. D_{eq} is an entirely empirical quantity but produces correlations of momentum thickness comparable to the real diffusion ratio.

Several years later Koch and Smith (1976) described a way of adapting D_{eq} to include the effect of Mach number and streamtube contraction (which means axial-velocity-density ratio) and approximately including the effect of changes in streamtube radius between inlet and outlet. The analysis is for minimum-loss incidence only and results in four empirical parameters.

There seems to be no particular reason for preferring either diffusion factor or diffusion ratio but the former appears to be more widely accepted and will

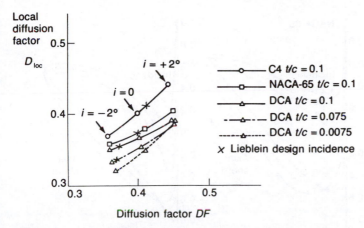

Fig. 4.16 Computed local diffusion factor versus overall diffusion factor for various profiles at incidences of $-2°$, 0 and $+2°$. Stagger 33.6°, camber 27.3°, $\sigma = 1.25$, $M_1 = 0.2$

be used here. Lieblein himself never referred to the diffusion factor at any incidence other than minimum loss but others since have used it widely in that way. To assess the effectiveness of the expression for *DF* given in equation 4.14 in terms of inlet and outlet velocities a number of inviscid calculations were performed for five cascades with each cascade at three incidences, $-2°$, 0 and $+2°$. The results are shown in Fig. 4.16 where the local diffusion factor D_{loc} (a true measure of the severity of the conditions on the suction surface) as ordinate is compared with the conventional diffusion factor *DF* from equation 4.14. The approximate or overall expression clearly does a good job at predicting the variation in the true severity of the blade loading as the incidence is varied. The overall diffusion factor does not include the effect of blade shape but the variation in the value of D_{loc} with blade type is substantial even at low Mach numbers, the C4 blade coming out significantly more severely loaded. The diffusion factor does not allow for blade thickness effects but Fig. 4.16 shows that these are appreciable. In Fig. 4.16 the crosses on each curve show diffusion factors at the estimated minimum loss incidence obtained by Lieblein's method, discussed in Section 4.4.

The result of inviscid calculation for the C4 and NACA-65 blades in cascade for an inlet Mach number of 0.6 and three values of incidences are shown in Fig. 4.17. The generally less satisfactory distribution on the C4 is very apparent with distinctly higher peak velocity and greater deceleration on the suction surface; on the pressure surface a wholly undesirable suction peak followed by a deceleration. The DCA blade is better than the NACA-65 or C4 blades so far as D_{loc} is concerned. In Fig. 4.16 the DCA was used to demonstrate the effect of thickness on D_{loc}; thinner blades have lower surface velocities when operating at design incidence but the design incidence itself changes as the thickness is altered.

In the simple expression for the diffusion factor in terms of inlet and outlet

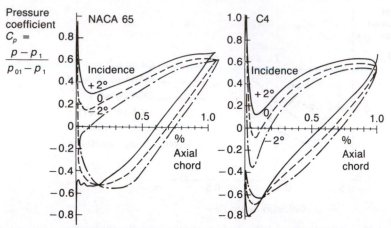

Fig. 4.17 Predicted pressure distributions on NACA-65 and C4 blades for identical inlet and outlet flow. Stagger 33.6°, camber 27.3°, $\sigma = 1.25$, $t/c = 0.1$, $M_1 = 0.6$

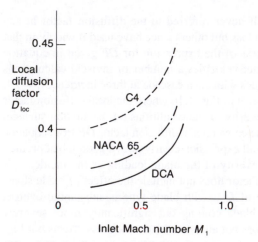

Fig. 4.18 Predicted local diffusion factor D_{loc} versus inlet Mach number for constant inlet angle and three different blade profiles. Zero incidence; stagger 33.6, camber 27.3°, solidity 1.25, thickness–chord ratio 0.10

velocities the Mach number is not considered. Figure 4.18 shows how an inviscid calculation method predicts the variation in local diffusion factor for the C4, NACA-65 and DCA blades as the Mach number rises whilst holding the flow directions and solidity constant, in other words holding the conventional diffusion factor DF constant. The increase in the amount of deceleration, as reflected in D_{loc}, with Mach number is large and should not be overlooked. The pressure distributions calculated on two cascades identical in every way, except for the thickness of one being twice the other, are shown in Fig. 4.19. The results are presented at two inlet Mach numbers, 0.2 and 0.7. It should be noted how the results at $M_1 = 0.7$ show a much greater

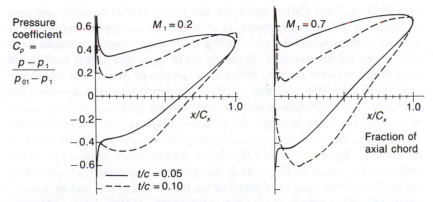

Fig. 4.19 Predicted pressure distribution for a NACA-65 blade at two inlet Mach numbers and two thickness—chord ratios. Stagger 33.6°, camber 27.3°, $\sigma = 1.25$, $i = 0$

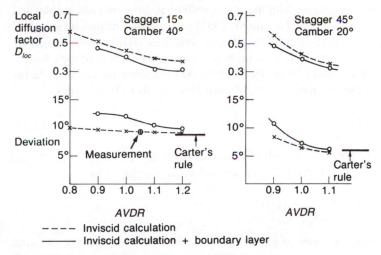

Fig. 4.20 Local diffusion factor and deviation as a function of axial velocity—density ratio. C4 sections, $\sigma = 1.0$, $t/c = 10$ per cent, incidence zero, inlet Mach number $M_1 = 0.3$

deceleration on the suction and pressure surfaces and also show much more the effect of blade thickness. The larger suction peak is the reason that the local diffusion factor is greater at higher inlet Mach numbers. The point to note is that both thickness and Mach number have quite large effects on the true (or local) diffusion factor which are omitted when the normal approximate diffusion factor, e.g. equation 4.14, is used. The simple forms for diffusion factor continue to provide a convenient design tool but are not to be regarded as precise estimates for blade loading.

Lieblein worked mainly with the cascade data obtained at NACA for which the axial velocity was equal into and out of the cascade ($AVDR=1.0$). Not all cascade data was taken with this constraint, particularly the British data used by Howell and Carter, and Felix and Emery (1953) demonstrated how significant *AVDR* is. Furthermore, in most compressor applications the axial velocity changes across the row. In Fig. 4.20 some results are shown from calculations for two cascades: deviation (to be discussed further below) and local diffusion factor D_{loc} are shown as a function of *AVDR*. For this figure two sets of calculations were performed, one termed inviscid, in which the boundary layer development was ignored, and the other in which the boundary layer was calculated and the effective blade profile modified by the displacement thickness before recalculating the inviscid flow. The effect of the boundary layer is to reduce D_{loc} by an amount which increases as *AVDR* falls. The magnitude of D_{loc} falls as the axial velocity ratio rises, reflecting the smaller overall amount of deceleration then taking place, and an approximation for the change in D_{loc} of 0.6 to 0.9 times ($AVDR - 1.0$) would seem to be appropriate.

Just as the blades in a compressor may operate with *AVDR* not equal to unity so too they may operate with mean streamlines at different radii at inlet and outlet. This was addressed by Smith (1954) not long after diffusion factor was first proposed. He recognized that it is the velocities relative to the blade row under consideration which are crucial in assessing the severity of the environment for the boundary layer. He rewrote the expression for diffusion factor in terms of the circulation Γ about each blade so that DF becomes

$$DF = \left(1 - \frac{V_2}{V_1}\right) + \frac{\Gamma}{2cV_1} \tag{4.17}$$

The mean radius and solidity are defined by $r_m = (r_{in} + r_{out})/2$ and $\sigma_m = (\sigma_{in} + \sigma_{out})/2$. The circulation around a stator is given by

$$\Gamma = \frac{2\pi}{N} \Delta(rV_\theta)$$

where N is the number of blades so for a stator row with change in radius the diffusion factor can then be written

$$DF = \left(1 - \frac{V_2}{V_1}\right) + \frac{\Delta(rV_\theta)}{2\sigma_m r_m V_1} \tag{4.18}$$

For the rotor row the circulation has a component due to the change in the absolute velocity and a negligibly small one due to the blade angular velocity so that in this case the diffusion factor becomes

$$DF = \left(1 - \frac{W_2}{W_1}\right) + \frac{\Delta(rV_\theta)}{2\sigma_m r_m W_1} \tag{4.19}$$

In concluding this section it may be pointed out that many designs have used the loading limits corresponding to Howell's nominal condition and it still finds some application today. There is therefore some interest in relating it to the Lieblein diffusion factor. It will be recalled that Howell chose his nominal

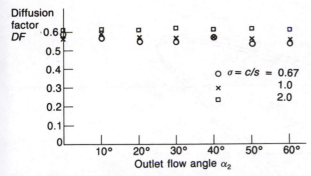

Fig. 4.21 Diffusion factor for 1.25 times Howell's nominal condition versus outlet flow direction for different solidities

condition to be at a deflection of 80 per cent of the stalling value: the correlation is shown here as Fig. 4.10. Figure 4.21 shows the computed diffusion factor for flow deflections equal to 1.25 times the nominal values given by Howell and so these should correspond to the stalling condition. It is significant how closely the curves derived from Howell collapse around a diffusion factor of 0.6, which is generally regarded as the value close to stall for a wide range of solidities and outlet flow directions. Put another way, the Howell nominal condition corresponds to a diffusion factor of about 0.45.

To summarize the diffusion factor DF, the diffusion ratio DR and the Howell nominal condition all provide a convenient and fairly reliable way of assessing the overall loading of blades for a wide range of blade geometries. This is also true at incidences other than those for design or minimum loss. It needs to be remembered, however, that the actual deceleration of the suction surface boundary layer, which is characterized by the local diffusion factor D_{loc}, varies quite significantly with blade thickness–chord ratio, blade shape and inlet Mach number. These are all effects neglected with the usual correlations for diffusion factor etc. referred to above and it indicates that great significance cannot be placed on their precise values.

4.4   The selection of incidence

In Howell (1942) the interchangeability of camber and incidence as a way of achieving deflection of the flow was emphasized but by Howell (1945) the optimum was specified to be at a small incidence close to zero. It was recognized by then that there is merit in having the stagnation point near to the leading edge, particularly at high inlet Mach numbers. The appreciable changes around the leading edge region for small changes in incidence are illustrated in Fig. 4.22 at an inlet Mach number of 0.7. The precise position of the stagnation point depends on several factors but the main ones are the mean inlet flow direction, the thickness in the leading edge region, the solidity c/s and the camber.

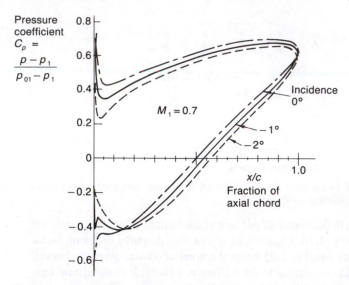

Fig. 4.22 Predicted pressure distribution about a blade at three incidences. NACA-65 blade, stagger 33.6°, camber 27.3°, $\sigma = 1.25$, $t/c = 0.05$

The leading edge region for a NACA-65 blade of 10 per cent thickness−chord ratio is drawn in Fig. 4.23. Shown on it are the stagnation points computed using an incompressible singularity method for three incidences ($-2°$, 0 and $+2°$) and at two values of solidity, 1.25 and 0.83. Increasing the incidence brings the stagnation point around onto the pressure surface; so too does reducing the solidity at a constant incidence. In each case the effect of the stagnation point being on the pressure side is to produce a negative peak (i.e. a high velocity spike) in the pressure distribution on the suction side, even though the distance of the stagnation point from the nose of the aerofoil is very small as a fraction of the chord. It will be seen that the stagnation points are very close together for the case of high solidity and positive incidence on the one hand and low solidity and negative incidence on the other. The pressure distributions for both these cases are also shown in Fig. 4.23, from which it can be seen how similar the pressure distributions are in the leading edge region though, as expected, they are quite different away from the leading edge. It will be realized that zero incidence based on the mean inlet velocities does not in general put the stagnation point on the nose of the aerofoil and obviate leading edge spikes.

There is now no fundamental difficulty in selecting blades for a given inlet flow direction to produce a pressure distribution in the leading edge region which is smooth and free from spikes. This does, however, beg the question as to whether the spikes have any deleterious effect. The results of Andrews (1949) have been referred to earlier and some are shown in Fig. 4.5. One of the things noticeable from these is the very wide range of incidence over which the loss remained low and nearly constant, particularly for the low Mach

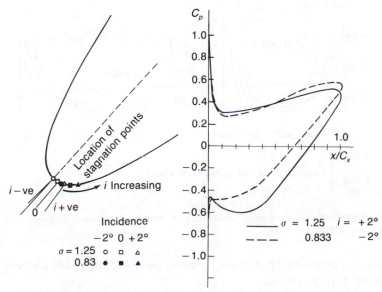

Fig. 4.23 The position of the stagnation point on the leading edge of an aerofoil for different solidity and incidence: also the predicted pressure distribution for two cases with stagnation points close together. NACA-65 blade, stagger 33.6°, camber 27.3, $t/c = 0.1$

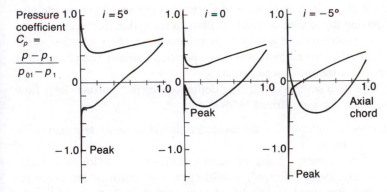

Fig. 4.24 Incompressible calculations of pressure coefficient about double circular arc blades of form tested by Andrews (1949). Stagger 47.5°, camber 25°, solidity = 1.33, $t/c = 0.105$

number cases. Figure 4.24 shows pressure distributions evaluated by an incompressible singularity method for the cascade of double circular arc blades tested by Andrews. The calculations were performed at incidences of −5°, 0 and +5°, all values well within the region for which the measured loss was low and virtually constant. Very large spikes are apparent in the predicted pressure distribution near the leading edge at both +5° and −5° incidence. Since there is good reason to expect the spikes to have been present in the tests it is not

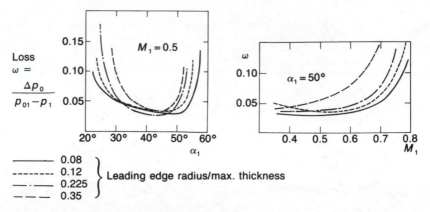

Fig. 4.25 The effect of leading-edge radius on loss: unpublished data of Rolls-Royce. (From Carter, 1961)

clear why they do not lead to massive boundary layer separation and a severe degradation in performance.

Carter (1961) showed previously unpublished test results from Rolls-Royce on the effect of leading edge radius which are reproduced here as Fig. 4.25. The conventional wisdom, founded on isolated aerofoil theory, would predict the narrowest operating range to have occurred with the sharpest leading edge because of the strong spike produced in such cases. The exact opposite was found and in the written discussion of Carter's paper two other people also reported having found similar trends. Although the spikes themselves do not seem to be a problem there is very clear evidence, e.g. Fig. 4.5, that as the Mach number rises the range of low-loss operation narrows sharply and it is important to get the incidence correct.

There have been several conflicting definitions of the 'correct' inlet flow angle and these are summarized below.

> Howell (1945): 'optimum' — the incidence should be small, less than about 5°, perhaps zero;
> Carter (1950): 'optimum' being the maximum lift–drag ratio for the cascade;
> Emery *at al*. (1958): 'design', a subjective assessment of the pressure distributions measured in low-speed tests such that there is a smooth distribution of pressure on the suction surface;
> Lieblein (1956, 1960): 'minimum-loss' incidence angle.

Carter's choice of optimum is theoretically the most correct: minimum loss is not a sufficient goal but rather minimum loss consistent with a high blade loading. He found the lift/drag ratio by an approximate method, with several rather questionable assumptions, and led to results summarized in Fig. 4.26 for blades of 10 per cent thickness–chord ratio (this figure also shows curves corresponding to the optima of Emery *et al*. and Lieblein). Now although Carter's method may predict the incidence to give the highest lift–drag ratio

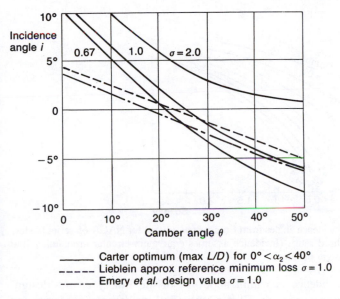

Carter optimum (max *L/D*) for 0° < α_2 < 40°
Lieblein approx reference minimum loss $\sigma = 1.0$
Emery *et al.* design value $\sigma = 1.0$

Fig. 4.26 Different optimum incidence angles

for a cascade with a given camber, it does not give the highest lift–drag ratio
for a given flow deflection. There are, for example, good reasons for suspect-
ing that the large values of 'optimum' incidence at low camber are very far
from ideal combinations.

Emery *et al.* chose the design condition to have a smooth distribution of
pressure to allow the low-speed tests to have validity when the inlet Mach
number was high. In this they showed a proper appreciation of the real problem: flow with inlet Mach numbers high enough for supersonic patches to form
on the blades. They found a dependence on solidity and camber but no effect
due to stagger or flow inlet angle and because they used only NACA-65 series
blades of 10 per cent thickness their results are strictly only valid for these.
Their results are summarized in Fig. 4.27. (In fact the results of Emery *et
al.* were given in terms of the angle of attack $A = \alpha_1 - \xi$ and to present it in
terms of the familiar incidence it has been necessary to assume a circular arc
camber line.) The results for a solidity of one have been overlaid on the curves
of Carter's optimum (Fig. 4.26) and the agreement is quite good in the region
of moderate or large camber. The agreement is poor at low camber but it is
expected that Carter's approach is erroneous there.

Lieblein's minimum loss is easy to determine from measurements but, as
noted above, it is philosophically wrong as a way of optimizing blades. In
Lieblein's method for predicting the minimum-loss incidence, described in more
detail below, the solidity, camber, thickness–chord ratio and thickness distribution are taken into account. For comparison with the other optima in Fig. 4.26,
an average of values given by Lieblein for a solidity of one but a range of
stagger angles (from Fig. 139 of NASA SP-36) is presented. Generally the

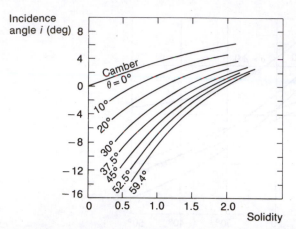

Fig. 4.27 Design incidence angles from Emery *et al.* (1958) for NACA-65 series blades, 10% thickness−chord ratio. (Incidence assumes equivalent circular arc camber line to convert from angle of attack)

minimum-loss incidence is within about 1° of the Emery *et al.* 'design' incidence and quite close to the Carter 'optimum' incidence at moderate and large cambers. It therefore seems that the differences in the definitions are relatively unimportant.

It is worth elaborating on the Lieblein method for estimating the minimum-loss incidence because it brings out rather clearly the opposing effects at work and because it provides a useful and versatile method for choosing incidence. Lieblein began with two theoretical results, one due to Weinig (1935) showing the effect of camber for infinitely thin plates in cascade and the other due to Stanitz (1953) for cascades of thick flat plates. These are shown as Fig. 4.28 and Fig. 4.29 respectively and in each case the ordinate is the incidence required to put the stagnation point exactly on the leading edge to produce so-called 'impact-free entry'. It should be noticed that camber requires a negative incidence and thickness a positive incidence to produce this condition. The direction of the blade also has a marked effect, particularly for the thickness. From this Lieblein assumed a correlation of the form

$$i = i_0 + n\theta$$

where i_0 is the incidence angle to put the stagnation point on the leading edge for zero camber and n is the slope of the variation in incidence with camber. i_0 is a result of the thickness and this was assumed to depend on the shape near the leading edge and on the magnitude of the maximum thickness−chord ratio so that it can be written in Lieblein's notation as

$$i_0 = (K_i)_{sh} (K_i)_t (i_0)_{10}$$

The term $(i_0)_{10}$ is the minimum-loss incidence for a NACA-65 cascade of zero camber and 10 per cent thickness−chord ratio; the form recommended for this by Lieblein is shown in Fig. 4.30. The $(K_i)_{sh}$ accounts for the shape of the thickness distribution and $(K_i)_{sh}$ is by definition equal to unity for the

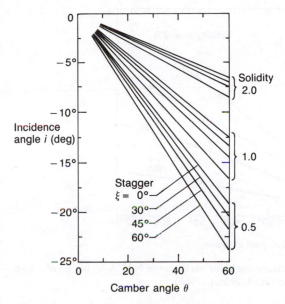

Fig. 4.28 Theoretical incidence to put stagnation point on leading edge of thin cambered plates in cascade. (From Weinig, 1935)

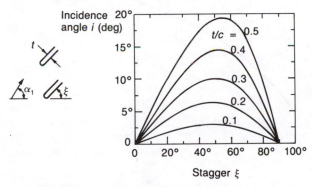

Fig. 4.29 Theoretical incidence to put stagnation point on leading edge of thick uncambered plates in cascade. (From Stanitz, 1953)

NACA-65 aerofoil. Because of its greater thickness near the leading edge Lieblein proposed $(K_i)_{sh}=1.1$ for the C4 and because of its sharper leading edge 0.7 for the double circular arc blade. $(K_i)_t$ is to account for the different magnitudes of thickness–chord ratio and the recommended variation is shown in Fig. 4.31. Finally the rate or change of minimum-loss incidence with camber, n, is shown in Fig. 4.32.

Because the thickness and camber act in opposite directions, the resultant is often a small difference between two fairly large numbers. Take for example the cascade for which calculations are shown in Fig. 4.23, NACA-65 blades with a stagger of 33.6°, camber of 27.3° and thickness–chord ratio of 10 per cent. Table 4.1 below shows the components for three different solidities, including the two used in Fig. 4.23.

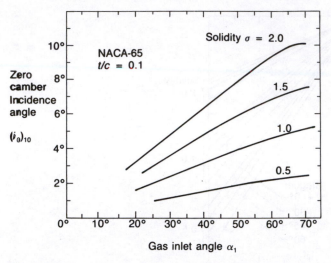

Fig. 4.30 Minimum-loss incidence angle for *uncambered* NACA-65 blades of 10 per cent thickness—chord ratio. (From Lieblein, 1965)

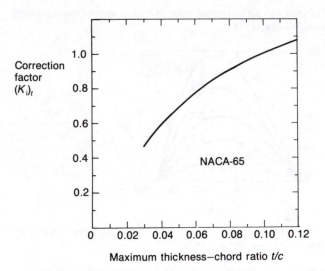

Fig. 4.31 Correction factor for thickness on minimum-loss incidence. (From Lieblein, 1965)

Table 4.1

Solidity $=0.83$	$i_0=3.2°$	$n\theta=-0.190\cdot27.3=-5.2°$	$i_{\text{min loss}}=-2.0°$
$=1.00$	$=3.7°$	$=-0.165\cdot27.3=-4.5°$	$i_{\text{min loss}}=-0.8°$
$=1.25$	$=4.65°$	$=-0.130\cdot27.3=-3.55°$	$i_{\text{min loss}}=+1.1°$

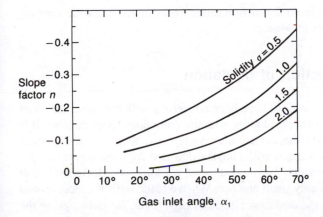

Fig. 4.32 Dependence of minimum-loss incidence angle on blade camber. (From Lieblein, 1965)

From the calculated position of the stagnation points shown in Fig. 4.23 the position of the stagnation points for $\sigma=0.83$, $i=-2°$ and $\sigma=1.25$, $i=+1°$ would be very nearly in the same place, quite close to the nose of the aerofoil and reasonably consistent with the Lieblein correlation. Because there is greater thickness near the leading edge of the C4 than the NACA-65, thickness contributes more in the case of the C4 cascades; thickness requires positive incidence to put the stagnation point near the nose so C4 blades are therefore expected to operate better at more positive incidence than the NACA-65. Similarly the DCA is expected to operate better at lower incidence. The PVD blades of the supercritical type generally have rather thick leading edges and may in this respect be expected to be more like the C4. They do, however, have the camber near to the leading edge and this would probably bring about an increase in the parameter n.

As already noted, the loss is so insensitive to incidence over a wide range at low inlet Mach numbers that the only justification for a procedure like Lieblein's described above is to get conditions right at high inlet Mach numbers. There are no satisfactory methods, other than the full calculation of the flow, for estimating the correct incidence at high subsonic inlet Mach numbers. The main evidence here is that of Andrews (1949); for the C4 profile the minimum-loss incidence remained more or less constant as Mach number was raised but for the DCA, with the fairly sharp nose, the incidence for minimum loss increased by about 4° between M_1 of 0.4 and 0.8.

To summarize, it would seem that the different specifications of the optimum or design incidence lead to very similar predictions (with the exception of Carter's method for low camber blades). The method by Lieblein gives a fairly convenient way of arriving at this for a wide range of combinations of shape, thickness, camber and solidity. For low Mach number applications the performance is remarkably tolerant to a range of incidence, but at Mach numbers for which sonic velocity on the suction surface is achieved it becomes important to get the correct incidence.

4.5   The prediction of deviation

Whereas the previous sections have been concerned with the specification of the input, this one is concerned more with the consequences of choices. It is conventional to specify the deviation as $\delta = \alpha_2 - \chi_2$, where χ_2 is the blade outlet angle and α_2 is the flow outlet angle calculated from the mass averaged outlet velocity components. The advantage of working with deviation is that it is normally a reasonably small number. When a blade stalls the most serious effect is usually that the deviation rises to a high level, far outweighing the effect of the rise in loss.

There is in fact no reason to suppose that the flow would leave the cascade in the direction of the blade outlet and the discrepancy is predominantly a potential flow effect for which the explanation is as follows. Around mid-chord the potential flow may reasonably be thought of as following the camber line and the centripetal acceleration requires a pressure difference across the passage. At the trailing edge, however, the blade loading must go to zero so that the Kutta−Joukowsky condition can be satisfied there and as the trailing edge is approached there must be a gradual reduction in the pressure gradient across the passage to make this possible. The chordwise distance over which the effect of the unloading will be felt upstream of the trailing edge will be proportional to the blade pitch. If the blade camber line remains curved right up to the trailing edge, as it normally does with the simple forms used with conventional profile families, it is clear that the flow can no longer follow the camber line because the loading is reduced below the level needed to turn it and balance the centrifugal acceleration. This means that the flow, averaged across the passage, can no longer have the outlet direction of the blades, in other words a deviation is produced.

For blades which are not stalled the boundary layer development makes a relatively small contribution to the deviation. This is demonstrated by the calculated values of deviation shown in Fig. 4.20 for two different cascades. The axial velocity ratio has a fairly weak effect on the inviscid calculation of deviation but quite a marked effect on the deviation calculated to include the development of the boundary layer.

Results of inviscid calculations of deviation for three types of blade at incidences of $-2°$, 0 and $+2°$ are given in Fig. 4.33. (Shown by an arrow on each curve is the minimum-loss incidence angle calculated by Lieblein's method.) As well as the three blade sections of 10 per cent thickness−chord ratio the figure also shows deviations calculated for an infinitely thin blade of the same camber and stagger; clearly the thickness has quite a significant contribution to the deviation and so too does the shape, with the very thin trailing edge of the NACA-65 or double circular arc (DCA) blades producing less deviation than the C4.

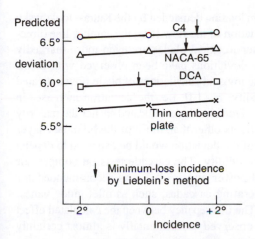

Fig. 4.33 Predicted deviation by an inviscid method versus incidence for various blades. The arrow shows Lieblein's predicted minimum-loss incidence. Stagger 33.6°, camber 27.3°, solidity 1.25, $t/c = 0.10$, $M_1 = 0.2$

With the advent of fast methods of computing the flow around cascades there is strictly speaking no need for correlations for deviation, but for many applications it will remain very useful indeed to have a method of obtaining a reasonable estimate of the deviation with a minimum of computation. The correlation which is still most often used, but with local modifications and adaptations, is known as Carter's rule and was a modification of one proposed earlier by Constant (1939). It takes the form

$$\delta = m\theta \left(\frac{s}{c} \right)^n$$

Carter (1950) introduced the empirical curves for m, reproduced here as Fig. 4.34. The parabolic arc camber line gives less curvature over their rear portion, which means that the blade shape is such as to unload gradually over the rear.

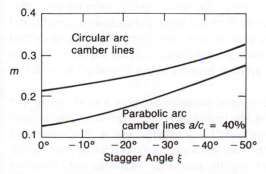

Fig. 4.34 Carter's rule; m is empirical parameter in the deviation rule. (From Carter, 1950) (a/c denotes position of maximum camber)

As a result much of the reduction in loading demanded by the Kutta–Joukowsky condition is achieved without deviation of the flow from the camber line direction. The current generation of supercritical blades are even more markedly straight to the rear and very low deviations have been observed with them. When drawing up Carter's rule the inviscid calculations of blade flow indicated a first-power dependence on solidity, $n = 1.0$, i.e. the deviation increases in linear proportion to the pitch for a given chord. As indicated earlier the majority of the deviation is produced by effects other than those of the boundary layer and on physical grounds this part of the deviation would be expected to depend on the ratio of pitch to chord, or solidity. The cascade tests on compressor blades agreed better with a half-power and $n = 0.5$ was recommended for decelerating cascades. For accelerating cascades, such as inlet guide vanes, a first power was recommended. The discrepancy between the calculated effect of solidity on deviation and that observed experimentally is almost certainly due to the boundary layer growth on the suction surface.

Carter's rule was intended for application at his 'optimum' incidence but as already noted this is quite close to the minimum-loss incidence and to the other design incidences. Furthermore as the calculations presented in Fig. 4.33 and many sets of measurements show, the deviation is actually quite a weak function of incidence until blade stall is approached. This is just as well, because Carter's rule is applied widely at incidences other than the optimum.

Lieblein (1956 and 1960) produced a correlation with a systematic approach similar to the one he used for the prediction of minimum-loss incidence. The deviation at minimum-loss incidence was assumed to be given by an expression of the form

$$\delta = \delta_0 + m\theta$$

where δ_0 is the deviation for a similar cascade of uncambered aerofoils but the same inlet direction, θ is the camber and m is an empirical slope. The parameter m was assumed to be expressible in the form $m = m_{\sigma=1}/\sigma^b$ so that empirical relations had to be found for $m_{\sigma=1}$ and b. The index b varied continuously from about 0.97 when the inlet flow was axial (an accelerating cascade) to about 0.55 for an inlet angle of 70°. This continuous variation of the index supports the much cruder and discontinuous form for the power of the solidity proposed by Carter. With the additional variable it contains and the large amount of NACA data with constant axial velocity the Lieblein method appears more accurate and reliable than the much simpler Carter's rule. The Lieblein correlation typically predicts between about 1° and 2° more deviation than Carter's rule.

Although the inviscid predictions of deviation are only a weak function of the axial velocity–density ratio, the boundary layer growth does lead to quite significant increases in deviation as *AVDR* is reduced, Fig. 4.20. Carter based his correlation on data from cascades in which the axial velocity rose significantly between inlet and outlet, and the measured deviations were therefore smaller than they would have been for *AVDR* = 1.0. The data measured at

NACA, such as Emery *et al.* (1958), had been obtained with careful attention to maintaining $AVDR=1.0$; Felix and Emery (1953) showed that up to 3° more turning i.e. 3° less deviation in the tests for $AVDR>1.0$ could be attributed to this. Pollard and Gostelow (1967) made measurements in which the axial velocity ratio was systematically varied. The expression

$$\delta - \delta_1 = 10(1.0 - AVDR)$$

correlated their data quite well, where δ_1 is the deviation for $AVDR=1.0$, the angles being expressed in degrees. The general validity of this expression is not established; the calculated results in Fig. 4.20 support a variation like $8(1-AVDR)$ for the low stagger blade and $20(1-AVDR)$ for the high stagger. A method widely used in industry is to add 2° to the prediction of Carter's rule which compensates for the use of data with $AVDR>1.0$ in its formulation. It is worth emphasizing that although an axial velocity ratio of unity is a very good choice to use for the systematic testing of cascades, it does not follow that blades in a compressor will be operating at this value. The precision of a method at $AVDR=1.0$ may be less useful than at first sight it appears if there is no comparably accurate method for introducing the effects of $AVDR \neq 1.0$.

To summarize, the deviation can be predicted using inviscid calculation methods. Including the effect of the blade boundary layers improves the estimate, although most of the deviation is not connected with the boundary layer at all. Lieblein produced a correlation procedure which gives accurate estimates for deviation of the common profile families but only for $AVDR=1.0$. There is no satisfactory general correlation for predicting deviation for axial velocity–density ratios different to unity. Carter's simple deviation rule was based on data taken when $AVDR$ was allowed to vary but was generally greater than unity. Perhaps because of its simplicity, Carter's rule continues to be widely used, albeit with modifications: the modifications may be very simple, such as the addition of 2° to the predicted deviation, or may be quite complicated and use extensive proprietary data.

4.6 The determination and prediction of losses

The losses produced by the blades away from the endwall, the type of loss measured in cascades and often referred to as profile losses, is not the most serious source of loss in the compressor provided the blades are operating within the normally acceptable limits of incidence, diffusion factor and Mach number. This point was recognized and stated clearly a long time ago by Constant (1939). The losses attributable specifically to high Mach number effects will be addressed in a later section; what will be concentrated on in this section is the losses which are present at low as well as at high speed.

The mass averaged loss across a blade row, evaluated a short distance downstream of the blade row, will be denoted by

$$\Delta p_0 = \frac{\int_0^s \rho V_x (p_{01} - p_{02})\mathrm{d}y}{\int_0^s \rho V_x \mathrm{d}y} \qquad (4.20)$$

where subscript one refers to conditions at inlet to the blade row and the integration is in the pitchwise direction along the cascade, in the circumferential direction for a blade row in a machine. The blade pitch is denoted by s. It is usual to nondimensionalize the loss in stagnation pressure with respect to the inlet dynamic pressure $p_{01} - p_1$ to give $\omega_1 = \Delta p_0 / (p_{01} - p_1)$. For low Mach number flow the incompressible flow relation $p_{01} - p_1 = \frac{1}{2}\rho V_1^2$ may be used.

The loss in total pressure is not constant downstream of the blades but rises as the wakes mix out. This was discussed in Chapter 1 and a thorough account is given by Lieblein and Roudebush (1956) for incompressible flow and the modifications for high-speed flow by Stewart (1955). A measure of the degree to which the wake has mixed out is the form parameter H defined by $H = \delta^*/\theta$, where δ^* is the displacement thickness and θ the momentum thickness. For blade rows with 'healthy' boundary layers this mixing takes place rapidly and a short distance downstream is nearly complete: Lieblein and Roudebush show measurements which indicate that 30 per cent of a chord downstream H will be down to between 1.1 and 1.2. At these sort of values it is possible to approximate the wake quite accurately with a power law and the expressions can then be manipulated algebraically fairly easily. It is also then a good approximation to take the energy displacement thickness δ_3 to be equal to twice the momentum thickness θ. For such conditions it is easy to show that a very satisfactory approximation to the loss coefficient at low Mach numbers is given by

$$\omega_1 = \frac{\Delta p_0}{\frac{1}{2}\rho V_1^2} = 2 \frac{\theta}{c} \frac{\sigma}{\cos\alpha_2} \left(\frac{\cos\alpha_1}{\cos\alpha_2} \right)^2 \qquad (4.21)$$

It must be remembered that this is an approximation, which will be valid only when the velocity decrement in the wake is small, such as occurs at least a short distance downstream of blades with fully attached flow. It could be very misleading for blades which are 'overloaded' and have large regions of separated flow. It should also be recognized that the specification of loss coefficient in terms of the momentum thickness θ is a convention, but a rather misleading one. The concept of momentum leads one immediately to think of losses in terms of the drag on the blades but this is not the mechanism of loss, as was discussed more fully in Chapter 1. Also discussed in that chapter was the additional loss produced by the flow mixing out to a uniform condition. This is normally a fairly small addition once the initial mixing in the first 20 to 30 per cent of chord downstream has taken place. As discussed below the prediction of loss is not at all precise, so a few percent change due to mixing is not very important. Furthermore it is not clear in a compressor stage whether the fully mixed out condition has any special merit or significance, other than providing a consistent standard, since in a compressor

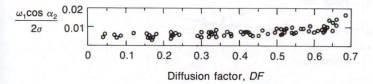

Fig. 4.35 Measured loss downstream of cascades of NACA-65 blades at minimum-loss incidence as a function of diffusion factor. (From Lieblein, 1965)

the next blade row will normally be no more than a small fraction of a chord downstream and there is almost no opportunity for the flow to mix before entering it.

In connection with the introduction of the diffusion factor Lieblein (1956) showed the variation in momentum thickness as a function of diffusion factor, DF, at conditions of minimum-loss incidence and this was reproduced here as Fig. 4.14. He went on to plot the loss coefficient measured by the NACA for the 65-series aerofoils at minimum-loss incidence against diffusion factor and this is replotted here as Fig. 4.35. What is most remarkable about this is the small variation in the value of $(\omega_1\cos\alpha_2)/2\sigma$ as the diffusion factor varies. If it is accepted that blades are likely to be designed with diffusion factor in the range $0.35-0.6$ it is evident that to assume a value of $(\omega_1\cos\alpha_2)/2\sigma=0.007$ would give as satisfactory an estimate as the scatter in the data would allow. By way of comparison the results shown earlier of measurements by Andrews, Fig. 4.5, demonstrate that the loss varies remarkably little over a significant range of incidence, at least for the low Mach number cases. The minimum losses measured by Andrews give values of $(\omega_1\cos\alpha_2)/2\sigma$ equal to about 0.005 and 0.007 for the C4 on the circular and parabolic camber lines respectively and 0.006 for the double circular arc. With the exception of the C4 on the parabolic arc there was no significant rise in minimum loss up to an inlet Mach number of 0.8. In fact some of Andrews' measurements, not reproduced in Fig. 4.5, show markedly higher loss at very low Mach numbers which is because these tests were carried out at low Reynolds number for which loss is known to rise, see Section 4.7. (For Andrews' measurements $Re=V_1c/\nu\approx2\cdot10^5$ at $M_1=0.5$.)

The loss for blade sections forming part of a compressor seems to behave quite differently to that in cascades. Figure 4.36 reproduces measurements reported by Robbins *et al.* (1956) for the loss across rotor and stator rows in a large number of single stage compressors. All the measured points shown were obtained while operating at the minimum-loss incidence and the inlet Mach number was subsonic in each case. The broken line shows the correlation drawn by Lieblein for two-dimensional cascades, essentially a line through the measured losses shown in Fig. 4.35. The higher loss near the hub and tip is not unexpected, since this region is likely to be affected by the endwall boundary layer, but the high loss compared with the cascade at mean height is a surprise. The solid line (and shaded region for the rotor tip) are the recommendations

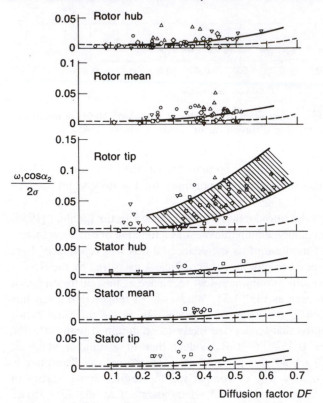

Fig. 4.36 Measured loss downstream of rotors and stators for minimum-loss incidence. Broken lines show correlation based on 2-D cascades. Solid lines are design recommendations based on the data by Robbins *et al.* (From Robbins *et al.*, 1965)

given by Robbins *et al.* for application to design. With this level of uncertainty for the losses attributable to the blades it is perhaps fortunate that the blade profile losses are not more important to the efficiency of the compressor. Unfortunately the other more important loss sources are even less well understood!

Koch and Smith (1976) looked at all losses in the axial compressor and described a method for predicting the blade profile loss based on a boundary layer calculation method. The predictions were in quite good agreement with the measurements of wake momentum thickness reported by Lieblein. The method allowed the inclusion of the effects of compressibility, in terms of inlet Mach number, of Reynolds number and the contraction of the streamtube (essentially the axial velocity–density ratio) through the blade row. Also included is the effect of profile surface roughness k_s on the boundary layer development. The criterion for hydraulic smoothness was selected to be $V_1 k_s / \nu$ less than or equal to 90. To predict efficiency, however, it was still necessary to make an empirical adjustment to the predicted momentum thickness by adding 0.0025 times the blade chord.

The similarity of the loss from various profile sections at low to moderate

Mach numbers has already been noted, as has the general desirability of having sharp leading edges. The trailing edges of blades are not made very thin for reasons of mechanical strength and ease of manufacture. Trailing edge thickness appears, however, to have little effect on the loss and Carter (1961) reported that thicknesses as great as 25 per cent of the maximum thickness will offer no disadvantage. Many of the PVD or supercritical blades have quite thick trailing edges and take advantage of this effect.

It seems appropriate to conclude this section on losses by referring to the recent designs of prescribed velocity distribution (PVD) blades. One of this class is the supercritical blade and there is no doubt that at inlet Mach numbers high enough to produce supersonic flow in the passages the supercritical type of blade offers advantages of lower minimum loss and higher useful inlet Mach numbers. The advantages in wholly subsonic flow are less certain; nevertheless, it would indeed be amazing if one of the original profile families were the optimum and it seems much more likely that a cascade with sections designed specifically for the duty would be better. In the example shown in Fig. 4.9, tested by Hobbs and Weingold (1984), the controlled diffusion blades gave a slightly higher loss at low Mach numbers, although they did give a substantially wider operating range at these conditions. These cascades all had the same solidity and, as already noted, a principal parameter determining loss and the maximum turning at low Mach numbers is the solidity. These results show that the controlled diffusion cascade tested by Hobbs and Weingold was of too high solidity to show the lowest possible loss at low Mach number, but instead it showed much wider operating range. In fact if the lower solidity needed to give minimum loss were chosen for a compressor (as opposed to a cascade with endwall suction) the endwall regions would almost certainly be in serious difficulty and any benefit near mid span would be undone. The expected narrowing of range with forward loaded blades, referred to in Section 4.2, was not observed. It seems very likely that such blades may be preferred at low Mach numbers more for this wider operating range, obtained at slightly higher solidity than that for minimum loss, than for any small reduction in minimum loss they may be capable of. An additional advantage with supercritical or controlled-diffusion type blades at all Mach numbers is the greater predictability; this shows in the smaller deviation and the greater likelihood of accurate boundary layer predictions. The prediction of boundary layer properties is of greatest importance at off-design conditions, since it is then that the boundary layer most affects the performance. The high Mach number properties of this type of blade is discussed further in Section 4.8.

To summarize, the loss from different profile sections, including those of the supercritical type, is very nearly the same until the Mach number is high enough for strong shocks to be formed in the passages. The loss is a weak function of loading, as indicated by the diffusion factor. For a useful working range of diffusion factor, $0.35 < DF < 0.6$, a good approximation for cascade tests at high Reynolds numbers, low to moderate Mach numbers and a wide range of profiles shapes is $(\omega_1 \cos\alpha_2)/2\sigma \approx 0.007$. Over quite a wide range of incidence the loss varies very little, except when the Mach number is high enough to produce high velocities and strong shocks. The comparable measurements made by the NACA in compressors show much higher profile losses even at mid-span and minimum-loss incidence. An important variable often omitted is the blade surface roughness.

4.7 The effect of Reynolds number on blade performance

The blading designed for all compressors relies on there being a turbulent boundary layer on the suction surface to allow the flow to decelerate without major separation. Many blades do in fact operate with some regions of separated flow (the double circular arc blades seem normally to be separated some way upstream of the trailing edge) but if this is too extensive there is a marked loss in turning (i.e. rise in deviation) and an increase in the loss.

It is very common for the laminar boundary layer formed near the leading edge or just downstream of the region of maximum velocity to separate to form what is known as a separation bubble. In most bubbles the shear layer between the nearly stationary fluid in the bubble and the fast moving fluid outside it is very unstable, so that the separated shear layer undergoes transition and the flow reattaches as a turbulent boundary layer. There is good reason to think that the presence of a small separation bubble does no harm, at least

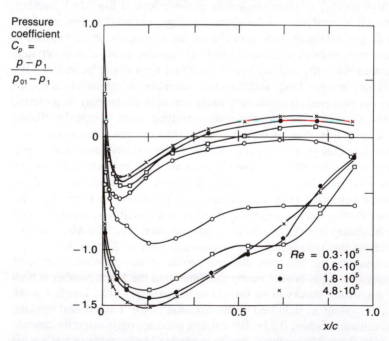

Fig. 4.37 Measured pressures around cascade of C4 aerofoils for various Reynolds numbers. Incidence $i = -1°$. Stagger 36.5°, camber 30°, solidity 1.0, thickness–chord ratio 0.10. Inlet Mach number $M_1 < 0.15$. (From Rhoden, 1956)

for subsonic flows. The harm is real, however, if the flow is unable to reattach, which may be the case at low Reynolds numbers. By convention the Reynolds number used is based on the inlet relative velocity, the blade chord and the kinematic viscosity at the static conditions for the entry flow, $Re = V_1 c / \nu_1$.

Figure 4.37 shows pressure coefficients measured by Rhoden (1952) around a C4 blade for a range of Reynolds numbers, though for clarity not all the Reynolds numbers which were measured are shown. The thickness—chord ratio of the blade is 10 per cent and the cascade is of solidity one. For each of the Reynolds numbers shown the incidence is constant at $-1°$ which is the value predicted by Lieblein's correlation for minimum loss. The diffusion factor DF for the cascade, using equation 4.14 and the measured outlet flow angle at high Reynolds number, is the reasonable one of 0.44. At the lowest Reynolds number there is clear evidence of a major separation on the suction surface with no reattachment and a massive reduction in the blade force. An approximate doubling of Reynolds number to $0.59 \cdot 10^5$ shows the flow still separating but reattaching at about 67 per cent chord. Further increases in Reynolds number produce a reduction in the size of the separation bubble but it is not until $Re = 4.8 \cdot 10^5$, the highest value at which Rhoden carried out the test, that evidence of the separation bubble is no longer discernible in the static pressure distribution.

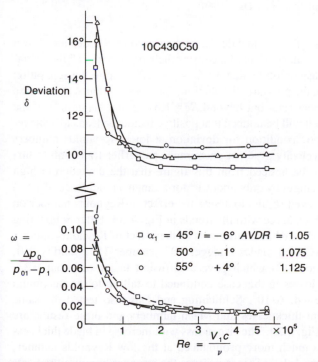

Fig. 4.38 Deviation and loss versus Reynolds number for C4 blade in cascade. Incidence $i = -1°$. Stagger 36.5°, camber 30°, solidity 1.0, thickness—chord ratio 0.10. Inlet Mach number $M_1 < 0.15$. Axial velocity—density ratio quoted at $Re = 3.10^5$. (From Rhoden, 1956)

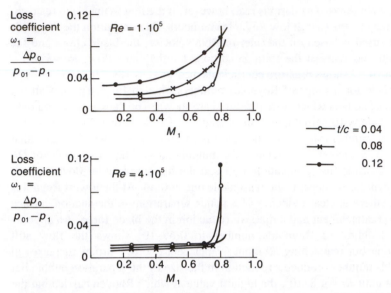

Fig. 4.39 Loss versus Mach number for NACA-65 cascades of different thickness at two Reynolds numbers. Stagger 40°, approx. camber 15°, solidity 1.0, flow inlet angle $\alpha_1 = 50°$. (From Schlichting and Das, 1969)

For the same cascade the loss and deviation measured by Rhoden are shown in Fig. 4.38 for three values of incidence. For the case of $i = -1°$ the deviation has reached its asymptotic value by $Re = 1.8 \cdot 10^5$ despite the separation bubble still evident in the pressure distribution. The loss continues to decrease as the Reynolds number rises but beyond $Re = 1.8 \cdot 10^5$ the magnitude of the change is very small. It will be noticed that positive incidence leads to achievement of the asymptotic condition for deviation at lower Reynolds numbers because the steeper velocity gradient encourages an earlier transition to turbulence. (It may also be noticed from this figure that the deviation at high Reynolds numbers changes by only about 1° for a range on incidence of 10°.)

Lieblein (1956) showed results to assess the effect of Reynolds number on loss and deviation which agree with the trends in Fig. 4.38. Later Schlichting and Das (1969) also showed measurements of the effect of Reynolds number on a cascade of NACA-65 blades (stagger 40°, camber 15°, solidity 1.0, thickness–chord ratio 0.12) which gave a diffusion factor of about 0.33, somewhat low. The losses in this case continued to fall up to the maximum Reynolds number tested, $6 \cdot 10^5$. Schlichting and Das also tested the same cascade with different thicknesses and Mach numbers and these results are summarized here as Fig. 4.39. The rise in loss with increase in blade thickness and Mach number is much more pronounced at the low Reynolds number, $1 \cdot 10^5$ than at the high, $4 \cdot 10^5$. This is because the stronger pressure gradients produced at high Mach numbers or with greater thicknesses are more readily surmounted by the boundary layer at the higher Reynolds number.

A lot of controversy surrounds the effect of turbulence in the inflow and in the early days of cascade testing attempts were made to calibrate the flow by measuring the drag on a small sphere in it to derive an effective Reynolds number (see for example Howell, 1942). Inside a compressor the turbulence levels can be very high: Schlichting and Das (1969) reported measurements of the turbulence level rising to 6 or 8 per cent, depending on the flow coefficient into the compressor, and de Haller (1953) reported values rising slightly above 6 per cent. More recently Wisler *et al.* (1987) measured turbulence levels well inside a highly loaded but very efficient multistage compressor. Away from the endwalls and outside the wakes of the immediate upstream blades the level was about 5 per cent at design and 10 per cent at reduced mass flow, to the left of peak efficiency. Near the endwalls or in the wakes the levels were much higher, around 20 per cent. Some very clear results of the effect of grid turbulence and a trip on the blades were shown by Schlichting and Das with the same NACA-65 cascade as that referred to above. At $Re = 1 \cdot 10^5$ the trip, and even more the turbulence grid, produced a reduction of loss and

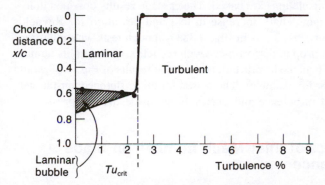

Fig. 4.40 The position of separation, reattachment and transition (inferred from surface flow visualisation) versus turbulence level for a cascade of NACA 65 blades. $Re = 1.6 \cdot 10^5$. Stagger 40°, camber 15°, $\sigma = 1.0$, inlet flow angle $\alpha_1 = 50°$. Turbulence produced by oscillating grid. (From Schlichting and Das, 1969)

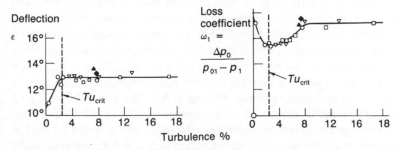

Fig. 4.41 Deflection and loss versus turbulence for a cascade of NACA-65 blades at $Re = 1.6 \cdot 10^5$. Stagger 40°, camber 15°, $\sigma = 1.0$, inlet flow angle $\alpha_1 = 50°$. Turbulence produced by oscillating grid. (From Schlichting and Das, 1969)

widening of operating range, the reduction of loss being more noticeable at
$M_1 = 0.7$ than at 0.5. At the higher Reynolds number, $Re = 4 \cdot 10^5$, the effects
oı both the trip and the turbulence grid were much smaller and each produced
an increase in loss and no great change to the operating range. The loss was
everywhere lower at the higher Reynolds number, with or without trip or grid,
and the magnitude was about 0.015.

Later tests by Schlichting and Das used an oscillating grid to introduce turbulence of more carefully controlled scale and frequency and the result of these
is shown reproduced as Figs 4.40 and 4.41. At the modest Reynolds number
of $1.6 \cdot 10^5$ a turbulence level of 2.5 per cent seems sufficient to eliminate the
separation bubble and simulate the effect of very much higher Reynolds number
on the deviation. They found that the effect does not appear very sensitive
to the frequency or scale of the turbulence in the range tested.

As noted, there are very high levels of turbulence inside multistage compressors and these are likely to make the blades behave more like those in
the tests of Schlichting and Das than in the tests with low inlet turbulence on
which most of the correlations are based. The cascade results obtained in low
turbulence show a quite sharp increase in loss and deviation for Reynolds
numbers below about $1.5 \cdot 10^5$, as in Fig. 4.38. Although tests with multistage
compressors show a drop in performance with reduction in Reynolds number
it is much less steep than the low turbulence cascade results of Fig. 4.38 would
suggest (see Fig. 1.8, for example). The reason for this difference must largely
be the high level of turbulence and unsteadiness inside the compressor.

4.8   The effect of inlet Mach number on blade performance

Conventional blades

This section is concerned only with flows for which the inlet flow is subsonic
even though the velocities are high enough for compressibility to play a significant role inside the passage, possibly with supersonic patches being formed.
Around 1940, in the early study of compressor blading, it was not appreciated
how high the velocities into and through the blades would be. With the thick
blade sections then being used high inlet Mach numbers would certainly have
led to poor efficiency. Since they did not forsee the high Mach numbers there
was little incentive to use thin blades and most of the work was done with
blades of about 10 per cent thickness−chord ratio. There was, moreover, a
tendency to see the optimization of blading at low Mach number as a goal
in itself and this may be seen as the background to the Carter (1950) specification of optimum conditions in which quite large incidence might be used at
low cambers. It also helps to explain the failure to accept fully the superiority
of the double-circular-arc blade over the C4 at all but the lowest Mach numbers,
as was demonstrated so clearly by Andrews (1949). The programme of low-

speed testing which had the high-speed application most clearly in mind was that at NACA Langley with their choice of design condition described in Section 4.4.

An approximate method for calculating the effect of compressibility is the Prandtl—Glauert rule, which is valid when the perturbations of the flow from the uniform inlet conditions are small. It has been described in its application to cascade flows in some detail by Scholz (1977) and more briefly by Schlichting (1954). The compressible flow in a cascade is related to a cascade in incompressible flow with higher stagger and higher solidity, the extent of the scaling depending on the inlet Mach number. The pressures (and likewise the velocities) at points on the blades in compressible flow are then related to those in incompressible flow by relations of the form

$$(c_p)_{\text{compressible}} = \frac{(c_p)_{\text{incompressible}}}{\sqrt{(1 - M^2)}}$$

Given the ease with which compressible flows can now be calculated, the Prandtl—Glauert approach no longer has much practical utility other than to emphasize the sense in which compressibility exaggerates effects and the fact that increasing the Mach number into a blade row has an effect on the overall flow akin to a change in cascade solidity and stagger. Wisler (1985), describing the use of low-speed compressors to simulate high-speed multistage compressors, notes that to obtain comparable velocity distributions around the low-speed blades they generally need to have several degrees extra camber in the leading edge region and somewhat more thickness.

Figure 4.11 shows the result of an inviscid calculation for blade row with an inlet Mach number of 0.7. In the case with 2.5° positive incidence a supersonic patch is formed in which the Mach number exceeds 1.2 and the patch would be terminated with a shock. The comparison for the same cascades at $M_1 = 0.3$ gives a peak Mach number near the leading edge of about 0.4 with high camber and $-2.5°$ incidence but only 0.44 for lower camber and $+2.5°$ incidence. As the Prandtl—Glauert scaling shows, one effect of the compressibility then is to exaggerate differences. This can be thought of as one reason for the narrowing of the operating range as the Mach number rises, as illustrated by the results shown in Fig. 4.5.

As the inlet Mach number is increased there comes a point at which the velocity on the blade surface reaches the local sonic velocity. This condition is often referred to as the critical condition and the corresponding inlet Mach number as the critical Mach number. The critical Mach number will depend on many variables including the overall blade thickness and camber (and the distribution of these along the chord, particularly near the leading edge) and, most important, the angle of incidence. The achievement of critical conditions is not in itself likely to have much effect on the performance of the cascade. However, if the inlet Mach number is raised the peak Mach number on the blade increases more rapidly and the supersonic patch is likely to be terminated

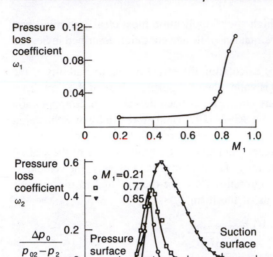

Fig. 4.42 Loss versus inlet Mach number and wake total pressure profiles at three Mach numbers. NACA-65 profile, 30° camber. (From Lieblein, 1956)

by a shock wave of increasing strength. The shock wave brings an alteration in the pressure distribution but neither this nor the losses across the shock itself are likely to be of major concern.

The crucial change in performance occurs if the shock is strong enough to lead to the boundary layers separating and not reattaching, in which case there is a major alteration in the pressure distribution and a big increase in loss. Figure 4.42, taken from Lieblein (1956), shows the rise in loss coefficient as the inlet Mach number is increased for a NACA-65 blade of 30° camber and 10 per cent thickness−chord ratio operating near to the minimum-loss incidence. The reason for the rapid rise in loss is evident from the lower part of the figure which shows the distribution of pressure-loss coefficient in the pitchwise direction. At $M_1=0.77$ the wake is noticeably wider on the suction side than at low speed; at $M_1=0.85$ it has become enormous, clear evidence that a shock wave has produced a massive separation. The rapid rise in loss and drop in deflection at positive incidence for the C4 sections shown in Fig. 4.5 are caused by the same effect. The corresponding double circular arc blade in Fig. 4.5 does not show this rapid deterioration in performance at positive incidence because with the much thinner and sharper leading edge region the peak Mach number does not become high enough with an inlet Mach number of 0.8 for the resulting shock to cause the suction surface boundary layer to separate drastically.

If the inlet Mach number is increased beyond the critical value there comes a point at which the flow is sonic or supersonic across the entire passage. At this condition the flow is choked, the corrected mass flow cannot be further

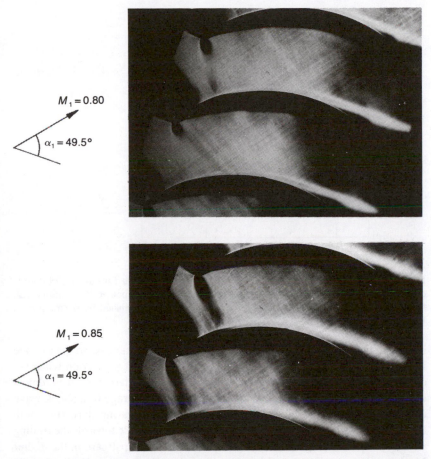

Fig. 4.43 Schlieren photographs of double circular arc blades in cascade at two inlet Mach numbers. Stagger 19.2°, camber 56.8°, solidity 2.2, thickness–chord ratio 0.07, blade inlet angle 47.6°. (Pictures by Hoheisel published by permission of Rolls-Royce plc)

increased and the conditions in the inlet region and upstream cannot be modified by a reduction in the downstream static pressure. In fact, lowering the pressure causes the flow to accelerate in the diverging part of the blade passage to higher supersonic velocities which terminate in a strong shock bringing loss and almost certain boundary layer separation.

Two schlieren photographs, taken by Dr. Hoheisel for Rolls-Royce, are shown in Fig. 4.43. The cascade is of double circular arc blades of thickness–chord ratio 7 per cent, camber 56.8°, stagger 19.2° and with the high solidity of 2.2. The inlet flow direction α_1 for the case shown is 49.5°, giving a positive incidence of 2°. At the inlet Mach number of 0.8 a dark patch from the leading edge indicates the presence of high velocities somewhere in the vicinity; at $M_1 = 0.85$ the dark region extends right across the passage,

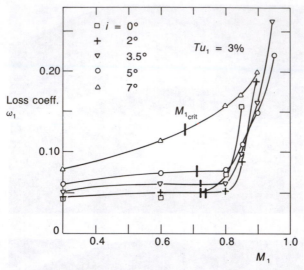

Fig. 4.44 Measured loss coefficient versus inlet Mach number for cascade of double circular arc blades at various incidences. Stagger 19.2°, camber 56.8°, solidity 2.2, thickness−chord ratio 0.07. (Measurements by Hoheisel published by permission of Rolls Royce plc)

indicating that the flow is sonic or supersonic all the way across, so the cascade is choked at this higher speed. It is just possible to see a separation when the dark region intersects the suction surface and so the dark region is almost certainly a shock wave or, more precisely, the time average of a shock wave or waves which are moving around in the region appearing dark. For both the Mach numbers shown there is a separated region visible towards the trailing edge; the separated region is larger and begins further upstream in the higher Mach number case, presumably because of the effect of the shock wave on the boundary layer. Quantitative measurements of loss for the same cascade

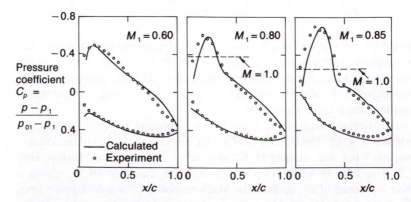

Fig. 4.45 Measured and predicted pressure distributions about a DCA blade in cascade, $\chi_1 = 47.5$, $\chi_2 = -9.2$, $\alpha_1 = 49.5°$, $\sigma = c/s = 2.2$, $AVDR = 1.20$. (From Hoheisel & Seyb, 1986)

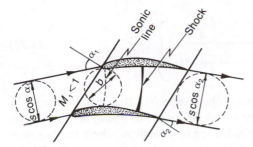

Fig. 4.46 Schematic of choked flow in a blade passage: inlet and outlet flow subsonic

are given in Fig. 4.44, the case for $i = 2°$ being shown with the crosses. The loss for this inlet flow angle is almost doubled between $M_1 = 0.80$ and 0.85, consistent with the schlieren pictures which show a large change taking place in this interval. It should also be noticed that at $i = 0$ the loss rise due to choking occurs at lower inlet Mach numbers. (The flow at a positive incidence of $7°$ is clearly stalled at all Mach numbers.) Also shown on Fig. 4.44 is the critical Mach number, i.e. the inlet Mach number at which the flow is first sonic on the blade surface; as anticipated, there is no discernible discontinuity in performance attributable to this.

The static pressure coefficients measured and calculated around the same cascade by Hoheisel and Seyb (1986) are shown in Fig. 4.45. Some supersonic flow is evident at $M_1 = 0.80$ but the extent and the peak level is much greater at $M_1 = 0.85$. The shock which must be present at the higher Mach number does not show up as a steep rise in pressure for the reason given above: the shock wave is almost certainly unsteady so that its effect appears spread out in time-averaged measurements. It is significant that even though the flow can be clearly seen in the schlieren pictures to be separated there are still large rises in static pressure along the suction surface more or less to the trailing edge, indicating that the simple ideas about performance with separated flow are inaccurate for high solidity cascades.

The usual diagram for choked flow in a blade passage is drawn in Fig. 4.46. The sonic line (along which the velocity is equal to the local velocity of sound) is drawn nearly straight and a circle of radius b is drawn to indicate the cross-sectional area of the throat. In the usual idealization the flow is assumed uniform across the passage and the sonic line is exactly straight. With this idealization conservation of mass flow between the uniform conditions upstream and the uniform conditions across the throat can be written

$$\rho_1 V_1 s \cos\alpha_1 = \rho^* V^* b$$

where * indicates conditions at sonic velocity. Thus

$$b/(s\cos\alpha_1) = \rho_1 V_1 / \rho^* V^*.$$

By simple one-dimensional gas dynamics it can be shown that for reversible flow up to the sonic line

$$\frac{\rho_1 V_1}{\rho^* V^*} = M_1 \left\{ \frac{1 + \dfrac{\gamma - 1}{2}}{1 + \dfrac{\gamma - 1}{2} M_1^2} \right\}^{\frac{\gamma + 1}{2(\gamma - 1)}} \tag{4.22}$$

where γ here is the ratio of specific heat capacity. Hence $b/(s\cos\alpha_1)$ is a unique function of the M_1. Figure 4.47 is a plot of this one-dimensional relation on which is superimposed a shaded region to indicate where the measurements lie (the shaded region was drawn on a similar figure given by Scholz, 1977). It will be noticed that the measurements always lie below the line so that for a given inlet flow angle α_1 the blades choke at a lower inlet Mach number than that calculated assuming that the flow is uniform. Alternatively for a given M_1 the blades choke at a higher α_1, that is at a lower mass flow, than the uniform flow case. In fact, in the expression $b/(s\cos\alpha_1)$ only α_1 is a variable for a given cascade so it follows that in choked flow

$$\alpha_1 = \alpha_1(M_1)$$

It is clear that a large mass flow or high inlet Mach number requires the largest possible throat area, i.e. b/s as large as possible. One way of achieving this

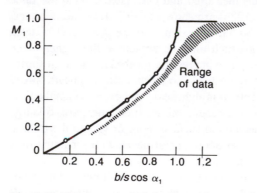

Fig. 4.47 The theoretical choke line with range of measured data; b denotes the blade throat width. (After Scholz, 1977)

is to have low solidity, but the problem encountered if this is taken to extremes is that the blade loading becomes too large. A large throat is also favoured by thin blades, particularly those with sharp leading edges and the maximum thickness well back: the C4 with maximum thickness at 30 per cent chord and a stubby nose is less satisfactory than the NACA-65 with maximum thickness at 40 per cent which in turn is bettered by the double circular arc (DCA) at 50 per cent. The throat area is larger if the camber can be greater towards the front of the blades, a trend which is very apparent with the modern supercritical sections but was also recognized in the 1940s as an advantage of blades on a parabolic camber line with maximum camber well forward.

The mass flow can never be greater than the theoretical one-dimensional value, that is, none of the measurements can be above the curve in Fig. 4.47. This is because the principal discrepancy in treating the flow as one-dimensional is the fact that real flows are always to some extent non-uniform. For a fixed inlet flow angle the mass flow will be higher, or the inlet Mach number greater, the more nearly uniform the flow is in the forward region of the blade passage. This suggests the desirability of having the least possible camber in the forward part, which is exactly contrary to the trend for throat area with camber position. Uniformity is also favoured by thin blades with the smoothest possible distribution of thickness and small leading edge radius and this again points to the suitability of the DCA. This is not, of course, to suggest that the DCA is immune from choking as is evident from the rapid fall in deflection and rise in losses at negative incidence for the higher Mach numbers in Fig. 4.5 and Fig. 4.4.

A common way of displaying the performance of blades at high subsonic inlet Mach numbers is given in Fig. 4.48 with inlet flow angle as ordinate and inlet Mach number as abscissa. The lines labelled 'choke' are the function $\alpha_1(M_1)$ for particular blade row geometries and indicate the absolute minimum possible inlet flow angle for a given inlet Mach number. The particular blades chosen for Fig. 4.48 are the double circular arc blades tested by Andrews (1949) of 10.5 per cent thickness—chord ratio and another set with the same angles and solidity but 5 per cent thickness—chord ratio, a value more typical of modern practice. The other line on the figure, shown broken, is a constant value of local diffusion factor, D_{loc}, defined earlier. This is expected to give a reasonably realistic estimate for the severity of the loading

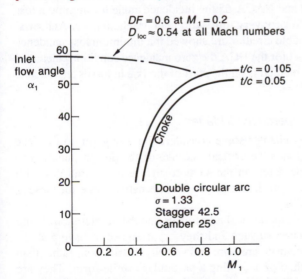

Fig. 4.48 Inlet flow angle versus inlet Mach number for cascades of DCA blades. Lines of choke at different thickness—chord ratio. Line for constant local diffusion factor for the thicker blade.

on the suction surface and for a constant D_{loc} the normal diffusion factor DF from equation 4.14 must decrease as Mach number rises, accounting for the downward slope of the line in Fig. 4.48. Whereas the choke lines are absolute limits, the constant D_{loc} line is more a line of guidance, indicating that above it the loading is very high.

With the high thickness–chord ratio tested by Andrews it can be seen that the working range between choke and the limit on D_{loc} narrows sharply to disappear altogether at about $M_1 = 0.8$. In fact as the results in Fig. 4.5 show, the blades operate quite efficiently at this Mach number with significantly greater inlet flow angles (and therefore apparently higher D_{loc}) very probably because the axial velocity–density ratio was allowed to rise. The measured rise in loss on the negative incidence side does agree well with the predicted choke line in Fig. 4.48. The choke line with the smaller thickness–chord blades is significantly lower, giving a very worthwhile increase in the operating range at inlet Mach numbers around 0.8.

There is evidence that the turning and deviation of the blades is not greatly affected by the inlet Mach number. This is even apparent for blades tested by Hoheisel for which the schlieren pictures are shown in Fig. 4.43, not-withstanding the clear evidence of flow separation on the suction surface. These blades were in a very high solidity cascade: it may not be true that deviation is hardly changed for low solidity blades when the shock causes the suction surface boundary layer to separate or greatly thicken. The double circular arc blade is slightly unusual in having a region of separated flow at most operating conditions and Mach number may affect the flow on some blade shapes more than others.

For blades like the C4 and NACA-65 the incidence angle for minimum loss was shown by Lieblein to vary very little with inlet Mach number. Andrews' measurements for the double circular arc showed the minimum loss incidence moving positive and in fact for the DCA the zero incidence condition calculated by Lieblein's low speed method is on the start of the rise in losses attributable to choking at almost $M_1 = 0.8$.

Prescribed velocity (supercritical) blades

In earlier sections of this chapter some consideration was given to the type of blade often described as supercritical because of the design intention to operate with a supersonic patch on the suction surface but to end this with an isentropic (i.e. shock-free) deceleration. An alternative name for these is controlled diffusion blading.

The essential feature of supercritical aerofoils is that the deceleration should be shock-free on the suction surface and they are designed to have a shape which will produce the requisite pressure and velocity distribution, rather than the more traditional method of accepting a particular profile form. They are a particular type of prescribed velocity distribution for which the avoidance of a suction surface shock is an important design constraint. The nature of the sort of pressure distribution chosen was discussed earlier in connection

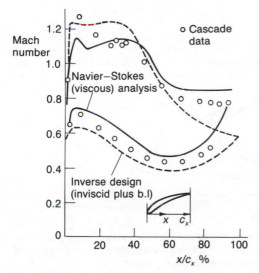

Fig. 4.49 Mach number distributions measured and predicted at the design point, (inlet flow angle 35.3°, inlet Mach number 0.75) for original design. (Design for axial outlet flow, actual $\alpha_2 = 6.9°$.) (From Schmidt *et al.*, 1984)

with Fig. 4.7 and the 'ski-jump' shape for the pressure on the suction surface is characteristic. This shape follows from specifying the pressure gradient to be as severe as possible without separating the boundary layer from the velocity peak to the trailing edge. An additional benefit with this pressure gradient is the higher blade loading which seems possible. This allows lower solidity and therefore a higher inlet Mach number before choking occurs. The lower solidity may contribute significantly to lower losses.

Although the inviscid calculations and the boundary layer calculations can be performed to choose the blade shape, the flow does not always behave as predicted. This is illustrated in Fig. 4.49, taken from Schmidt *et al.* (1984), for a supercritical cascade designed for an inlet Mach number of 0.75, inlet flow angle 35.3°, outlet flow angle 0° and solidity 0.91. When tested, the cascade exhibited a laminar separation near the leading edge which then altered the mainstream flow. After it reattached, the thicker boundary layer was unable to sustain the steep pressure gradient on the rear of the blade so that a major separation of the turbulent boundary layer occurred at about 60 per cent of chord with no subsequent reattachment. This brought a large reduction in blade loading and a stagnation pressure loss coefficient at design inlet conditions of 0.05, more than twice the normal value for conventional blades.

The original design intention for the blades shown in Fig. 4.49 was that the potential flow would be accelerated very rapidly from the stagnation point and then remain nearly constant at a high velocity 'plateau' from which it was to be rapidly decelerated, thus bringing about a rapid transition to turbulence of the boundary layer. It was concluded by Schmidt *et al.* (and independently by others too) that this is not realizable. More recent designs have dispensed

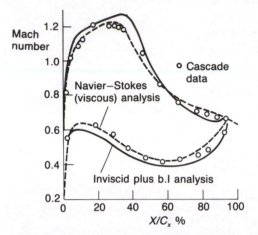

Fig. 4.50 Mach number distributions measured and predicted at the design point (inlet flow angle 35.3°, inlet Mach number 0.75) for redesigned blade. (Design for axial outlet flow, actual $\alpha_2 = 4.6°$.) (From Schmidt *et al.*, 1984)

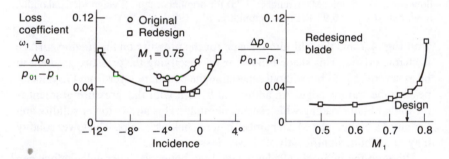

Fig. 4.51 Measured loss for original and redesigned supercritical blades. (From Schmidt *et al.*, 1984)

with the plateau and have a moderately favourable pressure gradient until the fairly abrupt start of the adverse pressure gradient, as shown diagrammatically in Fig. 4.7. Schmidt *et al.* redesigned their blade: a comparison of predictions and measurements for the Mach numbers along the redesigned blade at the design condition with this philosophy is shown in Fig. 4.50. In this case there is reasonable agreement between prediction and measurement and the design intent has been more or less realized. The corresponding loss distributions are shown in Fig. 4.51, with much lower losses and wider operating range being evident for the redesigned blade. These changes were achieved by modification to the forward part of the original blade with metal removed to the extent of no more than 1 per cent of chord. This points to one of the drawbacks with this type of blade: a *very* large change in performance was found for a small change in blade shape. It is important to recognize that for small blades the precision required to achieve this design goal almost certainly exceeds the normal precision of manufacture. Furthermore, in-service damage, particularly erosion, may undo any benefits achieved. If a design for the whole compressor is predicated upon achieving the design goals for such blades then the effect of in-service damage could be far more serious than with the more tolerant blades normally employed in the past.

Almost all the published verification of the supercritical type of blades has been performed in cascade tests. The tests in compressors have been less encouraging, partly, of course, because the profile losses make up only a small part of the total. There are many reasons to believe, however, that the prediction of transition in multistage compressors may be highly inaccurate. Tests by Dong (1988) have shown the transition on a supercritical type blade to be very different from that assumed in the design: in ordinary cascade tests there is a separation bubble whereas wakes or turbulence from upstream initiate transition earlier and the separation bubble is removed.

Although the 'ski-slope' pressure distribution characteristic of the supercritical blade has now found wide acceptance for flows which are entirely subsonic as well as those with supersonic patches, the shape originally came from designs to remove the shock wave terminating the supersonic patch. A recent paper by Weber *et al.* (1987) reported tests on a supercritical blade at design conditions and at positive and negative incidences. The design was for an inlet Mach number of 0.85, turning of 20°, solidity of 1.0 and an axial velocity—density ratio of 1.15. When the measured pressures were compared with the design values the level of agreement was sufficiently good to suggest that the design goals had been met. High-speed schlieren pictures made at the design condition showed moving shocks terminating the supersonic patch on the suction surface, Starken (1987). Because the shocks moved back and forth they were not apparent in the time mean surface pressure measurements and from these it appeared that the design intent of removing the shock had been achieved. The tentative conclusion is that the underlying idea behind the supercritical blade design, the removal of shocks, is not physically realizable. If the peak Mach numbers are prevented from exceeding 1.2 (perhaps even 1.3) however, the shock is of insufficient strength to separate the boundary layer on the suction surface and the loss attributable to them is not serious.

4.9 Concluding remarks

This has been a long chapter reflecting the large amount of work done and published over the years for subsonic inlet cascade flows. It needs to be remembered, however, that the isolation of the blade-to-blade flow so as to treat it like a two-dimensional cascade is a convenient model and will not always be an accurate description of what is really a fully three-dimensional flow. Three quite different examples will be offered to illustrate this.

The first has already been referred to and concerns the radial coupling of the flow so that each section of blading merely contributes to establishing the

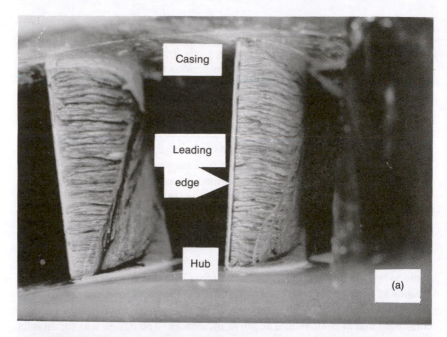

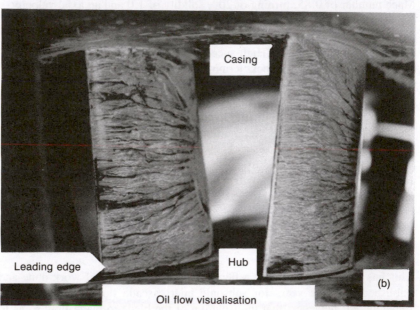

Fig. 4.52 Flow visualization on a stator blade with the stage close to stall (stall initiated in rotor): (a) shows flow with sealed hub; (b) shows flow with clearance of approximately 1 per cent of chord at hub. Equal flow coefficient for case (a) and (b). Flow visualization used pigment suspended in oil; the stator blades shown had their suction surfaces approximately horizontal to reduce effect of gravity. (From MacDougall, 1988)

whole flow. In the usual quasi-three-dimensional approach the coupling is provided by a solution to the flow in the meridional plane. If a highly swirling flow enters a stator row which turns the flow to axial it is inevitable that there will be a large rise in static pressure across the hub section and the *AVDR* will possibly be less than unity. The conditions at outlet from the stator near the hub are essentially fixed by the rest of the blade and local changes to the blade near the hub can have only limited effect.

A second example is for flows which are transonic, so that although subsonic at inlet there are supersonic patches of flow inside the passage. Close to sonic conditions the flow is extremely sensitive to changes in flow area and it cannot therefore be assumed that a blade-to-blade description will be accurate. Very small changes in adjacent blade-to-blade surfaces and boundary layer development on the hub or casing walls, can all lead to effective changes in the streamsurface thickness and hence to major changes in the flow.

The final example relates to the complicated boundary layer flow present in compressors. The nature of the boundary layer flows in compressors will be dealt with in another chapter but it is appropriate to show how the flow on a stator can be affected by the interaction of the endwall and blade boundary layers. Figure 4.52, from McDougall (1988) shows the flow visualization pattern on the suction surface of a stator for two cases, one with the tip sealed at the hub and the other with a tip clearance of about 1 per cent of chord. The aspect ratio of this blade is 1.33, the solidity 1.07 and the diffusion factor at the hub about 0.44 at the design point. Both photographs were taken at the same flow coefficient, fairly close to stall of the stage. With no clearance it can be seen that a separation line moves from the hub most of the way across the span, whereas with a small amount of tip clearance the spread of the separation is greatly reduced. The flow deterioration is not merely something which shows up in the flow visualization, but can be seen in the pressure distribution around the blades. Calculations with a three-dimensional Navier–Stokes solver by McDougall and Dawes (1987) predict the separation line remarkably well in both cases, indicating that it is gross features of the flow which are responsible for this pattern and not small-scale details of the turbulence or shear stresses, which are only included in a fairly approximate way.

In summary then, the blade-to-blade description of the flow provides an extremely powerful tool for designing and analysing the flow in axial compressors. However, it must not be forgotten that it is an incomplete model, whose possible validity must always be held open to question and which must be used with caution, particularly when three-dimensional effects are liable to be large.

5

Blade-to-blade flow for axial compressors with supersonic inlet flow

5.1 Introduction

The terminology for high-speed machines is somewhat inconsistent. Sometimes it is only if relative inlet flow is supersonic over the entire span that it is referred to as supersonic, whilst if the inlet velocity is supersonic in one spanwise region and subsonic elsewhere it is often referred to as transonic. This use of the word transonic is unfortunate because it is also normally used in other contexts to refer to flows where the velocity is everywhere close to sonic. The present chapter is concerned with flows which have supersonic relative inlet flow.

The supersonic flow into a blade row may be divided into several different classes set out in Bölcs and Suter (1986) and Starken (1986). At the present time the only practical choice is with a subsonic axial velocity at inlet. The shocks or expansion waves produced by the blades may therefore propagate upstream and the compressor is able to influence the incoming flow. Indeed there would be severe operational difficulties if this were not so. The axial velocity out of the blade row will be subsonic and in most cases the resultant flow at outlet is also subsonic.

If the use of supersonic inlet flow blading is restricted to cases with subsonic axial velocity it follows that the tangential component of velocity will be high — supersonic compressor blades are highly staggered, often by more than 60° to the axial direction. It is almost without exception that it is the rotor blades which operate with supersonic inlet velocity. Because of the high blade speed it follows from the Euler work equation that a very large work input can occur with a relatively small change in the absolute whirl velocity and a small deflection of the relative flow in the blades. With a compressible flow it is possible to change the absolute velocity with no change in direction in the frame of reference fixed to the moving blades, all that is required is a reduction in axial velocity as illustrated in Fig. 5.1. The deceleration of the relative velocity in the blade passage would normally be accomplished with one or more shockwaves. It is therefore another feature of supersonic blading that the camber is normally very small. The thickness is also kept very low, about

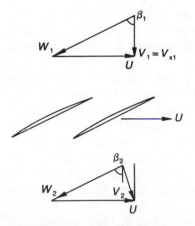

Fig. 5.1 Velocity triangles for a supersonic rotor with no turning in the relative frame $\beta_1 = \beta_2$, $V_{x2} < V_{x1}$

2 per cent of chord for the tip section of a transonic fan, and it is almost impossible to provide realistic sketches for illustrative purposes without exaggerating both thickness and camber.

The consideration of supersonic blade-to-blade flows is taken in conjunction with a meridional flow which determines the radius change of the mean streamsurface and the corresponding area reduction of the meridional flow across the blade row. The blade-to-blade flow is then expected to provide the turning and the loss and, since these affect the density and pressure change, they in turn affect the meridional flow. This is similar to the subsonic blade-to-blade flow, though the coupling between the two is much stronger with supersonic flow. The changes in density and pressure are normally larger, choking of the blades is more common and the losses can be large, up to 10 or even 20 per cent of the inlet relative dynamic pressure. There is a large effect of small area changes for transonic flow as the supersonic relative flow is decelerated to subsonic and one spanwise section can strongly affect others at different radii. The maximum mass flow capacity of the blade passage at choking is of major concern; not only must this give the desired overall mass flow for the whole machine but it should also give more or less constant values along the span. Under such conditions there is no well-founded reason for assuming that designs in which the meridional flow is calculated and this used as input to the blade-to-blade flow (the so-called quasi-three-dimensional method) should be at all satisfactory. Nevertheless it has to be admitted that such methods, chosen because there was no alternative, have worked well. To enable the methods to design machines with the desired aerodynamic performance has undoubtedly required a number of empirical inputs inferred from earlier tests. (The methods could be said to be consistent rather than correct; for example there is typically no unambiguous way of separating discrepancies in loss and in deviation from test data.) By their very nature such methods are proprietary. At the time of writing the designs are still quasi-three-dimensional although they may be checked with a full three-dimensional viscous calculation method before manufacture. The approach to design is discussed briefly at the end of this chapter.

When compressors were being designed and built in the 1940s it was generally believed that the inlet relative Mach number M_1 to each blade row must remain below unity. Given the thick blades which were then used this belief was not without foundation (by analogy with aeroplane flows it had been believed that the conditions around $M_1 = 1$ would be particularly bad). In 1949 the performance of a compressor with supersonic inlet flow was reported (see for example Wilcox *et al.*, 1959 or Bullock, 1961) and, although the efficiency was not very high, it operated best when the inlet relative Mach number was close to unity.

The early transonic and supersonic compressors used the blade profiles which had been used for subsonic conditions, such as double circular arc (DCA) blades, but as time went on the blades were modified in shape, with particular design goals in mind, so that the inlet part of the blades became flatter and the camber was moved rearwards. Once this trend became established it became almost impossible to think in terms of general profile families since each was different and was shrouded in the confidentiality of the company or organization concerned. Late in the 1960s DCA blades were still being used at design Mach numbers as high as 1.4 into the blade tip, but by then this profile was mainly a conservative choice used for comparison with more recent designs, Gostelow (1971).

For blades with supersonic inlet flow the greatest attention is directed to the forward or inlet region of the blades, ahead of a line from one leading edge drawn normal to the suction surface of the next blade. In most conditions this is where the majority of the pressure rise is produced and it is the region which fixes the maximum mass flow that the blades can pass. The inlet region has been recognized as paramount for a long time and has been the subject of many studies using linear cascades. The separate treatment of the inlet region is a convenience which can be rigorously supported for choked flow when the inlet region is truly independent of the downstream conditions. (By thinking of the inlet region separately, the ideas formulated for supersonic axial blades also find application to the inducer of centrifugal compressors with supersonic relative inlet flow.) Downstream of the inlet region the velocity is generally below sonic and information derived from subsonic blading has been used, for example diffusion factor and subsonic profile losses, though the justification for this is slight.

In the early days much was learned about subsonic compressors blading with linear cascades. A feature of supersonic blade-to-blade flow is that special problems and conditions exist, both experimental and theoretical, which are peculiar to linear cascades. The special topic of supersonic cascades is thoroughly covered elsewhere, for example Bölcs and Suter (1986) and Starken (1986), and will be given only superficial treatment here with the emphasis instead on blade performance in compressors. Experimentally it is much easier to operate with the cascade choked and much of the cascade work has been done at this condition. Much of the literature on supersonic blades is concerned with this type of flow even though compressors normally operate unchoked. The most common treatment is of a type of choked flow when the bow shock is attached to the leading edge and the choking occurs where the flow is still supersonic, ahead of the enclosed part of the blade passage. This is usually described as the 'unique incidence' condition even though it is really a minimum incidence (i.e. choke) condition and is neither the only nor even the most common choke condition. As will be discussed later the blades are more likely to choke inside the passage where the thickness is greatest and it is only at very high blade speeds that the 'unique incidence' condition prevails. (The 'unique-incidence' condition is described in Section 5.2.)

In the 1950s the NACA tested a large number of single-stage compressors and for a summary of this see Wright (1970) or Lieblein and Johnsen (1961). Each of the major companies involved in the manufacture of jet engines has probably designed and built literally scores of supersonic compressor stages, all at very high cost. (Only a few were reported in the open literature, for example Gostelow, 1971.) This trend was forced upon them by the inability of most two-dimensional supersonic cascades to model the flow at realistic operating conditions, as discussed above. The use of compressors rather than cascades did ensure that the results obtained were immediately applicable for design purposes. It did also, however, impose some severe constraints because until recently it was impossible to obtain any measurements inside the moving blade rows. Pressure transducers were installed in the casing wall over the rotor tips and hot-wire measurements were made downstream to infer what was happening in the moving frame of reference, see for example Miller *et al.* (1961). These did allow considerable insight but nevertheless such refinement in technique had come relatively late and had restrictions; in particular the pressure transducer measurements may well have been compromised by the endwall boundary layer producing conditions very close to the wall untypical of those only a short distance further in along the span. All this was accompanied by severe mechanical difficulties which had to be overcome to allow the very high speed operation. Finally there was and remains the additional complication that the blades deform under the exceedingly high centrifugal accelerations. A low hub—casing ratio fan is likely to experience up to 4° reduction in stagger and a significant change in camber between stationary conditions and maximum speed. The exact amount of untwist and uncamber depends on the blade design but will be higher for low aspect ratio blades and for blades with no part-span shrouds. Nowadays the deformation can be predicted reasonably accurately and methods exist for measuring the changes with considerable precision.

The shock pattern is not only affected by the geometry of the blades but also very strongly by the inlet Mach number, the inlet flow direction and the back-pressure behind the blade row. Figure 5.2, taken from Schreiber and

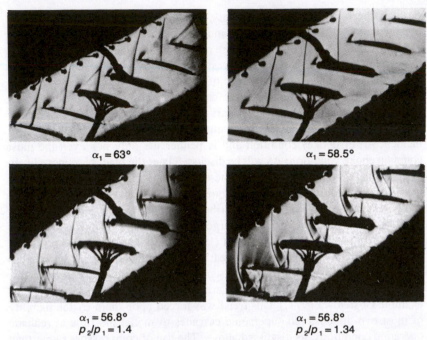

$\alpha_1 = 63°$ $\alpha_1 = 58.5°$

$\alpha_1 = 56.8°$ $\alpha_1 = 56.8°$
$p_2/p_1 = 1.4$ $p_2/p_1 = 1.34$

Fig. 5.2 Schlieren pictures of cascades with supersonic inlet flow. Stagger 48.5°, camber 14.9°, solidity 1.61, thickness—chord ratio 0.05. Cascade is choked for $\alpha_1 = 56.8°$. (From Schreiber and Starken, 1981) © American Institute of Aeronautics and Astronautics; reprinted with permission

Starken (1981), shows schlieren pictures for blades in a linear cascade with an inlet Mach number just above unity. Two of the pictures are for the choked condition but with different back pressures. The flow pattern is much simpler at the lower flow rates, i.e. higher incidences, with the shock at the leading edge dominating and creating the majority of the pressure rise. This is the common observation also found in rotors using either laser anemometry or holography. The oblique shocks which propagate upstream of the blade row, the bow shocks, are weak and these are essentially cancelled by the expansion waves — it is the passage shock or shocks stretching across from one blade to the suction surface of the next blade which produce the pressure rise and also the losses.

The next section considers the special 'unique-incidence' flow before moving on to other types of flow. A separate section is given over to losses in supersonic blading and a short section considers design aspects.

5.2 Choked flow with attached shocks — 'unique incidence'

With a subsonic axial velocity an idealized choked flow is shown in Fig. 5.3 with oblique shock waves and expansion waves emanating from each blade.

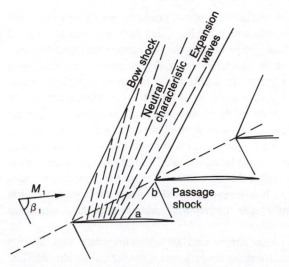

Fig. 5.3 An idealized shock and expansion wave pattern around the forward part of a supersonic rotor row at the 'unique incidence' condition

The shocks are attached to the leading edge and the type of flow is known as 'unique-incidence'. Because the geometry is axisymmetric the pattern must repeat and the flow into each blade is exposed to exactly the same number of shocks and expansions. Most of the pressure rise is produced by the passage shock. The bow shock is caused by the leading edge thickness and also by the expansion waves produced by the blade suction surface curvature; these turn the flow so that it is locally at a negative incidence to the leading edge of the next blade. The expansion waves interact with the shocks upstream of the blades and the further upstream one moves the weaker the shocks become. (Although the shock waves are drawn as straight they are in fact curved because of the intersection of the expansion waves.) In fact the shocks are sufficiently weak everywhere except directly ahead of the leading edge that the entropy rise is negligible and as a reasonable approximation the stagnation pressure relative to the blades may be taken as uniform and equal to the value well upstream. For flow regions such as this, where the characteristic lines run in only one direction and where the entropy is nearly uniform, the flow direction is constant along each characteristic. One expansion wave sketched in Fig. 5.3 does not interact with either the shockwave upstream or downstream of it and this is referred to as the neutral characteristic. Since the neutral characteristic extends to upstream infinity the flow crossing it has the flow direction of the uniform upstream flow and therefore intersects the blade suction surface at the point where the blade direction is equal to that of the uniform inlet flow. The flow sketched in Fig. 5.3 has one expansion wave, labelled ab, which passes from the suction surface of one blade to the leading edge of the next. and the flow is therefore supersonic which it enters the domain of the enclosed part of the bladed passage.

In an ideal flow with infinitely sharp leading edges and a flat suction sur-

face over the forward part of the blade there could be no shocks upstream
of the blades at the condition in which the incoming flow was exactly aligned
with the blade suction surface. (In producing the flat suction surface allowance
would have to be made for the boundary layer displacement thickness, so it
would not be geometrically flat. In addition the uncambering of the blade
because of the centrifugal acceleration would need to be considered.) Very
thin, sharp leading edges are normally unacceptable for reasons of mechanical
integrity and a leading edge thickness of about 2 per cent of the staggered gap
is typical. Any curvature of the blade suction surface in the forward region,
where the flow is supersonic, must bring about a Prandtl—Meyer acceleration
of the flow leading to a higher Mach number across the section ab and a
therefore lower mass flow. It is therefore usual to make the forward region
of the blades as flat as possible with the required camber put in the downstream
part of the blade chord.

It is possible to carry out some simple analysis which illustrates some aspects
of this flow and provides a basis for more general numerical treatment. The
leading edge is assumed to be sharp so that the shock is attached to it as shown
in Fig. 5.3. Along the characteristic ab which leaves the suction surface and
intersects with the leading edge of the next blade the flow direction and Mach
number are constant and will be denoted by β_e and M_e respectively. The flow
direction along the characteristic has to be the slope of the blade surface where
it intersects and, to be consistent with the usage elsewhere in the book, this
is denoted by χ_e. (More exactly the flow direction should be that of the blade
plus the displacement thickness but this is a refinement that can be ignored
here. The point a on the suction surface depends on the inclination of the Mach
wave and therefore on the local Mach number; for simplicity it will be assumed
that the position of a is known although in practice it would have to be found
by an iterative process.)

Since the leading edge is sufficiently sharp that the shock is attached and
the inlet Mach number M_1 for compressor applications is not too high, it
follows that the entropy rise in the bow shock is negligible. The oblique bow-
shock ahead of the blade row may therefore be treated as equivalent to an
expansion wave, but of opposite sign, so that the entire region ahead of the
line ab may be described by the Prandtl—Meyer relations.

For conditions far upstream, subscript 1, and conditions along ab, subscript
e, the Prandtl—Meyer relation is

$$\nu_1 + \beta_1 = \nu_e + \beta_e \tag{5.1}$$

where β is the relative flow direction measured from the axial and ν is the
Prandtl—Meyer function given by

$$\nu(M) = \sqrt{\left(\frac{\gamma+1}{\gamma-1}\right)} \tan^{-1}\sqrt{\left(\frac{\gamma-1}{\gamma+1}(M^2-1)\right)} - \tan^{-1}\sqrt{(M^2-1)} \tag{5.2}$$

and tabulated in textbooks on gas dynamics. For a given gas ν is a function only of the Mach number. The flow direction all along ab is the inclination of the blade surface at a, i.e. $\beta_e = \chi_e$, and the Mach number M_e along ab must therefore be such that equation 5.1 can be satisfied given that β_1 and $\nu_1 = \nu(M_1)$ are fixed by the inlet flow.

The Prandtl–Meyer relation is not the only equation which must be satisfied between upstream and the line ab. It is also necessary that the mass flow continuity be satisfied. The flow function, or corrected mass flow per unit area, is a unique function of Mach number.

$$F = \frac{m\sqrt{(c_p T_0)}}{A p_0} = \frac{\gamma}{\sqrt{(\gamma-1)}} M \left\{ 1 + \frac{\gamma-1}{2} M^2 \right\}^{-(\gamma+1)/2(\gamma-1)}$$

For the flow considered here the mass flow and T_0 are exactly constant and p_0 is approximately constant, since the shocks are assumed weak between upstream and the line ab. It follows from the expression for flow function that

$$A_1/A_e = F(M_e)/F(M_1).$$

where $A_1 = s\cos\beta_1$ is the incoming streamtube area and

$$A_e = s\cos\beta_e - (t_e + \delta_e^*)$$

is the streamtube area across ab, t_e being the thickness of the blade at the leading edge and δ_e^* the displacement thickness of the boundary layer on the suction surface at a.

Formally presented then one has two equations to satisfy:

Prandtl–Meyer $\qquad \beta_1 + \nu(M_1) \quad = \beta_e + \nu(M_e)$ $\qquad\qquad$ (5.3)

Continuity $\qquad\qquad F(M_1)s\cos\beta_\epsilon = F(M_e)A_e$ $\qquad\qquad$ (5.4)

For particular blades β_e and A_e may be regarded as given parameters, although each depends slightly on the position of a because of the effect of the blade slope and thickness and because of the development of the boundary

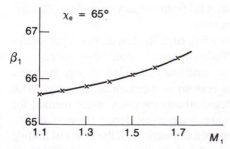

Fig. 5.4 Variation in inlet flow angle with Mach number for a cascade operating at the 'unique incidence' conditions. Effective leading edge thickness 2 per cent of staggered gap

layer. If M_e is taken as the variable then it must satisfy both equations 5.3 and 5.4, each of which is dependent on the two specified flow parameters, M_1 and β_1. It is clear that the two independent equations for one unknown, M_e, can only be satisfied for particular combinations of the parameters β_1 and M_1. This means that if M_1 is given then there is only one value of β_1 which can satisfy the equations, in other words the incidence is unique. If the incidence cannot change neither can the mass flow and the flow is choked.

The Prandtl—Meyer and continuity relations, equations 5.3 and 5.4 are solved iteratively to obtain the combination of β_1 and M_1 to satisfy both equations. (The variation in β_e and A_e as the line ab moves can also be included in an iterative scheme.) The value of β_1 does not need to change much in order to accommodate significant changes in inlet Mach number, as the results of simple calculations in Fig. 5.4 show for the case of blades of effective leading edge

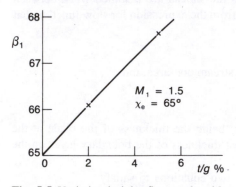

Fig. 5.5 Variation in inlet flow angle with ratio of effective leading edge thickness to staggered gap for a cascade operating at the 'unique incidence' condition

thickness equal to 2 per cent of the inlet staggered gap, g. The 'unique incidence' rises rapidly with the thickness, as Fig. 5.5 demonstrates.

With the conditions along the line ab fixed it is possible to calculate the remainder of the flow, in particular the pressure rise across the passage shock which runs from the leading edge to intersect the blade suction surface towards the trailing edge. This shock may be reflected from the suction surface, depending on the Mach number and configuration.

Real blades have, of course, finite leading edge thickness. For large blades this will be a very small fraction of blade chord: a leading edge thickness of perhaps 1 per cent chord is realistic. For small blades the leading edge thickness will be a larger fraction of chord, for reasons of mechanical integrity. With finite thickness the shock is locally detached and the basis of the method for including the leading edge radius in the 'unique-incidence' flow is due to Moeckel (1949). Starken (1986) has reported calculations of the effect of leading edge thickness for staggered cascades of flat plates. For an uncambered blade staggered at 60° with an inlet Mach number of 1.5, for example, increasing the ratio of leading edge radius to blade staggered gap from 0.5 per cent to

2.5 per cent increases the 'unique incidence' from about 0.7° to 3.3°. York and Woodard (1976) described a method for predicting the inlet flow direction corresponding to 'unique incidence' with finite leading edge thickness using the method of characteristics and including the effect of finite leading edge thickness. In a discussion to this paper Prince described the approach adopted by General Electric for the treatment of leading edge thickness. The predictions by Prince agreed almost exactly with those of York and Woodark and the methods must be presumed to be physically equivalent. Both methods show that the formulation following Moeckel (1949) leads to an underestimation of the true loss because it predicts a too-rapid attenuation of the shock strength away from the blunt leading edge.

As already remarked the 'unique incidence' condition corresponds to choking of the blade row and compressors generally operate at conditions other than this, with higher overall pressure ratio. The 'unique incidence' condition has attracted considerable attention because it is the condition which can be established most easily in linear cascades with supersonic inlet flow. If the back-pressure downstream of the linear cascade is raised there comes a point at which the shocks are no longer attached and then the flow is able to spill from one blade passage to the next; it is consequently given the name spill point. Mikolajczak *et al.* (1971) carried out systematic tests on three blade sections with supersonic inlet Mach numbers, testing them first in a linear cascade and then in a compressor rotor. All the blades had 10° camber, 60° stagger, solidity 1.25 and thickness–chord ratio of about 4.4 per cent. One was a double circular arc blade, DCA, with maximum thickness and camber at 50 per cent chord. Another was a multiple-circular-arc blade, MCA, which is a common way of specifying blades; for this the maximum thickness was around 60 per cent chord from the leading edge. The third section was termed

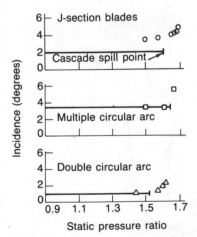

Fig. 5.6 Incidence versus static pressure ratio for supersonic blades. The solid lines show results for cascades and are all at the unique incidence condition, the points for results obtained in the rotor. (From Mikolajczak *et al.*, 1971)

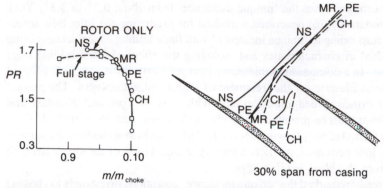

Fig. 5.7 The rotor and stage operating characteristic and measured leading edge shock structure at design speed (point PE is that for peak efficiency, CH for choke, MR for mid range and NS for near stall). (From Strazisar, 1985)

a J section with maximum thickness at 60 per cent chord and entirely straight for the first 50 per cent of chord. Results for the three blades are shown in Fig. 5.6. The cascade results denoted by a solid line were only obtained for the choked flow, believed to be the 'unique-incidence' condition and the lines are terminated at the spill point. For the rotor measurements, denoted by symbols, the DCA and MCA blades show the same incidence condition obtained at low static pressure ratios, when the blades are choked, but not at high values: the J section in the rotor choked at a different incidence condition at all pressure ratios perhaps because the rotor geometry at high rotational speed was not quite as specified for this blade. The total pressure loss was

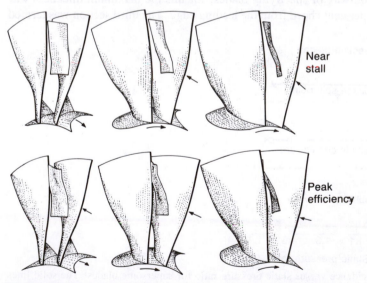

Fig. 5.8 Perspective views of the passage shock in a low hub–casing ratio rotor. For each flow rate each of the three views is rotated by 20°. (From Wood *et al.*, 1986)

not found to be significantly different for the choked ('unique incidence') or unchoked cases, although in the latter case the static pressure rise was greater.

More recently there have been abundant measurements with laser anemometers and flow visualization to confirm that rotors normally operate with the leading edge shock somewhat detached. Results published by Strazisar (1985), reproduced in part as Fig. 5.7, make it clear that even when the blades are choked the shock is detached and the pattern is not that of the 'unique-incidence' condition. A very clear demonstration of the detached passage shock and its three-dimensional character is given by the measurements of Wood *et al.* (1986) in a low hub—casing ratio fan and Figure 5.8 is reproduced from their work.

5.3   Operation with detached shocks

As stressed above the 'unique-incidence' condition, which is choked with super-sonic flow entering the closed portion of the blade passage is not that at which supersonic blades would normally operate. Instead the forward shock will be detached from the leading edge, rather as shown in Fig. 5.7. Whilst operating in this mode, questions of major concern are the mass flow capacity of the blading, the static pressure rise and the loss in stagnation pressure, all of which are determined primarily in the leading edge region.

A simple analysis by Freeman and Cumpsty (1989) considers the control volume around the forward part of the blade, Fig. 5.9. Flow enters at Mach number M_1 and inclination to the axial β_1. Inside the control volume there will be a shock wave which may be strong (there may indeed be several shock waves). The blades are assumed to be sufficiently thin and lightly cambered in the leading edge region that a single blade direction, χ, suffices to describe the blade there; for most supersonic blades this is likely to be quite accurate. The flow leaves the control volume subsonic with a Mach number M_2 and

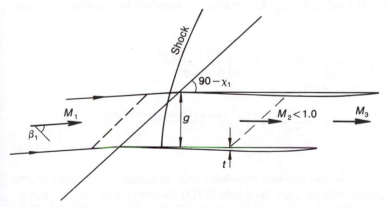

Fig. 5.9 A simple control volume for the analysis of unchoked supersonic flow

in the direction of the blade, χ. The blades do have thickness and at the downstream side of the control volume where the thickness is a maximum it is denoted by t, As noted earlier it is a general feature of supersonic blades that the thickness is small, typically about 2 per cent of chord which is between 6 and 8 per cent of the staggered gap, the distance between two blades measured normal to the chord line, denoted here by g.

The analysis consists in equating the mass flow, stagnation enthalpy and momentum into and out of the control volume. It is simplified if the component of momentum used is that parallel to the blades, i.e. in the direction χ, because it is assumed that with a thin leading edge, small blade thickness and with very small camber in the forward part, the blade force resolved along the blade will be negligible.

The equation for the conservation of mass is

$$\rho_1 V_1 g \cos\beta_1 = \rho_2 V_2 (g - t) \cos\chi \tag{5.5}$$

for conservation of relative stagnation enthalpy

$$T_1 \left[1 + \frac{\gamma - 1}{2} M_1^2 \right] = T_2 \left[1 + \frac{\gamma - 1}{2} M_2^2 \right] \tag{5.6}$$

and for momentum parallel to the blades

$$p_1 g \cos\chi + p_1 \gamma M_1^2 g \cos\beta_1 \cos(\chi - \beta_1) = p_2 g \cos\chi + p_2 \gamma M_2^2 (g - t) \cos\chi$$
$$+ \text{higher order terms} \tag{5.7}$$

It is easy to derive an equation for p_2/p_1 directly from the momentum equation. Another expression for p_2/p_1 can be constructed from the equations for the conservation of mass and stagnation enthalpy. Both these equations for p_2/p_1 must be satisfied simultaneously. It is, however, more convenient and more instructive to carry out some algebraic manipulation of equations 5.5, 5.6 and 5.7 to isolate M_1 and M_2 on alternate sides of one equation. The result is

LHS $\quad \left[1 + \dfrac{\gamma - 1}{2} M_2^2 \right]^{-1/2} \dfrac{\{1 + \gamma M_2^2 (1 - t/g)\}}{M_2 (1 - t/g)}$

RHS $\quad = \left[1 + \dfrac{\gamma - 1}{2} M_1^2 \right]^{-1/2} \dfrac{\{(\cos\chi/\cos\beta_1) + \gamma M_1^2 \cos(\beta_1 - \chi)\}}{M_1}$

$$\tag{5.8}$$

Figure 5.10 shows the left- and right-hand sides of equation 5.8 plotted against Mach number with the right-hand side (RHS) for the case of $\chi = 65°$. For no thickness, $t/g = 0$, the left- and right-hand sides become identical if the inlet

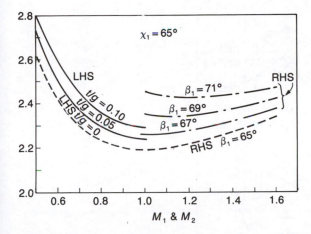

Fig. 5.10 The left-hand side (LHS) and right-hand side (RHS) of equation 5.8 plotted versus Mach number. Blade inlet angle $\chi_1 = 65°$

flow is aligned with the blades, $\beta_1 = \chi$, and the curve for this case, which is shown as a broken line, then describes the normal shock wave. A solution clearly requires that the ordinate of Fig. 5.10 is equal for both the LHS and RHS: thus for an inlet flow with $M_1 = 1.5$ and $\beta_1 = 67°$, for the blade inlet angle $\chi = 65°$ and thickness equal to 5 per cent of staggered gap, the curve of the LHS indicates that M_2 will be approximately 0.75. Once M_2 is known it is easy to find pressure ratios and loss in stagnation pressure.

For a non-zero thickness there is a minimum value of the LHS which exceeds the minimum value of the RHS. Consider for example a rather thick blade for which $t/g = 0.1$ and $\chi = 65°$. If the inlet Mach number is 1.5 it can be seen on Fig. 5.10 that the smallest inlet flow angle for which the LHS and RHS can be equal is $\beta_1 \approx 67°$. It is impossible to reduce β_1 further, i.e. to increase the axial velocity, and this represents a choking condition. The magnitude of 'unique incidence' increases with blade speed whereas this other type of choking occurs at smaller incidence as speed rises. For realistic values of blade maximum thickness and leading edge thickness the 'unique-incidence' type of choking does not become the limiting condition until the blade speed becomes quite high, $M_1 \approx 1.5$ for $\chi = 65°$ or about 450 m/s at standard-atmosphere conditions.

For the choice of $\chi = 65°$ the ratio of the downstream static pressure to the absolute stagnation pressure at inlet p_2/p_{01} has been calculated as if the blade row formed a rotor with no absolute inlet swirl. The results of this are shown in Fig. 5.11 for the idealized case of uncambered blades. The blade thickness is 7.5 per cent of staggered gap. The flow coefficient V_x/U is given as abscissa. The lines are non-dimensional corrected blade speed; $U/\sqrt{(c_p T_{01})}$ = 0.8 corresponds to blade speed of 430 m/s at $T_{01} = 288$ K, giving a relative inlet Mach number $M_1 = 1.42$ at $V_x/U = 0.4$. Also shown in Fig. 5.11 are lines for the relative stagnation pressure loss and overall isentropic efficiency at

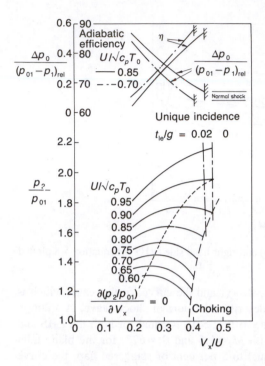

Fig. 5.11 Pressure ratio, loss and efficiency for a supersonic cascade of uncambered blades predicted by Freeman and Cumpsty (1989), $\chi = 65°$, $t/g = 0.075$

two speeds. No extra loss beyond that created in the inlet region is included. It should be noted that the loss at each speed is tending to the value correspon- ding to a normal shock which would be obtained when $\beta_1 = \chi$ but it is impossi- ble to reach this condition because the flow chokes at a lower value of V_x/U. (The mimumum possible loss is that of a normal shock at the inlet Mach number.) The highest efficiency also occurs at the highest possible flow. The subject of losses is discussed later in this chapter but it should be noted here that values predicted by this simple model agree quite well with those deduced from measurements.

The curves of p_2/p_{01} show the expected rise with increase in rotational speed. Superimposed on the curves of p_2/p_{01} are three loci, two show choking conditions and one is the locus of the peak value of pressure ratio, when $\partial(p_2/p_{01})/\partial V_x = 0$. At very high speeds the choking locus is that for 'unique- incidence': with zero leading edge thickness this incidence is zero but at a typical leading edge thickness of 2 per cent of the staggered gap the maximum value of V_x/U is reduced. At lower speeds the choking condition is that discussed in connection with Fig. 5.10 and the change occurs at $U/\sqrt{(c_p T_{01})} \approx 0.82$.

The significance of the peak of p_2/p_{01} is that stable operation is only pos- sible when $\partial(p_2/p_{01})/\partial V_x < 0$, see Chapter 9. As the speed rises the locus of the peak moves to higher values of V_x/U so that by $U/\sqrt{(c_p T_{01})} \approx 0.9$ there is no range between instability and choke. Although the precise value of blade speed at which the flow range disappears depends on the blade geometry used and the assumptions made, it does indicate a real limit on the useful range of blade speed for satisfactory operation. Choked blades always satisfy $\partial(p_2/p_{01})/\partial V_x < 0$ but away from the peak pressure rise, which is also the instability or stall line, the efficiency rapidly becomes low.

5.4 Shock structure and the nature of flow in supersonic rotors

There have been several papers describing the laser measurements or holographic pictures obtained in supersonic rotors, one of the earliest being by Wisler (1977), one of the most recent by Parker and Jones (1988). It is now quite clear that the shocks normally stand forward of the blade leading edge (i.e. are detached) and that the shock structure is really three-dimensional; Fig. 5.8 clearly demonstrated both these points. The blade-to-blade treatment is therefore strictly inappropriate, although it is the conventional way to view the flow and does still form the basis of most design.

Strasizar (1985) showed the extent to which the surfaces of both the bow shock and passage shocks are inclined to radial direction. This is illustrated

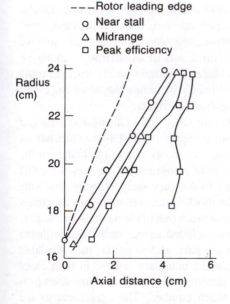

Fig. 5.12 The intersection of the passage shock with a surface from hub to casing midway between pressure and suction sides of blade passage. (From Strazisar, 1985)

by Fig. 5.12 showing the intersection of the passage shock (the forward passage shock if there are more than one) with a mid-pitch surface from hub to casing. The shocks are inclined backwards. In addition the streamlines are also inclined radially outwards in the meridional plane so the net effect is to produce a much more oblique shock than would seem apparent from a projection onto the blade-to-blade surface. The combined obliquity has marked effects on the conditions across the forward shock and the most pronounced effect was found by Strazisar at 30 per cent span in from the tip, the case illustrated in Fig. 5.7. The inlet Mach number to the shock at this position was measured to be 1.36. Treating the shock as two-dimensional in the blade-to-blade surface (i.e. neglecting the radial obliquity and the radial component of the velocity) Strazisar calculated the downstream Mach number was 0.76. Including the radial components the downstream Mach number was calculated to be 1.06, which was fairly near to the measured value of 1.2. The calculated shock loss at this section with the flow treated as two-dimensional was 0.048 but this was reduced to 0.016 with the radial shock obliquity and flow inclination included. Over most of the span the differences found by Strazisar were less than this.

There seems no doubt that understanding the flow does require the three-dimensional treatment, a point demonstrated by Prince (1980) and later by Wennerstrom and Puterbaugh (1984). Kerrebrock (1981) has stressed the need to think of the flow in three-dimensional terms and has used theoretical reasoning to show that around the radius where $M_1 = 1.0$ there will be a discontinuity if the three-dimensional nature of the structure is not taken into account. Modern calculation methods make this possible.

Figure 5.13 shows measurements and results of three-dimensional calculations[†] for a section near the hub of a supersonic rotor. The rotor, designated Rotor 33 by NASA, has a fairly high hub—casing ratio with supersonic relative inlet flow down to the hub. There are three sets of calculations: taking no account of the boundary layers on the blades or annulus walls; including the blockage effect of the boundary layers; finally including the blockage effect of the boundary layers and viscous dissipative effects.

The laser measurement in Fig. 5.13 were taken from Chima and Strazisar (1983) for an equal mass flow in the same compressor. Just as the calculations show smearing of the shocks the laser measurements also fail to bring out the sharpness evident with such methods as laser holography. One reason for this is the failure of the particles to respond to the very steep velocity gradients of the shock, another is the fact that the shock is not stationary but fluctuates in strength and position with time and also from one blade passage to another. The calculation with the viscous effects included agrees quite well with the measurements, particularly in the forward part, and so too do the calculated results including just the blockage from the boundary layers; in both cases the calculation predicts the correct flow pattern even though there are discrepancies in the downstream contours of Mach number. The agreement of the

† The calculations were performed specially by Dr J D Denton for this comparison using his time-marching method discussed further in Chapter 11.

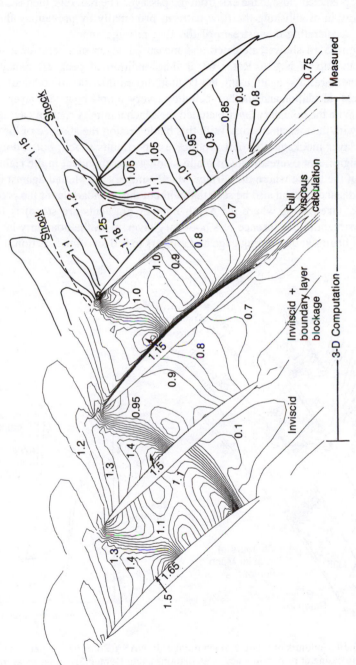

Fig. 5.13 Predicted and measured Mach number contours near the hub of a supersonic rotor, NASA Rotor 33. Calculations performed by Denton, measurements from Chima and Strazisar (1983)

calculations neglecting the boundary layer is not satisfactory and in this case a quite different flow pattern has been established with a strong, almost normal shock predicted close to the exit from the passage. The blockage then is a crucial quantity in establishing the flow pattern, principally by preventing the flow from accelerating downstream of the first passage shock.

Figure 5.14 shows calculated and measured Mach numbers for a section near the tip of NASA Rotor 67 at the condition of peak efficiency. The calculated results, again using the Denton three-dimensional code, are for an inviscid calculation and another including a very simple boundary layer prediction. With the boundary layer included the Mach numbers contours agree quite well with the measurements. With this blade section the absence of the boundary layer blockage does not produce a superfluous shock, but it does lead to a significant overestimate of blade loading, most evident in the right-hand part of the figure where the surface Mach numbers are plotted against chordwise distance. It can also be seen that without the boundary layer the predicted shock (taken to be where the Mach number gradients are steep) is further downstream in the absence of a boundary layer; with the boundary layer the shock begins at the leading edge and appears to reflect off the suction surface.

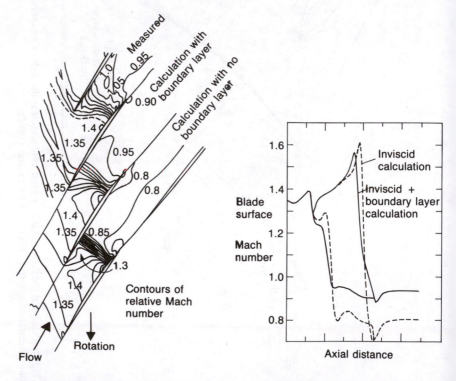

Fig. 5.14 Contours of relative Mach number in NASA Rotor 67, 10 per cent of span in from casing at peak efficiency. Calculations using Denton 3D code, measurements from Pierzga and Wood (1985)

The three-dimensional calculations shown here demonstrate that nearly all the pressure rise takes place across the passage shock or shock system. A good estimate for performance of supersonic blades, particularly those well away from the hub, may therefore be obtained from a method concentrating on the forward region of the blades. Just as most of the pressure rise is being produced in the leading edge region, it is also true that this is the region responsible for most of the losses. In early cascade tests by Starken and Lichtfuss (1970) it was realized that the shock could separate the boundary layer on the suction surface and that the loss then became sensitive to the cascade solidity. The explanation is that high solidity allows a separated boundary layer to reattach. The possibility of boundary layer separation introduces a quite strong influence of Reynolds number, because the ability to resist separation or to reattach as a turbulent flow after a laminar separation rises with the Reynolds number.

It was also clear from cascade tests that the strength of this shock was increased by the amount of convex curvature on the suction surface between the leading edge and the shock. By having a very nearly flat suction surface the expansion could be minimized; it was then a small step to decide to have curvature in the opposite sense, often referred to as negative camber, in the leading edge region so as to have a gradual compression along the suction surface. This may be thought of in two ways: the curvature of the blade surface towards the tangential produces compression waves which may coalesce into a series of weak oblique shocks; alternatively, but equivalently, the flow cross-sectional area is decreased in the flow direction by this negative camber and this leads to a deceleration of the supersonic flow.

An example of pronounced negative camber is shown by the blades and static pressure contours of Fig. 5.15, taken from a rotor designed by Prince (1980). These were obtained whilst operating very close to design speed (which was high, about 550 m/s) and on the design operating line. The efficiency was rather disappointing, $\eta_{\text{adiabatic}} = 0.79$. Downstream of the leading edge there is a rise in static pressure along the suction surface, as intended, prior to the shock across the passage. However, the flow seems to be off design because the static

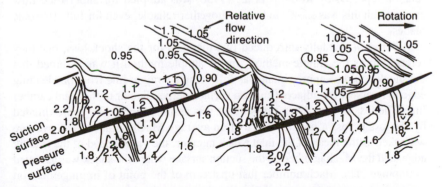

Fig. 5.15 Contours of casing static pressure beneath a high-speed rotor (550 m/s tip speed) with pronounced negative camber. (From Prince, 1980)

pressure decreases along much of the pressure surface before terminating in a strong shock. This is one of the dangers of high-speed blades with appreciable camber and area change — away from the optimum condition (ideally the design condition) a supersonic flow can be accelerated where it should decelerate and high loss be created in the ensuing shock.

Ginder and Calvert (1988) describe the design of a rotor for a tip inlet relative Mach number of 1.47. With a conventional blade the Mach number was predicted to be 1.5 in front of the passage shock, causing the suction surface boundary layer to separate and reattach just before the trailing edge. With negative camber the Mach number ahead of the shock was reduced to 1.4 and the boundary layer separated for only a short length and the predicted loss was substantially reduced. Ginder and Calvert point out that the success of the design depends critically on the response of the boundary layer to the shock, for with only slightly more blockage the negatively cambered blade could choke.

5.5 Losses in supersonic blading

From the beginning the estimation of loss in supersonic compressor blades has been a subject of major concern. For subsonic flow the blade-to-blade loss is usually of secondary concern compared to other sources of loss and with other aspects of blade performance. With supersonic inlet velocities, however, the blade-to-blade loss can be so large as to be of first-order importance.

Miller *et al.* (1961) were probably the first to describe a method specifically addressing the loss of supersonic blades and what they did has been followed by others. Essentially they assumed that the loss could be considered in two parts, the loss created by the shock across the passage from one leading edge to the next (the passage shock) and the profile loss. The profile loss is analogous to the loss in subsonic blading, described in Chapter 4, and the correlation reported by Robbins *et al.* (1956) was adopted for supersonic flow even though this was shown to be not altogether reliable, even for fully subsonic blades.

Miller *et al.* actually introduced two methods for the shock loss, one they referred to as the simple method. In the simple method it was assumed that the shock could be approximated as following a straight line from one leading edge to intersect the adjacent suction surface perpendicular to the local camber line. The loss in stagnation pressure, it was assumed, could be approximated by the loss across a normal shock at an average Mach number. The average was taken as the mean of the inlet Mach number, the value ahead of the leading edge, and the Mach number on the suction surface where the shock was assumed to impinge. The Mach number just upstream of the point of impingement on the suction surface is normally higher than the inlet Mach number but may be lower if there is negative camber near the leading edge. In the more detailed model the stand-off of the shock around the leading edge was considered and the expansion waves interacting with the shock were considered so that the mass-averaged loss along the shock could be found. A previous investigation had indicated that the mass-weighted losses associated with the bow shock were small compared with the passage shock. Generally the shock losses for the examples given were very similar for the two methods and by far the greatest uncertainty arose from the estimate of profile losses. Most of the examples given were for the design incidence case but one comparison was carried out with the detailed flow model for a compressor at a range of incidence. Whereas the predicted loss increased from about 0.15 to 0.17 for a change of incidence from about 3.5° to 8° the comparable measured loss rose from 0.12 up to 0.35. On this basis there can be little confidence that this method is correctly accounting for the losses.

Koch and Smith (1976) described a method having much in common with the methods of Miller *et al.* (1961). The shock loss was again thought of as an addition to the profile loss which could occur in subsonic flow. This time a relation for the entropy rise due to the leading edge thickness was an explicit term derived by Prince (see discussion to York and Woodward, 1976). Good agreement with measured loss was obtained at conditions near peak efficiency.

Most methods for loss estimation in supersonic blading stem from the method of Miller *et al.* (1961), with two separate loss sources being assumed, one due to the shock and the other the profile loss arrived at using the measurements made in cascades and in blade rows which are entirely or predominantly subsonic. This division is not physically plausible. A property of shock waves is their ability to bring about separation of boundary layers. Schlichting (1979) states that a turbulent boundary layer will separate if the pressure rise across the shock exceeds that for a normal shock with upstream Mach number of 1.3, a value likely to be exceeded on very many blades. (A more detailed analysis shows that the necessary pressure rise across the shock to produce separation decreases as the form parameter of the incoming boundary layer increases.) In many cases the boundary layer on the suction surface will be laminar ahead of the shock, particularly if there is significant acceleration between the leading edge and the shock, and with a laminar boundary a separation occurs at very low Mach numbers. It is probable that most supersonic blades have a region of separation beginning where the passage shock strikes the suction surface — a soundly based fluid dynamic description of loss needs to take this into account. It is also probable that decisions on blade design based on models of the flow which do not admit separated flow will lead to erroneous trends for major variables, such as solidity or blade shape. Starken and Lichtfuss (1970) in their tests in linear cascades found that at the highest solidity they tested ($\sigma=c/s=1.4$) the separation was further upstream (because the passage shock impinges further upstream) but the downstream width of the separated region was smaller than for the lowest solidity ($\sigma=1.0$). Experience in design has shown that when inlet Mach numbers are sufficiently high that the passage shock can separate the suction surface boundary layer, good efficiency requires that the solidity must be high to allow the flow to reattach.

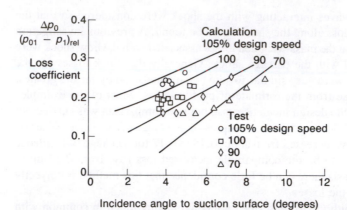

Fig. 5.16 Measured loss compared with predictions by Freeman and Cumpsty. Measurements by Sulam *et al.* (1970) at 90 per cent span: 100 per cent speed at $U/\sqrt{(c_p T_{01})} = 0.87$

At the present time the Navier–Stokes methods for three-dimensional flow are not reliable as predictors of the magnitude of loss, even though they seem able to give good indications of the overall flow pattern, including static pressure variations and the distribution of loss. The accurate prediction of loss probably requires a fairly accurate description of shear stresses (and therefore turbulence modelling) whereas the overall flow pattern is only weakly sensitive to this.

Freeman and Cumpsty (1989) predicted loss by their simple conservation model in the inlet region. Predictions were compared with losses deduced from measurements in two fan rotors and Fig. 5.16 shows a comparison for one of them, the fan tested by Sulam *et al.* (1970) at 90 per cent span. Not only is the absolute level of loss well predicted but the trends with speed and incidence are also well described. More generally it was found that for values of $U/\sqrt{(c_p T_{01})}$ above about 0.8 the total measured overall loss was quite well predicted except near the casing, where, as one would expect, the measured loss was much higher. As speed was reduced the loss from the inlet region prediction began to underestimate the total loss.

Freeman and Cumpsty found that the *minimum* loss was that of a normal shock at the inlet Mach number ahead of the blades but this could only be achieved for zero thickness blades at zero incidence. The entropy rise, equivalent to the relative stagnation pressure loss, across a normal shock increases as $(M_1 - 1)^3$ so the minimum loss rises rapidly with speed. (The efficiency falls more slowly than this implies because the useful pressure rise increases strongly with blade speed.) More generally, the calculations show loss is a strong function of incidence and blade speed. For a given incidence, however, the thickness has only a small effect on loss. Nevertheless thickness is indirectly most important because it determines the minimum incidence at which choking occurs. With the minimum loss at zero incidence (obtainable only from flat blades of zero thickness) being that of a normal shock $\{\Delta p_0/(p_{01}-p_1)\}_{n-s}$ Freeman and Cumpsty used parameter studies to derive the empirical relation for the loss

$$\Delta p_0/(p_{01}-p_1)=\{\Delta p_0/(p_{01}-p_1)\}_{n-s}+[2.6+0.18(\chi-65°)]10^{-2}(\beta_1-\chi)$$

This linear dependence on incidence, $\beta_1-\chi$, should be valid for values up to $5°$.

The satisfactory prediction of the loss using the simple method of Freeman and Cumpsty shown in Fig. 5.16 is in apparent conflict with the evidence now available from laser measurements and holography that the flow is three-dimensional and the shocks are oblique in the radial sense. As noted above Strazisar (1985) showed from his measurements of the shock structure that where the shock appears to be nearly normal in the blade-to-blade surface a two-dimensional estimate could overestimate the shock loss. Wennerstrom and Puterbaugh (1984) estimated the inclination of the shock in the radial sense and used the appropriate obliquity to estimate the loss by an appropriately modified method of Miller *et al.* (1961). In this case the largest differences were in the regions where the Mach number was highest, i.e. near the rotor tip, but the differences were much less than those found by Strazisar.

The explanation for the satisfactoriness of the loss prediction by the simple one-dimensional approach of Freeman and Cumpsty is that it follows from the imposition of conservation relations for mass, momentum and stagnation enthalpy; provided the radial fluxes of these are small and other assumptions of the method are justified, such as the blades of very small camber, the method must predict loss correctly. In this respect the flow is analogous to that in short, straight pipes or ducts with supersonic inlet flow and subsonic outlet flow: the flow pattern may be very complicated with many shocks and local separations but overall the static pressure rise and stagnation pressure loss are well predicted by the normal shock relations precisely because the same conservation relations are imposed. This type of flow is sometimes called pseudo shocks.

5.6 The design process for supersonic blades

As noted in the introduction to this chapter most designs of supersonic blading are performed as two separate steps, the axisymmetric meridional calculation and the blade-to-blade calculation. The term quasi-three-dimensional calculation is sometimes used to indicate the inclusion of the effect of streamsurface radius change and axial velocity—density ratio change (contraction of the meridional streamtube) but operating still in a blade-to-blade surface with the meridional flow given by a separate axisymmetric calculation. Fortunately, because the endwall boundary layers are normally thin in relation to the blade span, the endwall blockage is much less important than in late stages of multistage compressors and the prediction of the meridional flow can be quite accurate. At each spanwise position the blades must be matched to the predicted local meridional inlet flow velocity and as the relative Mach number rises the range between choke and flow instability becomes narrower. The mass flow swallowing capacity is one of the most important quantities to be obtained from a blade-to-blade calculation: not only does it affect the meridional calculation directly but it also is the the major influence on the loss generation.

One of the long-standing features of the design of high-speed blading is that each blade section is designed rather than chosen from a family of profiles (a practice now occuring with subsonic blading as well). This makes it difficult to generalize on the methods adopted because the crucial details of the procedure for successful designs are confidential. A rare recent example where the recent design philosophy of the blade profile is explained is by Wennerstrom (1984) who designed and tested a highly successful rotor. It would appear that many of the designs in common use are very simple — a straight wedge in the forward part and at a particular chordwise position a blend into a curved section at the rear to give the desired camber. Sometimes this is all achieved by a number of circular arcs. There is no unanimous view of the optimum but this may in any case be difficult to determine since so much depends on the radial matching of the blade sections as well as on the accurate prediction of the actual geometry at operating speed.

The pressure rise and mass swallowing capacity of supersonic or transonic blade rows can now be calculated numerically with some confidence. At each radial section the blade shapes are essentially chosen to be compatible with the axisymmetric flow field calculated on a meridional surface. The correct prediction of blade-to-blade flow requires the boundary layer to be included, but the effect of the blockage seems much more important than the correct inclusion of the loss producing processes in the boundary layer.

Because of the high stagger common with supersonic blades, the predictions of work input and pressure rise are exceedingly sensitive to outlet flow angle, i.e. deviation. Freeman and Cumpsty (1989) calculated the performance of a blade row for which measurements were given by Sulam *et al.* (1970). The general level of measured deviation was high but it serves to show the effect: at a particular spanwise position and blade speed an increase in the deviation from $7.4°$ to $8.9°$ reduced the predicted ratio of absolute stagnation pressure from 2.13 down to 1.81. In some design methods the estimate for deviation is still based on correlations similar to Carter's rule, derived for incompressible flow, incorporating the effect of camber and solidity. The estimate of deviation needs to take into account the tendency of supersonic blading to have the camber near the rear with the blades flat or reverse cambered near the front. It also needs to take account of the change in meridional streamline position and streamtube height across the blade row. Estimates for deviation of the necessary accuracy can only be obtained at the present time from tests of rotors similar to that under design. It seems plausible that changes in deviation are one of the ways in which blades can make the spanwise adaptations necessary to satisfy radial equilibrium. If separated or nearly separated flow exists on the suction surface very small changes in back pressure could effect significant changes in wake thickness and thus in deviation. If this is the case then no two-dimensional method will correctly predict the magnitude.

Despite the obvious neglect of three-dimensional features, such as the inclination of the shocks, the design methods can be remarkably successful. Most are proprietary but a recent description of such a method has been given by Ginder and Calvert (1987) where boundary layer growth on the blades is included. Solutions to the three-dimensional transonic flow which include the major viscous effects are now quite widely used, Denton (1986). Methods for solving the transonic Navier–Stokes equations are also available, for example Dawes (1987a and b), and the separated flow in a cascade with supersonic inlet flow at quite large incidence can be calculated. This is discussed briefly in Chapter 11. Full three-dimensional calculations are, however, likely to be reserved for the final checking of designs or for the investigation of particular regions suspected of being a problem. This is not just because of the cost of such calculations. The real problem is that there are few, if any, design strategies in three dimensions and there is likewise little experience from which to assess whether features of flow predicted in three dimensions are satisfactory.

6
The centrifugal impeller

6.1 Introduction

Unlike the axial compressor, where the rotor and stator blades are comparatively similar, the rotating and stationary parts of a centrifugal compressor, the impeller and the diffuser, are quite different. Because of this it is convenient to treat each in a separate chapter: this chapter is concerned mainly with the impeller and Chapter 7 with the diffuser.

As noted in Chapter 2 there are wide variations in the geometry of the impellers in use. Most of the highly loaded compressors are of the unshrouded type with inducers; since it is these that have attracted the most study and for which most has been published it is also about these that there is most to write. Shrouded compressors without inducers usually have heavily backswept blades (that is the impeller blades are inclined backwards in the circumferential direction so that the relative flow is in the opposite direction to the impeller) and, for reasons which will be elaborated, this relieves many of the aerodynamic problems.

Most centrifugal compressors produce a sufficiently large rise in pressure that there is a marked rise in density across each stage. However, the relative Mach numbers inside the passages of the impeller are often not very high. Away from the entry to the impeller the relative Mach numbers are normally well below one and the rise in pressure is caused predominantly by the centrifugal effect and not by deceleration of the flow. Accordingly many of the processes inside the impeller can be considered with an incompressible flow approach, although the flow area must be adjusted to take account of the rise in density arising from the centrifugal effect. This is not likely to be acceptable at inlet to the impeller, where the relative velocity is often supersonic, nor at outlet when conditions are considered in the stationary frame of reference. One of the features of many impellers is the very low average aspect ratio of the blades by the standards of the axial compressor. Perhaps more relevant is to say that the mean hydraulic diameter is small in relation to the passage length. There is therefore great scope for loss generation in the shear regions adjacent to the solid walls.

A common feature of many impellers is that the cross-sectional area frequently increases far too much between inlet and outlet. The streamtube outlet area is given by

$$A_2 = (2\pi R_2 b_2 - N b_2 t)\cos\chi_2,$$

where χ_2 is the blade backsweep, b_2 is the axial span of the blades at outlet, t their thickness and N their number. The inlet area of the relative flow streamtube for an impeller with an inducer is

$$A_1 = \pi[(R_{outer})^2 - (R_{inner})^2]\cos\beta_1,$$

with β_1 denoting the relative inlet flow direction and R the inducer radii. The inlet area is much smaller than it appears in a meridional section because the inlet flow is inclined at a large angle β_1 to the meridional direction, which is to the axial direction if there is an inducer. The outlet area A_2 is also often much larger than it appears on such sections because of the increase in radius between inlet and outlet.

On a one-dimensional basis what matters to the overall state of the relative flow is the ratio of the mean inlet and outlet relative velocities, W_1/W_2, or alternatively the equivalent ratio of Mach numbers. (Some prefer the ratio W_2/W_1 which is the same as the de Haller number used for axial compressors.) For most impellers the ratio of velocity or Mach number is not likely to be very different. If the flow were uniform at inlet and outlet the velocity ratio would be given by

$$W_1/W_2 = \rho_2 A_2/\rho_1 A_1.$$

Compressors can easily be designed, and are designed, with ratios for $\rho_2 A_2/\rho_1 A_1$ as large as three, whereas boundary layer theory, or the knowledge of diffusers or axial cascades, would suggest that a value of two would only be achieved for W_1/W_2 in the most favourable circumstances. If the density-area ratio for a diffuser becomes large the blockage is increased by means of thick boundary layers or regions of separation to limit the effective area ratio. Put another way, the flow becomes highly non-uniform so that the simple relation between density–area ratio and velocity ratio no longer holds. With the centrifugal impeller the problem is exacerbated by the large turning in the blade-to-blade and in the axial-to-radial directions. For very many impellers the basic outline ensures that there will be a high degree of flow non-uniformity, possibly with major separation of the flow, and that the flow will not decelerate in the way which the area ratio seems to indicate.

The reasons for the large $\rho_2 A_2/\rho_1 A_1$ lies in the choice of the axial span of the blades at outlet, b_2. This is often not reduced as much as the one-dimensional velocity ratio suggests it should be, because to do so would lead to exceedingly shallow flow passages towards the outlet of the impeller. For unshrouded impellers this would give tip clearances very large in relation to the span of the blades. This problem is more acute for high-pressure ratio impellers, because the density rise is large, or those of low specific speed, because the outlet diameter is normally large in relation to the inlet. In general, impellers with no inducer seem to have smaller area ratios and the problem is less severe, particularly since they normally have large amounts of backsweep at outlet which also reduces the amount of area increase.

Separation in the impeller therefore seems a necessary consequence of the stipulation of the basic geometry in many cases. The possibility of a separated flow, referred to as a neutral zone, was considered by Carrard (1923) and flow visualization and measurements in water pumps had shown the existence of separated flow even before this. Despite this evidence the designers of compressors had, until quite recently, failed to take full note of it. It was Dean who persistently drew attention to the need to take into account the separated nature of the flow and who has given the name to the type of flow, the jet-wake model. In this most of the flow leaves the impeller in the jet, which has the appropriate stagnation pressure for almost loss-free flow in the relative frame of reference; the wake has only a small velocity, with much lower relative stagnation pressure, but it may occupy a large part of the passage area. The idea seems to have been first aired in the paper by Dean and Senoo (1960) in which the flow was treated as two-dimensional (uniform in the axial direction) with the jet and wake dividing the passage in the circumferential direction, Fig. 6.1. The wake was on the side of the passage bounded by the suction side of the blade (i.e. the trailing surface). (The aspects of the Dean and Senoo paper relating to the behaviour downstream of the impeller will be considered in the context of the diffuser in Section 7.2.

In the next section mainly experimental evidence is used to describe the flow in radial impellers. Following this there is a description of the flow from the point of view of the equations of motion and simplifications thereof, followed by a brief account of the contribution of numerical methods. Of special interest to the radial machine is the estimation of slip factor and loss and a section is devoted to each of these. Finally a brief treatment of the aerodynamic design of impellers is given.

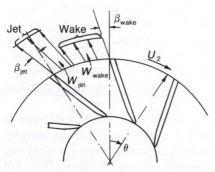

Fig. 6.1 The idealized jet-wake model. (From Dean and Senoo, 1960)

6.2   The flow pattern in impellers

When Dean (1971) discussed the existence of a region of separated flow in impellers the evidence was still somewhat circumstantial. There were four main pieces of evidence to point to the existence of the jet and wake:

1. Measurements of strongly unsteady flow downstream of the impeller with a strong repeating pattern.
2. Measurements of the unsteady pressure along the shroud showed that towards the outer part of the impeller the pressure was uniform across part of each passage, which could only occur if the throughflow velocity was very small in this region.
3. The time-mean or steady pressure distribution along the shroud was incompatible with deceleration of the relative flow in the outer part of the impeller.
4. Potential theories were unable to predict the variation in the slip factor observed, indicating that the flow was being quite incorrectly modelled by them.

There was, in addition, flow visualization carried out in a pump by Fischer and Thoma (1932). Some direct evidence was produced for the existence of the strongly nonuniform flow in impellers by the work of Fowler (1968) in which a huge impeller was rotated slowly with an observer sitting in the hub. Not long after, Moore (1973) published his work on a rotating diffuser. The experiment showed the existence of a separated region on the suction (or trailing) surface and provided support for the ideas on why the flow should choose this surface in a rotating diffuser, a topic discussed further below.

The most important information about the flow in impellers, mainly unshrouded impellers, has come from measurements at DFVLR in PorzWahn, Germany, using laser anemometry. The first measurements were by Eckardt (1975) and Eckardt (1980). These laser measurements were backed up with very careful and meticulous measurements of a more conventional kind. More recently Krain (1987) has published measurements also made in DFVLR on a modern and very efficient impeller.

With measurements available inside high-speed impellers most of the ambiguity of interpreting the measurements taken outside the impeller was removed. Figure 6.2 shows measurements of the meridional velocity at a number of sections inside a radial outlet (i.e. no backsweep) impeller presented by Eckardt (1976). (The impeller had a tip diameter of 400 mm, and was designed to give a stagnation pressure ratio of 3.0 with a mass flow of 7.2 kg/s at 18 000 rev/min.) Also shown is a meridional section through the machine indicating the position of the measurement planes. At plane 2, at the end of the inducer, there is no evidence of any irregularity in the pattern but some emerges near to the shroud at plane 3. By plane 4 the irregularity is recognizable as a region of loss, probably what is normally describable as a separated region, which has moved to the suction surface/shroud corner. The separation grows and flattens as the flow continues to the outlet plane 5. At this plane it can be recognized as the wake. These pictures show the time-mean flow averaged

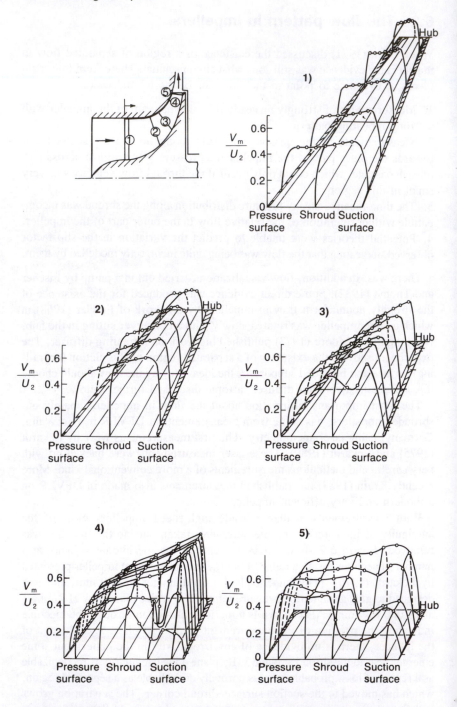

Fig. 6.2 Velocity measurements by Eckardt in a centrifugal impeller with no backsweep (14 000 rev/min, $m = 5.31$ kg/s, pressure ratio 2.1). (From Eckardt, 1976)

over all the blade passages but the flow is in fact very unsteady with average fluctuations of 25–30 per cent of the mean in the region of the wake close to the shroud.

The wake at impeller outlet shown in Fig. 6.2 is not quite like the two-dimensional region modelled by Dean and Senoo, but it is clearly a large important part of the flow, with a mean velocity very much less than the jet and positioned near to the suction surface. Dean had envisaged the wake beginning in the shroud-suction surface corner, migrating to fill the passage at outlet from hub to shroud near the suction surface; some details may therefore be inaccurate but the overall jet-wake model proposed is good. (Because the wake is close to the shroud it can be seen that measurements of static pressure on the shroud would tend to overestimate its size.)

Eckardt (1980) also made measurements in an impeller with the same shroud line and the same blade shape from inducer to 80 per cent of the outlet radius, but beyond which the curvature was changed to give 30° of backsweep. The hub contour was also moved outwards. The velocity profiles showed a similar pattern to those in the original impeller, with the separation beginning on the shroud and growing in size. The relative stagnation pressure distributions for this impeller are shown for two flow rates in Fig. 6.3. In this case there was much less migration to the suction surface and, at the mass flow for best efficiency, the wake had not really reached the suction surface of the blade even at exit. At the choke condition the wake can be seen to have migrated to the shroud/suction surface corner.

From his measurements Eckardt was able to estimate the pattern of the secondary velocities across the passage. (Secondary velocity means the components

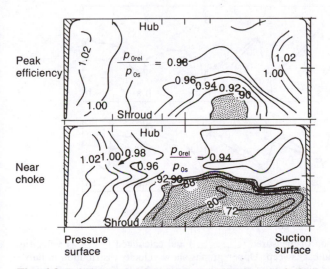

Fig. 6.3 Relative stagnation pressure contours at exit from an impeller with 30° backsweep (non-dimensionalized with isentropic relative stagnation pressure). Wake region shown shaded. (From Eckardt, 1980)

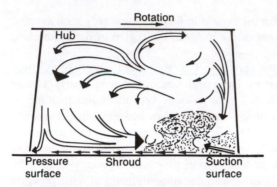

Fig. 6.4 Schematic of secondary flow pattern observed in impeller with no backsweep. (From Eckardt, 1976)

of velocity normal to the primary flow parallel to the solid surfaces forming the passage.) A sketch taken from his paper for the radial outlet impeller is shown as Fig. 6.4. The significance of the wake for calculation of the flow in the impeller was shown very clearly by Eckardt; Fig. 6.5 shows his comparison of measured relative velocities inside the impeller without backsweep with those calculated by methods which ignore the separated flow. The upper set show the hub—shroud distribution along a streamsurface midway between the blades, whilst the lower set show the blade-to-blade distribution along a surface midway between hub and shroud. (Calculation methods will be discussed below but it suffices to note that any inviscid method would have given a similar comparison with the measurements.) The agreement is reasonably

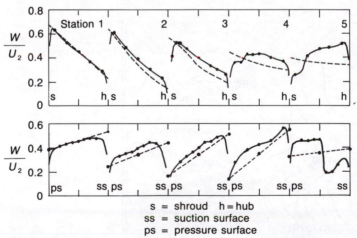

s = shroud   h = hub
ss = suction surface
ps = pressure surface

Fig. 6.5 A comparison of measured (————) and calculated (_ _ _ _) velocities in an impeller with no backsweep. Upper graphs show velocity distribution in hub—shroud direction at mid-pitch. Lower graphs show blade-to-blade distribution on surface midway between hub and shroud. Measurement stations are those shown in Fig. 6.2. (From Eckardt, 1976)

good until plane 3, although the boundary layer blockage causes the velocities calculated at planes 2 and 3 to be underestimated. By plane 4, where the separated region is a major part of the flow, the prediction is poor and by plane 5 the opposite trend is being predicted to that measured.

The measurements of Eckardt have provided a firmer foundation for the modelling of flows in centrifugal compressors as well as providing the first real test cases for numerical methods. Certain things should be said in reservation. Even at the start of the test programme the radial-bladed impeller was a rather old-fashioned design, see Moore (1976), and the impeller stagnation pressure ratio at the highest speed was only about 3:1, a very modest figure by modern standards for a compressor with an inducer. In fact tests were performed at only 78 per cent of this maximum speed. The results of Eckardt, exceptionally useful as they are, must not therefore be taken as an infallible model for all impellers, particularly modern ones of good design.

Whereas the laser measurements of Eckardt used a conventional design of impeller, Krain (1987) describes similar measurements in a wholly new impeller of his own design. A photograph of the impeller is shown in Fig. 6.6. Constraints adopted in the design was ease of machining, using a five-axis milling machine, and the description of surfaces by simple analytical expressions: there

Fig. 6.6 An impeller designed and tested by Krain with $30°$ backsweep. Overall diameter 400 mm, $U_2 = 468$ m/s at design, specific speed 0.62. (Reproduced by permission of DFVLR)

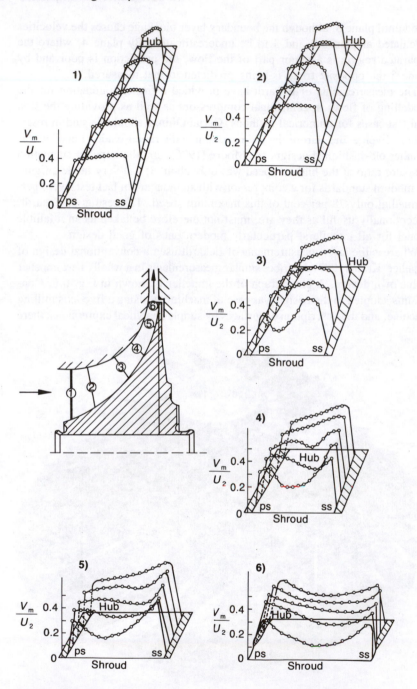

Fig. 6.7 Velocity measurements by Krain in an impeller with 30° backsweep. Measured at design speed and mass flow, 4.0 kg/s. Impeller stagnation pressure ratio approx. 4.7. (From Krain, 1987)

is absolutely no evidence that performance was compromised in any way by these constraints since the polytropic total-to-total efficiency for the impeller was measured to be 95 per cent, a very high value. At this condition the exit tip clearance was about 1.3 per cent of the tip width and the inducer tip clearance was about 0.4 per cent of the inducer radius at inlet. The overall pressure ratio was much higher than for the wheel tested by Eckardt, being 4.7 for the impeller itself at the design point when the mass flow was about 4 kg/s and the peripheral speed 468 m/s. The outlet diameter was 400 mm and there was 30° of backsweep. The specific speed was relatively low, about 0.62, implying a small inducer diameter in relation to the outlet diameter. A meridional section is sketched in Fig. 6.7 where the effect of the small inducer diameter can be seen to have been to give a large and smoothly varying radius of curvature on the shroud. The shroud shape is very different from that on the impeller tested by Eckardt, see Fig. 6.2.

The main part of Fig. 6.7 is the meridional velocity distributions measured with laser anemometry at design speed for a flow rate close to that for peak efficiency. The meridional velocity does not in this case show a feature describable as a wake at exit, section 6, but a region of reduced velocity near the shroud. At section 3 the flow at the shroud has clearly undergone a major change. Here, as in Eckardt's measurements, the measurements are ambiguous, since what is obtained is the average over many revolutions of all the blade passages; it is *possible* that the velocity is not reduced near the shroud as shown but that for some measurements the flow was wholly separated, with very low meridional velocity, whilst for others it remained fully attached. The big difference between these measurements and those obtained by Eckardt is that with Krain's impeller the low velocity region remains close to the shroud, does not grow or a deepen and does not migrate to the suction surface–shroud corner.

Separation from the shroud or massive thickening of the boundary layer on it is not surprising. Whenever a flow is made to follow a curved path the flow past the convex surface is accelerated and when the flow has to be decelerated downstream there is the likelihood of separation. At first sight it is surprising that the separations in Figs 6.2 and 6.7 occur well before the shroud wall straightens in the meridional sections shown. The explanation seems to be that once separation occurs the effective meridional curvature is removed earlier and the flow pattern is sufficiently altered that the deceleration occurs much sooner. (In this it seems analogous to the separation of the laminar flow about a circular cylinder, which can occur upstream of the point predicted by potential flow theory to give maximum velocity because once the wake is present the entire pressure field is altered.) If the separation from the shroud is a key process in making possible the jet-wake pattern it is clear that an important design feature is a large and smoothly varying radius of curvature of the shroud. In this respect it should be noted that Krain's design took advantage of the low specific speed (i.e. small inducer relative to the outlet diameter) to have a large radius of curvature along the shroud. Furthermore by choosing simple mathematical forms for the shape Krain ensured continuous meridional curvature. At higher specific speed this would be harder to achieve, yet higher specific speed is where optimum designs are conventionally expected to be, see Fig. 1.9. It seems likely that only when advantage is taken of the low inducer diameter to give a desirable shape to the shroud does the optimum choice of impeller move to lower specific speed. This does require greater axial length than has conventionally been the practice.

Separation from the shroud therefore seems to be a prime cause of the region of low meridional velocity and low relative stagnation pressure (referred to as the wake for convenience) in impellers of very different type. The extent of this, and its position, can have a big effect on the performance. It is now possible to calculate many of the features of the flow in impellers using three-dimensional methods to solve the Navier—Stokes equations, with suitable models for the turbulence. There is also a place for simpler methods which can give useful guidance: secondary flow theory, considering only the convection of low stagnation pressure fluid without considering its origin, can be used to indicate the position of the wake and to assess the relative effects at work.

Secondary flow

Secondary flow, which will be considered more generally in Chapter 8, refers to the flow at right angles to the primary flow. Secondary flow is generated when a primary flow with non-uniform stagnation pressure is subjected to accelerations perpendicular to the primary streamline direction. These sideways accelerations can be produced by curvature or, in centrifugal impellers, by Coriolis effects. In considering secondary flow in radial machines it is appropriate to use a relative stagnation pressure defined for the case of incompressible flow, by $p^* = p + \frac{1}{2}\rho W^2 - \frac{1}{2}\rho\omega^2 r^2$. The rate of increase in secondary vorticity can then be written in a form derived by Hawthorne (1976), for which the derivation is also given by Johnson (1978), as

$$\frac{\partial}{\partial s}\left[\frac{\Omega_s}{W}\right] = \frac{2}{\rho W^2}\left[\frac{1}{R_n}\frac{\partial p^*}{\partial b} + \frac{\omega}{W}\frac{\partial p^*}{\partial x}\right] \qquad (6.1)$$

Here Ω_s denotes the secondary vorticity (i.e. the component of absolute vorticity in the relative flow direction) and ω the angular velocity of the impeller about an axis in the x direction. R_n is the streamline curvature in the relative frame of reference and b is distance in the binormal direction, that is the normal to both the streamline direction and the vector of length R_n from the centre of curvature to the streamline. For a centrifugal impeller there is curvature in the meridional plane (characterized by the shroud and hub curvature) and curvature in the blade-to-blade surface (i.e. camber of the blades). These, as well as the rotation, all act on fluid of low relative stagnation pressure to produce secondary flow. Equation 6.1 predicts that secondary flow moves low stagnation pressure fluid to regions where the reduced static pressure, $p' = p - \frac{1}{2}\rho U^2$, is low. The low $p*$ fluid can be in the boundary layers or in a wake. If the relative stagnation pressure has a gradient in the axial direction, as the boundary layer on the shroud has in parts of the impeller downstream of the inducer, the rotation drives the low $p*$ fluid towards the blade suction surface. The blade-to-blade curvature is in the surface normal to the stagnation pressure gradients in the hub and shroud boundary layers and this curvature gives a secondary flow driving the low $p*$ fluid towards the suction surface. Likewise the meridional curvature acts with the gradient of $p*$ on the blade surfaces to give a secondary flow towards the shroud.

The ratio of the terms multiplying the differentials in equation 6.1 gives a Rossby number, $R_0 = W/\omega R_n$; if this is large there will be a tendency for the effect of curvature to dominate over the rotation. For the centrifugal compressor the blade-to-blade curvature tends to be large in the forward part of the impeller, in the inducer, where the boundary layers are very thin and regions of separated flow are likely to be small. The more important curvature is that in the meridional plane, drawing the low $p*$ fluid to the shroud. Once a region of low relative stagnation pressure $p*$ is formed it will therefore be drawn to the shroud if the Rossby number is large, whereas it will be drawn towards the suction surface if the Rossby number is low (implying rotation is the dominant term).

Johnson and Moore (1983a and b) studied the flow in a large radial impeller very similar to the radial outlet wheel tested by Eckardt. Although designed

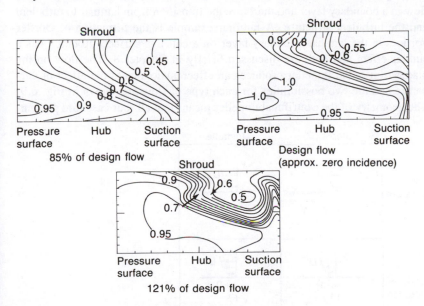

Fig. 6.8 Contours of non-dimensional rotary stagnation pressure, $p* = p + \frac{1}{2}\rho \ (W^2 - \omega^2 r^2)$, in a low-speed impeller with rotating shroud and no backsweep. (From Johnson & Moore, 1983b)

for high-speed operation it was operated for this work at very low speed, so the meridional velocity at outlet would be higher than that for which it was designed. The impeller was fitted for these experiements with a rotating shroud so the effects of tip clearance were not included. In their compressor, without tip clearance, the separation began in the shroud—suction surface corner. The size of the wake was determined by the overall flow deceleration and its position by the secondary flow, that is the velocity components normal to the main or primary flow. This is illustrated by Fig. 6.8 taken from Johnson and Moore (1983b) showing contours of relative stagnation pressure at outlet from the impeller. At the so-called design flow rate the area of low stagnation pressure fluid, the wake, is in the suction surface—shroud corner. On reducing the flow the size of the wake increases and moves around to the suction surface of the blade, showing that the rotation effect is increased, i.e. R_0 decreases. On increasing the flow, and therefore increasing R_0, the wake is squeezed down and moves around to the shroud, showing that the meridional curvature is relatively more important. In a compressor in which the meridional velocity had been reduced by density rise it is probable that the rotational effect would be proportionately even greater, at least in the downstream part of the impeller. The same trend with flow rate is evident in the measurements of Eckardt, Fig. 6.3.

Boundary layer stability

The flow over curved surfaces has been known to affect the stability of the flow in a boundary layer and therefore the transition from laminar to turbulent and the turbulent structure. A familiar example is the formation of Görtler vortices in the laminar boundary layer on a concave surface. Rotation has a similar effect and both are discussed briefly in Chapter 8. The flow on the blades of a radial impeller produces an effect analogous to the effect of curvature and the two possibilities for each type of flow are shown in Fig. 6.9. The geometry of the centrifugal impeller means that all four cases are present.

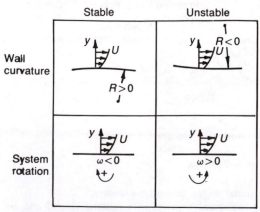

Fig. 6.9 The different regimes of stable and unstable flow near solid walls

The origin of the stability or instability can be understood heuristically by considering just one case, for example the stabilized flow in a rotating impeller. Consider a particle of fluid which is in a state of equilibrium in the boundary layer on the suction surface, i.e. the trailing surface. There is a pressure gradient normal to the blade surface balancing the Coriolis acceleration appropriate to the local radial velocity, $2\omega v_R$. Suppose now that the particle is displaced slightly away from the wall whilst retaining the same velocity (i.e. momentum) parallel to the wall. The particle now finds itself in a region where the particles which are at equilibrium have a higher velocity, $v_R + v'_R$, so that the local pressure gradient normal to the surface must be large enough to balance a Coriolis acceleration $2\omega(v_R + v'_R)$. The perturbed particle therefore experiences an acceleration $2\omega v'_R$ from the pressure gradient over and above that needed to keep it in its new position and in the opposite direction to that in which it was originally perturbed. The motion is therefore stable. For the corresponding case of flow over curved surfaces the argument is the same but the acceleration is the centripetal, v^2/R, instead of the Coriolis acceleration. In most applications stability is the desirable state but this is not the case when a viscous flow is to be decelerated, because it is the turbulent shear stresses which prevent or delay the separation of the boundary layer.

Earlier investigations focused on flow separation from the impeller blades, following the same concentration on the blade-to-blade flow considerations evident for axial compressors. The effect of rotation was to stabilize the flow on the suction surface, depress the turbulence in the boundary layer and therefore reduce the ability to resist deceleration. Analogous considerations apply to the flow along convex walls, such as the shroud. Since in the measurements of both Eckardt and Krain separation appears to begin on the shroud it seems that this is where attention to stabilization should really be concentrated.

The most obviously relevant experiments on the effect of rotation were performed by Rothe and Johnston (1976), in which the performance of straight diffusers was studied stationary and rotating about an axis normal to the flow — a good analogy to the flow in the radial part of simple centrifugal impellers. What was most different from the radial impeller was the absence of the curvature of the shroud. In contrast to the flow in stationary diffusers those with rotation were steady and relatively quiescent, with the region or regions of stall fixed on the suction side of the passage. At small area ratios there were fairly symmetric stalled regions in the suction surface/endwall corners; at larger area ratios this became a stalled region right across the passage, more or less straight except near the corners and referred to as two-dimensional stall. Further increase in area ratio led to the separation beginning at the minimum area plane and this is referred to as full stall. A diagram of the regions is shown in Fig. 6.10; here the ordinate is the diffuser area ratio and the abscissa is a parameter $\omega h_1/W_1$ (an inverse Rossby number) where h_1 and W_1 are the diffuser width and the mean velocity at entry respectively. The lines mark the approximate boundaries of the different types of separation and the points show the values

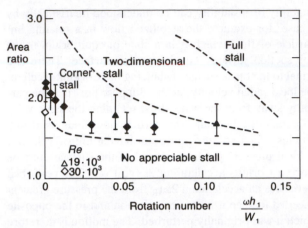

Fig. 6.10 The performance of two-dimensional diffusers with rotation. Uncertainty limits shown on each point with bar. (From Rothe and Johnston, 1976)

of area ratio for peak pressure rise; the dependence of this on Reynolds number was weak in the range investigated. For a stationary diffuser the line terminating the 'no appreciable stall' region would correspond to an area ratio of about 2.3, so the rotation clearly brings a marked reduction in the capacity to withstand deceleration, with all of this degradation occurring on the suction surface. The pressure recovery of the unstalled diffusers was found to be unaffected by rotation; the loss in performance with rotation was due to onset of separation at lower area ratios.

For the measurements referred to it appears that the first signs of separation in rotating diffusers or a shrouded impeller were in the suction surface/shroud corner but in the cases with tip clearance the first sign was on the shroud. The extent to which the separated region grows depends mainly on the ratio $\rho_1 A_1/\rho_2 A_2$ of the machine, whilst the position of the wake is fixed by secondary flow considerations, primarily the balance between the effect of meridional curvature and the effect of rotation.

Most of the discussion so far in this chapter has been concerned with impellers with inducers so the absolute inlet flow is approximately parallel to the axis of rotation. Most compressors used for multistage application have no inducer and have the flow entering as well as leaving the region of the impeller in the radial direction. A drawing of a model of such an arrangement, together with the meridional velocity distribution at entry to the impeller, from Casey and Roth (1984), is shown in Fig. 6.11. The large angle through which the meridional flow is turned and the small radius of curvature in the meridional plane means that the hub-to-shroud variation is extremely important. The measurements made in a low-speed rig, in agreement with predictions from a streamline curvature method, indicate that the velocity on the shroud is almost four times as large as that on the hub. This was a particularly severe example and a more typical inlet to an impeller without inducer would ex-

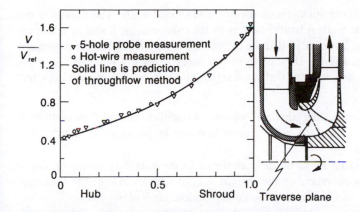

Fig. 6.11 The velocity measured and predicted in a model of a severe inlet to a shrouded radial impeller. Measurements on plane shown with broken line in the cross-section. (From Casey and Roth, 1984)

perience a meridional velocity at the shroud about twice that at the hub. In a high-speed application it is probable that supersonic velocity would be obtained near the shroud and possible choking across the entire section at some higher mass flow conditions.

Even with an inducer the flow is not normally uniform immediately ahead of the leading edge, mainly because the hub and shroud curvature exert an influence upstream of the impeller, but the variation is much smaller than without an inducer: with a long inducer the axial velocity at the inducer tip might be 1.1 times that at the hub and with a short inducer 1.3 times. In design it is quite normal to assume a linear variation in leading edge axial velocity between hub and tip.

In assessing the importance of curvature in the meridional plane for impellers with or without inducers, it should be born in mind that the extent to which the flow on the shroud is accelerated depends on the ratio of the hub and shroud radii of curvature in the meridional section and not on the shroud curvature alone. The acceleration will therefore be greatest when the hub radius of curvature is much larger than that on the shroud, as might be the case when the inlet cross-sectional area is large to accommodate a large mass flow. The effect that this high velocity has on the boundary layer on the shroud depends on the ratio of some measure of the boundary layer thickness to a characteristic length scale along the shroud; for example, the ratio of the overall boundary layer thickness to shroud radius of curvature would suffice. The greater the thickness of the boundary layer the further upstream the separation is likely to occur.

For multistage compressors the flow normally enters the stage radially inwards before turning to enter the impeller. The cross-sectional area of the flow therefore reduces very substantially as it approaches the impeller and therefore is accelerated. This can be expected to bring a large reduction in the boundary layer thickness on the hub and shroud walls. The flow near the hub and shroud walls is further helped by the change in the frame of reference as it enters the impeller so that the difference in stagnation pressure between the freestream and the flow close to the walls is greatly reduced.

To sum up then, it seems that some separation is inevitable in most impellers because:

1. Area ratios are often stipulated which, by one-dimensional flow considerations, can be recognized as far greater than can be permitted if separation is to be avoided.
2. The stabilizing effect on the turbulence in the boundary layers produced by the convex curvature on the shroud and, to a lesser extent, the rotational effect on the suction surface, make separation particularly likely.
3. Many impellers, particularly early ones of the unshrouded, inducer type, have quite unrealistic amounts of blade curvature (i.e. camber) attempting to turn the flow by 60° or more whilst there is simultaneously a 90° turn from axial to radial to be accomplished.
4. Many impellers have very strong meridional curvature (small radius of curvature) on the shroud wall and very often this curvature has discontinuities. This is particularly true of high specific speed impellers for which the inlet diameter is not much smaller than the outlet diameter.
5. Compressors in which the flow enters the region of the impeller in the radial direction and the meridional flow must be turned through 180° (such as many multistage compressors) experience a very high degree of deceleration of the flow near the shroud as it enters the impeller and there is a high risk of separation in most cases.

Once the separation has occurred, the size of the separated region is fixed mainly by the overall area ratio and the position of the wake by the balance of secondary flow effects.

6.3 Calculation methods and predictions of flow in impellers

Calculation methods are considered in general in Chapter 11 and in connection with meridional throughflow in Chapter 3. The development of ideas about centrifugal compressors was so related to the thinking about calculating the flow, largely because of the activities at the NACA Lewis Laboratory in the 1940s and early 1950s, that some particular treatment is appropriate here. This also provides an opportunity to discuss the different accelerations which the flow in the highly three-dimensional geometry of an impeller has to experience.

Equations of motion in the impeller

It seems natural to consider the flow in the impeller in a frame of reference rotating with the impeller itself. This introduces an axial component of relative vorticity given by $\Omega = -2\omega$, where ω is the impeller rotational speed. (Sometimes it is easier to solve the problem in stationary coordinates, so that the flow is unsteady but remains irrotational except in the boundary layers.) The equations governing the relative flow can be written, approximating the viscosity as constant, which is valid if the temperature and density gradients are modest, as follows for the continuity, momentum and enthalpy (i.e. rothalpy):

$$\frac{\partial \rho}{\partial t} + \nabla.(\rho \mathbf{W}) = 0 \qquad (6.2)$$

$$\frac{\partial \mathbf{W}}{\partial t} - \mathbf{W} \times (\nabla \times \mathbf{W} + 2\omega) + \nabla \tfrac{1}{2}(W^2 - \omega^2 r^2) + \tfrac{1}{\rho}\nabla p - \nu \nabla^2 \mathbf{W} = 0 \qquad (6.3)$$

$$h + \frac{W^2}{2} - \frac{\omega^2 r^2}{2} = \text{const} \qquad (6.4)$$

It is clear that some simplification to equation 6.3 is possible when the density variations are very small if the terms operated on by the gradient operator are combined, writing a relative stagnation pressure

$$p^* = p + \tfrac{1}{2}\rho(W^2 - \omega^2 r^2).$$

The solution of equations 6.2, 6.3 and 6.4 is the basis of many calculation schemes but unless one goes to a full numerical solution the complications are such that it is impossible to make much headway without major simplification. For most impeller flows an important term is that including the viscosity, μ, here taken to include turbulent eddy stresses, for without this quite misleading predictions can be made and insights gained are liable to be suspect.

An approximate method for blade loading

The blade-to-blade calculation is made more complicated by the fact that the flow has a component of vorticity $2\omega\cos\gamma$ normal to the plane, ω being the impeller rotational speed and γ the inclination of the blade-to-blade surface to the axis of rotation. For the inducer $\gamma = 0$, so, just as in an axial machine, there is no vorticity normal to the surface, but this changes along the flow path so that near the outlet $\gamma = 90°$ and there is a component of vorticity equal to twice the rotational speed. There is a further practical difficulty which is that the blade-to-blade surface is generally non-developable so pictures of the blade-to-blade flow cannot be shown without distortion.

If one is prepared to make a number of assumptions it is possible to gain some insights into the flow which are useful in deciding on the effect of major changes such as the effect of backsweep. Figure 6.12a shows a view along the axis of rotation of a backswept impeller (the corresponding view of the meridional plane is shown sketched in Fig. 6.12b). A packet of fluid is shaded and to one side it is shown enlarged with arrows attached to indicate the pressure

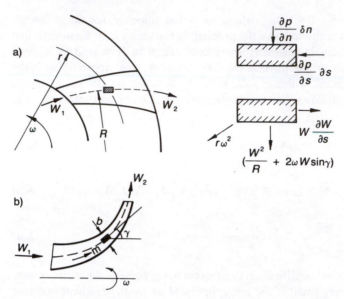

Fig. 6.12 The geometry, stresses and accelerations of inviscid flow in a centrifugal impeller

gradients and accelerations to which it is exposed in the blade-to-blade surface; r is the radius about the axis of rotation and R is the local radius of curvature of the streamline in the blade-to-blade surface. The impeller rotation gives a centrifugal acceleration $\omega^2 r$ radially inwards towards the axis of rotation. Normal to the local flow direction there is a component due to the streamline curvature W^2/R and a Coriolis acceleration $2\omega W \sin \gamma$, the sine term arising because the velocity vector is inclined at angle γ to the axis of rotation. In the absence of shear stresses due to viscosity or turbulent shear stresses the only way these accelerations can be balanced is by the pressure gradients on the element.

It is possible to make some assumptions about flow direction, equate the pressure gradients and accelerations and integrate from inlet to outlet. There is, however, a rather neater way to arrive at what is most useful, namely an approximate expression for the blade loading. This has its origins in the work at the NACA, see for example Stanitz and Prian (1951) but what follows is an approach given by Johnston (1986). Consider the view of the impeller in Fig. 6.13 in which an element is shown defined by two circular arcs about the axis of rotation and the blade surfaces. The flow will be taken as uniform in the spanwise direction normal to the plane shown in Fig. 6.13; this is not a good assumption when the spanwise width b is large, where the curvature of the meridional streamline is large or where the flow has been complicated by the effects of viscosity. The analysis is only valid well away from the blade ends, that is well inside the blade passage. For the present purpose the flow direction β from the meridional direction will be assumed uniform across the

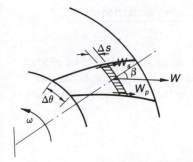

Fig. 6.13 A section through part of a centrifugal impeller

passage and equal to the blade direction on each side of the passage. β will be taken as positive in the direction of backsweep which is positive as drawn. The passage width is $r\Delta\theta$ with $\Delta\theta$ taken to be small, equivalent to having a large number of blades.

For an irrotational absolute inlet flow there is a constant relative vorticity of 2ω parallel to the axis of rotation. The shaded region of included area A therefore has a relative circulation given by

$$\Gamma_{\text{rel}} = (-2\omega \sin\gamma)A = (-2\omega \sin\gamma)r\Delta\theta\Delta m. \qquad (6.5)$$

The distance in the meridional direction Δm can be related to that in the streamline direction Δs and the radial direction Δr by

$$\Delta r = \Delta m \sin\gamma = \Delta s \sin\gamma \cos\beta$$

and the differentials are related in a similar manner.

The relative circulation can also be evaluated by considering $\oint v dl$ around the perimeter of the shaded region. Doing this in the same direction as the angular velocity of the impeller gives

$$\Gamma_{\text{rel}} = W_p\Delta s - W_s\Delta s - d/ds(W\sin\beta \, r\Delta\theta)\Delta s$$

or
$$(W_s - W_p) = - \Gamma_{\text{rel}}/\Delta s - d/ds(W\sin\beta \, r\Delta\theta) \qquad (6.6)$$

where W_s and W_p are the relative velocities on the suction and pressure faces and W is the mean relative velocity across the element. Replacing Γ_{rel} in equation 6.6. by that in 6.5 and rearranging slightly gives

$$\frac{W_s - W_p}{r\Delta\theta} = 2\omega \sin\gamma \sin\beta - \frac{\sin\gamma \cos\beta}{r\Delta\theta}\frac{d}{dr}(W\sin\beta \, r\Delta\theta) \qquad (6.7)$$

Now the mass flow through the passage, $\Delta m = \rho br\Delta\theta W\cos\beta$, is constant and may be used to remove $Wr\Delta\theta$ from equation 6.7. After differentiating Δm may be again removed to yield the final expression

$$\frac{W_s - W_p}{r\Delta\theta} = 2\omega \, \sin\gamma \, \cos\beta \; - \; W\left(\frac{d\beta}{dm} - \frac{\cos\beta \, \sin\beta}{\rho b}\frac{d\rho b}{dm}\right) \qquad (6.8)$$

The first term on the right-hand side is the loading which would be produced in an impeller with straight blades; in the wholly radial section of such an impeller $\gamma = 90°$ and for no backsweep $\beta = 0°$ so that the velocity difference is simply given by

$$W_s - W_p = 2\omega \, r\Delta\theta.$$

This is the simplest way of estimating the loading and obtaining the linear variation of velocity across the blade passage. The loading from this mechanism is clearly smaller when the flow is partly axial ($\gamma < 90°$) and when the impeller is backswept ($\beta > 0°$ in the sign convention adopted here). The first term in the brackets in equation 6.8 arises from the curvature of the blades and if the blades have been given backsweep it is quite probable that towards the exit the curvature will be increasing such that $d\beta/dm$ is positive (i.e. the suction surface is then concave.) As equation 6.8 shows, the effect of this is to reduce the difference $W_s - W_p$ and unload the blades. In the inducer region of an impeller (or for axial blades) the curvature is in the opposite sense, the suction surface is convex, and it is this term which produces the pressure difference across the blades.

The second term in the brackets of equation 6.8 comes from the acceleration of the mean flow. For this one-dimensional analysis the compressibility is adequately compensated for by the contraction of the passage width b and the combination ρb is appropriate. Having $\partial\rho b/\partial r > 0$ corresponds to an additional decrease in the mean velocity and the effect of this is to raise the loading in the sense that the velocity on the suction surface is increased relative to that on the pressure surface.

The importance of the loading, expressed here as the difference in the velocities on the suction and pressure surfaces at the same radius, is severalfold. Most important is that the two velocities are required to be equal at the trailing edge and if $W_s - W_p$ is large this requires that there is a large decrease in W_s towards the trailing edge with consequent boundary layer separation. In fact separation of some sort seem to be common, so an impeller designed to produce a large value of $W_s - W_p$ may have a larger wake than one designed for a low value. Moreover with large loading the difference between the design intent and the real flow will be even greater.

If the loading is thought of in terms of the pressure difference between the two blades it can be recognized that this is equal to the integral across the streamlines of the accelerations and in the outer part the main component is the Coriolis acceleration which is the product of the relative velocity and angular velocity. The loading, which is the Coriolis acceleration integrated across the blade passage, is therefore proportional to the mass flow and the angular velocity. In other words, the loading calculated assuming idealized fully attached flow will be reasonably accurate even if there is a significant region of separated flow. This was evident in the measurements of Prian and Michel (1951) in which they showed very good agreement between the measured pressure difference across the blades and that calculated by the approximate method of Stanitz and Prian, but very poor agreement in the value of the relative velocities along the heavily separated blade surfaces. (Similar reasoning for the blade loading arising from the blade curvature shows that this depends on the square of the mean velocity and not on the rotation.)

The pioneering work at NACA

In the 1940s and early 1950s the NACA Lewis Flight Propulsion Laboratory was a pioneer in the study of centrifugal compressors, and in particular in the calculation of the flow in them. What they did has affected the way people have thought about radial compressors to the present day. This work is summarized in a thorough manner by Traupel (1962) where a similar theoretical approach is brought to bear on a range of machines, including many of the industrial type. (This book has been largely overlooked in the English-speaking

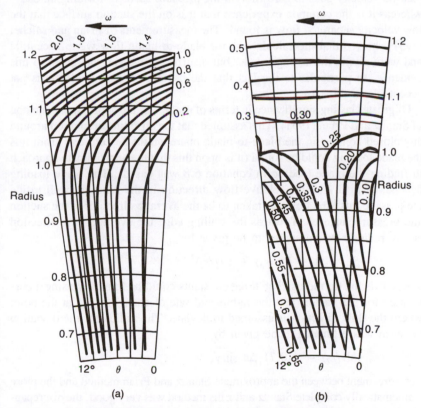

Fig. 6.14 Computed contours in a two-dimensional impeller passage: (a) shows relative streamlines; (b) shows relative Mach number. (From Stanitz, 1952)

world but has many of the ideas, largely theoretical, which have since become widely accepted.)

Figure 6.14 shows the calculated streamlines and contours of relative Mach number for an impeller with straight blades calculated by Stanitz and Ellis (1950) using a stream function method solved by relaxation. The streamlines show how even for this idealized inviscid case there is an overall inclination of the streamlines backwards at the trailing edge, giving rise to the slip velocity: slip, which is very important for estimating the work input, is discussed in a separate section of this chapter. It should also be noticed how the Mach number is low on the pressure side with an increase in relative Mach number to the tip, whilst there are calculated to be high Mach numbers on the suction surface with a very rapid drop to the tip. At the tip, velocities and Mach numbers must be equal to satisfy the Kutta–Joukowski condition of no load at the trailing edge. It is generally impossible to achieve this amount of deceleration on the suction surface; instead there would be a massive separation on the suction surface and this calculated flow is physically unrealizable. If the flow rate through such an impeller were reduced the calculation would predict that the velocity goes to zero first on the pressure surface, forming an eddy, whereas it is the common experience that it is on the suction surface that the low velocity separated flow is found. The measurements of Prian and Michel (1951) showed that the calculations for blade-to-blade flow were inaccurate and were giving the wrong trends, but insufficient weight was given to this evidence, in particular to the fact that the neglect of viscous effects is not acceptable in this case.

Of greater lasting significance in terms of usage was the approximate method of Stanitz and Prian (1951). This assumed that the absolute circulation around any closed contour in the blade-to-blade plane would be zero and from this the blade loading could be found (it is upon this that the approximate approach to finding the blade loading in equation 6.8 was based). The blade loading depends critically on the relative flow direction in the passage; well inside the passage this direction was taken to be the average direction of the suction and pressure surfaces. Towards the trailing edge the average flow direction at any radius r was assumed to be given by

$$\beta_{av} = A + B(r/r_2) + C(r/r_2)^2$$

where r_2 is the tip radius. The three constants could be found by using a correlation for slip factor, and the radius and rate of change of β_{av} at the point where the flow direction is assumed to deviate. The radius for the deviation to set in was estimated to be given by

$$\ln(r/r_2) = -0.71 \, \Delta\theta \, \sin\gamma.$$

The agreement between the approximate Stanitz and Prian method and the more mathematically complete Stanitz and Ellis method was very good, the discrepancies being far smaller than those arising from the real effects, mainly viscous, neglected in the procedures. To the present day the Stanitz and Prian approach continues to be used as a rapid way of getting an assessment of the blade-to-blade loading of an impeller. It is usually used in conjunction with a streamline curvature method for the hub—shroud flow. It is probable that the original treatment for the unloading at the tip can be omitted in favour of some very simple smooth curve so long as it leads to equal velocity on the suction and pressure surfaces at the trailing edge.

The methods of calculation devised for hand calculation are not generally suited to a computer and different methods, such as that known as streamline curvature, soon took over. Design is still almost invariably performed using two-dimensional methods with two intersecting surfaces, the throughflow calculation being performed on a meridional surface from hub to shroud and the blade-to-blade flow on a surface of revolution. For radial compressors it is essential to have a good prediction for the meridional flow, because of the large curvature of the hub and shroud, but a very simple method for the blade-to-blade flow often suffices. The whole topic of calculation in impellers has been reviewed by Adler (1980) but it is of course clear that the principal limitation is the neglect of important fluid dynamic features in the inviscid description of the flow, in particular the tendency to form a large region of separated flow, and no amount of numerical refinement can overcome this.

Since the limitation of the accuracy of prediction with inviscid methods is the model and not the method of solution, a simple throughflow calculation, probably streamline curvature, and a simple blade-to-blade calculation is not significantly worse than more elaborate methods and can give useful guidance over about the first two-thirds of the impeller length (Casey and Roth, 1984). With this simple approach it is possible

1. To predict the flow direction along the leading edge and choose the impeller inlet angle distribution.
2. To find the blade loading and velocity distribution in the inlet region and for about two-thirds of the way along the impeller. The velocities are normally highest in the inlet region and since loss is roughly proportional to the cube of the velocity, see Section 1.5, the inlet region is the most important one for loss generation.
3. To find the hub-to-shroud velocity gradients from upstream to about two-thirds of the way along the passage.

What it is not possible to do with inviscid calculations is

1. To calculate the velocity distribution or loading in the outer part of the impeller.
2. To calculate the losses without large empirical input in most regions.
3. To obtain rational limits on loading based on the prediction of the flow behaviour.

Inadequacies in the calculations in centrifugal impellers have been known for some time. When Dean (1971) set down his ideas on the nature of flow in centrifugal compressors and the importance of the jet-wake type of flow

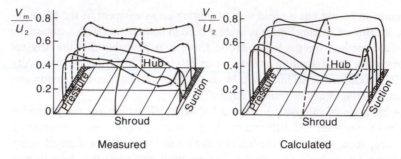

Fig. 6.15 Comparison of measured meridional velocity at outlet from an impeller with calculations by Dawes using a Navier–Stokes solver. Measurements by Eckardt (1978) on impeller B designed by Sulzer with 40° of backsweep. (Operating point 14 000 rev/min, 3.30 kg/s, close to surge line)

he wrote with great clarity as follows: 'Despite inklings of these difficulties, most designers and even research workers still fall back on the sophisticated potential flow prediction techniques. Given the insufficient understanding of internal flows of impellers, this tactic is understandable, but not necessarily reasonable.'

Nowadays full viscous transonic flow calculations can be accomplished. One of the first was reported by Moore and Moore (1980) and since then development has continued. Figure 6.15, produced by Dawes (1987b), shows the calculated and measured flow near the outlet plane of an impeller with 40° backsweep and no inducer, designed by Sulzer and tested by Eckardt (1978). The operating condition was one close to the surge line for which the agreement between measurement and calculation was closest. Although there are differences between the two there is much in common, including the low-velocity region near the suction surface–shroud corner. Dawes' calculation was performed using a very simple mixing length model for the turbulent shear stress and no attempt was made to include the suppression of turbulence on the suction surface and shroud. This leads to the tentative conclusion that these effects are less important in the complicated flow in impellers than has been believed. In other words the details of the shear stress are relatively unimportant and it is the gross features which the calculation method is able to address which also dominate the real flow.

It has not yet been worked out how such three-dimensional viscous methods should be used. It seems beyond most people to think in terms of three dimensions and presentation is inevitably normally going to be two-dimensional. What seems probable is that the design will continue to be largely two-dimensional but those with access to three-dimensinal viscous methods will check their designs and their ideas with them. What is certain is that the availability of three-dimensional viscous calculations has *not* brought the immediate understanding of the flow which seemed possible before they existed.

6.4 Slip and the estimation of slip factor

Slip factor is the means used to estimate the work input in radial compressors, as discussed in Section 2.3, and is one of the most important empirical inputs. In the most primitive assessment of impeller performance the flow would leave the impeller in the same direction as the blades are pointing. This does not happen, but instead there is a component of relative velocity in the opposite direction to the rotation that is known as the slip velocity or just the slip. In a hypothetical impeller with an infinite number of thin blades the flow would indeed leave in the blade outlet direction and the ideal relative flow direction would be

$$\beta_2 = \chi_2$$

For this idealized case the relative whirl velocity leaving the impeller can be written

$$W_{\theta 2i} = V_{R2} \tan \chi_2$$

where V_{R2} is the radial velocity at outlet, assumed uniform. The corresponding absolute whirl velocity is

$$V_{\theta 2i} = U_2 - V_{R2} \tan \chi_2.$$

The slip velocity is defined as the difference between the idealized and actual whirl velocities (the idealized flow being in the direction of the actual impeller blades at outlet) by

$$V_s = W_{\theta 2} - W_{\theta 2i} = V_{\theta 2i} - V_{\theta 2}$$

The slip factor is the ratio of the actual absolute whirl to the idealized case

$$\sigma = V_{\theta 2}/V_{\theta 2i} = (V_{\theta 2i} - V_s)/V_{\theta 2i}.$$

which can also be written in terms of the *absolute* outlet flow angles

$$\sigma = \tan \alpha_2 / \tan \alpha_{2i} = \tan \alpha_2 / \{U/V_{R2} - \tan \chi_2\}. \tag{6.9}$$

The simplest case to consider is when the blades are radial at outlet. In the ideal case the relative flow would have no whirl component at outlet and $V_{\theta 2i} = U_2$, but in practice there is slip so that $V_{\theta 2} < U_2$. There are other ways of defining slip factor: Wiesner (1967), for example, defines it by $\sigma = 1 - V_s/U_2$. For impellers with no backsweep this reverts to the earlier definition of equation 6.9.

There still exists considerable confusion as the cause of the slip factor, but the explanation is quite simple. It is analogous to the deviation in axial blades in the sense that it arises from the blade loading being required to diminish gradually to zero at the trailing edge; if the loading falls off there is not the means to impose the acceleration on the flow to make it follow the blade direction. To be more specific, for the radial impeller the rotation and flow together produce the loading of the blades so that the pressure is higher on the leading side of the blade (the pressure surface) and lower on the trailing (or suction) side. At the trailing edge the pressure difference must be zero, in order to satisfy the Kutta—Joukowsky condition, and the reduction in pressure difference has to be gradual, beginning some way upstream. Because the pressure difference across the passage must be reduced near the trailing edge it follows that there is no longer enough force to make the flow follow the blades. The average flow is therefore turned back to give a slip velocity in the opposite sense to the impeller rotation. The distance upstream from the trailing edge which the unloading affects is intuitively related to the passage width at outlet and there is correspondingly a large influence of blade number in the magnitude of the slip velocity.

A different but equivalent way to explain slip is to consider the relative vorticity $\Omega = -2\omega$. Inside the passage, well away from the blade ends, there can only be very small components of relative velocity normal to the blade surface. The relative vorticity is therefore generated by the gradients in the velocity parallel to the blades and, if this is assumed uniform,

$$\Omega = (W_p - W_s)/r\Delta\theta = -2\omega.$$

At the trailing edge the velocity on the suction and pressure surfaces must be equal and the vorticity can no longer arise from their difference. Instead the vorticity is set up by radial gradients in the relative whirl velocity, the component known as slip velocity.

The loading of axial blades is produced by turning the flow, usually by the camber. Deviation occurs primarily because of inviscid effects when the flow cannot follow the blade direction because the loading decreases towards the trailing edge. The deviation often increases very rapidly when a blade stalls because the means for applying the acceleration to the flow is sharply reduced. In the centrifugal impeller the loading comes primarily from the Coriolis acceleration and, integrated across the circumferential width of the passage, this is proportional to the mass flow and angular velocity. The loading of the radial impeller does not drop catastrophically when the flow becomes strongly separated and accordingly the slip factor does not alter much with flow rate.

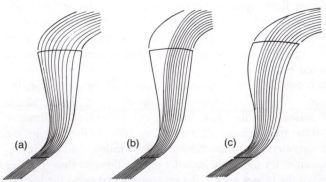

Fig. 6.16 Computed streamline patterns in an impeller for inviscid flow: (a) fully attached flow, no backsweep; (b) separated flow, no backsweep; (c) separated flow, with backsweep. (From Sturge and Cumpsty, 1975)

Low pressure on the suction surface implies high velocity and it is a consequence of the unloading towards the tip that flow near the suction surface must decelerate. The deceleration on the suction surface may cause the boundary layer to separate and this causes the flow to be deflected, in the sense opposite to the rotation, by the blockage of the separated region. However this is not the origin of most of the slip. Figure 6.16, taken from Sturge and Cumpsty (1975) shows the computed streamline pattern for a fully attached flow in an impeller without backsweep, a separated flow in the same impeller and a separated flow in a backswept impeller. The flow is highly idealized, with lossless flow in the unseparated region and zero flow in the wake, but it does show the formation of slip in all three cases. Because slip is fundamentally an inviscid flow effect it is calculable from the inviscid methods for blade-to-blade flow such as was carried out by Stanitz. Presumably because of the three-dimensional nature of the separated flow, the results of two-dimensional calculations for slip are less reliable than the correlations. It is possible that the recent three-dimensional viscous calculation methods will be able to achieve really accurate predictions of slip but they have not yet been applied sufficiently widely for their accuracy to be established.

Wiesner (1967) carried out an exhaustive examination of the existing correlations or expressions for slip in terms of the available data. He looked at methods due to Stodola (1927), Busemann (1928) and Stanitz (1952). Organizations designing impellers will have evolved their own methods for estimating slip factor and there are methods in the literature apart from those investigated by Wiesner. (Wiesner, it will be recalled, defined slip factor by $\sigma = 1 - V_s/U_2$.) The results of Busemann were approximated by Wiesner by the expression

$$\sigma = 1 - \frac{\sqrt{\cos\chi_2}}{N^{0.7}} \tag{6.10}$$

where N is the number of blades and χ_2 is the blade outlet angle to the radial. This simple expression was also compared with the measurements and with the other three methods.

The method of Stodola comes from considering the relative circulation of the flow at outlet from the impeller. It is outlined here because it shows very clearly in an approximate but quantitative way the formation of slip in an inviscid flow. Taking the inlet flow to the impeller to be irrotational and the flow inside to be inviscid, then at outlet (when the flow has no radial component) the vorticity $\Omega = -2\omega$ where ω is the impeller angular velocity. The width of the blade passage at outlet is approximately given by

$$h = (2\pi r_2/N)\cos\chi_2.$$

The slip velocity is related to the peripheral velocity v_p about a loosely defined eddy of diameter h and mean vorticity Ω,

i.e. $\qquad v_p\pi h = \Omega\pi h^2/4 \qquad$ or $\qquad v_p = \Omega h/4.$

Inside the blade passage the velocity normal to the blades is negligible so the peripheral velocity across the eddy is the slip velocity, V_s, i.e.

$$V_s = v_p = \Omega h/4 = \pi \omega r_2 \cos\chi_2/N.$$

Substituting this into the expression for slip factor, using the definition of equation 6.9, gives the Stodola expression

$$\sigma = 1 - \frac{V_s}{U_2 - V_{R2}\tan\chi_2} = 1 - \frac{(\pi/N)\cos\chi_2}{1 - (V_{R2}/U_2)\tan\chi_2} \tag{6.11}$$

or $\sigma = 1 - (\pi/N)\cos\chi_2$ by Wiesner's definition.

This form has found widest acceptance when the backsweep is large, for example greater than 60°.

The Busemann slip factors were calculated for log spiral impellers and the results actually used by Wiesner were taken from the diagrams given by Wislicenus (1947). For this the outlet to inlet radius ratio must not exceed

$$\frac{r_2}{r_1} \approx \exp\left\{\frac{8.16\cos\chi_2}{N}\right\}$$

The Stanitz form for slip factor was derived from numerical calculations for six radial bladed impellers and was found to be of the form

$$\sigma = 1 - \frac{1.98}{N} \tag{6.12}$$

The range of validity for this was for backsweep not exceeding 45° and at least eight blades. Stanitz found only a weak influence of compressibility on slip.

The Stanitz form for slip factor shows no dependence on backsweep; the Stodola form shows the slip factor to decrease slowly with backsweep ($\sigma = 0.88$ and 0.86 for $\chi_2 = 0°$ and $60°$ respectively in a 26-bladed impeller); the Wiesner form, based on Busemann's calculations, shows a rise in slip factor with backsweep ($\sigma = 0.90$ and 0.93 for $\chi_2 = 0°$ and $60°$ respectively for a 26-bladed impeller). Comparing all four methods and expressions with the measured values available Wiesner found that the simple expression derived from the curves of the Busemann calculations, equation 6.11, gave the most satisfactory agreement and in the absence of anything better this is probably the best expression to use.

The actual value of slip factor will depend on many features of the flow and the geometry. Dean (1971) showed from the measurements by Sakai *et al.* (1967) that none of the methods begin to predict the trend in slip with flow rate, which is not surprising since the essential physical behaviour has been largely omitted. Eckard (1977) showed the variation of slip factor with flow rate for three impellers. The impeller with no backsweep, that used for Fig. 6.2, gave an approximately linear *increase* in slip factor as the flow rate was reduced from choke to surge from about 0.86 to 0.885. The impeller with an inducer and 30° backsweep, used for Fig. 6.3, gave a linear *decrease* from about 0.97 down to 0.93. An industrial type impeller with no inducer and 40° of backsweep showed an approximately linear *increase* from 0.825 to 0.85

as flow rate was reduced, over most of the range. The corresponding values from Wiesner's expression, equation 6.9, are 0.88, 0.89 and 0.89 respectively. Very clearly the estimates for slip factor must be taken with caution.

The slip factor somewhat disguises the true level of ignorance in predicting the flow. The discrepancies would look much larger if the true measure of the flow divergence from the blades, the slip velocity, were used. Taking for example a 26-bladed impeller with no backsweep the, slip factors predicted by Stodola, Wiesner and Stanitz are 0.88, 0.90 and 0.92 respectively, a spread of only 4 per cent. The corresponding spread of slip velocity is 20 per cent of the impeller tip speed. Figure 6.17a shows a comparison taken from Dean (1971) of Wiesner's expression for slip factor, equation 6.9, with data; most of the measurements lie within the ± 5 per cent band. When the true parameter for describing the slip velocity, which is $1 - \sigma$, is used in Fig. 6.17b a more

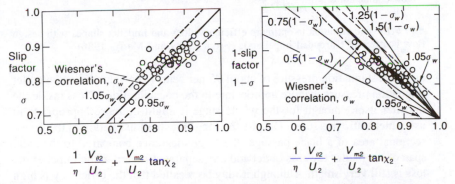

Fig. 6.17 A comparison of the Wiesner correlation for slip, σ_w, with measurements and similar comparison of $1 - \sigma_w$ (i.e. the slip velocity $\div$ the slip impeller tip speed). (From Dean, 1971)

reasonable estimate would be that most of the data lies within ± 25 per cent of the correlation. All the correlations were assessed at the maximum efficiency point. Fortunately it is the slip factor which is required in the prediction of the work input and the unsatisfactoriness of the estimate for slip velocity is therefore hidden.

6.5   Loss in impellers

The losses in good impellers cannot be very large because the efficiencies obtained near the optimum specific speed, which means when the duty is such that the geometry can be favourable, are normally very high. Figure 1.9 taken from Rodgers (1980) shows that for unshrouded impellers with an inducer, isentropic efficiencies of around 93 or 94 per cent are attainable, with the backswept impellers being about 2 per cent higher than those with radial ending blades. Krain's impeller achieved an impeller polytropic efficiency based

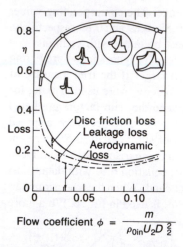

Fig. 6.18 The variation in optimum efficiency, loss and impeller shapes with design flow coefficient (shrouded impellers). (From Casey and Marty, 1986)

on outlet stagnation pressure of about 95 per cent. (The value based on outlet stagnation pressure is more appropriate to the consideration of loss inside the impeller itself, though for the whole stage it might be more appropriate to use outlet static pressure.) Despite the long flow path in relation to the cross-sectional area of a blade passage, the large clearance compared to the axial span of the blades towards outlet and any separated region in the impeller the loss is still very small. Although it may be recalled that the efficiency is high partly because there is a large contribution to the pressure rise from the centrifugal effect which is not subject to loss, it is nevertheless clear that the losses are not overwhelming.

Figure 6.18 shows the efficiency and estimated contributions to the loss for shrouded impellers, taken from Casey and Marty (1986). These are shown as functions of the design flow coefficient defined here by $\phi = m/(\rho_{0in}U_2D_2^2)$, where ρ_{0in} is the stagnation density at inlet and U_2 and D_2 are the impeller tip speed and outer diameter respectively. The efficiencies are generally lower than those quoted above for unshrouded impellers, but this is largely because these are stage efficiencies which are much lower than those of the impeller. (Krain's impeller, which has been referred to above, operating with a vaneless diffuser gave a *stage* total-to-total efficiency of about 84 per cent, to be compared with the *impeller* efficiency of 95 per cent.) At the very small design flow coefficients, corresponding to low values of specific speed, the aerodynamic losses do rise somewhat, but the overwhelming cause of lower efficiency is the disc friction and leakage loss.

Rodgers (1980) estimated the losses in a number of impellers, all with inducers and several with high inlet relative Mach numbers, using a correlation of pressure loss for pipes with a 90° bend, the diameter and length of the equivalent pipe having been chosen to represent the mean hydraulic diameter and length of the blade passage. He found that the viscous effects associated with turning the flow in these relatively long passages accounted for most of the loss, with only a small additional amount attributable to what Rodgers grouped as diffusion. The loss from the clearance flow was presumably included in the diffusion loss. It is normal for separation to occur in pipe bends and if the correlations were complete this loss source would have been allowed for.

Senoo and Ishida (1987) have examined the effect of tip clearance on loss experimentally and have re-examined the methods due to Eckert and Schnell (1980), Pfleiderer (1961) and Pampreen (1973). The effect is conventionally expressed for impellers as a loss in efficiency $\Delta\eta/\eta$. Pampreen used measurements at a range of clearances on six different compressors and found that the average decrement in efficiency was given by

$$\Delta\eta/\eta \approx 0.3t/b_2$$

where t is the axial clearance and b_2 is the blade height at the impeller outlet. At small clearance ratios there was a range of about ± 50 per cent in the measured values of $\Delta\eta/\eta$ about the correlation. Furthermore individual compressors did not follow this trend as the clearance was varied and an individual compressor was likely to show a more rapid increase in efficiency decrement with clearance.

Both Eckert and Schnell and Pfleiderer expressed the loss in the form

$$\frac{\Delta\eta}{\eta} = \frac{2at}{b_1 + b_2}$$

where b_1 and b_2 are the blade width (i.e. span) at inlet and outlet and t is the axial clearance at outlet. Eckert and Schnell recommend $a = 0.9$ whilst Pfleiderer recommends $a = 1.5-3.0$ Pampreen's expression agrees approximately with the Eckert and Schnell form for $b_1/b_2 = 4$ and $\eta = 0.8$.

Senoo and Ishida (1987) developed a simple model for the tip loss, considering momentum normal to the blades and Bernoulli's equation for the pressure difference. Remarkably good predictions of the loss in pressure ratio and loss were produced for a number of compressors. The decrement in efficiency was found to be almost proportional to the ratio of clearance-to-blade height at the impeller outlet (provided this ratio was less than 0.1) and the trend seems to be about

$$\Delta\eta/\eta \approx (t/b_2)/4$$

i.e. 1 per cent loss in efficiency for every 4 per cent tip clearance-to-height ratio. The loss in efficiency was found to be somewhat smaller as the flow rate was reduced. They also found that for the same clearance-to-height ratio at the outlet the high pressure ratio compressor tends to have smaller loss because at higher pressure ratios the inlet blade height is larger. The reduction in efficiency due to tip clearance for a high pressure ratio machine was smaller when operated at reduced speed.

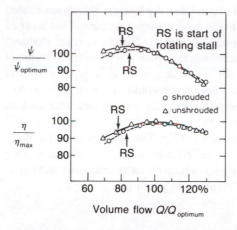

Fig. 6.19 The measured variation in loading and efficiency with flow rate for shrouded and unshrouded impellers. Optimum condition corresponds to the flow rate for maximum efficiency. (From Harada, 1985)

Although there is an increase in loss and a reduction in efficiency for unshrouded impellers as the clearance between the blades and stationary shroud increases, the efficiency may nevertheless be higher with an unshrouded impeller and small clearance than one with a shroud. This was demonstrated by Harada (1985), Fig. 6.19, and the explanation seems to be that a *small* clearance flow alters the main flow, particularly in the suction surface/shroud corner in such a way that the overall flow is improved. (It has also been observed with axial compressor rotors that optimum performance is with a small tip clearance and not with the rotor tips sealed.)

For multistage configurations axial location is such a problem that obtaining adequate control of clearance for unshrouded operation is virtually impossible and impellers for such applications are normally shrouded. The control of the ratio of clearance to blade height is also very difficult in small compressors, particularly for the automotive turbocharger where the requirement for long life at very high rotational speeds forces the design away from rolling contact bearings. Because the turbocharger must normally give a large pressure rise the tip speed is too high for a rotating shroud to be possible and such compressors probably lose several per cent in efficiency from tip clearance loss. For larger machines the ratio of axial clearance to blade height is likely to be much smaller; for example Krain's impeller was tested with the clearance of only about 1.3 per cent of the local blade height at exit.

In examining the source of loss in centrifugal compressors Moore *et al.* (1984) used the three-dimensional viscous calculation of Moore and Moore (1980) to examine the flow in a high-speed impeller with backswept blades. The design pressure ratio was 3.4:1 and the mass flow rate 1.0 kg/s. Because of geometric constraints, mainly the large hub diameter, this was not an optimum geometric configuration, but nevertheless the polytropic efficiency for the impeller was calculated to be 91 per cent. The impeller was much less efficient when viewed as a diffuser in the rotating frame of reference, the efficiency for this being only 60 per cent. The reason that this did not lead to low overall efficiency is that only 22 per cent of the static temperature rise in the impeller was produced by decelerating the relative flow and 76 per cent by the centrifugal effect. These results provide an example of an increase in rothalpy due to work input to the flow in the relative frame by friction: this gave a static temperature rise equal to about 2 per cent of the overall.

The calculations by Moore *et al.* showed that most of the loss came from quite localized regions. Loss production was high over most of the shroud. On the suction surface it was high at the end of the inducer, near the corner with the shroud and near the trailing edge of the blade. On the pressure surface loss generation was high only near the shroud corner. Loss generation was high on the hub wall downstream of the impeller where the highly skewed flow met the stationary wall. The calculations showed that 12 per cent of the inlet flow leaks through the clearance and the region of highest loss generation per unit volume was in the clearance gap; the total of this was calculated to be equivalent to 2 per cent of efficiency. (In the calculations the clearance was assumed to have varied linearly from the measured values of 1 per cent of blade height at inlet to 3.3 per cent at outlet.)

The calculations by Moore *et al.* (1984) cannot be taken to provide absolute quantitative estimates of the loss because comparisons with readily measured quantities were not given. Nevertheless they do show very clearly several aspects of the loss production. The clearance flow is a region of high entropy generation, though clearance flow may lead to improvements in the overall flow pattern. As anticipated from the shape of the flow passages the deceleration process is very inefficient. In most measurements of efficiency the inefficient deceleration process is hidden by the large pressure rise from the centrifugal effect. This pressure rise, though not capable of loss, is bought at the unavoidable price of high absolute exit velocity; as a result the inefficiency of the downstream diffuser is where most of the stage loss occurs (viz. Krain's impeller efficiency was 95 per cent compared with a stage efficiency of 84 per cent). Although the entropy generation per unit volume which Moore *et al.* calculated to occur inside the clearance gap is intense, the volume involved is small whereas the volume comprising the boundary layer over the shroud is, by comparison, large. With an unshrouded impeller the relative flow in the blade passage is moved past the stationary shroud with consequent high shear.

The mechanisms of loss generation inside the impeller are not really understood. It is not known, for example, whether the presence of a large wake is really detrimental, nor whether its position affects loss. It is known that when a flow is non-uniform, in the manner of the jet and wake, the loss will be created as the flow mixes out to uniform. In many cases this happens downstream of the impeller, so it is not accounted for in the measurement of downstream pressure and impeller loss (though it may make the measurement of pressure inaccurate).

Shrouded impellers without inducers usually have severe curvature on the shroud and there is the possibility of a massive separation. The mixing which takes place inside the impeller to return the flow to more nearly uniform is then likely to be a significant source of loss. It is also true that a shroud separation is likely to put the leading edge of blades of an impeller with no inducer at very large incidence, further complicating the flow and introducing more possible sources of loss. There is a great shortage of good measurements in shrouded impellers, at least in the open literature.

This above discussion begs the question as to what makes a good impeller; it is perhaps easier to say what will make a bad one. It does seem probable that small radius of curvature, particularly on the shroud, leads to an early separation and therefore a separated region which is unnecessarily large. Since, however, there have been many impellers with acceptable performance with very tight curvature on the shroud and large amounts of blade curvature (camber) this does not seem to be a complete explanation for the occasional catastrophe. Very large clearance gives a marked degradation of performance, but this is quantified, if only approximately.

What seem the most likely cause of occasional disastrous performance in the impeller are effects related to compressibility. The combination of shroud curvature, which increases the meridional component of velocity, and the blade curvature, can lead to high Mach numbers in the inlet region to the impeller. Blades with supersonic inlet flow have only a narrow range of incidence for which they perform adequately: too high incidence and a shock strong enough to separate the suction surface boundary layer is formed; too low incidence and the blade passage chokes. The effect of compressibility can also be felt in a different way at the impeller inlet due to changes or design errors at the rear. With a vaned diffuser the critical quantity is the cross-sectional area of the diffuser throat which must be specified on the basis of the estimated density rise in the impeller. If the density rise is smaller than that on which the design is predicated the diffuser is unable to pass the expected mass flow and the incidence into the impeller becomes unacceptably high. The seriousness of the high inlet incidence increases as the inlet relative Mach number rises so that high performance compressors are at risk in two connected ways. Choking of the diffuser gives a risk of high incidence into the inducer; high inducer incidence can lead to large loss creation in the impeller and consequently further reduction in the mass flow passed by the diffuser.

6.6 Design choices for the impeller

The impeller as a whole

The procedure used by most organizations to design impellers is shrouded in commercial secrecy. It is possible that some companies are designing very satisfactory impellers without properly understanding why or how. As long as the new designs are able to follow the previous ones without too radical changes the situation may be satisfactory. The design methods adopted may

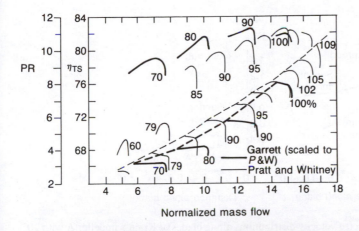

Fig. 6.20 A comparison of a Garrett and a Pratt and Whitney of Canada compressor designed for similar duty. The Garrett mass flow was scaled to the P&WC size. Total-to-static pressure ratio and efficiency are shown. (From Kenny, 1984)

have very different intermediate objectives, i.e. what distributions of area, curvature, velocity or pressure will lead to good performance. Very often inviscid calculation methods have been used to assess the performance of impeller blade shapes. As an illustration of the lack of agreement in design philosophy it is convenient to reproduce in Fig. 6.20 the pictures of two impellers designed by different companies for a similar duty, together with their corresponding performance maps for the entire compressors, taken from Kenny (1984). The sizes were different and for this comparison the results from the Garrett compressor were scaled. The P&WC compressor used a very efficient pipe diffuser, discussed in Chapter 7, whereas the Garrett compressor used a very simple cascade diffuser. Notwithstanding the very different impeller shapes the performance is similar at the design point. It can only be concluded that within certain limits a wide range of solutions to the design are acceptable.

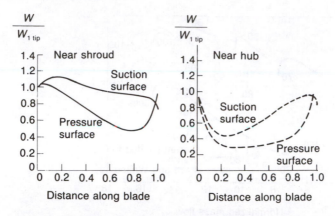

Fig. 6.21 Predicted velocity distributions along blades of Krain's impeller A with 20° of wrap and 30° of backsweep. (From Krain, 1985)

It was quite clear since the work of Dean in the early 1970s, but certainly since measurements by Eckardt, that the inviscid procedures are incapable of predicting the flow throughout the impeller because of the separated regions. Many designers have refused to believe that their exemplary designs would have separations; some may have been much better than that tested by Eckardt, but even the impeller tested by Krain showed some evidence of separation on the shroud. It should not be overlooked, however, that Eckardt's radial bladed impeller had a peak polytropic efficiency of over 90 per cent at the rotational speed of the laser measurements. In other words high efficiency is not incompatible with gross separation. Even were the separations to be suppressed by reducing the area at outlet, still the long narrow passage would introduce large viscous effects and possibly large effects from the tip clearance.

The traditional way of using calculation methods to design impellers has been to carry out a hub−shroud calculation, probably using a streamline curvature method, to find the meridional flow assuming that the flow is axisymmetric. The results of this are then used as input to blade-to-blade calculations, usually on hub and shroud surfaces but possibly on more. The hub and shroud line as well as the blade cambers can then be adjusted to optimize the velocity or Mach number or pressure distributions according to some empirically formulated rules. An example of this performed relatively recently by Krain (1985) is shown in Fig. 6.21. The impeller had 30° of backsweep and for the case shown the blades were arranged to have a wrap angle of 20°. (The wrap angle is that viewed along the rotational axis between the leading edge and trailing edge of a blade.) This wrap is fairly conventional and later designs are using much larger values; it is a desirable consequence of this that the blades along the shroud contour appear to have very little camber. The results in Fig. 6.21 show that this design has succeeded in having remarkably little deceleration of the flow on the suction surface towards the tip, even for the surface near the shroud. The pressure surface near the hub shows a large initial deceleration but, given the thin boundary layer near the leading edge, this may be acceptable.

(It may be remarked that if a massive separation can be avoided, as seems the case for the impeller designed by Krain, Fig. 6.6, the calculations have meaning right out to the trailing edge.)

It is a feature of some calculations in the blade-to-blade surface that the velocities in the inlet region are set equal or nearly equal on the suction and pressure surfaces; in fact this is unlikely. For subsonic inlet relative flow the optimum will almost certainly be achieved with some positive incidence and therefore an abrupt, large pressure difference; at supersonic relative inlet flow the leading edge shock structure puts high pressure on the pressure surface with the upstream static pressure (or even a lower pressure) on the suction surface in the leading edge region. So far as the blades are concerned, deceleration is likely to be tolerated without separation much better in the inducer/inlet region than further back in the impeller because the boundary layers are thin.

Many calculation schemes use the streamline curvature throughflow method to calculate the meridional flow and use an approximation such as Stanitz and Prian (1951) for the blade-to-blade flow to find the difference in velocity on the suction and pressure surfaces. Half the difference in velocity is added to and subtracted from the mean flow found from the throughflow calculation to give the velocity on the suction surface and pressure surfaces respectively. The loading is usually assumed to disappear at the trailing edge, which is a reasonable approximation to what happens. The simple approximation for the velocity difference between the two blade surfaces is probably adequate, given that the flow is often separated, and the true justification of method is as a way of getting some idea of the loading in relation to other designs.

Since many impellers have had significant regions of separated flow in them it is not really very clear what some inviscid calculations have achieved for the designers beyond allowing them to align the blades with the flow at the leading edge. There is one important exception to this where the usefulness is clear. All calculation methods, but streamline curvature methods in particular, are very sensitive to wall curvature. What the inviscid flow calculations have provided is an indirect route to achieving the maximum radii of curvature and the most smoothly varying curvature consistent with the overall layout of the machine. Wall curvature, and shroud curvature in particular, have a very large effect on the flow. Clearly the overall specification of the impeller geometry, in particular the inlet and outlet radii and the axial length, put very definite limits on what can be achieved.

Long ago it was realized that the overall specification of the impeller layout has a powerful effect on the performance, as is shown by Fig. 6.22 taken from Johnsen and Ginsburg (1953). The worst blade profile, the circular arc, has markedly lower efficiency but larger mass flow capacity. The best impeller is that with the parabolic blade curvature, and this looks remarkably like many modern impellers where the blades have lean at the trailing edge. ('Lean' is one of the terms given to the inclination of the blade trailing edge away from the axial direction in a view taken radially inward. There are many other names for it.)

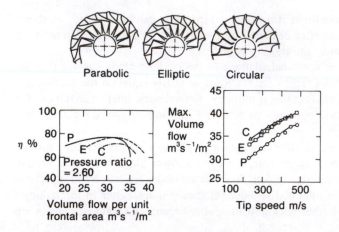

Fig. 6.22 The effect of blade shape on the efficiency versus flow rate at constant pressure ratio and on the maximum volume flow rate versus speed. (From Johnsen and Ginsbrug, 1953)

Leaning the blades creates backsweep with blades which retain purely radial fibres. The significance of this is that blades with radial fibres or generators experience no bending moments as a result of centrifugal loading. Blades which are backswept in the normal way have fibres which are not radial and there is a considerable increase in stress as a result. The contribution of lean to backsweep is explained by Casey and Roth (1984). The necessary definitions of the angles with which the blade surface intersects the planes is shown in Fig. 6.23: λ_R is the angle by which the fibres depart from the radial direction, λ_x is the local blade lean and γ is the inclination of the meridional streamline to the axial direction. Casey and Roth show that

$$\tan\chi = \sin\gamma \, \tan\lambda_R + \cos\gamma \, \tan\lambda_x$$

where χ is the local backsweep, the angle between the blade direction and the radial direction in a view directly along the compressor axis. At the impeller outlet $\chi = \chi_2$, and for an impeller which is fully turned to radial in the meridional sense, that is $\gamma_2 = 90°$, the backsweep is given by $\chi_2 = \lambda_R$. For many impellers the outlet meridional inclination γ_2 varies from hub to shroud but it is common for the hub surface to be virtually radial, that is $\gamma_2 = 90°$. The shroud surface may have considerable inclination to the radial, partly

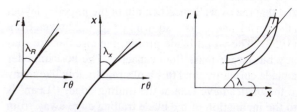

Fig. 6.23 The angles made by the blade surface in different planes

because this allows a larger radius of curvature on the shroud and also because it tends to give a reduced outlet area. Taken with blade lean this inclination of the shroud surface gives some backsweep. As a real example consider the impeller of an automotive turbocharger with blades which have radial fibres, i.e. $\lambda_R = 0$, but lean at outlet $\lambda_x = 40°$. At outlet the hub wall is radial but the shroud is inclined at $\gamma = 70°$ to the axial. As a result there is no backsweep at the hub but $\chi_2 = 16°$ on the shroud and so the overall effect will be a backsweep somewhere between.

As a means for assessing the loading of an impeller Morris and Kenny (1971) outlined a method of impeller design by analogy with a curved diffusing pipe and this has been adopted fairly widely. They proposed that $\Delta p/2q$ should not be allowed to exceed about 0.7 where Δp is the pressure difference between blades at the same radius and q is the mean relative dynamic pressure at that same radius. (In purely radial parts of an impeller with no backsweep it can be shown that $\Delta p = 2\pi r(2\omega W)/N$.) An example of the predicted velocity distribution along the shroud is shown in Fig. 6.24 taken from Morris and Kenny. The blade-to-blade loading has been assumed linear about the mean velocity distribution along the shroud and a guess has been made for the rate and position at which this decreases at the trailing edge. Splitters have been added to reduce the loading in the downstream parts of the impeller and prevent it from exceeding the loading criterion by very much. In reality the flow in such an impeller would have regions of separation in the downstream parts of the impeller as well as secondary flow and tip clearance effects. The significance of a diagram such as Fig. 6.24 and the loading criterion is that they provide a way of comparing designs. The most powerful design tool seems to be experienced with the design of previous machines and plots such as these provide a means of relating one design to another so as to abstract the salient differences.

The most common guide to impeller design is a diffusion ratio of some sort. This may be in terms of the mean relative velocities $W_2/W_1 = \rho_1 A_1/\rho_2 A_2$, effectively treating the flow as one-dimensional, which allows some ideas from

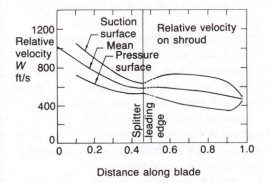

Fig. 6.24 Predicted velocity along the blades close to the shroud showing the effect of the splitter in reducing the loading. (From Morris and Kenny, 1971)

ordinary straight diffusers to be carried over. A typical realistic upper limit on diffusion is $W_2/W_1 \approx 0.6$. It is known that with incompressible flow a major separation is likely for stationary diffusers of area ratio above about 2, and Rothe and Johnston (1976) showed this to be much lower for rotating diffusers. (When there are significant changes in density it is the density–area ratio which matters.) As noted earlier in this chapter many impellers have been designed such that $\rho_1 A_1/\rho_2 A_2$ is significantly greater than 2.

Rodgers (1977) adapted the diffusion factor approach used in axial blade rows. In a discussion to that paper another approach to the diffusion ratio was presented by Young which avoids the one-dimensional assumption at inlet. In this method the ratio MR_2 of relative Mach number at the inducer tip to the average value at the impeller outlet is considered, the ratio being derived from measurement of static pressure at impeller tip. This is compared with an ideal ratio of Mach numbers MR_{2i}, for which the outlet value is calculated assuming that the outlet flow is uniform and in the direction of the blades (i.e. no slip) and the flow through the impeller is treated as isentropic. In fact the flow does not satisfy these assumptions and this has been investigated in great detail by Japikse (1987). Figure 6.25 shows comparisons of the Mach number ratios MR_2 and MR_{2i} given by Japikse for three impellers tested and reported on by Eckardt (1977). (Impeller O is the radial outlet impeller with inducer for which results were shown in Fig. 6.2, impeller A is the backswept impeller with inducer, see Fig. 6.3, and impeller B is an industrial backswept impeller with no inducer.) The region in Fig. 6.25 between the two broken lines is described by Japikse as the reference band and good impellers with inlet relative Mach numbers in the range 0.8 to 1.0 are expected to lie in this. Much of the Eckardt data exceeds this band and it must be concluded that, notwithstanding the large regions of separation found in them, these were indeed good impellers compared to others tested. An upper level of MR_2 of about 1.6 seems to be a reasonable expectation, even though the ideal value MR_{2i} is much larger, exceeding three in some cases. The reference band gives an estimate for the maximum deceleration likely to be achieved but it also provides a convenient way of estimating the value probable for a given geometry, i.e. for a specified MR_{2i}.

Japikse (1987), in this same paper, examined the methods available and the problems involved in deducing useful parameters related to compressor performance which are not directly measurable from the incomplete information available from the measurements. To do this some model of the flow is necessary and the problem is exacerbated when the flow contains large regions of separation. In evaluating the actual ratio MR_2 in Fig. 6.25 the outlet velocity inferred from the static pressure was that of the jet. In related papers Japikse and Osborne (1986) discuss the design of eight compressors with greater detail and openness than is usual from commercial organizations.

It is apparent from Fig. 6.25 that the actual values of MR_2 achieved are lower for the impeller with no inducer, impeller B, than for the other two. The impeller in question was without a shroud and was designed and tested

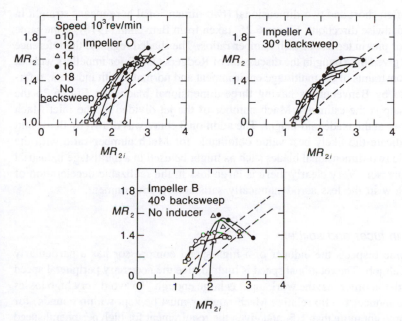

Fig. 6.25 Measured and ideal Mach number ratios MR_2 for three impellers tested by Eckardt (1977). Region between broken lines is the reference band. (From Japikse, 1987)

with the flow entering axially. Impeller B had blades which may be characterized as being of three-dimensional design, that is they were not straight in the spanwise direction. In many industrial compressors the conditions are much less favourable; the flow enters radially and is subjected to strong inlet curvature in the meridional plane, the inlet flow is often distorted and the blades

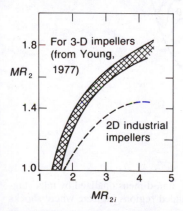

Fig. 6.26 A comparison of the measured and ideal Mach number ratios MR_2 for impellers with inducers (3-D impellers) and industrial impellers. (From Benvenuti, 1978)

are often short and two-dimensional (two-dimensional here means straight in the spanwise direction). Figure 6.26, taken from Benvenuti (1978), is the same type of plot in terms of Mach number ratios. The shaded region is the reference band given by Young in the discussion of Rodgers (1977) for impellers without the constraints of the multistage environment and normally with inducers, referred to by Benvenuti as having three-dimensional blades. In Fig. 6.26 the ordinate is the estimated Mach number of the jet divided by the inlet Mach number at the blade mid-height. The additional curve was drawn by Benvenuti to indicate the likely best value obtainable for Mach number ratio with the simple two-dimensional blades such as might be used in a multistage industrial compressor. Very clearly there is large loss in the realizable deceleration of the flow in the less aerodynamically satisfactory environment.

The inducer and splitters

In some respects the inducer of a high-speed compressor has a particularly difficult job. The rotational speed is high to give the necessary peripheral speed at outlet in order that the work input is large enough. To avoid very high losses in the inducer the tip relative Mach number must be kept within bounds, for example not more than 1.5, and given the requirement for high peripheral speed this effectively puts a restriction on the inducer outlet diameter. For many small machines the leading edge has to be relatively thick in terms of the blade pitch,

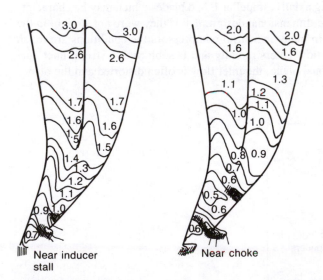

Fig. 6.27 Static pressure measured on the shroud, non-dimensionalized by inlet stagnation pressure, for an impeller with splitters. Shaded regions indicate where shocks are. (From Senoo *et al.*, 1979)

when compared to high-speed axial blades, because of the requirement to be mechanically robust and also because in some cases, such as automotive turbo-chargers, the expense of machining or hand finishing the leading edges of cast impellers would be too great.

At the rear of the impeller, in the radial part, there must be an adequate number of blades if the loading is to be kept within reasonable bounds. If this number were to be continued to the inlet of the inducer their blockage would be so great in some cases as to cause a serious constraint on the mass flow in high pressure ratio machines. The expedient adopted is the use of splitter blades which end somewhere downstream of the inducer. It is customary to design these so that they are identical to the full blades but cut back; in fact for research purposes the impeller may be made with a full set of blades going right down to the leading edge of the inducer and then every other one of these may be machined back to form the splitters. This is not a correct design pro-cedure, for the leading edge of the splitter will not be aligned with the flow approaching it in the middle of the passage. It is quite common to find that the two passages divided by such a splitter behave quite differently, which is illustrated in Fig. 6.27 by the shroud pressure contours measured by Senoo *et al.* (1979). There is great scope for variation in the way the splitters are ended, both as to the radial and axial variation in their leading edge. This is in addition to attention directed to changing the alignment and curvature of the splitters to conform to the flow direction midway between the blades. This is, in fact, an ideal application for numerical experimentation using three-dimensional or even two-dimensional methods.

Backsweep

There seems no doubt that impellers with backsweep generally have higher efficiency than those for which the blades are radial at outlet. This is true of the efficiency of the stage as well as the impeller itself; in fact it is easier to see why the stage should be more efficient as a result of backsweep than it is for the impeller. This is because the impeller is generally much more effi-cient as a decelerating device (or diffuser) than the diffuser downstream of it. Backsweep reduces the absolute whirl velocity out of the impeller and therefore raises the proportion of the stage static enthalpy and pressure rise occurring in the impeller, in other words it raises the degree of reaction. If the impeller is more efficient than the diffuser it follows that the efficiency of the stage should be higher.

It has already been observed that the impeller efficiency is high, at least in part, because a large part of the static enthalpy and pressure rise occurs as a result of the centrifugal effect $(U_2^2 - U_1^2)$ which is loss free. The other component of the static rise, the deceleration of the relative flow, is usually smaller and does incur losses. For the same stagnation enthalpy rise for the stage it is necessary to have a higher impeller tip speed U_2 for wheels with backsweep and thus there is more of the loss-free rise $(U_2^2 - U_1^2)$ to be had. There are, in addition, several possible reasons why backswept impellers should perform better than those without backsweep, although there is no unanimity as to which are most important.

1. Many impellers have streamtube area–density ratios, $\rho_2 A_2 / \rho_1 A_1$, which are too large for the avoidance of massive separations and therefore very large wakes. Backsweep reduces the outlet area without reducing the axial depth of the blades at outlet and thereby without raising losses due to tip-clearance flow or very narrow flow passages.

2. The inclination of the blades to the radial direction brings about a reduction of the loading, in the sense that the difference in the relative velocity between suction and pressure surfaces at the same radius is reduced. This reduces the secondary flow and the tip-clearance flow but most importantly reduces the extent to which the flow on the suction surface must be decelerated towards the trailing edge. The effect of this is probably to delay separation or to inhibit its growth.

3. Impellers with radial blades at outlet usually have extremely large amounts of curvature or camber by the standards of axial machines. Backsweep has the effect of reducing this so that the blades take on a much flatter appearance; most inducers are inclined at 60° or more to the axial direction at inlet so that 30° of backsweep halves the camber, a very large change. The inducer region is where the camber is usually most pronounced and it is particularly undesirable for separation to occur as far forward as this.

4. The large pressure ratio across the centrifugal impeller is possible because most of this large pressure rise is balanced by the large centrifugal acceleration due to the tangential velocity of the fluid particles. The fluid adjacent to a stationary shroud does not have a high tangential velocity and to prevent this flow from reversing and keep it attached there must be stresses applied to it. In the radial bladed impeller these can only be shear stresses from the body of the flow but in the backswept impeller there is a component of the blade force able to counteract the tendency of the fluid close to the shroud to decelerate in the meridional direction. (Such blade force effects are believed to be important in axial endwall flows.)

5. If the blades are curved near the trailing edge the effect of the backsweep increasing towards the trailing edge is to raise the pressure on the suction surface and thereby reduce the velocity difference and the loading.

6. Backsweep lengthens the flow passage in the impeller in relation to its width. The advantage of large length-to-width ratio has been documented for axial stages by Koch (1981) and is discussed in Chapter 9.

The advantages of backsweep listed in (1) to (6) above have consequences on the downstream diffuser. Any change which makes the flow more uniform at outlet from the impeller is likely to improve the performance of the diffuser. One of the main reasons that the diffusers do not work as effectively as one might hope is that the flow is highly non-uniform at entry; non-uniformity in the axial direction is the greatest problem and it is normally the high level of blockage (or thick boundary layer) on the shroud side of the diffuser passage which causes most of the difficulty. The blade-to-blade non-uniformity (the conventional two-dimensional jet and wake) seems to be much less of a problem, partly because of rapid mixing in the radial direction and partly because the *absolute* whirl velocity is more similar in the jet and wake than seems evident from the magnitudes in the relative frame.

Compressor stability and the avoidance of stall and surge will be addressed in Chapter 9 but it is relevant to the choice of backsweep. With backsweep the work input per unit mass flow rises as the flow rate falls. For an impeller with no backsweep the work input is very nearly independent of flow rate for a given rotational speed. For impellers with poor internal flow, and in particular poor flow in the inlet region, almost all the pressure rise comes from the centrifugal effect. For such impellers the maximum pressure rise will occur just about where the efficiency is a maximum. For a backswept impeller with negligible pressure rise by deceleration of the flow inside the impeller the maximum pressure rise can occur at a smaller flow than that for maximum efficiency. The stability limit for compressors is often taken to occur at the mass flow rate at which the pressure rise is a maximum, an issue discussed much more thoroughly in Chapter 9. The impeller with no backsweep therefore has its stability limit at the same flow rate as its maximum efficiency; operation must therefore occur at flow rates larger than that for optimum efficiency. With the backswept impeller operation is generally possible at maximum efficiency. With a reasonably good inducer or inlet it is often possible to get a quite appreciable pressure rise from the deceleration of the relative flow. This pressure rise increases quite steeply with incidence and contributes a stabilizing influence to the impeller flow. It is for this reason that impellers without backsweep have operated stably in such applications as turbochargers. In multistage compressors the inlet configuration is normally not very good from an aerodynamic point of view and since only a small pressure rise is likely to be generated by decelerating the flow this stabilizing influence is missing. Large operating range is of prime importance for industrial compressors and backsweep has been more or less universal for many years.

An overview

The topic of this book being aerodynamics it is not appropriate to consider mechanical aspects in detail. The effect of material properties and cost on impeller tip speed has already been alluded to in Section 2.3. There is no doubt that the consideration of stresses held back the use of backsweep for many years in high pressure ratio machines. With the use of better stress analysis methods and the determination to proceed down the backsweep route the original objections seem to be being overcome. It is nevertheless true that the stressing still imposes restrictions on the design of high-speed machines. Another aspect is the method of manufacture, which includes the methods used to specify the shape and to turn this into metal, see for example Casey (1983) and Krain (1984) for a discussion of different points of view. Certain restrictions on the geometry can have the effect of simplifying the manufacture of the machine very considerably and at the same time will exact no aerodynamic penalty.

7 The diffuser of the centrifugal compressor

7.1 Introduction

The flow generally leaves the impeller of a centrifugal compressor with a high absolute velocity, normally inclined at a large angle to the radial direction. It is necessary to decelerate this flow if the pressure rise and efficiency of the compressor are to be acceptable, and this is the role of the diffuser. As discussed in Chapter 6, the flow in the impeller is usually highly non-uniform and the outlet flow highly distorted. The diffuser is required to cope with flow velocity variations in the axial (i.e. hub-to-shroud) direction and in the circumferential direction.

There are broadly two types of diffuser: the vaneless is used when wide range or low cost are of prime concern, the vaned diffuser when maximum pressure rise and efficiency are needed. The vaneless diffuser is geometrically simple, being an annular channel surrounding the impeller. Usually the axial width b is constant, so that the diffuser is composed of two parallel plane surfaces, but it can also be designed with b varying in the radial direction.

Outside the diffuser there is often a volute, sometimes called a scroll, to collect the flow and decelerate it further. With a volute there is necessarily a circumferential discontinuity at one position, known as the tongue. It is possible to design the volute so that at one combination of V_θ and V_R from the impeller (in other words one mass flow for a given rotational speed) there will be no incidence of the mean flow onto the tongue and therefore a nearly uniform pressure distribution at the outlet of the vaneless diffuser. At other conditions, however, there will be a circumferential variation in static pressure which can have quite serious effects on the flow in the impeller as well as the vaneless diffuser. The non-uniformity may be noticeable upstream of the impeller and because of it compressors often surge at a higher flow rate than with a uniform downstream static pressure.

There are very many types of vaned diffuser. One common type consists of wedges and is illustrated in Fig. 7.1. This is sometimes known as the vane-island type of diffuser. The blunt end does not cause as much loss as it might seem because the velocities at this location are relatively low. The loss attributable to exit kinetic energy is easily quantifiable for a specific geometry and

266

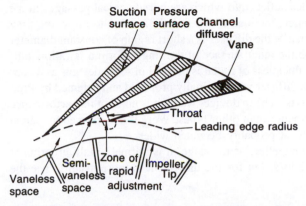

Suction Pressure
surface surface Channel
diffuser
Vane

Throat
Leading edge radius

Semi- Zone of Impeller
Vaneless vaneless rapid Tip
space space adjustment

Fig. 7.1 The geometry and notation of the vaned diffuser. (From Dean, 1971)

may be worth as much as 5 per cent of overall efficiency. The nomenclature generally used for vaned diffusers of all types is also shown in Fig. 7.1 (some of the terms and the diagram itself are taken from Dean, 1971). A second type of diffuser has cambered vanes which consist of thin plates or possibly aerofoil sections. This type has the advantage that the outside diameter of the diffuser is smaller than those consisting of more or less straight wedges, and a large outside diameter is often a serious drawback to the centrifugal compressor. On the other hand the curvature of the cambered vanes adds to the complexity of design and is thought to bring a reduction in potential pressure rise.

A rather special but successful type of vaned diffuser is the pipe diffuser, illustrated in the photograph in Fig. 7.2. It consists of a thick ring of metal

Fig. 7.2 A section of pipe diffuser. (Reproduced by courtesy of Pratt and Whitney of Canada). Impeller motion from right to left

surrounding the impeller outlet, into which are bored conical passages in the $r-\theta$ plane of the impeller, inclined to the radial direction to allow the flow to enter easily. The throat of the diffuser is a short region of constant diameter and this breaks through the inner surface of the ring to form a 'swallow-tail' type of geometry. The thickness of metal is utilized in some designs as a containment ring. The pipe diffuser was originally proposed and patented by Pratt and Whitney of Canada as a cheap device, which it is not, but its performance does seem to be rather better than other types. Were it not covered by patent its application would most probably be much wider.

When discussing the impeller, or the stage, it is normal to talk in terms of efficiency and pressure loss, but for the diffuser it is more usual to use the pressure coefficient

$$C_p = (p_3-p_2)/(p_{02}-p_2)$$

where station 3 is the outlet from the diffuser and station 2 is at inlet. It is not straightforward to define condition 2 because the conditions at diffuser inlet, impeller outlet, are highly non-uniform (the distribution often being described by a jet and wake) with quite large variations in stagnation pressure and velocity direction and magnitude. Quite separately, the flow field at entry to a vaned diffuser is strongly non-uniform in the circumferential direction because of the vane loading. To define conditions at either impeller outlet or vane inlet therefore requires some sort of averaging to take place. Because the impeller and vaned diffuser are so close, the averaging has to be done with proper care and it is common to define a hypothetical mixed-out state such as would exist if the impeller non-uniformity could mix out very close to the impeller outlet with no diffuser vanes present.

The region between the impeller and the diffuser vanes is known as the vaneless space and in this the flow is treated much like that in a vaneless diffuser. The performance seems to be best if the vane leading edge is only a short distance out from the impeller outlet, typically with a radius ratio of about 1.1. The radius ratio may be increased a little to reduce the vibration or to reduce the Mach number at inlet to the vanes. In fact the proximity of the non-uniform impeller flow and the non-uniform vane inlets makes it surprising that it is useful to consider the diffuser and impeller separately, but Inoue and Cumpsty (1984) showed that the conventional practice of doing this is normally quite accurate.

Over very short distances, such as the vaneless space, it is possible to treat the flow in inviscid terms to get some idea of the behaviour, but this must be used with caution because of the highly non-uniform nature of the flow. Apart from this, the inviscid approach to the flow in a vaneless diffuser has also affected the way thinking on the topic has developed. For axisymmetric, inviscid flow in a vaneless diffuser the moment of momentum rV_θ remains constant and the whirl velocity falls inversely proportional to radius. If the axial depth of the diffuser is constant and if the flow is incompressible, the radial velocity also decreases inversely proportional to radius. In this case the flow direction

$$\alpha = \tan^{-1}V_\theta/V_R$$

remains constant from inlet to outlet and the streamlines form a logarithmic spiral

$$\theta_2 - \theta_1 = \ln r_2/r_1.$$

If the width of the vaneless diffuser does not remain constant the tangential velocity is still inversely proportional to radius, whereas the radial velocity is modified by the variation in area. Many diffusers do maintain a constant axial width.

One of the effects of the vanes in a vaned diffuser is to alter the streamline pattern away from this logarithmic spiral, usually forcing it to follow a course which is more nearly radial so that the same radius is reached in a shorter path. The tangential velocity into the diffuser is usually around three times as large as the radial component and it is therefore the deceleration of the tangential component which gives most of the pressure rise. For the inviscid flow in a vaneless diffuser the only mechanism for decelerating the tangential velocity is the increase in radius.

It is a general feature of diffusers in centrifugal compressors that the axial depth b is small in relation to the impeller outlet diameter D_2 and typically $0.02 < b_2/D_2 < 0.12$. This means that for a vaned diffuser the aspect ratio of the flow passage is likely to be much less than one, a configuration known to be far from ideal in giving best diffuser pressure rise. For a vaneless diffuser small b/D_2 means that the flow path is long in relation to the axial depth of the diffuser and there is therefore ample opportunity for the viscous effects, including those associated with turbulent stresses, to diffuse across the flow. A treatment of the flow in the vaneless or vaned diffuser, on the basis of inviscid flow or inviscid core flow with boundary layers on the walls, is not likely to be very satisfactory. Inviscid analysis can be useful for giving qualitative guidance on trends but it is not to be trusted for quantitative estimates.

In what follows the behaviour of the non-uniform flow downstream of the impeller is first considered, beginning with the theory of Dean and Senoo (1960) and concluding with the detailed and comprehensive measurements of Eckardt (1975). This is relevant to the behaviour in vaneless and vaned diffusers. It leads on to a treatment of the vaneless diffuser, followed by the vaned diffuser and the chapter is concluded with a section on volutes. It is worth pointing out that the whole subject of diffusers in turbomachinery has been considered very thoroughly, with a very complete set of references, by Japikse (1984).

7.2 The non-uniform flow from the impeller

It is now firmly established that most impellers have extensive regions of separated flow in them and that their outlet flow is highly non-uniform. Dean and Senoo (1960) described the process downstream of the impeller in a vaneless diffuser with the jet-wake model illustrated by Fig. 6.1, taken from that paper. Their approximate analysis was incompressible and the flow direction, but not magnitude, was assumed uniform in the circumferential direction. Applying the equations of mass continuity and conservation of radial and tangential momentum separately to the wake and the jet regions they were able to set up a differential equation in the radial direction. By integrating this they could follow the development to uniformity. Skin friction at the wall of the vaneless diffuser and inter-region shear stress between the uniform jet and the uniform wake were included. In the process of reverting to uniform flow there were two quite different effects: a reversible work exchange and irreversible mixing. The reversible process arises from the different way in which the jet and wake *absolute* velocities change as they flow radially outwards and the simultaneous need to maintain compatible flow directions in the *relative* frame of reference.

Dean and Senoo compared their theory with measurements for a flow having a wake covering 72 per cent of the passage with very little flow in the wake. For a range of magnitudes of the skin friction and inter-region shear the theory predicted that the wake should disappear by a radius ratio (radius divided by impeller tip radius) of 1.05 and the measurements were able to bear this out. The rise in static pressure and fall in stagnation pressure were also well predicted without great sensitivity to the actual magnitude of skin friction or inter-region shear stress, although setting these to zero gave unsatisfactory predictions.

Calculations with non-zero relative velocity in the wake gave similar trends for static and stagnation pressure to those calculated with no flow in the wake, but showed a marked difference in the rate at which the wake decayed. Not only was the initial decay much slower if there were flow in the wake but in fact the decay became so slow that the wake never disappeared completely.

Johnston and Dean (1966) represented the mixing of the non-uniform impeller flow by a sudden expansion of the jet and wake to create a uniform flow. This was plausible, since the Dean and Senoo calculation had shown most of the reversion to uniform takes place very rapidly. There was no reversible mixing with the Johnston and Dean method. Loss was calculated from the mixing process and from skin friction on the faces of the vaneless diffuser in the downstream uniform flow. In fact the losses predicted by this process are almost identical with those calculated by the Dean and Senoo theory; this is shown by Fig. 7.3 taken from Inoue (1983). (In their original paper Johnston and Dean made numerical errors in their calculation of the Dean and Senoo losses and showed a large discrepancy for values of $\lambda_2 = V_{\theta 2}/V_{R2}$ less than about 1.5.) Inoue was able to show that although the reversible work does bring about a transfer of energy from the jet to the wake in the very early stages, this contribution is very small and can be ignored — hence the good agreement between the Dean and Senoo and the Johnston and Dean theories.

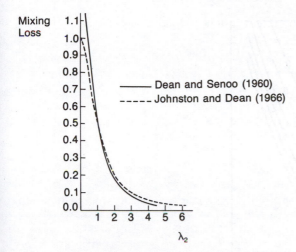

Fig. 7.3 A comparison of the predicted mixing loss versus diffuser inlet swirl parameter $\lambda_2 = V_{\theta 2}/V_{R2}$ by Dean and Senoo and by Johnston and Dean method

(In fact the transfer of energy from jet to wake is in the wrong direction for creating uniformity, for in most configurations the wake has higher absolute stagnation pressure and temperature than the jet.) Inoue was also able to show that either skin friction or inter-region shear stress can bring about this mixing and that if the magnitude of either one is reduced the loss by the other mechanism will rise to compensate. Because the predictions of mixing loss from the two theories are so similar the simpler Johnston and Dean method, which is very easily generalizable to include changes in geometric layout and compressibility, is quite adequate.

Figure 7.4 is taken from Johnston and Dean and shows their predicted loss in stage efficiency resulting from mixing, scaled by the non-dimensional impeller work input, as a function of wake width and swirl parameter λ_2. The comparison is for no flow in the wake, a vaneless diffuser of constant axial width and incompressible flow. (Dean, 1971, showed a result for compressible flow which was not significantly different). Many high-performance compressors have values of $\lambda_2 \approx 3$ and from Fig. 7.4 it can be seen that with this value the mixing loss is about 1 per cent for a wake with a width 0.3 times the pitch, but rises rapidly with increase in wake width to about 12 per cent at a wake width of 0.6. Figure 7.4 shows a strong dependence on the swirl parameter, the loss decreasing as the value of λ_2 rises. This occurs because the absolute tangential velocity is almost equal for the jet and wake and most of the loss comes from the mixing out of the radial component. For $\lambda_2 = 1$ the dynamic pressure of the radial and tangential components are equal but when $\lambda_2 \approx 3$ the dynamic pressure of the radial component is an order of magnitude smaller and the mixing loss correspondingly reduced.

Eckardt (1975) carried out very detailed measurements of the flow in the discharge of a centrifugal compressor, the machine with no backsweep for

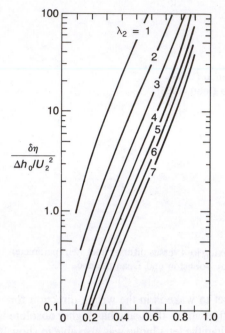

Fig. 7.4 Mixing loss in terms of loss in stage efficiency, scaled by stage loading, as a function of wake width and impeller outlet swirl $\lambda_2 = V_{\theta 2}/V_{R2}$. Zero relative velocity in wake. (From Johnston and Dean, 1966)

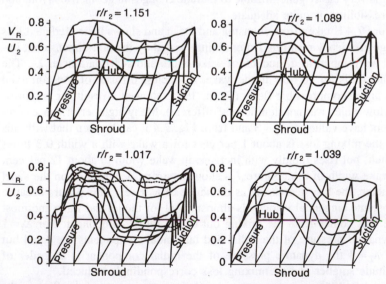

Fig. 7.5 The evolution of the flow pattern measured in a vaneless diffuser from an impeller with no backsweep. (From Eckardt, 1979)

which results are shown in Fig. 6.2. The vaneless diffuser width decreased
linearly with radius. Figure 7.5, taken from Eckardt (1979), shows laser
measurements of the radial velocity distribution across the vaneless diffuser
passage at four radius ratios for the flow coefficient giving the highest efficiency at this rotational speed. The general pattern at impeller outlet is for
the flow to have a wake region, but not one with zero radial velocity. The
wake is near the shroud/suction surface corner, as discussed in Chapter 6 and
over the short distance between the measuring stations there is a substantial
move towards uniformity, with the steep gradients greatly reduced. The pattern
for the stagnation pressure is similar. At other flow rates the level of distortion
is rather greater.

Figure 7.6 compares some measured quantities in the vaneless diffuser by
Eckardt compared with calculated values. Eckardt extended the Dean and Senoo
theory to include compressibility and this is shown as case 1, using the measured
inlet flow as the starting condition. In similar comparisons he varied the possible
assumptions, in particular comparing cases with uniform flow instead of the
jet-wake (case 2), with the flow treated as incompressible instead of compressible (case 3) and with no friction at the walls or jet-wake interface (case 4).
The method including all the effects, case 1, predicts the variation very well,
whilst omitting one of the effects leads to appreciable discrepancies. Also shown
in Fig. 7.6 is the measured variation in the fraction of the total area occupied
by the wake ϵ and the ratio of the average wake velocity to the average jet
velocity γ: these are compared with the calculation including all the effects,

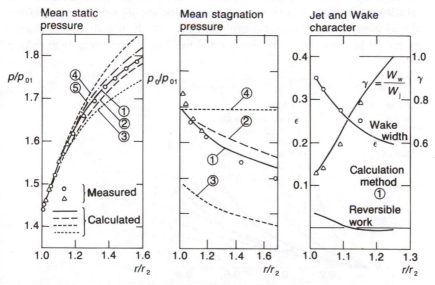

Fig. 7.6 Eckardt vaneless diffuser: comparison of measurements at 5.31 kg/s and
14 000 rev/min with calculations by extension of Dean & Senoo. 1: full analysis; 2:
ignoring jet & wake; 3: ignoring compressibility; 4: ignoring viscous effects; 5: ignoring
wall displacement thickness. (From Eckardt, 1979)

case 1. It is noteworthy that the wake disappears by the velocity rising to equal that of the jet, i.e. by γ tending to one, and not by the area shrinking to zero. The calculated reversible work exchange is also shown, positive being to the wake from the jet. As predicted by Inoue the effect is small and essentially confined to very small radius ratios. The level of agreement between the calculation and the theory is remarkable, bearing in mind its simplicity and the fact that the relative flow direction for the jet and wake were taken as equal whereas in fact at this flow coefficient the wake was inclined at about 10° to the radial and the jet at about 20°.

Eckardt used his calculation method to carry out parametric studies of the loss in total-to-static efficiency due to the non-uniform outlet flow from an impeller with no backsweep; the results are reproduced here as Fig. 7.7 with the abscissa being the area of the wake at diffuser inlet as a fraction of the total area, ϵ_2. The loss becomes large only when the wake is large and the ratio of the velocity in the wake to that in the jet γ_2 is small, i.e. the case shown with $\gamma_2 = 0.2$. The losses are generally small at high values of λ_2 for reasons given above in connection with Fig. 7.4. For the conditions of Eckardt's experiments using the impeller without backsweep the wake was quite large, $\epsilon_2 \approx 0.4$, and quite deep, $\gamma_2 \approx 0.5$, but even so only about a 1 per cent loss in total-to-static efficiency for the entire compressor was attributable to the mixing. The strong dependence of the loss on the ratio of wake-to-jet velocity, γ_2, means that the curves of predicted loss given by Johnston and Dean for the case of $\gamma_2 = 0$, reproduced here as Fig. 7.4, represent major overestimates of the likely level of mixing loss, because for well-designed impellers the velocity in the wake is a substantial fraction of the jet velocity. One effect of backsweep is to reduce the magnitude of λ_2 which, if nothing else changed, would increase the mixing loss; it follows that if backsweep is to reduce

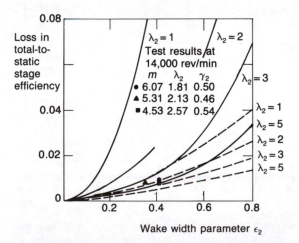

Fig. 7.7 Comparison by Eckardt of measured loss in stage efficiency with parameter study using modified Dean & Senoo method (method 1 in Fig. 7.6). Solid lines $\gamma_2 = 0.2$, broken lines $\gamma_2 = 0.6$ (From Eckardt, 1979)

the mixing loss it must do so by reducing the severity of distortion, i.e. reducing the size of the wake and increasing the velocity in it.

The original treatment of the non-uniformity of the impeller outlet flow by Dean and Senoo envisaged a pattern which had the jet-wake variation in the circumferential direction but uniform flow in the axial width of the passage. In fact the jet-wake is now known to be more complicated, with the wake usually near the shroud. The circumferential variation, which is perceived by the downstream diffuser as an unsteady effect sweeping by, does not seem to affect the aerodynamic performance of the downstream diffuser, although it is important as a vibration exciter and as a source of noise. What is far more serious for the diffuser is the nonuniformity across the passage in the axial direction. Figure 7.8, taken from Benvenuti (1978) shows the axial variation in radial velocity at a radius ratio of 1.08 for three different values of specific speed. High specific speed corresponds to a wide impeller, with much greater curvature on the shroud and hub, and it is therefore no surprise that this case has a much more distorted profile. The axial distortion may be better or worse as stall is approached: Fig. 7.9 is taken from Rodgers (1982) for different flow rates in a range of unshrouded impellers with different ratios of outlet width to outlet diameter b_2/D_2 (essentially different specific speeds). The distortion is worse for the impellers which are wide at outlet for the same reason as was given in relation to Fig. 7.8.

To sum up, the jet and wake mix out quite rapidly in the radial direction, but much more slowly than would be predicted by an analysis assuming no relative velocity in the wake. The decay of the strongly non-uniform pattern at impeller outlet seems to be well predicted by a method which is an extension of the Dean and Senoo method, although this neglects physical boundary conditions at outlet, treats a three-dimensional problem as two-dimensional and includes many assumptions known to be only approximately true, see Inoue (1983). The rate of decay of the non-uniformity from the impeller is such that at a radius ratio of even 1.2 there is likely to be still some irregularity in the circumferential direction, so that in most practical cases with vaned diffusers there will be a strongly unsteady flow into the vanes. The reversible work does not make a major contribution; although there is the reversible work transfer between the jet and wake identified by Dean and Senoo, this occurs near the impeller outlet, is generally small in magnitude, is not necessarily favourable and can usually safely be ignored.

The loss is well predicted by the simple Johnson and Dean sudden expansion theory. The loss from the mixing of the non-uniform impeller outflow decreases rapidly as the velocity in the wake rises and loss is seriously overestimated if no flow is assumed in the wake. The loss also becomes small for large swirl parameters; for $\lambda_2 \geq 3$ and a velocity in the wake only 20 per cent of that in the jet the wake would need to occupy about 40 per cent of the outlet area to give a loss in stage total-to-static efficiency of 1 per cent. Too much attention has been directed at the mixing loss of the jet and wake, probably because attention was directed at cases calculated with no flow in the wake.

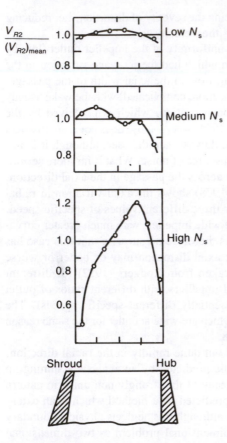

Fig. 7.8 Measured radial velocity at radius ratio 1.08 downstream of industrial (shrouded) impellers. Results for low, medium and high specific speed impellers at maximum efficiency operation point. (From Benvenuti, 1978)

7.3 The vaneless diffuser

Although the vaneless diffuser is one of the simplest components of any turbo-machine, the aerodynamic behaviour is complicated by the non-uniformity of the flow entering and by the dominant effect that viscous stresses have along the long flow path between inlet and outlet. Because the streamlines are curved there will be a large amount of secondary flow or, expressed in a different way, the flow direction will not be uniform across the width of the diffuser. The width of the diffuser, b, is normally constant but is sometimes allowed to decrease in the radial direction.

The geometric variables of the vaneless diffuser, assuming that it discharges into a uniform region, are the diameter ratios from inlet to outlet D_2/D_1 and the ratio of the inlet axial width to the inlet diameter b_2/D_2. An additional

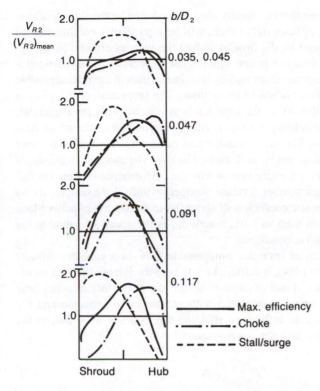

Fig. 7.9 Measured radial velocity profiles at radius ratio 1.1 downstream of unshrouded impellers. Results shown for range of diffuser axial width to diameter ratio b/D_2. (From Rodgers, 1982a)

ratio or angle is needed if the width is not constant. The flow variables are the mean inlet flow angle α_2 and the mean inlet Mach number M_2. An important feature of the impeller outlet flow is the non-uniformity, with that across the width being most important.

Making the axial width of the diffuser decrease outwards has a direct effect on the mean radial velocity through the equation of continuity, but only an indirect effect on the tangential velocity. (In an inviscid flow there would be no effect of width on the tangential velocity since rV_θ=constant.) By making the radial velocity decrease more slowly than the tangential, the mean flow is inclined more towards the radial direction and the distance followed by the mean streamline from inlet to outlet is reduced. Because of the shorter distance the losses are reduced. The most significant effect then of changing the axial width of the diffuser is the effect it has on the frictional effects at the wall, Rodgers (1982b). An additional benefit which accrues with a contracting diffuser is that by making the flow less inclined towards the tangential there is less tendency for instability, a topic discussed below.

Compressibility plays a part in determining the diffuser performance in real

machines. Suppose first that the density change produced by the flow deceleration is negligible: at diffuser outlet there will be a particular combination of
$V_{\theta 3}$ and V_{R3} determined by the flow including the viscous effects. Now suppose that the density does rise in line with the rise in pressure. The radial velocity must therefore decrease more rapidly with radius than in the incompressible
case to satisfy the conservation of mass flow. The tangential velocity is not
directly affected by density so the flow tends to turn towards the tangential.
One effect of compressibility is to increase the inclination of the mean flow
to the radial direction. The static pressure rise obtained in a vaneless diffuser
decreases with inlet flow angle, as discussed below. The increase in this angle
due to compressibility is a major reason why the performance of vaneless diffusers falls off as Mach number is raised. Rodgers (1982), for example, shows
a reduction in the pressure coefficient of about 0.1 in raising the impeller Mach
number, U_2/a_{01}, from 0.65 to 1.01, where a_{01} is the speed of sound at the
impeller inlet stagnation conditions.

Some considerations of inviscid compressible flow in a vaneless diffuser
were given by Taylor (1964); it suffices here to note the impossibility of supersonic radial flow in a diffuser of constant width, the possibility of supersonic
tangential velocities being decelerated without shocks to subsonic and the
possibility of downstream influences affecting the impeller even though the
resultant flow into the diffuser is supersonic.

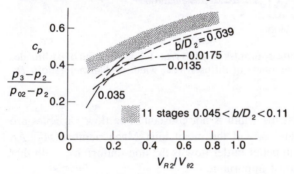

Fig. 7.10 Static pressure recovery measured in vaneless diffusers versus mean inlet
$V_{R2}/V_{\theta 2}$, for diffusers of various width-to-diameter ratios b/D_2. Radius ratio $R_3/R_2 =$
1.71 in all cases tested. (Adapted from Rodgers, 1982b)

Rodgers (1982b) described tests and presented results for a programme in
which 15 configurations of vaneless diffuser were tested for a range of impellers, all unshrouded and backswept, with impeller Mach numbers
$M_u = U_2/a_{01}$ ranging from 0.6 up to 1.2. The axial width of the diffuser was
constant in the radial direction and $D_2/D_1 = 1.7$. The results are summarized
in Fig. 7.10 in terms of pressure coefficient and the ratio of average inlet radial
velocity to tangential velocity, $V_{R2}/V_{\theta 2} = 1/\lambda_2$ (note that $1/\lambda_2 = \cot^{-1}\alpha_2$ where
α_2 is the mean flow direction into the diffuser). The tangential velocity
changes relatively slowly with mass flow and depends most strongly on the

impeller tip speed, so the abscissa *approximately* shows V_{R2}/U_2. Two things are immediately noticeable, the first being that most of the diffusers produce very nearly the same pressure coefficient for the same inlet flow angle, that is the same value of $V_{R2}/V_{\theta2}$. No systematic trend of pressure coefficient with diffuser width ratio b/D_2 could be seen until this was very small, less than 0.045, when a pronounced drop in performance occurred. For diffusers which are wider than this value a pressure coefficient of about 0.5 at an inlet swirl parameter $\lambda_2 = V_{\theta2}/V_{R2}$ of about 3 seems a reasonable estimate, with no clear trend to allow this estimate to be refined.

A second important point is that the pressure coefficient rises with increase in the ratio of $V_{R2}/V_{\theta2}$ — in other words, the diffuser becomes less effective as the flow becomes more nearly tangential. (Some of the results at very low values of V_{R2} were made with the flow in rotating stall and are less reliable.) The simple explanation is that with lower $V_{R2}/V_{\theta2}$ the path between inlet and outlet is longer, so there is more opportunity for the wall shear stresses to act, but the ideal no-loss pressure rise depends only on radius ratio. As noted above, $V_{\theta2}$ is mainly fixed by the blade speed U_2, so Fig. 7.10 shows approximately the variation of c_p with V_{R2}/U_2. A pressure rise which increases with flow rate tends to be inherently unstable and implies that a centrifugal stage with vaneless diffuser is kept from stall or surge by the impeller pressure rise—flow rate characteristic sloping in the other direction.

Experiments have shown that the static pressure continues to increase as the diameter ratio of the vaneless diffuser increases, but at a progressively slower rate. At the same time the stagnation pressure continues to fall. There is normally considerable incentive in keeping the outer diameter of a compressor small to prevent it becoming too cumbersome. Since in the outer part of the diffuser the velocities are low the additional rise in static pressure with an increase in diameter can become small and not worth the geometric inconvenience. It is usually preferred to keep the diameter ratio D_3/D_2 less than about two, and to attempt to achieve the additional pressure rise through the use of a volute around the outside of the vaneless diffuser. The volute flow does not seem to be well understood but some aspects of the flow in volutes are discussed in the last section of this chapter.

The radial velocity profiles measured by Rodgers (1982a) and shown in Fig. 7.9 were the inlet flow into the diffuser of a number of the stages for which the pressure coefficients are shown in Fig. 7.10. Since no obvious trend was evident for most cases, it must be tentatively concluded that for a vaneless diffuser the inlet velocity profile is not important, with the proviso that this must be for quite large ratios of outlet to inlet diameter so that the flow has an opportunity to develop into its own equilibrium state.

Some measured radial velocity profiles in a vaneless diffuser downstream of a shrouded impeller are shown in Fig. 7.11, taken from Benvenuti (1978). The profiles are modified quite considerably as the flow goes outwards and the apparent ability to overcome the distortion at inlet is particularly noticeable for the maximum efficiency case.

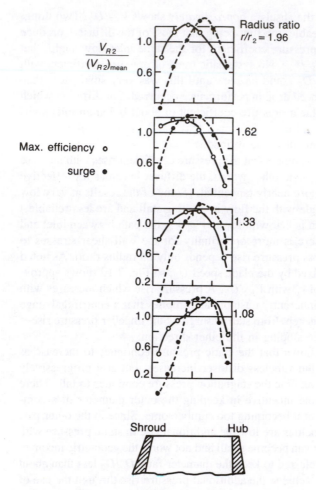

Fig. 7.11 Measured radial velocity at various radius ratios downstream of shrouded impeller of high specific speed. Results shown for operation at maximum efficiency and close to surge. (From Benvenuti, 1978)

Calculation methods for the vaneless diffuser

There have been a number of methods devised to calculate the flow in the vaneless diffuser. The simplest is to treat the flow as one-dimensional and apply momentum and mass conservation along a streamtube, including the effect of skin friction at the wall. This is the approach adopted by Johnston and Dean (1966), by Jansen (1964a) and more recently by Rodgers (1982). The difficulty comes in specifying the correct value for the skin friction. Deriving this from measured pressure distributions indicates that the value necessary to give the correct balance is higher near the inlet. This is partly because the non-uniform impeller flow is mixing out there and the mixing loss is incorrectly attributed to skin friction, but also because the skin friction will be higher where the flow is being established, just as it is at entry to a pipe. Rodgers derived the necessary skin friction coefficients to match his data for the 15 stages he examined and found a spread in value exceeding 2:1. The highest values were obtained for the impeller with the largest value of b/D_2, the highest specific speed machine. In all cases but one the skin friction coefficient equalled or exceeded the value of 0.003 which may be taken as a typical value for a turbulent boundary layer. Japikse (1984) recommended the expression

$$c_f = k(1.8 \cdot 10^5/Re)^{0.2}$$

for use with one-dimensional analysis. He tabulated the values of k deduced from tests at different flow rates on a turbocharger and from Eckardt's measurements. The highest value of k was 0.025, the lowest 0.0074; a suggested working value was around 0.01, but the range emphasizes the physical inaccuracy or incompleteness of the model.

Given the range of values for c_f implied by the variation in k, it seems hard to find any real use for such a one-dimensional analysis except perhaps to indicate trends. Figure 7.12 show the pressure coefficient and loss coefficient as a function of radius ratio at a range of values of inlet swirl parameter $\lambda_2 = V_{\theta 2}/V_{R2}$ calculated by Johnston and Dean (1966). These are all for one value of the parameter $c_f D_2/b$, which comes out of the one-dimensional analysis as the relevant variable to include the effect of skin friction. The reduction in static pressure coefficient with λ_2 is the counterpart of the result measured by Rodgers, shown in Fig. 7.10.

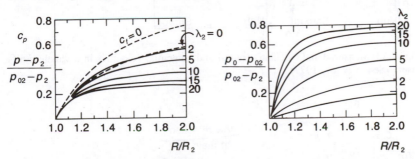

Fig. 7.12 Static pressure rise coefficient and loss coefficient calculated for a vaneless diffuser as a function of radius ratio and swirl parameter $\lambda_2 = V_{\theta 2}/V_{R2}$ for $c_f D_2/b = 0.36$. (From Johnston and Dean, 1966)

A step towards a more complete description is to treat the flow on each wall as a three-dimensional boundary layer. The simplest case is when the boundary layers do not meet and there is a core flow with uniform stagnation pressure. The streamwise direction is defined by this core but in the boundary layers the flow adopts a different direction with the component of the local velocity normal to the core direction known as the crossflow. Senoo *et al.* (1977) applied such a method to vaneless diffusers, making the usual assumptions for boundary layer flow. They assumed a power law profile for the streamwise component of the boundary layer of overall thickness δ given by $u/U = (y/\delta)^{1/n}$ with the index chosen as $n = 2.67(U\delta/\nu)^{1/8}$ and a crossflow profile of the form

$$v/U = u/U(1 - y/\delta)^3 \tan\beta$$

where β is the angle between the limiting streamline at the wall and the streamwise direction. An empirical expression was also used for the skin friction. The equations could then be solved for the core and the boundary layer thickness. When the boundary layers from each wall were thick enough to

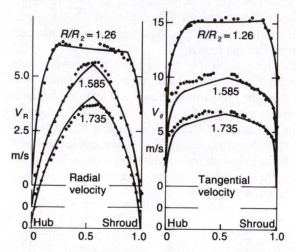

Fig. 7.13 Measured and calculated radial and tangential velocities in a vaneless diffuser. (Senoo *et al.*, 1977)

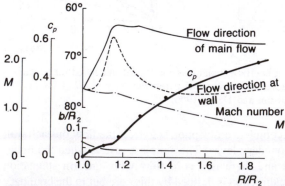

Fig. 7.14 Predicted flow direction, and Mach number of main flow and flow direction at wall in a vaneless diffuser as functions of radius ratio; measurements of pressure coefficient compared with predictions, (Senoo *et al.*, 1977). b/R_2 is diffuser width over inlet radius

meet it was necessary to define equivalent velocity profiles such that the overall momentum was conserved and the mass flow rate was the same. The application to the case of merged boundary layers can only be regarded as very approximate. Senoo *et al.* were able to calculate the radial velocity distributions, which agreed quite well with those measured in vaneless diffusers after fitting the initial velocity profiles to the measurements; Fig. 7.13 shows a comparison with the measurements by Jansen (1964) of the radial and tangential velocities. Figure 7.14 shows a comparison by Senoo *et al.* for the overall parameters for a high Mach number configuration. Both Figs. 7.13 and 7.14 demonstrate the good agreement possible with suitable empirical inputs.

One of the intriguing effects in the radial diffuser is the appearance of negative radial velocity close to the wall. Senoo *et al.* (1977) found this; so too did

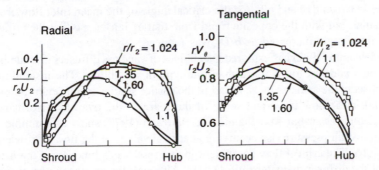

Fig. 7.15 The development of radial and tangential velocity profiles in a vaneless diffuser downstream of an impeller with no backsweep. (From Inoue and Cumpsty, 1984)

Inoue and Cumpsty (1984) and an example from the latter is shown in Fig. 7.15. The flow first reverses, i.e. goes radially inward, near the shroud at a radius ratio of 1.1 then recovers on the shroud and reverses near the hub at a radius ratio of 1.35. (The flow in the tangential direction does not reverse and this component is much larger than the radial velocity.) This occurred with the compressor operating near its peak pressure rise and there was no evidence of stall. Senoo *et al.* (1977) predicted reverse flow first on the shroud side and then on the hub. The presence of one patch of reversed flow can be envisaged by a secondary flow approach; the streamlines are curved concave inwards and the low stagnation pressure fluid near to the walls is turned inwards more. It can also be envisaged by thinking of the flow in its radial and tangential components; in this approach one accepts that most of the radial pressure gradient is created by the centrifugal acceleration arising from the tangential velocity but in regions where the tangential velocity is low the pressure gradient can drive the flow radially inwards. It appears that the blockage created by the separation on one wall drives the flow towards the other wall but the physical explanation for the reattachment and then separation on the other wall is not firmly based.

Stall and stability in the vaneless diffuser

The range over which a compressor can operate is of great concern and vaneless diffusers are often chosen because they have wider operating range than the vaned type. There has therefore been considerable interest in prediction of point of instability in the vaneless diffuser flow. Jansen (1964) identified a condition for rotating stall as that at which the radial velocity on one of the diffuser walls became negative. He predicted that the mean inlet flow angle for the flow to breakdown, α_{2crit}, would decrease (i.e. the flow would be more nearly radial) as the ratio of the diffuser width to the inlet diameter b/D_2 was reduced. Since it is now known that the diffuser can operate with negative radial velocity at some radii it is not clear what this criterion represents. The problem was later addressed by Senoo and Kinoshita (1977) who looked at a more general velocity profile across the width of the diffuser. They predicted values between 81° and 69° for the critical angle of the mean inlet flow from the radial, but with the opposite trend from that of Jansen, so that the critical inlet flow angle increased as b/D_2 was reduced.

Abdelhamid (1982) has explored instabilities of vaneless diffusers in a number of papers and has found more than one type of instability. The instabilities do not seem the same as those found in the earlier studies of Jansen or Senoo. It is not clear how important some of these are to the overall performance of the compressor but Abdelhamid and Bertrand (1979) show an example of loss in overall pressure rise occurring at critical values of the flow coefficient, Fig. 7.16. The critical flow coefficient in this case was a function of the b/D_2 and of the diffuser diameter ratio D_3/D_2. Although the existence of a change in flow character was unmistakable, it is not clear how this brought about the loss in pressure rise in the diffuser. Abdelhamid (1982) was able to delay the instability to lower flow coefficients by putting an axisymmetric throttling ring at exit to the diffuser, which indicates that it is not a boundary layer type phenomenon. He was also able to use an inviscid theory to predict two classes of instability; the critical flow angle was a strong function of outlet to inlet diameter ratio ranging from about 84° to the radial direction at $D_3/D_2=1.4$ down to 79° at $D_3/D_2=2.0$. Another investigation of instabilities in the

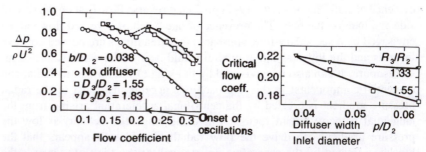

Fig. 7.16 Pressure rise flow coefficient characteristics for a centrifugal compressor with vaneless diffuser and the flow coefficient at which oscillations begin as a function of diffuser width. (From Abdelhamid and Bertrand, 1979)

vaneless diffuser was reported by Frigne and Van den Braembussche (1983). Two sorts of rotating pattern were found in the diffuser, with different propagation speed and number of lobes. The critical angle was found to vary with width of the diffuser between about 76° and 85° to the radial.

In summary, it is clear that there is great confusion about the nature of instability and flow breakdown in vaneless diffusers. The overall variation of c_p with flow rate is such that the device is to be expected to be inherently unstable, but other effects including the impeller may ensure that the entire compressor stage operates without breakdown into a rotating stall or surge. Because of the coupling between components, the stability and breakdown of the flow in the vaneless diffuser should not be studied theoretically or experimentally in isolation. The occurrence and significance of the rotating disturbances referred to above are the subject of considerable confusion but they are not necessarily indicators of breakdown into rotating stall or surge. Nor is the presence of reverse flow on one of the walls of the diffuser an indicator of stall. For a moderate or high specific speed machines (relatively large b/D_2), with α_2 chosen to be not more than $\tan^{-1}3$, in line with normal practice, it would seem that this type of disturbance is avoided. As a general design guide this seems a safe upper limit to adopt for α_2.

7.4 The vaned diffuser

When the requirement from a radial compressor is for high efficiency, a vaned diffuser is normally indicated. As noted earlier there are many types available which are capable of doing a good job. This diversity is indicative of the very different philosophies which lie behind many designs. Some of the diversity springs from considerations of ease of manufacture and cost but some from a lack of understanding of the underlying physics.

Matching the impeller and diffuser

In Chapter 2 the matching of different stages of axial compressors was discussed and the problems that arise from the change in density which takes place. In the case of the centrifugal compressor the pressure and density rise in the impeller alone is normally sufficiently large that it must be allowed for in the design of vaned diffusers for even a single-stage machine. Just as with the axial it is difficult to estimate the crucial quantity, the blockage, and there are serious problems when changes in rotational speed alter the density into the diffuser.

When operated at low speed, so that the compressibility of the gas is relatively unimportant, the diffuser vanes can suffer two serious failures. As the mass flow is reduced from the design value the radial velocity decreases but the tangential velocity will increase and the mean absolute flow into the diffuser is inclined at a larger angle to the radial: the reduced mass flow is perceived by the diffuser vane as a positive increase in incidence. Conversely a larger mass flow than that chosen for the design gives a flow direction which is more nearly radial and a negative incidence onto the diffuser vane. If these incidences become too large the loss in pressure in the diffuser vanes rises rapidly and one would describe them as stalled. (Unlike axial blades the diffuser vanes are usually sufficiently long in relation to the pitch that loss of turning is not one of the effects to be associated with their stall.)

At high impeller tip speeds and high absolute Mach number into the diffusers the range of positive incidence which can be tolerated is normally much smaller than at low speeds. The Mach number approaching the vanes of high pressure ratio machines is often close to unity and small amounts of positive incidence can lead to very high Mach numbers on the vane suction surface to be followed by a strong oblique shock. The incidence is determined by the mean radial and absolute tangential velocities out of the impeller. If an impeller is operated at constant incidence as the speed is increased it is easy to show that because the density at outlet increases with impeller rotational speed the radial velocity will rise by less than the tangential velocity. In consequence the absolute flow angle into the diffuser $\alpha_2 = \tan^{-1}(V_{\theta 2}/V_{R2})$ rises with impeller speed. The diffuser vanes therefore tend to find their incidence increasing as the impeller speed rises; conversely the vanes for a high-speed build of compressor need to be inclined at a larger angle to the radial than one for low speed.

When the impeller tip speed becomes high enough for the compressibility of the gas to be significant, negative incidence stall does not normally occur but instead the flow chokes across the narrowest section of the diffuser, the throat. (At still higher impeller speeds it tends to be the impeller which chokes.) Choking occurs in axial stages, of course, but the high Mach number stages are usually of relatively large span in relation to radius so that choking is approached gradually as the speed is increased and the sonic region extends along the blade. With the centrifugal diffuser the span (i.e. the axial width) is very small in relation to the other dimensions and choking can be relatively sudden.

Treating the flow into the diffuser as locally one-dimensional there is a particular value of the flow function, $m\sqrt{(c_p T_{0t})}/A_t p_{0t}$, (or equivalently the corrected mass flow) corresponding to sonic velocity across the throat. The conditions for choking therefore depend not only on the mass flow itself but also on the stagnation pressure and temperature and the effective cross-sectional area A_t. The stagnation temperature depends on the inlet conditions, the rotational speed and, to a small extent, the mass flow rate. The stagnation pressure depends on all of these but also on the efficiency of the processes ahead of the throat as well. The effective area must include the effect of the flow blockage, which is mainly that of the viscous layers (loosely the boundary layers) on the hub and shroud wall. It is conventional to define the blockage as a way of summing all the boundary layer contributions (more specifically the sum of the displacement thicknesses) as

$$B = 1 - \{\text{effective flow area}\}/\{\text{geometric flow area}\}.$$

The blockage is affected by the flow upstream of the throat and cannot be assumed constant. The blockage can become quite large, up to 20 per cent in some cases, and affects not only the mass flow capacity of the throat but also the pressure rise in the diffuser, a point discussed below.

The effect of speed can be made rather more specific by considering the variation in $m\sqrt{(c_p T_{02})}/A_2 p_{02}$ at impeller outlet with rotational speed. The mass flow is assumed proportional to the impeller speed, which is a good approximation for the impeller operating at constant incidence. Thus

$$m = gU_2/\sqrt{(c_p T_{01})}$$

where g is a constant and $\sqrt{(c_p T_{01})}$ provides a convenient non-dimension-alizing parameter, T_{01} being inlet stagnation temperature. For an impeller with little or no backsweep and no swirl at inlet, the ratio of the absolute stagnation temperature into and out of the impeller is well approximated by

$$T_{02}/T_{01} = \sigma U_2^2/c_p T_{01} + 1$$

The corresponding pressure ratio is

$$p_{02}/p_{01} = \{\sigma U_2^2/c_p T_{01} + 1\}^{\eta\gamma/(\gamma-1)}$$

where η here denotes the polytropic efficiency. Taking, for example, the efficiency to be 0.9 and $\gamma = 1.4$ it is easy to show that

$$\frac{m\sqrt{(c_p T_{02})}}{A_2 p_{02}} = \frac{g}{A_2} \frac{U_2/\sqrt{(c_p T_{01})}}{\{\sigma U_2^2/(c_p T_{01}) + 1\}^{2.65}}$$

At low rotational speeds, $U_2^2/c_p T_{01} \ll 1$, the flow function or corrected mass flow rises in proportion to the wheel speed. In contrast at high values of $U_2^2/c_p T_{01}$ the denominator dominates and the flow function falls with wheel speed. The maximum flow function occurs for the parameters chosen here, when $\sigma U_2^2/c_p T_{01} = 0.23$ which, for a slip factor of 0.85 would correspond to an impeller Mach number $U_2/a_{01} = 0.83$ and a stagnation pressure ratio of 2.1. Different values would be obtained with different parameters, or for an impeller with backsweep, but the underlying form of the variation would be the same.

The flow function derived is for the impeller outlet in terms of the absolute pressure and temperature. To get the corresponding value at the throat of the diffuser the loss in stagnation pressure must be allowed for, as well as the decrease in area resulting from the vane thickness and flow blockage. Assuming for the present purpose that loss and blockage are independent of rotational speed it can be seen that the diffuser will be most liable to choke when the flow function at impeller outlet is a maximum. Conversely the throat area will be a maximum for a compressor with stagnation pressure ratio of about 2.1 (for the parameters of the above example) and if the same impeller were to be used for a higher pressure ratio the throat area could be reduced. A compressor matched with a small diffuser throat area for high pressure ratio operation will be liable to encounter diffuser choking problems at low rotational speed, being worst around that for a pressure ratio of about 2 in the case of an impeller without backsweep. A choked throat represents an absolute limit on the flow rate and what is inclined to happen is the impeller inlet flow separates with corresponding additional loss in pressure rise and efficiency. Furthermore the impeller may go into a rotating stall which can lead to blade or diffuser vane failure by fatigue.

There is an obvious need to be able to design the diffuser vanes so that in the leading edge region the incidence is not high enough to give large losses and the throat is large enough that the choking of the diffuser will not impose a mass flow on the impeller which is too low for its efficient operation. When the velocity into the diffusers is supersonic the problem is most acute, for their operating range is quite narrow, and there have been attempts to use the same approach adopted with supersonic flow into axial blades. The circumstances are very different, with a highly non-uniform distorted flow in the case of the centrifugal diffuser vanes, and it is not realistic to expect accurate predictions with inviscid methods which neglect this strong variation in the axial direction.

Some of the effects referred to above are revealed in tests reported by Stiefel (1972) with three different vaned diffusers downstream of a single impeller with no backsweep at a wide range of rotational speeds. For clarity, results from only two diffusers, with axial widths 8.0 mm and 10.7 mm, are shown in Fig. 7.17. (The impeller diameter is believed to be about 125mm.) The vane shape was the same for each and the leading edge was inclined at 71.5° to the radial direction. The two compressor stages shown in Fig. 7.17 behave very differently. At the low rotational speed the maximum mass flow was much greater for the large diffuser; the maximum mass flow for each was in about the ratio of the diffuser heights and it is inferred that the smaller diffuser was choking. At low speeds the pressure ratio at stall was very similar, implying that both the diffuser and the impeller were working within their effective envelope at the Mach numbers involved and that higher incidence implied by the larger diffuser was tolerated without excessive loss.

At high speeds the variation in mass flow at choke shown in Fig. 7.17 was much smaller, but the difference in the pressure ratio at surge was large. The difference in the pressure ratio was probably because the higher density at impeller outlet reduced the radial velocity so much that with the larger diffuser the incidence was large enough to produce very high losses. (Note that in Fig. 7.17 what is shown is the ratio of stagnation pressures, so differences represent loss and not failure to raise static pressure.) The small difference

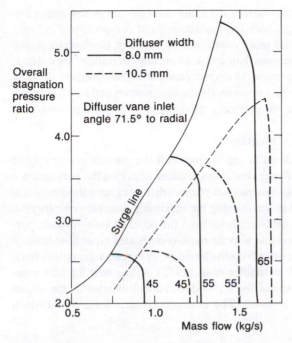

Fig. 7.17 Overall stagnation pressure ratio—flow rate characteristic for the same impeller operating with vaned diffuser of different axial width. (Numbers refer to impeller speed in thousands of revolution per minute.) (From Stiefel, 1972)

in mass flow at choking for the highest speed is most interesting. The choking cannot be occurring in the impeller for both machines, since if it did the mass flow at choke would be identical within the measurement accuracy. (In fact the original measurements showed the same choking mass flow rate, 1.7 kg/s, at 70 000 rev/min for diffusers with width 8.0 and 8.6 mm, i.e. the impeller choking, but results for the 10.7 mm diffuser were missing at this condition.) The conditions in the impeller inducer must therefore have been very near to choke at 65 000 rev/min and it is reasonable to assume that the impeller with the larger diffuser was choked, whereas with the smaller diffuser it was the diffuser itself which was choked.

Stiefel's results illustrate the importance of the correct choice of relative flow area for the matching of impeller and diffuser. The impeller he used could, for example, form a satisfactory compressor delivering a pressure ratio of 4.5 or 5 with the 8 mm diffuser. If, however, the requirements were for a pressure ratio of only 3 a wider diffuser would offer more flow rate and more operating range. At the low-speed end there is a sufficiently wide operating range that matching is easy, it is at the high-speed end, when the range of incidence between choke and stall shrinks, that the problems become considerable. Figure 7.17 shows very clearly how a wrong choice of size, e.g; a diffuser which is too big in relation to the size of the impeller, can produce very poor performance at high speed, whereas the same combination at low speed can be very satisfactory. Just as with the multistage axial compressor it is probable that mismatching is far more common as a cause of poor performance with high pressure ratio machines than details of impeller or diffuser vane shape. Just as with the axial compressor the central problem is that details of the flow need to be known in order to size correctly the components and some of these, notably the blockage, are not generally predictable with great precision.

Pressure rise in diffuser passages

Blockage has been considered in connection with the mass flow capacity of the throat of the vaned diffuser, but it has another effect on the performance of diffusers. Studies of two-dimensional channel diffusers have shown the importance of inlet blockage in determining the maximum possible pressure rise. A thorough treatment of diffusers, both from the point of view of basic performance of simple diffusers and also the more complicated types used in compressors, with a comprehensive list of references, is given by Japikse (1984). Figure 7.18, taken from Rundstadler *et al.* (1975), shows how the maximum pressure rise coefficient varies for a two-dimensional diffuser as the aspect ratio and the blockage are varied. The example given is for a throat Mach

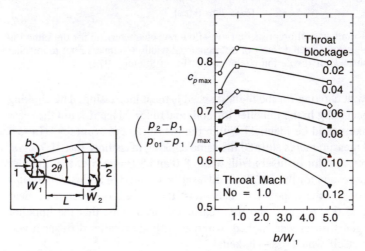

Fig. 7.18 Maximum pressure rise coefficient for flat diffusers as a function of aspect ratio, b/W_1, and throat blockage. Throat Mach number on centre-line = 1.0. (From Rundstadler and Dean, 1969)

number of 1.0 and the blockages also refer to the throat. It should be noted that the high throat Mach number is not in itself deleterious but lowering the back pressure, so that strong shocks exist after the flow becomes supersonic, would, of course, lead to very poor diffuser performance. The aspect ratio, the ratio of the vane height to the throat width, is a relevant variable but Rundstadler *et al.* found only a small variation in the maximum pressure coefficient with aspect ratio for values less than one in the range likely to be encountered in the vaned diffuser of a centrifugal compressor. The geometric variable which does have a marked effect is the ratio of the passage length to the width at entry. In the range normally encountered in compressors the effect of Mach number and Reynolds number variation is small and the dominant flow variable in determining the maximum pressure rise is the blockage. As Fig. 7.18 illustrates, for any given diffuser the pressure rise drops rapidly as the blockage is increased.

Blockage correlations and effects on vaned diffusers

Figure 7.19 shows results given by Kenny (1984) for the effectiveness of cambered-vane diffusers and pipe diffusers. Effectiveness is a parameter common in the assessment of diffuser performance and is the ratio of the achieved static pressure rise to that which would be produced by an ideal uniform flow experiencing the same increase in passage area. In Fig. 7.19a the measured effectiveness of pipe diffusers is shown compared with a correlation curve. The effectiveness is around 80 per cent for small blockage falling to 55 per cent at the maximum blockage of about 0.20. As well as showing the dramatic effect of the blockage on the realizable pressure rise the figure also indicates that a high degree of uncertainty exists, with a scatter of about 30 per cent of the correlation value. The performance of the pipe diffuser approaches that of an isolated diffuser tested by Sprenger (1959) for some values of blockage. Figure 7.19b compares correlation curves for the effectiveness of the pipe diffuser and a cambered vane diffuser, which for a given level of blockage at the throat shows superior performance for the pipe diffuser.

The blockage at the throat is crucial in determining the pressure rise in the

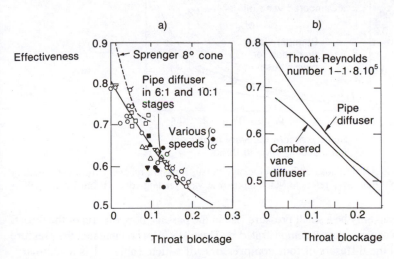

Fig. 7.19 Plots of diffuser static pressure effectiveness versus blockage at throat. Figure on left compares measurements in pipe diffuser with correlation; figure on right compares correlations for pipe diffuser and cambered vane diffuser. (From Kenny, 1984)

diffuser passage. It is desirable not only to keep the blockage small, but also to be able to estimate its extent; the level of blockage is required to estimate the pressure rise of the diffuser and to size the diffuser throat as part of the matching of diffuser to impeller. The throat blockage is normally estimated on the basis of correlation, adopting an approach first suggested by Kenny (1972), but since supported by Dean (1974) and Tramm in the discussion of Baghdadi and McDonald (1975). The type of correlation used is shown in Fig. 7.20, taken from Kenny (1984). The correlating parameter is the pressure rise coefficient from the mixed out state downstream of the impeller to the throat of the diffuser where the pressure is p_t,

$$c_p = (p_t - p_{2*})/(p_{02*} - p_{2*}).$$

(The mixed out condition is a hypothetical one but it provides a consistent means of assessing the stagnation and static pressure, p_{02*} and p_{2*}, after the impeller.) If the throat blockage is well correlated by the pressure rise coefficient it also follows that the shape of the inlet velocity profile (the outlet profile from the impeller) is not of overwhelming importance since this varies widely from one configuration to another.

A most important conclusion can be drawn from Figs 7.19 and 20. If there is to be a large pressure rise between the impeller and the diffuser throat there will be large blockage at the throat; if there is large blockage at the throat

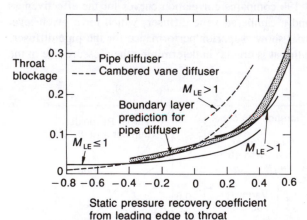

Fig. 7.20 Correlations of throat blockage versus pressure recovery from vane leading edge to throat. M_{LE} refers to Mach number at vane leading edge. (From Kenny, 1984)

there can only be a small pressure rise in the passage downstream of the throat. This was very clearly demonstrated by Dean (1974) who tabulated the pressure rise in the diffusers of four compressors, of which Table 7.1 is a summary.

From this one may safely conclude that to a large extent the pressure rise of the inlet region or the channel are alternatives with no clear indication that one is more efficient than the other. Rodgers (1982a) found in a wide range

Table 7.1 Diffuser static pressure rise coefficients

Compressor	Overall coefficient	Entry region fraction	Channel region fraction
Boeing RF-2, 10:1	$c_p = 0.69$	45%	55%
Creare 6:1	$c_p = 0.53$	23%	77%
Creare 4.5:1	$c_p = 0.73$	1%	99%
Creare 5.5:1	$c_p = 0.67$	34%	66%

of tests that the maximum static pressure recovery was in the order of 0.5 to 0.6 in the vaneless space compared to an overall recovery between impeller tip and diffuser vane exit of 0.7 to 0.8; nearly three-quarters of the static pressure rise was occurring ahead of the throat. (Rodgers also found a strong dependence on blockage). This is similar to Dean's conclusion that the pressure rise in region ahead and behind the throat are more or less alternatives, the sum varying much less than the parts, with the overriding input variable being the blockage.

The effect of vaned diffuser geometry

Great care sometimes goes into the design of the vane shape but there is little evidence that this has a significant effect on the flow. Sometimes the vanes are designed as cascades using a transformation to convert the radial geometry into an axial one for which there is extensive data, Pampreen (1972). Given the very low aspect ratio of the diffuser vanes and the highly distorted inlet profile (shown in Figs 7.8 and 7.9 for the radial velocity, but also present in the angle distribution) this does not seem a very profitable approach. Dean (1971) has observed that the shaping of the suction surface of the vane in the semi-vaneless space did not seem to matter very much and that the vane setting angle could be changed by several degrees without any effect on performance of a very high pressure ratio machine. The measurements of Rodgers (1982a) also show fairly small changes in performance of a variation in vane inlet angle of almost 5°. The shape and setting of the vanes are not crucial but, as indicated above, it is the throat area which is the determining quantity.

A design choice is the ratio of the vane leading edge radius to the impeller tip radius. One factor here is the need to reduce the vibration and noise generation which normally implies a minimum radius ratio of about 1.05. There is also very often the need to keep the overall size of the machine small and this gives a strong incentive to keep the radius ratio as small as possible. If the absolute flow from the impeller is supersonic some designers use an extended vaneless region so that the flow is subsonic before meeting the vane leading edge, although this may require radius ratios of as much as 1.25. There seems no particular merit in this because it is now widely accepted that in axial geometries inlet flows can be in the low supersonic region without serious increase in loss with high static pressure rise. An alternative is to keep the vaneless space small, since the vaneless diffusion process is generally accepted as less efficient, and then adopt diffusers which are thin and sharp in their inlet region to make best use of the high dynamic pressure. Figure 7.21, taken from Rodgers (1982a), shows the performance of four stages with different radius ratios for the diffuser inlet. The best performance was obtained with a radius ratio of 1.125, with performance significantly down at the lowest value of 1.035, but it is not clear how general this result is.

It is not easy to say what geometry is most suited to the vaned diffuser since the performance depends so critically on the matching and on the flow distribution from the impeller. It does seem that at high Mach numbers the vanes

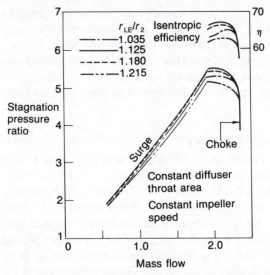

Fig. 7.21 The effect of vane leading edge radius ratio r_{LE}/r_2 (r_2 is impeller outlet radius). (From Rodgers, 1982)

should be thin in the region ahead of the throat and sharp at the leading edge. The pipe diffuser seems markedly better, particularly at Mach numbers above one. The explanations for the better performance of the pipe diffuser are largely conjecture but they may contain some truth. The shape generated when the pipe breaks through on the inside of the ring has characteristic 'swallow tails' which protrude forward and may inhibit secondary flow in the vaneless and semi-vaneless regions.

The number of vanes is another important variable which seems able to influence the performance very markedly — the reason for this is either not properly understood or at least not published. Figure 7.22, taken from Japikse (1980) shows very clearly how large this effect can be. The stage with 17 diffuser pipes had a markedly larger operating range but suffered from fatigue and produced slightly less pressure rise around its design point. From static

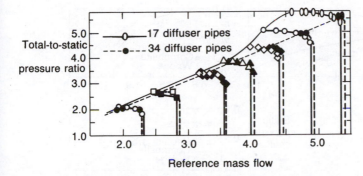

Fig. 7.22 A comparison of pressure ratio — mass flow characteristics for two centrifugal stages differing only in the number of diffuser passages (small differences in throat area account for differences in choke mass flow rate). (From Japikse, 1980)

pressure measurements on the endwall Japikse concluded that the wider range came from the presence of strong shocks such that at the higher speed the flow from impeller to diffuser throat would traverse two shocks with an accelerating region between, a configuration known to be stable. Others have found that reducing the number of vanes is detrimental to performance; Dallenbach, for example, in the discussion of Baghdadi and McDonald (1975) remarked that reducing the number of vanes had been found in his experience to reduce the flow range from choke to surge. Came and Herbert (1980) quote a test by Rolls-Royce for which the operating range was considerably wider with 13 diffuser vanes than 29. In their own tests Came and Herbert found that with 13 vanes the operating range was smaller than with 37 and the efficiency lower by about 2 per cent. Clearly this is not a well understood topic.

In the case of the pipe diffuser the number of diffuser pipes actually fixes the axial width of the diffuser at impeller outlet: the fewer pipes the greater the width. The choice adopted by Pratt and Whitney of Canada is such that the width is greater than the impeller tip and there must be a transition region where the axial width increases abruptly.

Tests in stationary rigs and the importance of unsteadiness

In almost every aspect of the design and performance of the diffuser it has been found that the evidence is confusing or incomplete or with a lot of scatter. One problem is the difficulty of getting a common base on which to compare measurements if the performance of the diffuser depends strongly on the upstream impeller. Baghdadi and McDonald (1975) built a simple stationary rig which would produce a radial swirling flow free from wakes (the non-rotating wakes from the swirl vanes had been criticized in earlier tests of diffusers made without an impeller). They examined three types of vanes. The Mach number at the vane leading edge was subsonic but the Mach number at inlet to the vaneless space, equivalent to the absolute Mach number out of the impeller, went as high as about 1.3. They first established that the velocity

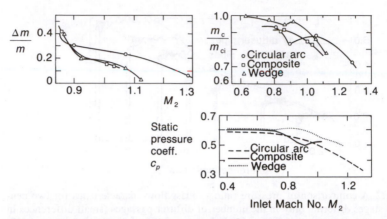

Fig. 7.23 Measured performance of three vaned diffusers versus Mach number M_2 at inlet to vaneless space. $\Delta m/m$ is range of corrected mass flow from surge to choke over corrected mass flow; m_c/m_{ci} is choking mass flow over choking mass flow for ideal flow. $c_p = (p_3 - p_2)/(p_{02} - p_2)$. (From Baghdadi and MacDonald, 1975)

profile into the diffuser was fairly similar to that measured downstream of an impeller. Some of their results are shown in Fig. 7.23. It should be noted that the wedge-type vane showed a slight increase in pressure recovery around sonic inlet conditions. The range of mass flow between choke and stall decreased with increase in inlet Mach number, but more slowly with the circular arc vanes. The ratio of the actual mass flow at choking conditions to the ideal value decreased as the inlet Mach number was raised showing that the blockage was increased.

The Baghdadi and McDonald paper attracted nine written discussions, many of them long. Much of the discussion addressed the stability aspects for which this facility was not most useful, but some criticized the nature of the experiment since it omitted the rotating jet-wake flow from the impeller and the axial profile at inlet was said to be unrepresentative. In his discussion Tramm showed comparative results from the test rig with a different diffuser vane made to match one tested in a compressor. He was able to show very similar behaviour for the rig and compressor, with a trend for the blockage at the throat as a function of inlet pressure recovery in line with that given by Kenny (1972). In their closure this point was also addressed by Baghdadi and McDonald, who plotted all their measurements as throat blockage against inlet static pressure coefficient and, with some considerable scatter, found reasonable agreement with the Kenny correlation. The scatter is most pronounced at the low pressure recovery end, which is not surprising since there other factors than pressure rise must be determining the blockage. In a later paper Baghdadi (1977) gave a more extensive comparison of the vaned diffuser performance in the stationary rig and in a compressor and, within the experimental spread, the agreement is remarkably good. Some of these results are shown in Fig. 7.24 from which the intermixing of the somewhat scattered results of both the rig and the compressor can be seen.

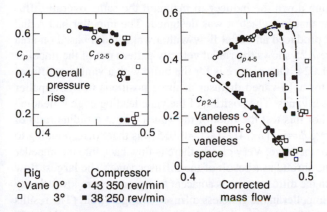

Fig. 7.24 Comparison of vaned diffuser pressure rise coefficients for test rig and real compressor (vane angle refers to inlet swirl vane setting). (From Baghdadi, 1977)

As far as stability is concerned, it is the composite pressure rise from diffuser inlet to outlet which matters, since the components are closely coupled. For a compressor the impeller should be included too. The pressure rise of the entire diffuser shown in Fig. 7.24 has a negative slope with respect to flow rate throughout the flow regime and is therefore stable. The combined vaneless and semi-vaneless region ahead of the throat is strongly negatively sloped, conveying a significant degree of stability. The pressure rise in the channel is mainly positively sloped, undoing some of the stabilizing effect of the inlet region. The positive slope for the channel is almost certainly a result of the increased blockage at entry to the channel resulting from the greater pressure rise ahead of the throat.

The work of Baghdadi is strong evidence that the unsteadiness of the flow seen by the diffuser as a result of the circumferential non-uniform flow out of the impeller is of secondary importance. As part explanation it should be noted that the circumferential velocity leaving the impeller is normally about three times the mean radial velocity and the circumferential velocity is proportionally much more uniform. The non-uniformity out of the impeller therefore gives swings in the flow angle but relatively small swings in the stagnation enthalpy or pressure. The feature of non-uniform flow which reduces the pressure rise in diffusers, the regions of low stagnation pressure, is thus mainly removed by the relative motion of impeller and vanes. The absolute stagnation temperature and pressure for flow in the ideal wake will be higher than in the jet, so the usual experience of a diffuser with distorted inlet flow is not relevant. The position is less clear, however, when the wake is collected on the shroud wall.

The effect of impeller outflow non-uniformity

Inoue and Cumpsty (1984) investigated the interaction between an impeller and the diffuser vanes consisting of circular arcs. The impeller was run at low speed after machining down the inducer to maintain the same overall diffusion at low speed as that for which it was designed. The impeller had radial blades at outlet and produced an outlet flow with a wake width based on area of about 0.4. Figure 7.25 shows the radial velocity pattern out of the impeller at a radius ratio of 1.024. One picture is for the build with a vaneless diffuser and the others are for a 10-vaned diffuser at three positions of the impeller relative to the diffuser vanes. The position of the vane leading edge is denoted by the letters DV. For this test the inlet of the vanes was at a radius ratio of 1.04 and the flow rate $\phi = m/\{\rho\pi/4D_2^2 U_2\} = 0.055$ was that corresponding to peak pressure rise for the stage. Very clearly there is flow back into the impeller occurring near to the shroud as a result of the diffuser vane. The largest flow reversal occurs when the diffuser vane is adjacent to regions where low velocity occurred out of the impeller in the vaneless diffuser configuration. (The results shown were all ensemble averages so that instantaneous flow reversals might be much larger). With the vaned diffuser at a flow coefficient of 0.045 there was slightly more reverse flow near the hub than the shroud. The amount of backflow, and all other evidence of the interaction, decreased as the radius ratio of the vane leading edge was raised to 1.1 and also as the number of vanes was increased to 20 and 30. No systematic trend between the pressure rise for the impeller or stage and the occurrence of backflow could be detected.

Inoue and Cumpsty resolved the instantaneous velocities and pressures into the time-mean distribution observed by a stationary observer and the unsteady distribution from the pattern rotating with the impeller. The time-mean velocities at mid-height are shown in Fig. 7.26; the variation in velocity is largely explainable by the diffuser vane loading. Surprisingly, in view of the

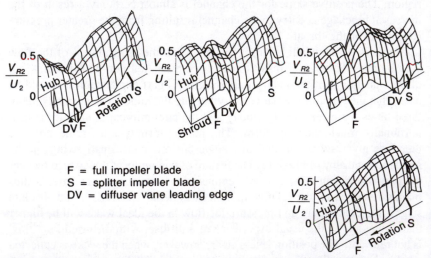

F = full impeller blade
S = splitter impeller blade
DV = diffuser vane leading edge

Fig. 7.25 The radial velocity pattern at a radius ratio $r/r_2 = 1.024$. Three top traces for vaned diffuser (10 vanes, leading edge at radius ratio 1.04), bottom right-hand trace for vaneless diffuser. Measured at flow rate for maximum pressure rise with vaned diffuser. (From Inoue and Cumpsty, 1984)

non-uniformity in the thick viscous endwall regions, the velocity distributions at mid-height can be predicted quite well by a method assuming inviscid irrotational flow (Fisher and Inoue 1981). The time-mean velocities, averaged circumferentially, are virtually identical in their distribution across the passage at the radius ratio of 1.024 for the vaned and vaneless diffusers. However, quite large differences are locally apparent at the circumferential position near to the vane leading ledge. The corresponding unsteady parts of the velocities measured by Inoue and Cumpsty were compared with the unsteady velocities measured downstream of the impeller with the vaneless diffuser. These generally agreed very well, showing that the total velocity field can be reasonably well represented as the sum of the steady field from the vanes and the unsteady field from the impeller.

The evidence from this is clear and convenient, supporting the findings of Baghdadi (1977) discussed above, that the diffuser can be studied (and probably modelled) whilst ignoring the blade-to-blade variations of the impeller.

Static pressure rise and stagnation pressure loss

Morishita (1982) continued Inoue's work with a slightly different diffuser con-

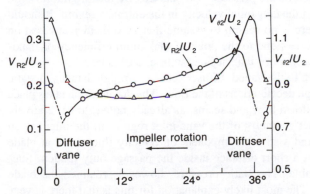

Fig. 7.26 The velocity pattern in the vaneless space measured at mid-passage height. Diffuser with 10 vanes, leading edge at radius ratio 1.04. Flow coefficient for maximum pressure rise. (From Inoue and Cumpsty, 1984)

figuration. Of particular interest are the measurements of pressure. Figure 7.27 shows the contours of static pressure in the 20-vaned diffuser with an inlet angle of 70° for the flow rate giving the largest pressure rise. This corresponds to a positive incidence of 6°. It is very noticeable that most of the static pressure rise occurred near the inlet in the semi-vaneless space. The isobars were nearly normal to the flow at entry to the passage whereas further upstream they were nearly circumferential. Similar effects have been shown at high speeds, for example Japikse (1980). The small rise in pressure downstream of the throat is because of the blockage build-up which keeps the velocity in the passage high despite the area increase.

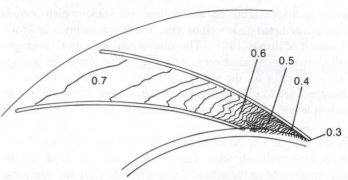

Fig. 7.27 Mean static pressure contours, normalized by ρU_2^2, in a vaned diffuser at flow coefficient for peak stage pressure rise. Diffuser with 20 vanes, leading edge radius, ratio 1.04, vane inlet angle 70°, mean flow inlet angle 76°. (From Morishita, 1982)

The large pressure rise in the entrance region, the zone of rapid adjustment in Dean's terminology, makes clear that it is sometimes more appropriate to think of the diffuser in terms of blades or vanes than as diffuser passages, with a large proportion of the pressure rise obtained by turning the flow, in fact removing some of the tangential velocity in the entrance region. It should also be noted that there is no reason to assume that zero incidence based on mean velocity will correspond to the point of maximum efficiency or maximum pressure rise, any more than it does with axial blades.

The entrance region to the vaned diffuser sees very rapid changes in static pressure and very large decelerations of the flow. The deceleration takes place with a highly non-uniform flow and seems, as already noted, to be relatively insensitive to the exact geometry of the vane inlet region. In the inlet region Morishita (1982) found a strong component of velocity fluctuation at blade passing frequency but a short distance inside the passage only random fluctuations were detectable. The static pressure, however, retained the blade passing noise signal. The most likely explanation for this is that there is very rapid mixing in the zone of rapid adjustment and that this is responsible for the removal of the vortical components of unsteadiness, i.e. the impeller wakes and any other separation, and also makes the large rise in static pressure possible.

Although the inlet region is where a large part of the pressure rise takes place, it is also where much of the loss is generated. From his measurements Morishita estimated the total-to-total efficiency of this stage was 86 per cent and the efficiency of the impeller 96 per cent (the absolute accuracy of these values is not great, and the estimates seem rather high, but it is the breakdown which is most relevant here). The impeller was therefore only converting around 4 per cent of the work input into loss. The losses in the vaneless space $(1.024 \leq r/r_2 \leq 1.04)$ were 2 per cent, in the inlet region $(1.04 \leq r/r_2 \leq 1.175)$ 7 per cent and in the downstream region $(1.175 \leq r/r_2 \leq 1.633)$ only 1 per cent. High loss is what would be expected of a region of high mixing but it is perhaps surprising that this region contributes so much of the stage loss, much more than the impeller. These figures need to be treated with some reservation, particularly since it is a low-speed test, but there seems no reason to doubt that it is in the region where the very high velocity from the impeller meets the diffuser that the most loss is created.

Summary and conclusions for the vaned diffuser

The evidence suggests that the conventional method of dealing with components separately, for example the impeller without the diffuser and the diffuser without the impeller, is reasonably accurate and will suffice for most purposes. It supports the results obtained by Baghdadi. Not only is the jet-wake mixing relatively unimportant as a loss source for reasonably well-designed impellers, but any detrimental effect of the wake in perturbing the flow in vaned diffusers seems much less important than was supposed. The inlet region, the semi-vaneless space, is important for the static pressure rise it produces, the large proportion of the total stage loss generated and the creation of blockage. A high pressure rise upstream of the diffuser throat gives high blockage there and then only a small pressure recovery downstream.

The calculation of the flow in the vaned diffuser should not be beyond the capacity of modern three-dimensional viscous methods. It remains to be seen if special levels of shear stress in the inlet region are needed to represent the mixing in the flow. It would seem that useful solutions could be obtained whilst treating the flow as steady, that is ignoring the rotating pattern from the impeller. The calculations that have been carried out so far have usually been of the boundary layer superimposed on a core flow type, but, given the highly three-dimensional nature of the flow and the fact that flow reversal is common, it is hard to see what such calculations can yield except in comparative terms between one configuration and another. Estimates of blockage arrived at by such methods should be regarded with caution.

The wide variety in diffuser types, in inlet radius ratio and in number of vanes all point to the underlying fact that the flow in vaned diffusers is not understood and the designers are forced to use methods which are not based on sound physics. (A similar conclusion can be drawn concerning the design of impellers!) Success seems to come to those organizations which carry out many systematic and carefully thought-through designs leading to systematic testing; the 'secret' is probably the following of design hunches and the continuance along directions which are found to lead to improvements – there seem to be no very effective short cuts. In this connection it should be stressed that the single most important design step, after the initial choice of the design point, is the correct matching of components: at high speed an error in this will almost certainly lead to serious loss in pressure ratio, flow capacity and efficiency.

7.5 The volute

The geometry of a volute or scroll surrounding a compressor with a vaneless

diffuser is shown in Fig. 7.28. Such arrangements are very common with industrial compressors, particularly single-stage ones, and with turbocharger compressors. They are used with vaned diffuser but more commonly with vaneless
ones. The geometry shown in Fig. 7.28 is for a rather large flow area, corresponding to a highly backswept impeller, and the tongue and the off-take pipe
are arranged so as to give a rather smooth transition; where space is at a
premium, such as on automotive turbochargers, the transition can be much
more abrupt.

The volute in Fig. 7.28 is shown with an approximately circular cross-section,
the section being symmetric with respect to the vaneless diffuser and wholly
outside it. These features tend to increase the overall diameter. The configuration with the volute at a larger diameter than the vaneless diffuser is known
as an external arrangement. If the flow path after the diffuser is curved or
wrapped around so that the volute outer diameter is smaller than the outside
of the diffuser it is known as an internal arrangement — many designs lie
between these two extremes.

Quite substantial additional pressure rises can be obtained from the volute
but appreciable losses can also be produced. For an industrial compressor stage
there is typically an efficiency loss for a well-designed volute of between 2

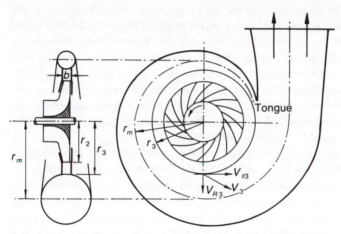

Fig. 7.28 The geometry of a volute surrounding a centrifugal compressor with a vaneless
diffuser

and 5 per cent, based on compressor inlet and outlet stagnation pressure, nearly
all arising from the inability of the volute to use the radial kinetic energy out
of the diffuser. It is remarkable that so little has been done, at least as reflected
in the number of papers or articles, to understand the flow in volutes. In a
recent paper Weber and Koronowski (1986) remarked that a computer search
had found only 28 references on the subject of volutes and most of these per-
tain to pumps. A comprehensive reference list has been given by Sideris and
Van den Braembussche (1987) but there are still few referring to compressors.

The consideration of the volute and the design method are different when it is used with a vaneless or vaned diffuser. Most of this section will be directed to the vaneless case but some remarks about the vaned case will be made at the end.

Volute with vaneless diffuser

It is not only the consideration of efficiency and static pressure rise of the volute itself which is important, for the sizing of the volute can exert an enormous influence on all aspects of the performance for a compressor with a vaneless diffuser. Figure 7.29, taken from Stiefel (1972), shows the stagnation pressure ratio versus mass flow characteristic for the same impeller and vaneless diffuser with two different volutes. That on the left had a volute cross-sectional area some 30 per cent larger than that on the right. The striking difference is the performance at high pressure ratios, above about 3.5, where the smaller volute is very much better. One of the most interesting aspects of these performance maps is that the performance was much better at pressure ratios (and rotational speeds) much lower than that for which the volute was designed. The larger volute was designed for 3.8:1 and gave best performance at about 2:1 whereas the smaller volute was designed for 6:1 and gave best performance at about 3.5:1; this will be discussed further in connection with the design

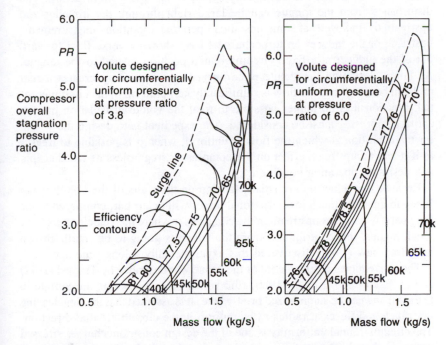

Fig. 7.29 Measured pressure ratio versus mass flow characteristic for one impeller and vaneless diffuser but two different volutes. Lines of constant speed in thousands of rev/min and contours of isentropic efficiency shown. (From Stiefel, 1972)

method. With the larger volute the pressure rise characteristics perform some strange undulations at the high-pressure end which are amplified in the efficiency contours, but no explanation is available for these. The large changes in performance corresponding to the undulations point to quite significant changes in the overall flow, it being very probable that the work output from the impeller changed for there to be such significant changes in the pressure rise.

From Fig. 7.29 it is clear that the matching of the volute to the impeller and diffuser is of great significance but it is a different type of matching from that which was discussed in connection with the vaned diffuser. There the matching was essentially a compressible flow phenomenon which could be reduced to choosing the flow function (corrected mass flow) out of the impeller to be acceptable to the diffuser vanes on a purely one-dimensional basis. The conventional layout for a compressor, though not for pumps, has the volute downstream of a diffuser so that the velocities in the volute are relatively small and choking is not an issue.

The matching of the diffuser vanes could be handled one-dimensionally because there are normally a large number of them and their periodic effect decays rapidly upstream. With the volute it is very different because there is normally only one off-take and one tongue. Any non-uniformity generated by these is therefore going to contain a large harmonic at the fundamental, once per revolution, which will decay only slowly in the upstream direction. The disturbance from the tongue can be large right through the impeller and upstream of it as well; at some unstalled operating conditions measurements upstream of the inducer on turbochargers have shown reverse flow forward out of the inducer at a fixed circumferential position related to the tongue. As far as designing the volute is concerned non-uniformity creates a real problem because once the flow is significantly non-uniform in the circumferential direction it is no longer possible to treat the volute in isolation but the entire compressor must be considered. The expedient is to design the volute for the condition at which the flow is uniform, when it is possible to treat it as if it had no upstream effect on the compressor, regardless of whether this is a desirable operating condition.

Because the velocities are small, friction on the walls of the volute is not likely to generate much loss; frictional losses will occur but, compared to the long narrow passages upstream in the compressor where the velocities are very much higher, this contribution from the volute is going to be small. Brown and Bradshaw (1949) discovered this experimentally a long time ago.

The conventional design theory of the volute was described by Traupel (1977) and by Eckert and Schnell (1980) (the volute being known as die Spirale in German) and can be summarized briefly here. It is assumed that the flow leaving the vaneless diffuser at radius r_3 is uniform in the circumferential direction. The circumferential uniformity is one of the design constraints but, as stressed earlier in this chapter, the flow leaving a vaneless diffuser will have a large variation in velocity across the width of the diffuser. For the purpose of the analysis mean velocities out of the diffuser in the radial and tangential directions V_{R3} and $V_{\theta 3}$ are used. The flow will be assumed to be frictionless from the exit of the diffuser to the pipe leading from the volute. With uniform static pressure and no wall friction it then follows from the conservation of the moment of momentum

$$r_3 V_{\theta 3} = r V_\theta$$

throughout the volute. This means that regardless of the shape of the volute the circumferential velocity is fixed by the local streamline radius about the axis of the machine; in particular at the mean radius of the volute r_m

$$r_3 V_{\theta 3} = r_m V_{\theta m}. \tag{7.1}$$

Once the mean radius of the volute is prescribed it follows that for a given tangential velocity out of the diffuser the mean velocity through the volute is fixed.

The radial component of velocity out of the diffuser will normally be much less than the tangential component. Since the kinetic energy of the radial velocity is not used at the design condition it does not affect the efficiency of the scroll whether the radial velocity vanishes entirely by turbulent mixing or by a large scale eddying motion before finally being dissipated by turbulence and viscosity. The radial velocity enters the conservation of mass flow around the circumference, which for a small arc around the circumference, can be written

$$\mathrm{d}/\mathrm{d}s\left(\int \rho V_\theta \mathrm{d}A\right)\mathrm{d}s = \rho_3 V_{R3} b r_3 \mathrm{d}\theta \tag{7.2}$$

where the exit of the diffuser at radius r_3 has an axial width of b, $\mathrm{d}A$ denotes an element of the volute cross-sectional area at the position under consideration and s here denotes distance around the volute. The length $\mathrm{d}s$ can be replaced by $r_m \mathrm{d}\theta$ where r_m is a mean radius.

It is appropriate to replace the integral in equation 7.2 by the product $\rho_m V_{\theta m} A$, where ρ_m and $V_{\theta m}$, the values at the mean radius, are taken to give a reasonable average for the whole cross-section. Introducing equation 7.1 into 7.2, and using $V_{\theta 3}$=constant, leads to

$$r_3 V_{\theta 3} \mathrm{d}/\mathrm{d}s\left(\rho_m A/r_m\right)\mathrm{d}s = \rho_3 V_{R3} r_3 b \mathrm{d}\theta \tag{7.3}$$

Since ρ_m is uniform circumferentially this gives on rearranging

$$\mathrm{d}/\mathrm{d}s\left(A/r_m\right) = \left(\rho_3/\rho_m\right)\left(V_{R3}/V_{\theta 3}\right)\left(b/r_m\right) \tag{7.4}$$

showing how the geometric quantity which characterizes the volute, the ratio A/r_m, must vary along its length to maintain V_θ constant. The right-hand side of equation 7.4 contains a density ratio ρ_3/ρ_m but, since the difference in velocity at r_3 and r_m is likely to be small compared to the local velocity of sound, the density ratio will be nearly one. The principal variable in the specification on the volute shape is the ratio $V_{R3}/V_{\theta 3}$ and for a particular value the variation in A is fixed once r_m is chosen. Many aspects of the shape are probably not critical, since the loss is mainly from the kinetic energy of the radial velocity, and a natural choice is the log-spiral as described in Section 7.1. Major design decisions are the shape of the tongue, the inclination this makes to the local tangent and radius at which the tongue is placed.

Once a volute is specified the design condition of circumferentially uniform flow can only be achieved for one particular value of the ratio $V_{R3}/V_{\theta 3}$. This means that at a given corrected rotational speed there will be one and only one inlet corrected mass flow which will give the uniformity designed for. Figure 7.30 shows on the left-hand side the stagnation pressure ratio—mass flow characteristic for a compressor built and tested by Stiefel (1972). One line shows the locus of conditions for which the uniform static pressure should be achieved; the design condition was a point on this line at which the pressure ratio was 2.0. Shown on the same diagram is the line of peak efficiency, which does not in general coincide with the line for uniform pressure. The diagrams on the right of Fig. 7.30 show the circumferential variation of static pressure around the annulus along a single constant speed line. Point 1 is close to the locus for uniform static pressure and the very nearly uniform static pressure measured indicates that the design process has succeeded in its goal. Point 2 is near to the stall or surge line; for this case the radial velocity will have been smaller than at design and the scroll therefore too large. The flow is therefore decelerated in the θ-direction and the pressure rises around the volute. Point 3, by contrast, has a flow rate which is too large and a scroll which is too small, so that the flow is accelerated in the tangential direction. For both point 2 and point 3 the variation in static pressure is very large and larger at impeller exit than in the volute itself. For point 3 the fall in static pressure around the circumference at the impeller exit is about 0.47 bar compared with the overall rise in stagnation pressure for the stage of about 0.65 bar. To produce a circumferentially repeating pattern the tongue is required to produce a pressure difference, a rise in pressure in the case of point 3 requiring that the stagnation point has to be on the impeller side of the tongue.

The variation in the entry flow direction to the volute as the flow rate or speed are changed is essentially very similar to that discussed in connection with the entry to the vaned diffuser earlier in this chapter. As the impeller speed rises the flow into the diffuser inclines more and more towards the tangential. For impellers with large backsweep the absolute tangential velocity at outlet increases rapidly as the radial velocity falls, exacerbating the tendency for the flow to incline towards the tangential as speed increases. The friction in the vaneless diffuser works so as to reduce the tangential momentum so that the mean tangential velocity decreases *more rapidly* than inversely proportional to radius. The friction in the radial direction only drops the static pressure (and therefore also the density) so the mean radial velocity decreases slightly *more slowly* than inversely proportional to radius. The result is that

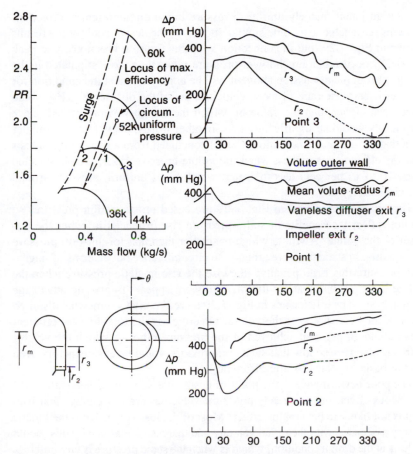

Fig. 7.30 Overall performance and static pressure variation around circumference for compressor with vaneless diffuser and volute. Volute designed to give uniform static pressure circumferentially for pressure rise of 2.0. Static pressure variations shown all measured at 44 000 rev/min. (From Stiefel, 1972)

comparing outlet from the vaneless diffuser, station 3, and impeller outlet, station 2

$$V_{R3}/V_{\theta 3} > V_{R2}/V_{\theta 2} \text{ or } \alpha_3 < \alpha_2.$$

The difference is normally small, only a few degrees once the initial mixing to uniform is over, see Fig. 7.14, and as a reasonable working approximation the mean flow direction out of the diffuser can be taken to be that of the mixed-out flow from the impeller.

Given an impeller and vaneless diffuser it is therefore possible to design a volute for a particular combination of inlet corrected mass flow and pressure ratio. From pressure-versus-angle trace for case 1 in Fig. 7.30 it is clear that Stiefel had succeeded there in producing a flow at the design point which met the design intent, namely uniform pressure in the circumferential direction. It seems reasonable to assume similar success in the design goal for the results shown in Fig. 7.29, but in connection with this figure it was noted that peak efficiency occurred at much lower pressure ratios than those stipulated in the design. From the discussion above it will be apparent that at the condition for peak efficiency the ratio of mean velocities out of the diffuser $V_{R3}/V_{\theta 3}$ would have been higher than the value on which the design was predicated, since the density would be significantly less, in other words, the volute was smaller than that for the design condition. Steifel concluded from a wide range of tests that the efficiency is highest when the volute cross-sectional area is selected to be 10 to 15 per cent less than that which gives uniform pressure around the circumference.

The efficiencies shown in Fig. 7.29 are based on stagnation pressures so do not reflect the benefits of static pressure rise produced in either the diffuser or the volute. A volute which is smaller than that for uniform pressure gives a drop in static pressure around the circumference but in terms of stagnation pressure this is no penalty; likewise the rise in static pressure when the volute is larger than that for uniform pressure appears to give no advantage on that basis. The efficiencies in Fig. 7.29 therefore give a somewhat distorted view of the best choice of the volute; nevertheless the total-to-total efficiencies around 70 per cent achieved at the pressure ratio for which the volutes were designed are so low that such compressors would be unsatisfactory even on the basis of total-to-static efficiency.

The poor performance at the point of design has sometimes been attributed to viscous effects but as already noted the velocities are low enough that friction is not likely to be a major effect. Most of the loss comes from the kinetic energy associated with the radial velocity and there is no reason why this should be least at the design condition, which is when the static pressure is circumferentially uniform. This condition constrains the velocity around the volute since it must satisfy the conservation of moment of momentum, equation 7.1. Under this constraint the kinetic energy associated with the radial velocity leaving the diffuser cannot be redirected to the circumferential direction around the volute and must therefore be dissipated.

All this is changed once the condition on uniform pressure in the circumferential direction is relaxed. If the pressure decreases in the circumferential direction the flow out of the diffuser can be turned towards the tangential (i.e. V_θ allowed to increase in the circumferential direction) and the radial component is no longer wasted. It is this which makes operation so much more efficient when the volute is smaller than that chosen to set up the uniform conditions. If, on the other hand, the volute is oversized, so that the static pressure increases around the annulus, conditions are exacerbated with the flow turned back towards the radial and a greater amount of kinetic energy wasted.

The difficulty of designing the volute to have the desired fall in static pressure comes when it is realized that the consequent non-uniformity is not just at the outlet from the vaneless diffuser but throughout the entire diffuser, see Fig.

7.30, and through the impeller as well. A proper calculation needs to include the whole compressor stage. To complicate matters further, the conditions in the impeller passages are unsteady. Some progress has been made with this, Sideris and Van den Braembussche (1987), though it is still some way from being in a position to directly affect the design of compressors.

To sum up it would seem that a problem in the design of the volute has been the failure to recognize that the condition which makes the flow circumferentially uniform, which can be calculated easily and which essentially allows the impeller, diffuser and volute to be considered separately, is not the one to give best efficiency. The reason for this is that it fails to take advantage of the kinetic energy associated with the radial velocity out of the diffuser. It is therefore a more serious problem for compressors for which the radial velocity is high in relation to the absolute tangential velocity, for example those with highly backswept impellers. Given the confusion which has arisen with the nature of loss and the difference between the condition for design and the optimum operating condition, it is not surprising that wide discrepancies have occurred in the literature for estimating the losses, for example Weber and Koronowski (1986).

A falling pressure around the circumference is clearly not ideal for the generation of high static pressure rise; a more successful strategy might well use the falling static pressure to maximize the stagnation pressure at the plane level with the tongue and then use a conical diffuser to decelerate the flow.

Volute with vaned diffuser

The effect of the diffuser vanes is to prescribe the diffuser outlet flow direction so that more or less regardless of the design of the volute or the compressor operating point the ratio $V_{R3}/V_{\theta 3}$ is fixed. The volute is designed for the single value of inlet angle that this corresponds to.

The volute is normally designed so that the static pressure is uniform around the circumference, just as with the vaneless diffuser. If the sizing of the volute is such that the pressure around the circumference is non-uniform, the constraint on the flow angle means that the consequent variation in velocity magnitude must be communicated along the diffuser vanes to the impeller and beyond; the effect of the vanes is therefore to couple the volute and impeller more closely.

As a first approximation the mean flow angle in a vaneless diffuser of constant axial width does not change much from inlet to outlet, and this means that the radial kinetic energy is usually small at outlet from the diffuser. A vaned diffuser usually turns the flow more towards the radial to obtain the highest possible static pressure rise and the lowest possible stagnation pressure loss. As a result a larger part of the total kinetic energy at outlet from the vaned diffuser is in the radial direction. The radial kinetic energy which is lost in the volute is therefore likely to be higher with a vaned diffuser and the loss in efficiency correspondingly greater.

8 Viscous effects in compressors

8.1 Introduction

The term viscous effect is used here and elsewhere as a shorthand to include the shear stresses generated by turbulence. In most compressor flows the existence of the turbulent shear stresses is essential for the flow to surmount the pressure gradients without separation. Generally the performance of compressors improves as the turbulent stresses get stronger relative to the laminar viscous stresses, that is as the Reynolds number increases. The regions in which the viscous effects are largest are often referred to as boundary layers. For brevity and compatibility with other literature this term is also used in this book but, for reasons discussed in this chapter, the boundary layer model is potentially misleading in compressor flows.

Viscous effects have three major influences on the flow in compressors.

1. Viscous effects put a limit on the pressure rise which can be produced. Attempts to exceed this lead to a flow instability and then rotating stall or surge, as discussed in Chapter 9.
2. Viscous effects lead to blockage, an effective reduction in the flow area, which affects the work input, particularly for axial compressors, and which can have an enormous effect on mass flow capacity.
3. Viscous effects in shear layers are ultimately responsible for all loss generation apart from the small losses generated in shock waves. Although loss is undoubtedly very important the estimation of the maximum pressure rise and the prediction of blockage are normally more worrisome to the designer.

Blockage has been introduced in connection with the stage matching in Chapter 2, the meridional flow calculation in Chapter 3 and the sizing of centrifugal vaned diffuser throats in Chapter 7. It may be defined by

$$B = 1 - \text{(effective flow area)/(geometric flow area)}$$

This can be rewritten in terms of the sum of the displacement thicknesses

$$B = 1 - (A - \Sigma\delta^*)/A$$

where A is the total cross-sectional area and the displacement thickness

$$\delta^* = \int (1 - \rho v / \rho_\infty V) \mathrm{d}y$$

is the integral of the velocity–density deficit across the passage, ρ_∞ and V being the free-stream density and velocity in the usual boundary layer terminology. With a uniform flow region outside the viscous region, the simple case on which boundary layer theory is based, the evaluation of blockage is unambiguous, but with the complicated non-uniform flow across the whole passage, which is normal in compressors, this is not the case and some arbitrariness exists in defining the conditions corresponding to the freestream. A general form useful for turbomachines is

$$B = 1 - \left(\int \rho v \mathrm{d}A \right)_{\text{actual}} / \left(\int \rho v \mathrm{d}A \right)_{\text{no viscous regions}}$$

The blockage is perhaps the most critical quantity in high-speed compressor design but, as will be shown, its creation is not well understood nor its magnitude accurately predictable.

The nature of viscous flows in compressors

A complete description of the flow in compressors can only come from the full Navier–Stokes equations. Even these equations require some empirical input to accommodate the turbulent shear stresses, and this takes the form of Reynolds averaging so that the turbulent velocities appear only in the time-mean products giving rise to stresses. Solutions can now be found for the Navier–Stokes equations, considering only one blade row at a time, and this is discussed briefly in Chapter 11. For most purposes, however, solution of the Navier–Stokes equations is still not a practical method of analysis although it is now used as a way of understanding special flows − with the current rate of progress its use may be much more general in the near future. The use for several blade rows at one time seems quite a long way off.

Because of this the conventional way of considering viscous flows is to use the boundary layer methods, calculating the flow separately on the blade surfaces, Fig. 8.1a and for the meridional flow on the annulus walls, Fig. 8.1b, drawn in each case for an axial compressor. Boundary layers have a large literature and it is quite inappropriate to summarize it here. The classic text is Schlichting (1979) but many textbooks give good treatments; Duncan, Thom and Young (1970) in particular gives a highly informative description. An extensive account specifically devoted to the calculation of boundary layers in turbomachines is given by McDonald (1985). The boundary layer model and approximations have been enormously successful in many applications, in external aerodynamics in particular. This has given confidence in their application to compressors but, as will be shown below, this is in many cases misplaced. On the other hand designers have needed some tools and the boundary layer calculation methods have often been all that is available. A

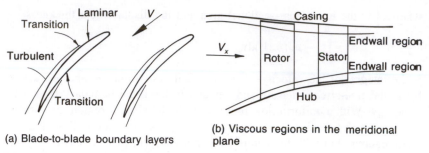

(a) Blade-to-blade boundary layers

(b) Viscous regions in the meridional plane

Fig. 8.1 Schematic of blade-to-blade and meridional viscous regions

description of the methods for compressors was given by Horlock (1969); at that time it seems to have been generally supposed that such methods would have a major influence on the design of compressors. Examining the literature one finds, however, that there is very little effective comparison of prediction with measurements inside compressors, as opposed to cascades, or with a range of different compressors without alteration in the empirical parameters introduced to obtain a satisfactory match with the data. In the following paragraphs some reasons are given why the accurate prediction of the viscous flows in compressors is so difficult and why boundary layer methods have such general problems.

The special problems of viscous flow in compressors

The grounds for caution in the application of boundary layer methods stem from the complicated nature of the flow in the compressor and from the severity or magnitude of the effects present. The following are some of the special features of compressor flows:

1. There are very steep pressure gradients in the flow direction in relation to what can exist without separating a two-dimensional boundary layer, both on the blades and on the annulus walls. The flow on the blades is often taken close to or beyond the point of separation and in this respect is like a heavily loaded wing. More serious is the flow near the annulus walls. In assessing the severity of a pressure gradient it is necessary to have a length scale relevant to the viscous layer, such as the displacement thickness δ^*. For an axial compressor the thickness of the viscous layer on the annulus walls (the region in which the viscous effects are large) is typically comparable to the blade chord and the displacement thickness an order of magnitude less: measured in these terms the pressure gradient along the annulus walls is very high indeed.

2. The stagnation pressure and temperature are not uniform in the freestream, i.e. the region outside the viscous region. This is particularly important for the flow in the meridional plane for which there is very often no properly defined freestream (which is the underlying idea behind the spanwise mixing discussed in Chapter 3). In consequence there is no proper edge to the viscous regions on the endwalls analogous to the boundary layer edge.

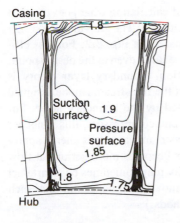 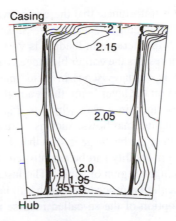

(a) Design point   (b) Increased loading

Fig. 8.2 Contours of stagnation pressure, non-dimensionalized by $\frac{1}{2}\rho\,U_{tip}^2$, downstream of a third stage stator. (Rotor moves from left to right.) (From Wisler *et al.*, 1988)

3. The turbulence levels are very high in the free-stream. This partly overlaps with the previous paragraph. High levels of turbulence are thought to be most significant for their effect on transition.

4. The flows are highly unsteady because of the passage of blades past one another. The unsteadiness takes the form of potential interactions between the pressure fields and also wake interactions, the latter being particularly important for transition.

5. There are large gradients in every direction including along the span of axial blades. As a result the static pressure is not constant across the viscous region. For blades which are long in relation to δ^* on the endwalls it may be possible to treat the flow as two-dimensional near the middle of the blade, since the variation in the blade shape is normally relatively small; near the endwalls and for low aspect ratio blades there may be no region for which the two-dimensional approximation is realistic. Figure 8.2 shows traverses of normalized stagnation pressure downstream of the third stage of a very efficient compressor, Wisler *et al.* (1987) and clearly there is a wide region at the design point for which the flow along the blade span is sensibly two-dimensional. As the pressure ratio is increased the influence of the endwall extends along the span of the blades and the region of two-dimensional flow decreases. Figure 4.52 also shows the extent to which a compressor blade can be operating without any region that can properly be called two-dimensional.

6. The dominant regions in terms of limiting the pressure rise, creating blockage and generating loss, are the corners, mainly those between the suction surface of the blades and the endwall. These are highly three-dimensional flows, normally with regions of separation. The flow is further complicated by the presence of tip clearance and by the skewed non-uniform flow at entry to the blade. (Since the corners are where most of the deleterious effects are

created it is unfortunate that there is no way of calculating what goes on short of the full three-dimensional Navier–Stokes methods.) The growth in the corner blockage with increase in loading is very apparent in Fig. 8.2. For less efficient compressors the corner blockage could be large even at the design point.

7. There is strong viscous-inviscid interaction. Boundary layer theory is predicated on the layer being thin enough that the freestream can be assumed to be prescribed in a way which is unaffected by the boundary layer. For decelerating internal flows, such as diffusers and compressors, this is not so. The growth of the blockage is such that the freestream velocity is altered appreciably and instability can be established. This becomes more serious at flow velocities in the region of sonic. (The instability may sometimes be an artefact of the calculation procedure which can be removed by changes in this, such as the adoption of the so-called inverse methods.)

The list of problems is considerable and serious. No method, not even those that solve the Navier–Stokes equations, can include all the effects. Inevitably what is used is to some extent an approximation.

Chapter organization

Despite its attraction, the division of the flow into endwall and blade regions is not genuinely satisfactory, in particular because the regions which have most effect on the flow in terms of blockage, loss and inception of stall are the corner regions between suction surface and endwall. Nevertheless to relate what is written here to the body of work which has been performed, and to give some order to it, the usual divisions have been retained. The major part of this chapter will therefore be in three main sections. The first will be devoted to the axial blade-to-blade flow for which the boundary layer is often a reasonable approximation. The boundary layer on the blades is normally of crucial importance only for special cases, for example when the blockage is critical in transonic flow or for the supercritical type of blade. In contrast the prediction of the flow in the endwall region is absolutely vital for all multistage axial compressors as well as flow for high pressure ratio centrifugal compressors. The endwall blockage is usually large enough to affect the performance of most compressors and the endwall and corner regions are responsible for most of the loss generation. The next section of the chapter therefore looks at the endwall regions in axial compressors, where the assumptions of boundary layer theory are not good. The final section looks at the viscous flow in centrifugal compressors. Before looking at the blade boundary layers a more general introduction to three-dimensional viscous flows in compressors is presented.

8.2 Three-dimensional viscous flows in compressors

Most flows are not two-dimensional in the sense that the flow over a surface is very often curved in the plane of the surface and this curvature of the

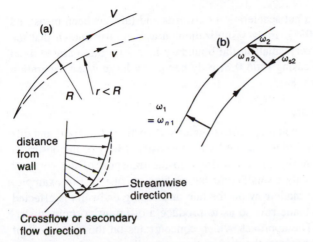

Fig. 8.3 Schematic representation of the formation of crossflow or secondary flow by curved flow over a surface. (a) shows streamline trajectory, solid line in freestream, broken line in boundary layer. (b) shows generation of streamwise vorticity by curvature of streamlines in plane of surface

freestream gives rise to three-dimensional boundary layers. To make the freestream follow a curved path there must be a pressure gradient parallel to the surface but normal to the streamline direction. The magnitude of this pressure gradient is sufficient to produce the centripetal acceleration of the freestream, V^2/R, where R is the radius of curvature of the freestream. The flow closer to the wall, inside the boundary layer, has lower velocity, $v < V$, and the pressure gradient set up by the freestream exceeds that needed to make the boundary layer fluid follow the same radius of curvature. The boundary layer therefore follows a tighter curve of smaller radius r, as sketched diagrammatically in Fig. 8.3. Components of velocity perpendicular to the local freestream direction, known as crossflow, are then produced inside the boundary layer. The mechanism sketched in Fig. 8.3a can be visualized to occur whenever there is curvature of the flow in the plane of the surface and the flow has a velocity which is non-uniform in the direction normal to the surface. This is very important for the flow on the casing and hub, as a result of the turning of the flow by the blades, and can also occur on the blades as a result of strong meridional curvature.

There are other ways in which three-dimensional flows in the viscous regions can be established in compressors. Crossflow can be produced whenever the centrifugal acceleration of the fluid in the viscous region differs from the pressure gradient perpendicular to the freestream direction. A simple example of this occurs in wakes from stator blades in an axial compressor. The freestream flow will have a component of swirl velocity V_θ which sets up a radial pressure gradient, to a first approximation given by $\rho(V_\theta)^2/r$. Inside the wake there is a smaller circumferential velocity v_θ so there is a net acceleration of the wake fluid radially inward. The boundary layer fluid on the stator blades is likewise driven inwards. For rotor blades the wake and boundary layer fluid will have a net acceleration outwards and this has been measured by Pougare *et al.* (1985). Flow visualization, however, has shown that the radial velocities are very small in the boundary layers on unseparated axial rotor blades of hub—casing ratio 0.8, only becoming large when separation occurs, McDougall (1988).

Secondary flow theory

The viscous layers may be thought of as regions of vorticity transport and diffusion, the latter being brought about by the viscosity. Many of the observed effects in turbomachinery can be put down to the transport of the vorticity already present, with only a small contribution from diffusion. For example, the inlet vorticity to a blade row on the hub and casing endwalls is affected by the turning in the blade row so as to produce a different vorticity distribution downstream. The approach which concentrates on the movement of existing vorticity and neglects the diffusion is normally referred to as secondary flow theory. It can give quite useful qualitative indications in regions where the shear stresses are not large, i.e. in regions where the diffusion is small, which means some distance out from the solid surfaces. Secondary flow in broad terms means flow at right angles to the intended primary flow, so the secondary vorticity is therefore parallel to the streamwise direction. The definitions of the primary flow can be slightly different, as explained by Horlock and Lakshminarayana (1973). The secondary flow nearly always has the connotation of being undesirable and the term 'secondary flow analysis' is reserved for that which is inviscid. A review of the field of secondary flow was given by Hawthorne (1965) and another by Horlock and Lakshminarayana (1973).

Figure 8.3b gives an explanation in terms of secondary flow for the same crossflow generated by a curved free stream in Fig. 8.3a. The flow is assumed to enter with normal vorticity ω_1 and no streamwise vorticity. If there is no diffusion or creation of vorticity the vortex lines are merely convected by the flow; the velocity on the inside of the curve is higher than that on the outside so the vector ω_2 is tipped up relative to ω_1. The vorticity ω_2 can be resolved into two components, one the downstream normal vorticity ω_{2n} and the other the secondary vorticity ω_{2s}. The process can be shown to be equivalent to that used in explaining Fig. 8.3a.

The first application of secondary flow to the theory of turbomachinery can be traced to the paper by Squire and Winter (1951). They looked at a cascade of vanes in the corner of a wind-tunnel and were able to predict remarkably well the crossflow velocity pattern produced by the endwall boundary layers being turned through 90°. The type of geometry and notation is shown in Fig. 8.4. The vorticity normal to the streamlines, convected in as the boundary layer on the endwall, was given by $\eta = \partial u / \partial z$. Although the flow was turned the primary or streamwise velocity U was equal in magnitude and directed along lines of constant height z at inlet and outlet. The undisturbed nature of the freestream was equivalent to assuming that the secondary flow was sufficiently weak that its effect on the primary flow could be neglected.

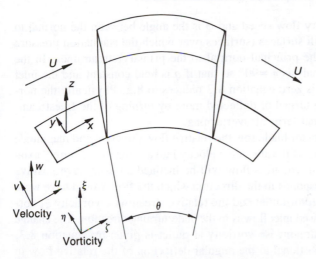

Fig. 8.4 Curved duct as used for Squire and Winter (1953) secondary flow analysis

The most widely quoted result, which Squire and Winter obtained assuming that the shear is weak, the freestream velocity at the edge of the shear layer is uniform in the streamwise direction and the primary flow is turned by a small angle θ, is the expression for the streamwise vorticity ξ

$$\xi = -2\theta\eta = -2\theta\partial U/\partial z \tag{8.1}$$

The secondary vorticity is inclined so that the crossflow is directed towards the low static pressure region, in the direction of $-y$ in Fig. 8.4. The crossflow velocity pattern was obtained by assuming that far downstream the flow quantities did not vary with streamwise distance so the equation for the crossflow streamfunction ψ is a Poisson equation

$$\frac{\partial^2\psi}{\partial y^2} + \frac{\partial^2\psi}{\partial z^2} = 2\theta\frac{\partial U}{\partial z} \tag{8.2}$$

where $w=\partial\psi/\partial y$ and $v=-\partial\psi/\partial z$ are the crossflow velocities. This equation is readily integrable.

Soon after Squire and Winter published their paper Hawthorne (1951) published the first of many papers on the subject of secondary flow. This was a more general treatment expressing the vorticity vector Ω in terms of the gradient in stagnation pressure p_0

$$V \times \Omega = \text{grad}(p_0/\rho)$$

The secondary vorticity could then be written

$$\left(\frac{\xi}{q}\right)_2 - \left(\frac{\xi}{q}\right)_1 = -2\int_1^2 |\text{grad } p_0/\rho|\sin\phi\,\frac{d\theta}{q^2} \tag{8.3}$$

where q is the primary flow speed and ϕ is the angle between the normal to the so-called Bernoulli surfaces (surfaces over which the stagnation pressure p_0 is uniform) and the principal normal to the primary streamlines. In the Squire and Winter analysis $\phi = 90°$ so that if q is held constant and the inlet streamwise vorticity is zero equation 8.3 reduces to 8.2. Physically the normal vorticity is being turned or deflected more by turning of the freestream, giving rise to the usual term of overturning.

At inlet to a compressor blade row the relative flow direction and magnitude vary near the endwalls. If the absolute velocity into a rotor row is purely axial close to the endwall the relative flow will be inclined so as to have positive incidence. This corresponds to the direction which the flow would have were it underturned at the rotor outlet and the relative streamwise vorticity ξ_1 corresponding to this skewed inlet flow is in the underturning direction. The change in the total relative streamwise vorticity at outlet is given by equation 8.3, where $\xi_2 - \xi_1$ is proportional to the angular deflection of the relative flow in the blade row and is in the sense to produce overturning. Hunter and Cumpsty (1982) showed that for an axial rotor tip the streamwise vorticity from the inlet skew is many times larger than that produced by the flow turning and the net vorticity at outlet is in the underturning sense. This is likely to be a general result at the tips of rotors, because the high stagger implies high inlet streamwise vorticity in the underturning sense and most rotor tips sections have only small camber. The relative contributions must be assessed for each particular geometry and in some cases, such as the rotor hub, the overturning may be larger than the underturning from inlet skew. In cascade tests there is conventionally no inlet skew and therefore no inlet streamwise vorticity; the conditions in a cascade are therefore very different from a blade row in a compressor and measurements or theories for flow near the endwalls of cascades must only be generalized to compressors with the greatest caution.

The range of geometries and coordinate systems lends itself to many different formulations for the secondary flow and the mathematics is sufficiently involved that the connections may not be obvious. Lakshminarayana and Horlock (1973) gave generalized expressions using intrinsic coordinates. Soon after, Hawthorne (1974) gave a range of equations specific to flows about an axis of symmetry, an obvious special simplifying feature of compressors. One of the simplest and most convenient expressions for use in turbomachinery derived by Hawthorne has already been used in Chapter 6 for the flow in centrifugal impellers and is

$$\frac{\partial}{\partial s}\left(\frac{\Omega_s}{W}\right) = \frac{2}{\rho W^2}\left(\frac{1}{R_n}\frac{\partial p^*}{\partial b} + \frac{\omega}{W}\frac{\partial p^*}{\partial z}\right) \tag{8.4}$$

where Ω_s is the streamwise vorticity in the direction s, W is the relative velocity, ω is angular velocity of the machine, z is the coordinate in the direction of the axis of rotation and b is the binormal coordinate which is normal to both the streamwise direction s and the radius of the streamline R_n. The variable $p*$ is the stagnation pressure equivalent to the rothalpy and is defined by

$$p* = p + \rho/2(W^2 - \omega^2 r^2).$$

Surfaces of constant $p*$ are the rotating system equivalent of the Bernoulli surfaces. The physical meaning of equation 8.4 is that the fluid with low $p*$ (i.e. low relative stagnation pressure) migrates to the region where the static pressure is low. This is the inside of a bend of radius R_n or the trailing side of a rotating passage, gradients of $p*$ in different directions being involved in each case.

Equation 8.4 is for incompressible flow and when variations in density are permitted the relation between pressure and temperature must be specified. However, the usefulness of secondary flow theory is more qualitative than quantitative and for most compressor applications the incompressible formulation suffices.

One of the aspects of secondary flow theory which made it attractive was apparent from Hawthorne's approach. Once the expressions for vorticity and circulation conservation have been obtained, the subsequent analysis is kinematic: it is not necessary to consider pressures but only to keep track of the vorticity and circulation of the flow, e.g. Preston (1954) and Came and Marsh (1974). The kinematic description seems to offer an easier way of analysing very complicated problems, such as the flow along the endwall of a cascade or the annulus wall of a compressor. For this to offer a real simplification it is necessary that the velocities generated by secondary flow are small so that the transport of vorticity is overwhelmingly by the primary flow. It is also necessary that the flow is effectively inviscid. In practice this means that the secondary flow methods are reasonably applicable in regions where the stagnation pressure is slightly non-uniform and the shear stresses are small in relation to the pressure gradient — it is not valid close to solid surfaces or in regions where loss is mainly created.

At one time secondary flow methods offered the only approach to calculating the three-dimensional flow in compressors and considerable effort went into including effects originally left out of the formulation, see for example Horlock and Lakshminarayana (1973). (It is, however, remarkable just how few cases were actually calculated by secondary flow methods.) Now there are more quantitative approaches and secondary flow methods are not to be recommended for accurate prediction, especially in the regions very close to the walls. Secondary flow methods do still find an application in the qualitative consideration of flows, for example the work of Johnson and Moore (1983) in centrifugal impellers referred to in Chapter 6.

An original goal of the study of secondary flow was the estimation of the losses attributable to it. If it is assumed that the kinetic energy associated with the secondary velocity field is unrecoverable, or cannot be used by subsequent blade rows, then the secondary kinetic energy is a measure of the secondary loss. When this has been evaluated it has been found to be far too small and it is reasonable to assume that the secondary motion, in regions where secondary flow theory is applicable, is not in itself a major source of loss. Inviscid secondary flow analyses are thus not suited to investigations of loss sources, most of the loss is created in just those high-shear regions which secondary-flow theory ignores.

8.3   Axial blade boundary layers

This section connects with Chapters 4 and 5, but with changed emphasis. The consideration of boundary layers on blades is nearly always in terms of two-dimensional flow such as that found near mid-span of a carefully controlled cascade experiment or well designed compressors, for example Fig. 8.2. The real flow is often much more complicated, as indicated in Fig. 4.53. Nevertheless the boundary layer model does have real applicability on the blades of axial compressors.

   In the design of blades in the past the boundary layer was rarely calculated on the blades, but overall behaviour obtained from correlations. One reason for this was the common observation of separation on the suction surface, with the consequent unreliability of calculation methods in this vicinity. Two things have changed in the last few years, in addition to the comparatively small improvements in boundary layer calculation methods. One is the greater use of high Mach number blades; at conditions close to sonic the blockage of the boundary layers has a very large effect on the free-stream pressure and velocity field and cannot be ignored — this was demonstrated in Chapter 5. The other change is the trend to use blades with a prescribed velocity distribution so that the freestream velocity distribution is specified to avoid boundary layer separation. One of the effects of the supercritical design is to permit higher blade loading by choosing that the boundary layer is near to (but not beyond) separation over a large part of the suction surface. This means that errors in the boundary layer calculation used in the design process can lead to serious discrepancies in performance.

   Most of the loss and the breakdown of the flow at high loading is controlled by the suction surface boundary layer and with few exceptions it is this boundary layer which will be discussed here. The boundary layer is conveniently considered in the different flow regimes: laminar flow near the leading edge, transition, the turbulent boundary layer and separation.

Laminar boundary layers

There is evidence to suggest that the laminar flow on the forward part of the blades exists even in the extremely turbulent conditions prevailing in the multistage compressor. Quite simply the Reynolds number is too low to per-

mit turbulence. Very often there is leading edge 'spike' in the freestream velocity, Mach number and static pressure distributions close to the leading edge
because the stagnation point is not exactly located on the nose of the aerofoil,
as discussed in Chapter 4. The 'spike' may lead to local freestream velocities
twice those immediately downstream and its length along the surface is typically
no more than 1 per cent of chord. With such a 'spike' the deceleration on
the downstream side is far greater than that to exceed the separation criterion
for a laminar boundary layer but there is no evidence from the overall performance of the blades that it leads to premature thickening of the boundary layer,
see Figs 4.5 and 4.24. It seems likely that the 'spikes' at the leading edge
are so sharp and the changes are so rapid in the streamwise direction that one
of the fundamental assumptions of boundary layer theory is violated. Briley
(1971) compared Navier–Stokes solutions with boundary layer solutions for
rapid decelerations and found that unseparated flow was predicted by the
Navier–Stokes method at stronger adverse gradients than were possible with
the boundary layer method. Moreover the Navier–Stokes solution showed that
even when the flow separated it could reattach as a laminar layer. In other
words the boundary layer method was inadequate for the very severe conditions. Recent experiments in Cambridge by Addison and Storer have shown
that with the relatively sharp leading edge of a double-circular-arc blade separation occurred at 0° incidence (about 2° negative incidence is needed to remove
the spike). The separation occurred at about 0.5 per cent chord and reattachment was between 2 and 5 per cent of chord. At 5 per cent chord the reattached
flow was turbulent and the Reynolds number based on momentum thickness
was about 320. This supports the common calculation strategy which is to treat
the flow as turbulent from the leading edge.

The laminar region may persist to the trailing edge on the blade pressure
surface but on the suction surface this would occur only for exceptionally lightly
loaded blades at Reynolds numbers based on chord length less than about 10^5.
The crucial factor in determining whether the flow undergoes transition or
whether it separates is the pressure gradient or, equivalently, the freestream
velocity gradient. A strong favourable pressure gradient, that is a flow in which
the free-stream velocity increases in the flow direction, can delay transition
to quite high Reynolds numbers, whereas even a weak adverse pressure gradient, where the free-stream velocity decreases in the flow direction, can induce transition at markedly lower Reynolds number. For a laminar boundary
layer even quite a weak adverse pressure gradient can induce separation so
whether separation or transition comes first depends critically on the Reynolds
number and pressure gradient in a far from straightforward manner. The non-
dimensional parameter which affects both transition and separation is

$$\lambda = \frac{\theta^2}{\nu} \frac{dV}{dx}$$

where $\theta = \int v/V(1 - v/V)dy$ is the momentum thickness written here in

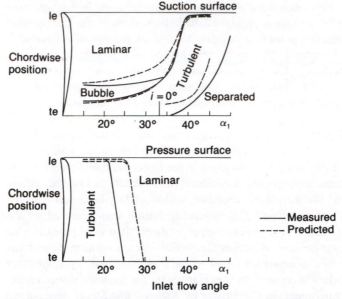

Fig. 8.5 The variation in boundary layer condition on the surfaces of a blade in cascade for a range of inlet flow angles. Reynolds number $1.5 \cdot 10^5$, turbulence intensity 2.6 per cent, C7 section, solidity 1.3, stagger 12.6°, camber 41.8°, thickness–chord 0.1; Lieblein minimum loss incidence $i = 0°$ at $\alpha_1 = 33.5°$. (From Seyb, 1972)

incompressible form. Separation of a laminar layer is usually assumed to take place when $\lambda < -0.09$.

Whether the laminar flow first undergoes transition or separates depends on the level of free-stream turbulence and the Reynolds number based on boundary layer thickness, for example $V\theta/\nu$, as well as on λ. Blade roughness also plays a part.

Figure 8.5, taken from Seyb (1972) shows the character of the boundary layer for a single cascade at different inlet flow angles. The cascade was tested with inlet turbulence intensity of 2.6 per cent, about half what it would probably be in a multistage compressor, and a modest Reynolds number based on chord length of $1.5 \cdot 10^5$. On the pressure surface transition was avoided for all incidences greater than about $-8°$; although Seyb's prediction gave the correct trend it was incorrect by about 5° of incidence. On the suction surface the true complexity is apparent. For incidence less than about 1° there was a separation bubble with turbulent reattachment and no subsequent separation. For more positive incidence transition occurred before laminar separation and there was no bubble. For incidence greater than about 4° the turbulent boundary layer separated before the trailing edge. The complication arises because increasing the incidence strengthens the pressure gradient but also causes it to begin further upstream; if the pressure gradient begins earlier the boundary layer is thinner and the magnitude of λ and Reynolds number correspondingly reduced. The prediction scheme used by Seyb was quite good on the suction surface, the main surprise being the relatively poor prediction of the laminar separation close to zero incidence which one would have expected to be quite accurate. The error may well reflect an important aspect of the unsteadiness of the bubble in the presence of fairly high turbulence and current thinking is chary of defining separation lines too precisely, preferring to acknowledge the fluctuating position at which the separation (and also the reattachment) actually takes place. Quite clearly the predictions from this method, which is similar to most other approaches for the blade boundary layer, must be used with considerable caution.

An approximate calculation method such as Thwaites (1949) seems adequate for the laminar flow downstream of the stagnation point, with the existence of the ambient turbulence and unsteadiness ignored. Any inaccuracy in the laminar calculation is dwarfed by uncertainty about transition. This method, like many boundary layer methods, was designed for an incompressible flow using information collected at low speed. At high Mach numbers some transformation of coordinates and velocities is usually made to allow the low-speed methods and correlations to be used. A widely used method, which assumes that the surface is adiabatic and the Prandtl number is unity, is known as the Stewartson—Illingworth transformation (see for example Schlichting, 1979).

Transition

Transition on compressor blades can come about in two ways. Either the attached laminar flow can undergo direct transition or the laminar flow can first separate. Once separated, the laminar shear layer can undergo transition and the flow reattach as a turbulent boundary layer, the separated low region between separation and reattachment being known as a separation bubble. The chordwise length of the separation bubble is typically up to about 10 per cent of chord, with the last 1/5th of this being the turbulent part, though the length depends on the freestream turbulence level, the Reynolds number and the blade pressure distribution.

Direct transition

The process of transition is an area where much is still not known, even for simpler configurations than turbomachines, and it continues to receive active study, Narasimha (1985). In a compressor transition is particularly important because the Reynolds numbers of most blades are such that the transition process can be expected to occupy a significant proportion of the chord length. This has been brought out by the studies reported by Walker (1987); the measured transition zone on a compressor blade suction surface, tested in cascade and also as a stator downstream of a rotor, extended over between 10 and 20 per cent of the chord. Walker's measured results are shown in Fig. 8.6 and indicate that the transition region on the suction surface moves forward as the incidence is increased, that the length of the transition region does not change very much with incidence. Walker also found that, although the

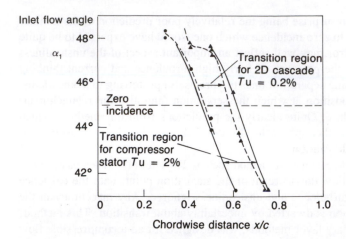

Fig. 8.6 The transition region on the suction surface of the same blade profiles in a cascade and as a compressor stator (C4 section, solidity 1.0, camber 31°, stagger 30°, Reynolds number $2 \cdot 10^5$). (From Walker, 1987)

starting point of transition was very similar in the cascade and stator, the transition was complete earlier in the cascade. Similar observations have been made elsewhere, including those of Dong (1988). The operation of incoming wakes on the blade boundary layers is complicated but it has been clearly demonstrated that they can inhibit the natural transition and thereby leave patches of laminar fluid further downstream on the blades. For Walker's blades the prediction by the method of Dhawan and Narasimha (1958) gives transition extending over 80 per cent of the chord. The experiments and most of the thinking behind Narasimha's approach are for flows in zero or very weak pressure gradients. Walker's comparison, with observed transition occupying 20 per cent of chord and the prediction by Narasimha's model occupying 80 per cent, shows that this model is not relevant to the transition found on compressor blades.

Because of the high levels of turbulence or unsteadiness likely to be present in all practical cases it seems that transition by the amplification of the small amplitude Tollmien–Schlichting waves in the attached boundary layer can be neglected. The effect of freestream turbulence on transition has been referred to in Chapter 4 where the results of Schlichting and Das (1969) were shown. Around the same time Evans (1971) carried out an investigation of transition using a cascade of blades with 44° of camber, exploring both the effect of turbulence intensity and turbulent length scale on transition. The turbulence intensity was varied from 0.25 per cent up to 4 per cent, three different length scales were used and tests were carried out at five Reynolds numbers between 2 and $6 \cdot 10^5$. The separation bubble type of transition occurred at the lower Reynolds numbers and lower levels of turbulence; the bubble was 11 per cent of chord in length at the lowest values of Reynolds number and turbulence level, but as these were raised it reduced to only 2 per cent before finally disappearing. The length scale of the turbulence has an effect on the position of

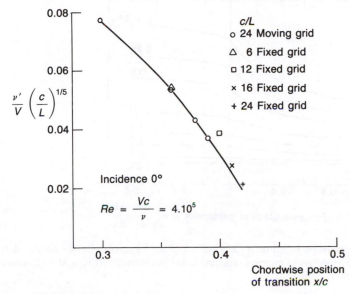

Fig. 8.7 The effect of $(v'/V)(c/L)^{1/5}$ as a correlating parameter for the effect of turbulence on transition (C4 section, solidity 1.4, camber 44.4°, stagger 13.5°, thickness–chord ratio 0.10). (From Evans, 1971)

transition and Evans found that the combined effect of turbulence intensity and length scale was correlated well by the parameter originally proposed by Taylor (1936) for the transition of flow on spheres,

$$(v'/V)(c/L)^{1/5}$$

where here c is the blade chord and L the integral length scale of the turbulence. The effect of using this is shown in Fig. 8.7. Although the generality of this correlation has not been proven it is probably safe to conclude that a given level of small-scale turbulence, typical of that from upstream blade wakes, is far more effective in promoting transition than large-scale turbulence from upstream obstructions or grids.

A study of the combined effects of turbulence and pressure gradient was made by Abu-Ghannam and Shaw (1980), a figure from which is reproduced here as Fig. 8.8 with the individual measured points omitted. The lines show the local Reynolds number, based on momentum thickness, at the start of transition, $R_{\theta S}$, corresponding to various levels of turbulence intensity. The lines are actually plots of the empirical relations deduced from their measurements

$$R_{\theta S} = 163 + \exp[(6.91 - Tu)F(\lambda)]$$

in which $F(\lambda) = 1$ for zero pressure gradients, $(1 + 1.84\lambda + 9.21\lambda^2)$ for adverse gradients and $(1 + 0.36\lambda + 1.78\lambda^2)$ for favourable pressure gradients. *Tu*

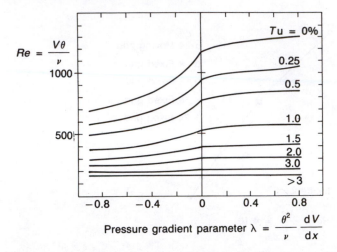

Fig. 8.8 A correlation based on measurements of the Reynolds number for the start of transition as a function of pressure ratio and turbulence intensity. (From Abu-Ghannam and Shaw, 1980)

denotes the turbulence intensity in per cent of the freestream mean velocity. For low levels of turbulence transition starts at a higher value of Reynolds number, $R_{\theta S}$, when the pressure gradient parameter, $\lambda = \theta^2 / \nu \, \mathrm{d}V \, \mathrm{d}x$, is positive, i.e. the pressure gradient is favourable, but drops rapidly for adverse pressure gradients. At higher levels of turbulence intensity the Reynolds number at start of transition is much lower for all pressure gradients and the difference between favourable and adverse pressure gradients markedly less. For a turbulence level of about 3 per cent, the effect of pressure gradient on the Reynolds number for the start of transition has become negligible and there seems to be no further reduction in $R_{\theta S}$ for turbulence intensities above about 5 per cent. For compressors, where the turbulence level is probably between 5 and 10 per cent, this means that even the strong pressure gradients on the compressor blades are not going to effect the start of transition; transition will begin when the boundary layer Reynolds number $V\theta / \nu$ exceeds about 160.

Abu-Ghannam and Shaw correlated the Reynolds number for the length L of the transition process $R_L = VL/\nu$ with the Reynolds number R_{XS} based on the distance x_S along the surface at the start of transition by

$$R_L = 16.8 (R_{XS})^{0.8}$$

Using the value of R_L they found the Reynolds number based on momentum thickness at the end of transition $R_{\theta E} = \theta_E V / \nu$ could be related by

$$R_{\theta E} = 540 + 183.5 (10^{-5} R_L - 1.5)(1 - 1.4\lambda).$$

The correlations for transition provide the starting point for the calculation of the downstream turbulent boundary layer. Because of the uncertainties associated with the prediction of the end of transition it is very often assumed that direct transition takes place instantaneously at a point on the blade with constant momentum thickness. With this instantaneous model the form parameter $H = \delta^*/\theta$ can be reassigned a typical turbulent value like 1.5 after transition instead of the laminar value prior to transition (typically above 2.5). This is sometimes unnecessary with the turbulent calculation stable enough to accept an 'improper' starting value and allow the turbulent prediction to relax to a more realistic value.

Separation bubble transition

As noted above and shown in Fig. 8.5 transition often occurs by means of a laminar separation bubble. The bubble and its effect on the pressure distribution are shown schematically in Fig. 8.9. The sketch shows what is known as a short bubble, occupying a fairly small fraction of the blade chord, typically less than 10 per cent, and having only a local effect on the surface pressure distribution. (In a long bubble the pressure distribution is radically altered, not only in the vicinity of the bubble; these are common on wings at high incidence but it is not clear that they are important in the conditions of a compressor.) The significance of the bubble is that free shear layers, such as that over the bubble, are very unstable and become turbulent at an earlier chordwise position than would have been possible for an attached boundary layer.

   The current modelling of the short bubble is based on the model proposed by Horton (1969) for wings with low ambient turbulence. Horton envisaged two regions of shear layer over the bubble. The first zone is a laminar shear layer of length l_1 in which region the surface static pressure remains more or less constant; the constant static pressure over the laminar part is how one normally notices separation bubbles, as sketched in the pressure plot in Fig. 8.9. The second region is the reattachment zone of length l_2 where the flow becomes turbulent and reattaches. The static pressure is assumed to rise over the length of the reattachment zone to rejoin the pressure distribution around the unseparated aerofoil. It is a premise of the model that the separation bubble

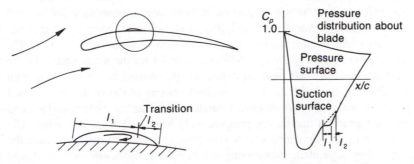

Fig. 8.9 A schematic diagram of a separation bubble and the corresponding pressure distribution. Broken line shows pressure distribution with no bubble

is small enough that the overall pressure distribution about the aerofoil is un-changed, except in the immediate vicinity of the bubble, so the turbulent zone is terminated on the undisturbed pressure distribution.

The start of the laminar zone of the bubble is the separation of the laminar boundary layer and this is taken to occur when, for example, $\lambda = \theta^2/\nu \mathrm{d}V/\mathrm{d}x = -0.09$. To calculate the start of transition along the shear layer Horton, with wings in a low turbulence stream in mind, assumed the Reynolds number based on the length l_1 of the laminar region and the free-stream velocity at separation would be constant and equal to 40 000. Dunham (1972) and more recently Roberts (1980) proposed an empirical expression for l_1 to take account of the high levels of turbulence in compressors using the data of Evans (1971).

The constant static pressure over the laminar region and the negligible wall shear stress which will occur under the bubble means that the momentum thickness of the shear layer at the start of transition is assumed equal to that at the point of separation. In the reattachment zone the shear layer momentum thickness is not constant because the static pressure rises. The momentum thickness has been calculated from the value at separation by applying momen-tum integral and energy-deficit integral equations over the turbulent reattach-ment zone shear layer, but other approaches could be used. Once the bubble shear layer starts to become turbulent the effect of freestream turbulence will be much smaller and therefore the correlation derived by Horton to obtain l_2 has been accepted by both Dunham and Roberts. At the point of reattach-ment it can be assumed that the non-dimensional velocity gradient $\theta/V \, \mathrm{d}V/\mathrm{d}x$ is constant at the theoretically determined value of 0.0059, together with $H = \delta^*/\theta = 3.5$, has been found to give the best prediction. The distance along the surface to find the appropriate level of non-dimensional velocity gradient then fixes the length l_2 of the reattachment zone. If there is no value of l_2 at which a solution can be obtained the flow is assumed to be unable to reattach and the bubble is said to have burst.

The most thorough published investigation of transition in turbomachinery to date was by Dunham (1972). Using his correlation scheme he was able to predict the occurrence of transition of Evans' blade reasonably well for the cases with low inlet turbulence, when in most cases a separation bubble was produced. The agreement was much less good, however, when direction tran-sition took place, but this predates the work of Abu-Ghannam and Shaw (1980).

More recent work by Dong (1988) casts doubt on the applicability of the bubble model when blade rows are exposed to the moving wakes from upstream rows. Dong worked with modern prescribed velocity blades of the supercritical type, designed to have an attached boundary layer on the suction surface and a pressure gradient decreasing progressively towards the trailing edge. The turbulence in the incidence wakes was able to induce transition before the laminar flow separated, suppressing the bubble. Similar effects were found with a high level of inlet turbulence from a grid. Dong found that although the transition models like those of Horton and Roberts are inaccurate, the resulting error in the momentum thickness at the suction surface trailing edge was only about 6 per cent. In the case which Dong considered the predicted flow did not separate towards the trailing edge but if flow separation had occurred in one case and not the other a much larger discrepancy could have resulted. In other words there is a need for the accurate prediction of transition and the present methods are inadequate. If blades are to be designed satisfactorily with particular boundary layer properties in mind it is essential to make estimates for the transition, since variations in this far outweigh discrepancies in the calculation methods for laminar or turbulent flow.

Rotation and curvature effects

There are known to be important effects on transition arising from the blade surface curvature and system rotation which are addressed more generally later in this chapter in connection with centrifugal compressors. On the concave surfaces a laminar boundary layer experiences a destabilizing effect. Similarly on the convex surface of a blade the flow is stabilized and transition is inhibited. The crucial quantity in the case of flow on curved surfaces is the local Richardson number

$$Ri_{\text{curv}} = \frac{2v/R_c}{\partial v/\partial y}$$

where R_c is the local surface radius of curvature. As a way of assessing the effect on overall boundary layer stability the velocity gradient can be expressed as V/δ, where δ is the overall boundary layer thickness, so that the crucial ratio is simply δ/R_c. For axial compressors the region where transition occurs on the suction surface is usually sufficiently far behind the leading edge that the radius of curvature is very large in relation to the boundary layer thickness and the Richardson number is low. In axial compressors there is no firm evidence that the stabilizing or destabilizing effects alter the laminar nature of the compressor blade boundary layer, nor that they affect the turbulent boundary layer once transition has occurred. The effect of curvature on stability is probably much more important for turbine blading because of the much greater curvatures encountered.

For compressors the details inside the transition region are normally unimportant and there is usually no need to resort to methods such as McDonald and Fish (1973) for calculating transitional flows. (The details are important for turbines because the local heat transfer rate is crucial.) Errors in the prediction of transition are serious primarily because the accuracy of the subsequent turbulent boundary layer prediction is strongly affected.

Turbulent boundary layer

The calculation of two-dimensional boundary layers has attracted an enormous amount of work and interest. The effectiveness of the methods has been assessed at a special conference organized in Stanford (Kline, Cockrell and Morkovin, 1968) and it is now known that there are a wide range of methods capable of predicting two-dimensional turbulent boundary layers with reasonable accuracy. There is no obvious advantage in terms of accuracy for differential methods over integral ones, but those based on the partial differential equations are generally more suited to the introduction of extra effects such as freestream turbulence, wall heat transfer, roughness and curvature effects. Because of their greater speed and ease of application the integral methods have tended to find favour as the companion to inviscid calculation methods on blades.

Although this section is about two-dimensional boundary layers, in the sense that variations in flow direction through the boundary layer are not considered, there is an effect of the convergence of the free-stream streamlines which does sometimes need to be included, particularly as separation is approached. In the case of an axial compressor this can be thought of as the narrowing of the flow passage because of the hub and casing convergence or because of the growth of the endwall boundary layers. An extra term must appear in the two-dimensional boundary layer equation and the continuity equation, which then become analogous to those over a body of revolution with varying radius, see Schlichting (1979). Because the flow calculations for axial blades are normally carried on what are approximate streamsurfaces this effect is often referred to as streamtube contraction. This has been considered for the specific case of an axial compressor blade row by Dunham (1974) and it is clear that where streamtube contraction is large in relation to the boundary layer growth it must be included if accurate predictions are to be produced.

It is a basic assumption of boundary layer theory that if the velocity field is known in the free-stream the boundary layer can be calculated up to the point when separation occurs. Many compressor blades have regions of separated flow: at high speeds separation can arise from shock-boundary layer interactions and for wholly subsonic conditions separation is caused by the general blade loading. To be useful a boundary layer method must be able to accommodate this and remain stable despite the singularity which occurs at the separation point (once downstream of the separation point the boundary layer method is no longer singular).

In fully attached flow a change in freestream velocity usually produces a small change in the downstream boundary layer displacement thickness and the process is stable, but when there is a region of separated flow this is sometimes no longer the case. A small change in freestream velocity can produce a very large change in displacement thickness, which in turn has a pronounced effect on the pressure distribution, possibly making the normal procedure unstable. This behaviour is well known on isolated aerofoils and in diffusers as well as in compressors. At transonic conditions the large changes in the freestream velocity due to small changes in the blockage (i.e. boundary layer displacement thickness) can also lead to instability with fully attached flow. Herbert and Calvert (1982) have had success in working in what is termed an inverse mode close to and in separated regions: the same equations are used but the displacement thickness is varied to give the required velocity distribution. An example where this has been successfully applied to the calculation of the flow in a high Mach number axial fan is given by Ginder and Calvert (1988).

An overview for blade boundary layers

Compared to other viscous regions in compressors the understanding and prediction of axial blade boundary layers is good. Nearly all the interest is with the suction surface. The laminar region is no problem and approximate methods or numerical methods are adequate. The sharp low-pressure 'spikes' found near the leading edge of blades when the stagnation point is not on the nose of the blade do not seem to be a serious disturbance to the overall blade performance although they may lead to immediate transition.

Transition can lead to substantial uncertainties and little information exists for this on compressor blades and none well inside a multistage compressor. The level of turbulence inside compressors is so high that transition can normally be assumed to start when the Reynolds number based on momentum thickness exceeds about 160. Separation bubbles, with laminar separation and turbulent reattachment, can occur when the turbulence level is low but are likely to be suppressed once the turbulence has become high, probably after the first stage. (Tests in cascades have often given a very misleading view of the whole problem.) Transition is much less predictable when the turbulence level is lower, for example on the outlet guide vane behind a fan rotor, and the process is modified by the passage of the rotor wakes.

The turbulent boundary layer can be calculated with a number of methods, all based on significant amounts of empiricism and all able to give reasonably accurate estimates of boundary layer growth. None are to be treated as generally reliable close to separation. The separation line very often adopts a strikingly three-dimensional form, see Fig. 4.53, and the two-dimensional basis of the calculation is called into question.

8.4 Flow in the endwall regions of axial compressors

The experimental background

The endwall regions are the most important and, unfortunately, the least well understood parts of the compressor. At the present time there is no really satisfactory way of predicting blockage or loss other than by correlation.

Theoretical approaches to the problem generally treat the flow as uniform in the circumferential direction, after carrying out some sort of averaging process, and use normal boundary layer methods to compute the development in the axial direction, usually assuming the flow to be fully turbulent at inlet. These have not proved very successful. The inaccuracy comes not from the averaging but primarily from applying methods and models which are unsuited to the flow. The averaged velocity profiles upstream and downstream of a blade row *do* look like conventional two- or three-dimensional ones. This does *not* mean that it is at all appropriate to use the conventional relations such as skin friction coefficient and entrainment rate inside the blade passage, nor does it mean that the velocity profiles are at all satisfactorily described by the conventional families inside the passage.

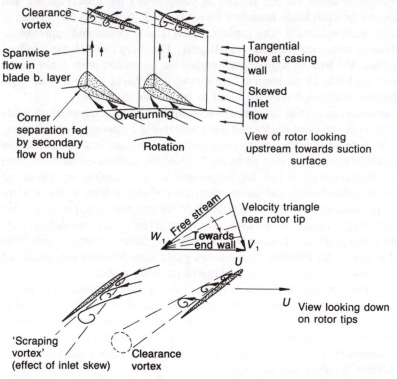

Fig. 8.10 Schematic representation of some viscous effects in a rotor passage

A pictorial representation of some of the effects to be included in a description of the endwall region is given in Fig. 8.10. In every sense the flow is highly three-dimensional and the tidy division of the flow into endwall and blade boundary layers is not supported by observation. The most important effects occur in the corner regions. Figure 8.11, from Dong *et al.* (1987), shows flow visualization in the hub region of a stator operating close to the design condition (the behaviour is similar to that sketched in Fig. 8.10 for

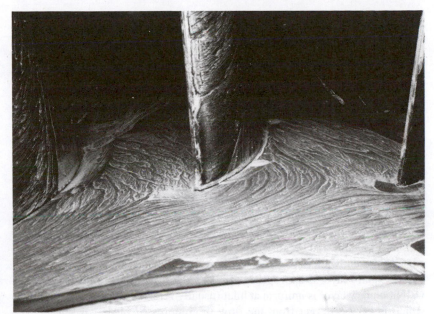

Fig. 8.11 Flow visualization on a non-rotating hub in the vicinity of the stator leading edge. (From Dong *et al.*, 1987)

the rotor hub). The high incidence close to the hub because of the inlet skew is apparent, so too is the overturning inside the passage so that the flow moves into the suction surface corner where there is a separated region. The separation affects not only the endwall but much of the blade; the two regions need to be considered together.

For axial compressors one way of identifying the major problem is to consider the rise in static pressure across the blade row in relation to the flow dynamic head. A typical value might be $\Delta p \approx 0.3\rho U^2$, where U here denotes blade speed. The corresponding flow coefficient of the freestream at the edge of the endwall viscous region might be $\phi = V_x/U \approx 0.5$. In terms of the axial velocity the pressure rise coefficient $\Delta p \approx 2.4(\frac{1}{2}\rho V_x^2)$ which is about four times as large as could be achieved in the best diffuser. The reason that it is possible to operate satisfactorily like this is that most of the pressure rise is supported by the blade force and the effective pressure difference to which the endwall region is exposed is much less than this. Unfortunately this means that small changes in the blade force can have disproportionately large effects on the endwall flow and there is no way of predicting what the blade force is with any accuracy inside the viscous region. The difference between the actual blade force and the value obtained assuming that the freestream force persists right down to the surface has been used to define a quantity known as the 'force deficit thickness'

$$\delta_F = \frac{1}{rF_{\text{edge}}} \int^{\delta} (F_{\text{edge}} - F) r \; dr$$

where F is the blade force and F_{edge} is the blade force at the boundary layer edge. (In considering the integral thicknesses in the endwall region the integration is understood to be on the hub or casing wall with the sign of the limits adjusted accordingly.) Although the force deficit is usually positive, that is the blade force per unit area is smaller in the endwall region than at the edge, on occasions it can be negative. Also the force deficit can first decrease when a small tip clearance is introduced and then increase again when the clearance is further increased.

The linear cascade

The simplest case to envisage is a compressor blade row in a cascade. This is the flow which has tended to be considered theoretically, because it is much simpler to handle and because experiments could be easily performed. At the edge of the endwall viscous region there is uniform flow so the description as a boundary layer is appropriate at inlet. The flow direction inside the endwall boundary layer is uniform at inlet (usually referred to as 'collateral flow') which is very different from the flow in compressor blade rows where the flow close to the walls is highly skewed. As was noted earlier in this chapter, Hunter and Cumpsty (1982) showed that the inlet vorticity to a rotor tip, resulting from the inlet skewing giving a flow which was not collateral, was much larger than the secondary vorticity produced inside the row and of opposite sign. By and large endwall experiments in linear cascades have little relevance to compressor blade rows.

There is an additional difficulty with cascade testing. For realistic blade loading the relatively thin turbulent boundary layer growing on the blade suction surface will be close to separation at or near the trailing edge. The same pressure gradient imposed on the much thicker endwall boundary layer will probably lead to some separation, usually in the endwall-suction surface corner. To complicate matters further, the endwall boundary layer flow approaching the leading edge of each blade in the cascade separates and rolls up into the familiar horse-shoe vortex found around all blunt obstacles standing out of a shear flow. The leg of the vortex on the suction side is moved by the flow close to this surface; the leg on the pressure surface stretches across the passage to join the suction surface of an adjacent blade so that the endwall-suction surface corner has two vortices rotating in opposite directions in or near a region which is separated.

The tendency for the crossflow inside a boundary layer to move the fluid near the walls towards the region of low static pressure (equivalently, for a secondary flow to be set up with streamwise vorticity such that the flow in the boundary layer is towards the low pressure wall) has been discussed briefly in Section 8.2 of this chapter. This was demonstrated in very elegant flow visualization pictures by Herzig *et al.* (1954), some of which are reproduced by Horlock (1958).

Axial stator rows

The stator row adjacent to a stationary wall is the compressor geometry most similar to the cascade of blades referred to above but there are important differences. One difference is that the flow into a stator in a compressor is highly unsteady because of the upstream rotor, but there is no evidence of how this alters the flow. It is more significant that in the endwall region the flow is rarely collateral: the flow direction changes in a direction normal to the wall, in a way similar to that sketched in Fig. 8.10 for the rotor, so the flow is highly skewed. The variation in direction may arise from having a moving hub, from the secondary flow in the rotor, or simply from the change in the frame of reference from rotor to stator in conjunction with variation of magnitude in the rotating frame. A particularly acute problem occurs with shrouded stators for the flow near the hub passes over a rotating surface in the rotor row and then abruptly changes to flow over a stationary surface in the stator. As explained in Chapter 2 the low velocity flow in the rotating frame of reference may actually have a higher magnitude in the stationary frame than the main flow — the freestream then has lower stagnation pressure (and temperature) than parts of the endwall boundary layer. If the endwall fluid has higher stagnation pressure the reason for the horseshoe vortex disappears. In Fig. 8.11 there is no evidence of the clear horseshoe vortex often seen in cascade experiments though there is evidence of a three-dimensional separation point upstream of the leading edge.

The direction of the inlet skew to a stator is generally that of positive local incidence such that the flow is towards the pressure surface of the blade, as in Fig. 8.11. This is in the opposite sense to the secondary flow likely to be produced by turning the flow in the row itself. The nature of the non-uniform flow leaving the stator blade depends on the difference between the two effects; if the inlet skew is larger the flow is described as underturned (since the flow has been locally turned less than in the freestream) but if the passage secondary flow is larger then it is described as overturned. High stagger blades generally have small camber and for these the contribution to secondary flow by the turning is small, so the inlet skew is likely to predominate giving underturning in the endwall region. Conversely when the stagger is small and the camber large, as is common near the hub, the contribution from the passage turning is likely to be large and overturning found. (The same arguments apply to the secondary flow in axial rotors.)

Even when viewed as a steady problem, essentially averaging out the circumferential variation from the upstream rotor, the stator flow is very different from the cascade flow. The concentration on modelling endwall flows in cascade passages has probably held back the understanding of the flow in axial compressors.

Axial rotor rows

With a rotor row the particular interest is with the tip, the hub being considered much like a stator passage. (For stators with hub clearance, however, there are features more similar to the rotor tip.) The rotor tip region is particularly complicated and some of the features are sketched in Fig. 8.10, the dominant feature being the tip clearance. Most rotor tip sections have relatively high stagger and low camber. The low camber means that little secondary flow towards the suction surface is produced in the passage. The high stagger means that the relative stagnation pressure of the low axial velocity fluid in the incoming endwall flow is high and may even reach a maximum for the fluid adjacent to the endwall with vanishing axial velocity. The mechanism for generating a horseshoe vortex is no longer present. Apart from the tip clearance flow the pattern tends to be dominated by the incoming endwall flow collecting in the endwall-pressure surface corner as a result of the relative inlet skew. This pile-up of endwall boundary layer fluid on the pressure surface is sometimes referred to as the 'scraping vortex'.

The inlet flow direction in the endwall region as a result of the inlet skew is equivalent to increases in the incidence and the angles involved can be quite large. In fact the local incidences are sufficiently large that, viewed as a two-dimensional strip, the blade suction surface boundary layer would be certain to separate at the high incidences involved. Some separation certainly takes place and from flow visualization there appears to be an unsteady and turbulent three-dimensional bubble in the endwall-suction surface corner. When there is a tip clearance the pressure field is modified very close to the clearance and the separated bubble is either removed or reduced in size. Flow visualization on the endwall beneath the rotor tip has shown that the flow close to the surface is moving axially upstream even at the design condition, the dividing lines between forward and backward flow occurring close to the leading edge and close to the trailing edge. The whole question of the flow in the region of tip clearance is so important that it is appropriate to consider it in a separate section below.

Repeating stages and fully developed flow in multistage axial compressors

Following the paper by Adkins and Smith (1982) the idea of spanwise or radial mixing has become accepted in axial compressors: this was addressed briefly in Section 3.6. The regions of high loss production, the endwall regions, experience an outflow of gas with raised stagnation temperature and reduced axial momentum whilst gaining an equal mass with lower stagnation enthalpy and higher axial momentum. Such a process is necessary if the entropy near the endwalls is not to build up to very high levels in a multistage compressor. All the evidence points to the velocity and temperature profile in well designed axial compressors settling down to an approximate equilibrium condition after two or three stages, as shown in Fig. 3.15, taken from Smith (1969) for a 12-stage compressor.

Gallimore and Cumpsty (1986) found that there was no clear edge to the viscous region on the annulus walls inside a multistage compressor and that the mixing and the variation in stagnation pressure extended right across the annulus. Wisler *et al.* (1987) found that in a very efficient four-stage compressor a clear spanwise demarcation did exist between flow which could be regarded as the annulus wall boundary layer and that which is the freestream, though quite strong mixing was present right across the annulus. Nearer to stall Wisler *et al.* found the high level of turbulence characteristic of the boundary layer to extend right across the annulus. Many of the test cases and much of the data used to derive or 'calibrate' annulus boundary layer calculation methods will have been obtained in conditions for which there is no clear edge to boundary layer and a fully developed flow might provide a better model.

Empirical approaches to estimating endwall parameters in axial compressors

It was realized early on that the loss at the annulus walls was important and Howell (1945) put some quantitative estimate on this for the axial compressors of the time. The combined endwall losses then as now exceeded the profile loss in the likely design regime. The endwall loss was assumed to be made up of two parts, a so-called annulus loss and a secondary loss. The former was just the frictional loss and was assigned the constant value of drag coefficient

$$C_{Da} = 0.20 s/h$$

s being the pitch and h the blade height. The secondary loss was derived by analogy with the trailing vortex loss behind aircraft; trailing vortices shed by the blades due to circulation variation along the length would induce additional incidence on the blades over the whole span, just as the trailing vortices do for an aircraft. Secondary loss was therefore taken to be related to the reduction in blade circulation in the blade end region and estimated to be

$$C_{Ds} = 0.018(C_{L})^2$$

where C_{L} is the blade lift coefficient outside the endwall region. Although there is no reason to think that the separation of loss into two components was accurate, nor to believe the model for secondary loss, this did provide a way of introducing the loss of approximately the right magnitude and later work confirmed that the secondary losses were proportional to the square of the lift coefficient.

In addition to the losses in the endwall region it was also necessary to know the work input. The approach adopted by Howell was the 'work done factor', defined as the ratio of the actual stage work input to the theoretical work with uniform flow entering and leaving the blades.

A simple analysis for a boundary layer along the endwall was given by Stratford (1967) and the geometry is shown in Fig. 8.12 for the case of a blade

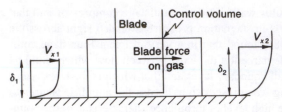

Fig. 8.12 Control volume around the tip region of a blade. (From Stratford, 1967)

row with clearance. The view shown is an axial section through the machine. If F_x is the axial component of blade force at the edge of the boundary layer and δ_{Fx} denotes the deficit thickness for the axial force, then applying conservation of axial momentum to the control volume drawn in Fig. 8.12 gives

$$F_x \delta_{Fx} = \rho \{ V_{x2}^2 \theta_{x2} - V_{x1}^2 \theta_{x1} + (V_{x2} - V_{x1}) V_{x2} \delta_{x2}^* \} \tag{8.5}$$

where the skin friction, the higher order terms and the effect of compressibility have been ignored. Here θ_x and δ_x^* denote the axial momentum and displacement thicknesses with velocity components in the axial direction. If the axial velocity is equal at inlet and outlet the equation simplifies to

$$F_x \delta_{Fx} = \rho V_x^2 \{ \theta_{x2} - \theta_{x1} \} \tag{8.6}$$

and so if the force deficit is small the momentum thickness should not change across the blade row. Hunter and Cumpsty (1982) found that this was true for a rotor tip until very close to stall, provided the tip clearance was significantly smaller than the inlet boundary displacement thickness, but when the tip clearance was similar in magnitude to the displacement thickness a rise in momentum thickness across the blade row occurred. (The displacement thickness was taken to be the relevant length scale of the boundary layer since most of the change in velocity takes place over this distance from the wall.) When the inlet boundary layer was thickened a reduction in the axial momentum thickness occurred across the blade row, indicating a negative axial blade force deficit. This suggests that for a given blade geometry and clearance there is a 'natural' boundary layer thickness depending only on the pressure rise. The results obtained by Hunter and Cumpsty agree with those measured in multistage compressors, giving support for their generality. The axial momentum thickness is of less direct concern to the designer than the displacement thickness δ_x^* since it is this on the hub and casing which determines the blockage. Further empiricism in the form of a relation for the form factor $H = \delta^*/\theta$ is necessary to get this.

The neglect of the skin friction is worth examining in a little detail. The skin friction coefficient is normally of order 0.003 for turbulent boundary layers at fairly high Reynolds numbers, so the component of wall shear stress in the axial direction is given by $\tau_w \approx 0.003(\rho V_x^2/2)$. The pressure rise may be approximated, as above, by $\Delta p \approx 2.4(\frac{1}{2}\rho V_x^2)$. Taking the measurements of Hunter and Cumpsty the axial chord of the blades c_x is some ten times the casing boundary layer displacement thickness so that comparing forces in the axial direction

$$\Delta p \cdot \delta^* \approx 80\tau_w c_x.$$

If the overall boundary layer thickness had been used for this comparison the smallness of the skin friction would appear even more overwhelming. Very clearly the model for the skin friction on the endwalls is largely irrelevant to the accuracy of the modelling. (Many methods for the endwall boundary layer have adopted an auxiliary equation for turbulent boundary layers based on entrainment by Head (1958); similar reasoning would show that the axial length is so short in relation to the boundary layer thickness that entrainment by the usual boundary layer processes cannot be important here.)

By far the most significant collection of measurements of endwall flows in multistage compressors was given by Smith (1969). Results for δ^*_x and $\delta_{F\theta}$ were originally given in graphical form with separate plots for hub and casing but in Koch and Smith (1976) the results for hub and casing were added to give the combined blockage and force deficit with rather smaller scatter. This is shown in Fig. 8.13, with g denoting the staggered spacing between the blade rows at outlet. Even with hub and casing values added it can be seen that there is a large scatter in the tangential force deficit.

Hunter and Cumpsty compared their results for δ^*_x and $\delta_{F\theta}$, obtained across an isolated rotor, with the extensive results published by Smith (1969) obtained in multistage compressors. These are shown non-dimensionalized with the tip

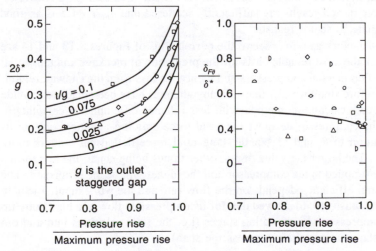

Fig. 8.13 Displacement thickness and tangential force deficit thickness for hub and casing regions combined, obtained in repeating stages (different symbols denote different compressors). (From Koch and Smith, 1976)

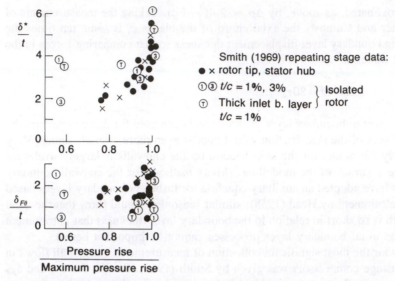

Fig. 8.14 Endwall displacement thickness and tangential force defect thickness versus pressure rise. Isolated rotor data measured by Hunter and Cumpsty. (From Hunter and Cumpsty, 1982)

clearance in Fig. 8.14, the abscissa being the ratio of pressure rise to maximum pressure rise, as in Smith's plots. Rather surprisingly the results for the isolated rotor agreed well for moderate or high blade loading with those obtained in a multistage compressor with repeating stages. Plotted in this way the overwhelming importance of the tip clearance is emphasized, although clearly it cannot be valid for stages with very small or vanishing tip clearance. The force deficit results are sufficiently scattered that $\delta_{F\theta}/t \approx 1.5$ is as good an estimate as seems realistic.

Unsatisfactory as it may seem, the correlations of Figures 8.13 and 14 are at present the most reliable basis for the prediction of blockage and tangential force deficit in axial compressors and thence the loss from the endwall region. It is therefore small wonder that so many attempts have been made to provide an adequate calculation method with less reliance on empirical correlation.

For the compressor designer the axial force deficit δ_{Fx} matters only in its contribution to θ_x and δ_x^* but the tangential force deficit $\delta_{F\theta}$ matters more directly. The larger the value of $\delta_{F\theta}$, other things being equal, the less torque must be supplied to the compressor and the higher will be the efficiency. The net overall efficiency depends on the flow rate and the work input. Lieblein and Roudebush (1956) showed that for incompressible flow in a high hub—tip ratio compressor with repeating stages (i.e. the flow velocities into and out of the stage are equal right across the span)

$$\eta = \bar{\eta} \frac{1-(\delta_{xh}^*+\delta_{xc}^*)/h}{1-(\delta_{F\theta h}+\delta_{F\theta c})/h} \tag{8.7}$$

In this equation $\bar{\eta}$ is the stage efficiency without the inclusion of the endwall loss, h is the blade span, $\delta_{F\theta h}$ and $\delta_{F\theta c}$ are the tangential blade force deficit thicknesses at hub and casing and likewise the displacement thicknesses are δ^*_{xh} and δ^*_{xc}.

Boundary layer calculation methods for the endwall region

Before summarizing the methods devised for calculating the endwall flow it is useful to set out some of the factors that make it so difficult. As discussed above there is a very large pressure rise in the axial direction, far too large to be sustained by the dynamic pressure of the incoming flow, and the endwall flow can only remain attached because of the blade force. This makes the flow development highly dependent on the magnitude and direction of the blade force in the endwall region, but the blade force itself is determined by the flow in that region. The blade force is felt as pressure gradients which are very severe when expressed in terms of the length scale appropriate to the boundary layer.

The intense pressure gradients and the large overall rise in pressure produce steep spatial variations in the endwall flow in the blade passages. It cannot be reasonably assumed that correlations for such properties as skin friction, fluid entrainment rate and velocity profile shape, derived for boundary layers in the relatively mild conditions normal in external flow, will be even approximately valid inside the blade rows.

Almost all the calculation methods that have been developed have treated the endwall flow as a boundary layer and used the boundary layer momentum integral equations. All can be described as control volume analyses with several unknowns; blade force deficits, skin friction, velocity profiles and other parameters such as entrainment. These are applied through the blade row, taking the freestream flow field to be prescribed. All average the flow in the circumferential (i.e. pitchwise) direction; sometimes the averaging is before integrating the partial differential boundary layer equations in the direction normal to the endwall and sometimes after, but the effect of such differences is small compared to the other underlying assumptions necessary.

Earlier in this chapter a simple order of magnitude argument showed skin friction to be negligible in comparison with the pressure forces involved, because the axial length over which the pressure rise takes place is so small, and similar arguments could be brought to bear for the entrainment. Surprisingly this does not seem to have been widely recognized, yet it implies that the estimates for skin friction coefficient and entrainment rate are essentially irrelevant to the final prediction.

Amongst the earliest of the methods was one by Mellor and Wood (1971) where the equations of motion and the momentum integral equations were set up in a coordinate system with axes in the axial, tangential and radial directions. The velocity profiles were prescribed with simple empirical expressions specific to turbomachinery passages and the great importance of the force defect was recognized.

There have been many other boundary layer calculation methods since Mellor and Wood, most resting on very much the same principles and assumptions and it is appropriate to mention only very few. Horlock and Perkins (1974) used a formulation in the streamwise and crossflow directions rather than the axial and tangential; the paper is unusual because of the thorough discussion and assessment it contains and because of its careful and detailed treatment of the defect issue. The paper by De Ruyck and Hirsch (1981) is of interest because of the thoroughness of the investigation into the applicability.

Recent papers include Lindsay *et al.* (1987) and Falchetti and Brochet (1987). The former sets out clearly some of the reasons why such methods should *not* be expected to be satisfactory; to this list should be added that the empirical relations used for such properties as skin-friction, entrainment, streamwise and crossflow profile shape are not suited to the violent conditions in the end-wall region of blades nor can they be reasonably be estimated by empirical relations using circumferentially averaged boundary layer properties. The one prediction shown for a compressor cascade by Lindsay, Carrick and Horlock does not give a satisfactory prediction downstream.

The recent experience reported by Falchetti and Brochet highlights the problems of all such methods. By varying one parameter, in their case the direction of the force defect in the endwall region, it was possible to adjust the method to give reasonable agreement with measurements. It is thereby possible to compensate for some of the errors introduced in the flow description and approximations used. This does not, of course, either justify the assumptions of the method or ensure its general applicability. As Figs 8.13 and 14 show, the tip clearance has a dominant influence on the displacement and force defect thicknesses for values of practical interest. The boundary layer calculation methods have either ignored the tip clearance or have modelled it very crudely: for this reason alone their validity is highly suspect. The continued development of new methods is evidence that a genuinely successful one has not been produced. Contrary to what is often assumed or claimed they in no sense model the flow behaviour.

If there is a place for boundary layer calculation methods such as the ones mentioned in this section it is that they give a consistent and fairly systematic method of including many of the geometric and flow variables in a correlation scheme. By using a wide and varied data base it may be possible to develop a fairly reliable scheme. Along these lines Freeman (1985) described how a method owing much to Hirsch has been used. The boundary layer method has been used in conjunction with an optimization procedure to find the defect forces in the streamwise and normal directions which give the best fit with measured data from twelve different compressors. Because measurements were available from choking mass flow to surge at a wide range of speeds there was a large body of data. Half the measurements were reserved for test cases after the optimum defects had been derived. Although such a method may give good predictions for variables it should not be taken to describe the flow. What has been achieved is at best an empirical fit and care must be exercised if it is applied outside the range of cases for which it was developed.

Given that the endwall blockage has a crucial effect on the design of multistage axial compressors, one questions how very efficient compressors can be achieved with methods and correlations which are so suspect. The first part of the answer is that few organizations will design a multistage axial compressor nowadays for the first time, most have large in-house data bases with which to guide the design of the next machine. (This is precisely the situation which Freeman described with his optimized calculation method; in essence the equations are used as a way of incorporating the measurements in a way which can be readily utilized.) An approach which relies on data from earlier machines is particularly reliable when the new machine bears a strong resemblance to one of its predecessors. The second part of the answer is that the prediction is sometimes *not* sufficiently accurate and an expensive and time-consuming development programme may then be required.

The tip clearance region

A separate section to examine the effects of clearance is somewhat artifical because the effects of tip clearance are overwhelming on the endwall flow development and on the blade-to-blade flow in regions near the tip as well. In justification it can be said that the investigation of tip clearance has often been carried on as a separate activity and it is useful to highlight the special features of it.

The effect of tip clearance on even the overall performance is not simple. Figure 8.15 shows total-to-static pressure rise−flow rate characteristics

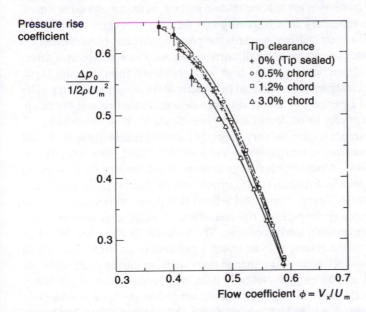

Fig. 8.15 Pressure rise−flow rate characteristics of a single stage compressor. Hub−casing ratio 0.8. At tip clearances less than or equal to 1.2 per cent of chord stall is initiated at the hub, at the larger clearance it is initiated at the casing. (From MacDougall, 1988)

measured by MacDougall (1988) at a range of rotor tip clearances in a low-speed single stage of hub—casing ratio of 0.8 with geometry similar to a modern high-pressure compressor stage. Generally the pressure rise increases and the stall point moves to lower flow as the clearance is reduced, but the performance with the tip sealed or with a clearance equal to 0.5 per cent of chord is below the optimum. Increasing the clearance from 1.2 per cent up to 3 per cent gave a reduction in peak pressure rise of about 8 per cent. With a high-speed single-stage compressor (rotor tip relative Mach number 1.66) Wennerstrom (1984) showed higher efficiency with a tip clearance of 0.68 per cent chord than the other values tested, 0.42 per cent and 0.94 per cent. Inoue *et al.* (1986) varied the clearance between 0.43 and 4.3 per cent of chord at the tip and found that the optimum must be less than 1 per cent of pitch. Lakshminarayana found the optimum to lie between 1 and 1.5 per cent of chord for rotors, compared with between 3 and 5 per cent for cascades. Unfortunately the optimum clearance for rotors is generally smaller than that which can be tolerated for mechanical reasons. Wisler (1985) reported that for a low-speed four-stage compressor the effect of increasing rotor tip clearance from 1.6 to 3.4 per cent of chord was to reduce the efficiency by 1.5 per cent, the flow range by 11 per cent and the peak pressure rise by 9.7 per cent.

Flow visualization studies have shown that with no clearance there is a major three-dimensional separation in the endwall-suction surface corner but a small clearance, of the order of 1 per cent of chord, seems to reduce this. The benefit of clearance is sometimes explained as opposing the secondary flow, which is assumed to be moving towards the suction surface; in the tip regions of rotors, however, the secondary flow is usually less than the inlet skew and the flow in the endwall region is driven towards the pressure surface. Nevertheless the flow in the suction surface-endwall corner does tend to separate as a three-dimensional region because of the high local incidence from the inlet skew. Although the clearance flow seems to be beneficial when very small, typically not more than 1 per cent of chord, at larger clearances the loss and the inception of stall appears to be dominated by the clearance flow behaviour.

The tip-clearance region has very properly received considerable study and this was summarized in two papers by Peacock (1982) and more recently one by Senoo (1986). A paper with a large amount of information on flow in the tip region is given by Freeman (1985), much of it obtained in high-speed compressors by Rolls-Royce. Figure 8.16 from that paper shows the effect of varying clearance on the pressure rise and efficiency of the high-pressure compressor of a three-spool aircraft engine. The variation in clearance, which is the same for all the rotors, is from about 1 per cent of the final rotor chord to 3 per cent and is typical of the change that can occur during rapid acceleration of a civil aircraft engine. Not only does the pressure rise and efficiency fall as the tip clearance is increased but the surge line moves a considerable way to the right, that is to higher mass flows. All of these effects are highly detrimental.

In a high-speed compressor the tip clearance will change during the operating cycle. The biggest problem in controlling tip clearance in the case of aircraft

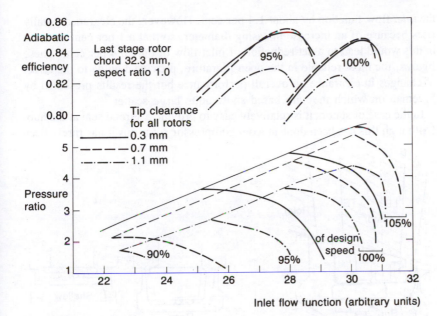

Fig. 8.16 The effect of tip clearance on the pressure ratio, surge line and efficiency of a six-stage, high-speed compressor. (From Freeman, 1985)

gas turbines is that the compressor disc temperatures rise more slowly than the relatively thin casing during acceleration (and likewise fall in temperature more slowly during deceleration). To accommodate this it is necessary to operate transiently with larger clearances than are desirable for aerodynamic considerations. Furthermore, the casing may deform so that it is no longer concentric; Freeman (1985) showed that the pressure rise is more or less the same for the concentric and non-concentric builds of higher clearance, the detrimental effect on pressure rise being from an integrated clearance around the annulus. However, the stall margin is made much worse for the non-concentric build, clearly showing that it is the worst clearance over an extensive sector which determines this.

In a multistage compressor the precise effect of tip clearance becomes confused because the performance of each blade row effects all those downstream. If, for example, increased tip clearance produces extra blockage and extra loss in the front stages, the matching of the downstream stages is altered. As another example, when the extra tip clearance is produced by an increase in casing diameter, as can occur during a transient, the mass flow capacity of the compressor is increased. This would move the stall line to right, that is to larger mass flow at stall, even with no deterioration of the flow. For clearances greater than the aerodynamic optimum Freeman's results suggest that a 1 per cent increase in tip clearance produced by increasing the casing diameter will give approximately 1.4 per cent loss in efficiency and 1 per cent increase in *exit* flow function; an increase in flow function moves the stall point to the right. An increase in clearance of 1 per cent of chord produced by a rotor crop reduces the *inlet* flow function by about 1.4 per cent. However, the clearance usually rises because of an increase in casing diameter, so that a 1 per cent increase in this would lead to a net reduction of inlet flow of only 0.4 per cent. These figures, the most reliable in the open literature, provide a guide to the effect of changes in clearance on overall performance but the results presented by Freeman on which they are based show quite large scatter.

In the case of stators it is relatively easy to provide a shroud seal at the hub. (Although this has been done at axial compressor rotor tips it has rarely been

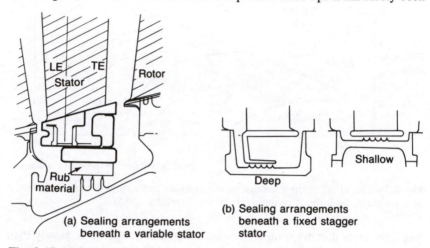

(a) Sealing arrangements beneath a variable stator

(b) Sealing arrangements beneath a fixed stagger stator

Fig. 8.17 Different arrangements of hub seals beneath stators

found to be worthwhile.) An arrangement of shroud seal under a variable stagger stator and two possible geometries under a fixed stator are shown in Fig. 8.17. This is a quite large topic and the reader is referred to the AGARD (1978) Conference Proceedings or to Wisler (1988) for a summary. It might seem that a stator shroud, be greatly reducing the leakage flow, would be entirely beneficial but in fact the picture is less straightforward. At small clearance the unshrouded stator, that is cantilevered stators with clearance at the hub, was shown by Freeman to perform better, Fig. 8.18, in line with the benefits found for rotor tips. (The shrouded stator does have significant mechanical advantages because of increased mechanical stiffness so that vibration problems are usually eliminated.) Particular tip clearance problems also arise with variable stators which are unshrouded and at some attitudes there may be very large clearances over part of the chord, particularly at the hub.

Most of the interest is with the tip clearance of rotors and there are now many measurements in the rotor frame of reference, particularly for low-speed machines. Figure 8.19, from Lakshminarayana *et al.* (1986), shows the evolution of the relative loss pattern through a rotor passage for a condition close to peak efficiency. The collection of the loss near the pressure surface towards the rear of the passage is very noticeable. Inoue *et al.* (1986) showed the

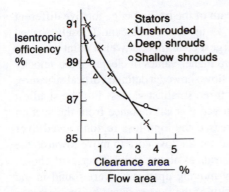

Fig. 8.18 The effect of stator hub-clearance on efficiency, unshrouded and with shrouds. Two-stage, 50 per cent reaction compressor. (From Freeman, 1985)

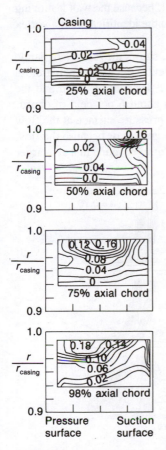

Fig. 8.19 Contours of loss in relative stagnation pressure, non-dimensionalized by $\frac{1}{2}\rho U_{\text{tip}}^2$, in a rotor passage at conditions of peak efficiency. (From Lakshminarayana *et al.*, 1986)

contours of relative loss just downstream of the trailing edge for five different levels of tip clearance from $t/c = 0.43$ up to 4.3 per cent. Closely spaced contours were evident in all but the largest clearance, at which point the clear pattern seems to disintegrate. At the two higher values of clearance the velocity vectors in the plane normal to the main flow showed a definite vortical structure, but this was not really distinct at the three smallest values. For the smallest clearances the maximum loss only moved a small distance from the suction surface but as the clearance was increased the high-loss region moved over towards the pressure surface; the trend for the loss core to move towards the pressure surface was reversed for clearances above about 2 per cent chord. Negative axial velocities, that is flow moving upstream, were found in the region of the maximum loss. The clarity of the vortex found by Inoue *et al.*, seems to be somewhat unusual and it is more common for it to burst or become blurred before the trailing edge. Plots of the relative crossflow vectors must show large components of velocity near to the wall, because the wall is moving relative to the blades, and this has sometimes led to the identificaton of a vortex when one is not present.

Figure 8.20, from Hunter and Cumpsty (1982), shows the relative dynamic pressure contours at three different flow coefficients, the lowest being close to stall. In all three cases the tip clearance was 1 per cent of chord and the inlet boundary layer displacement thickness was about 6 per cent of chord. The greater collection of high-loss flow near the pressure surface at the flow rate near to stall may be because the inlet skew is greater or because the flow through the tip clearance is stronger. Hunter and Cumpsty found that at a larger

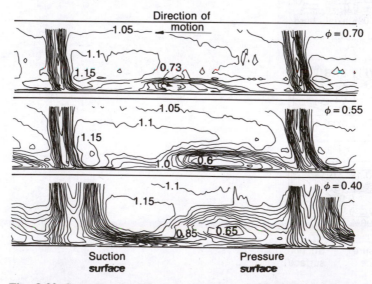

Fig. 8.20 Contours of relative dynamic pressure measured 20 per cent of chord downstream of trailing edge non-dimensionalized by mean value. $\phi = V_x/U = 0.4$ is just prior to stall. Tip clearance $t/c = 1$ per cent. (From Hunter and Cumpsty, 1982)

tip clearance, 5 per cent of chord, the systematic pattern disappeared, similar to the finding of Inoue *et al*. As well as showing the tendency for the high relative loss fluid to collect towards the endwall—pressure surface corner and to be strongly affected by the tip clearance flow, the contour pictures also show very clearly the strong circumferential variation which is present in the region between the blades. They provide clear evidence of the need for caution in using annulus boundary layer calculation methods.

Some high-speed evidence of the tip-clearance flow is presented in Fig. 8.21, showing photographs from holograms of the tip region from Freeman (1985). The two pictures were taken at different flow coefficients so that the position of the passage shock is different in each. In each a tip clearance vortex is evident, springing from the blade suction surface where the passage shock meets the pressure surface, which is just where the pressure difference across the blade row would be expected to become large and therefore able to drive flow through the gap.

Much earlier, Rains (1954) had used a three-stage axial water pump to examine tip flows, looking at effects of different clearance and also varying the flow rate through the machine. There were significant reductions in pressure rise and efficiency as the clearance was increased up to the maximum value of 1.4 per cent of chord. The tip clearance vortex appeared to spring from the leading edge of the rotor blade, being evident by its cavitation; similar tests with a stationary blade with clearance showed the corresponding vortex to spring from about the 1/4 chord point, around where the pressure difference across the blade might be greatest. The difference here is probably because the inlet skew to the rotor moved the peak loading forward to near the leading edge.

Rains also produced a simple model for the clearance flow which still may have application in conjunction with numerical methods for the remainder of the blade passage. For relatively large clearances this assumed an inviscid flow through the clearance gap to produce a jet normal to the blade on the suction

Low back pressure   Near surge

Fig. 8.21 Photographs from a hologram showing the tip region of a supersonic fan, rotor moving from right to left. Near surge the passage shock is slightly ahead of the leading edge; at low back pressure the passage shock is some distance down the passage. The vortex can be seen springing from suction surface at chordwise position of shock impingement on pressure surface. (From Freeman, 1985)

side but when the clearance gap was very small a simple one-dimensional viscous flow was assumed. The mechanism of loss generation was that the kinetic energy of the jet normal to the chord was assumed lost. The pressure difference was assumed to be that of the blades with no endwall and clearance effects present, so the effect of the clearance flow on the blades was neglected. (Pressure measurements on rotor blades with tip clearance have shown that the blade loading and pressure distribution do remain sensibly unchanged until very close to the blade tip indeed, see Smith and Cumpsty (1984), so this may be one of the better assumptions.) Harder to quantify is the exact behaviour in the clearance region when the gap is small enough for the viscous effects to be important and the wall may be moving at high velocity relative to the blades. Recent measurements by J.A. Storer in Cambridge support many aspects of Rains' model: the crucial change which occurs with small clearance gap is the reattachment of the clearance jet to the blade tip with consequent reduction in stagnation pressure and velocity perpendicular to the camber line.

The effect of clearance was investigated in an ingenious experiment by Lakshminarayana and Horlock (1967) which sheds some light on the phenomenon. A linear cascade was used with the blades cut in two at mid-span so that, with the two parts of the blade cantilevered out from the endwall, an adjustable clearance gap could be introduced. In the first set of experiments the inlet flow was uniform, in another a weakly non-uniform inlet flow was superimposed, modelling the outer part of the boundary layer, and in the third experiment a solid surface was introduced on which an inlet boundary layer grew. Most of the experiments were carried out with the uniform inlet flow. A vortex was shed from each side of the slot, see Fig. 8.22, springing from about the 1/4 chord position, just as in Rain's experiments for the isolated aerofoil, but unlike the rotor. With a clearance of 1 per cent of chord the pressure distributions even within 1 per cent of chord of the gap were very similar to the two-dimensional ones with no gap. By 4 per cent clearance the effect of the gap was quite marked even 8 per cent of chord along the span from it, with a large drop in pressure on the suction surface towards the rear, just where the vortex might be assumed to affect it. This effect was predicted quite accurately using the measured position of the vortex core and the deduced

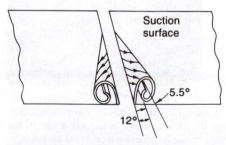

Fig. 8.22 The vortex observed above a gap in a blade in a two-dimensional cascade. (From Lakshminarayana and Horlock, 1967)

vortex circulation; for the case with uniform inlet flow the disturbance to the two-dimensional flow was mainly a result of the vortex. As a result of the low pressure on the suction surface the total blade normal force was increased by the presence of the gap up to a clearance of 6 per cent of chord but decreased at still higher values of clearance. Right at the end of the blade the normal force did decrease, the value being about 50 per cent of the two-dimensional value for a clearance gap of 3 per cent of chord. This indicates that a fraction of the bound circulation around the blade was being shed in the tip gap, the amount depending on the clearance gap. The shed vorticity, which is assumed to all come from the clearance, sets up an induced drag which Lakshminarayana and Horlock approximated by

$$C_D = 0.7(C_L)^2 t/h$$

where C_L is the two-dimensional lift coefficient, h is the blade span and t the tip clearance. The dependence on the square of lift coefficient is, of course, the same general form as proposed by Howell (1945) for the secondary loss.

Lakshminarayana (1970) extended the earlier work to include the kinetic energy of the spanwise secondary flow in the reduction of efficiency to derive the following approximate expression for the effect of clearance

$$\Delta \eta = \frac{0.7 t/h \psi}{\cos \beta_m} \left[1 + 10 \sqrt{\left(\frac{\phi}{\psi} \frac{t/c}{\cos \beta_m} \right)} \right] \qquad (8.8)$$

with β_m being the mean relative flow direction, $\psi = \Delta h_0 / (1/2 U^2)$ the non-dimensional stagnation enthalpy rise (written here with the half in the denominator) $\phi = V_x / U$ and h the blade span. The comparisons shown by Lakshminarayana indicated that this gives quite a good estimate for the loss. Others have found less satisfactory agreement, for example Inoue *et al.* (1986), although as a method for getting an estimate within a factor of two it seems to be reasonable for clearances of up to about 3 per cent of chord. Rather better agreement seems to come with the simple approximate method given by Senoo and Ishida (1986).

Lakshminarayana (1970) went on to consider the finite size of the vortex core in predicting the flow angle variation: Fig. 8.23 compares his predictions of the pitchwise average outlet angle with measurements made in the cascade with the endwall and thick boundary layer reported by Lakshminarayana and Horlock (1967). It can be seen that a good prediction of the average flow direction with clearance was produced by the Lakshminarayana model considering only the clearance flow effect. However most noticeable and most important is the great difference between the measurements with and without clearance. Without clearance there was a large region of separated flow in the endwall-suction surface corner, evident by the high outlet flow angle extending out for about a chord length in the spanwise direction. The clearance had a dramatic effect on the flow over a spanwise

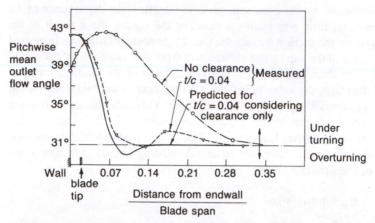

Fig. 8.23 Outlet flow direction averaged in pitchwise direction versus distance from the endwall with and without clearance. For the cascade well away from the endwall the outlet flow direction is 31°. (From Lakshminarayana, 1970)

distance equal to many times the tip clearance, reducing the region of separated flow to a fraction of its former size. The effect of the clearance flow in reducing the extent and moving the location of separated flow has been discussed by Lakshminarayana and Horlock (1967) for the case of endwall flow with clearance, based on flow visualization and quantitative measurements. It must be emphasized that the effect of the clearance cannot be thought of as a linear addition to the no-clearance flow, but as a total alteration of the whole flow structure. These results were obtained in the relatively simple cascade flow with a collateral boundary layer at inlet and no moving wall. Measurements, calculations with Navier–Stokes solvers and flow visualization (shown in Fig. 4.52) show this same large effect of tip clearance for a stator row: experimental evidence and the results of calculations point to a similar very large effect at the rotor tip.

Measurements have always been difficult for the endwall region, particularly in the case of rotors. The advent of the Navier–Stokes solvers opens up the possibility of numerical investigations with a view to understanding the flow behaviour. Hah (1986) reproduced quite accurately the flow in the tip region measured by Hunter and Cumpsty (1982). More recently McDougall and Dawes (1987) have calculated the flow on a stator with and without clearance at the hub. The flow was shown in Fig. 4.53 and is interesting in a number of ways, including the large extent of separated flow without the clearance and the greatly reduced separated flow with a clearance of about 1 per cent of chord. The calculation was capable of reproducing this in both cases. With any calculation procedure it is often unclear how far the results represent *pre*dictions and what extent *post*dictions with parameters adjusted to give best agreement. Calculations of the former type are clearly much more demanding but most of the boundary layer calculations have been in the latter category.

Axial blade modifications in the endwall region

Recently it has become fairly common to design the ends of blades for axial compressors with the intention of reducing the loss, even though the mechanism of loss generation is not properly understood. These have a variety of names but 'end-bends' is one of the more common. It is intuitively reasonable that locally high-incidence in the endwall region is undesirable and one of the aims is to align the inlet to the blades so that flow enters with only a small local incidence. To cope with skewed inlet flow it is therefore possible to bend the tips to incline the forward part of the blade towards tangential, i.e. increase χ_1. Very close to the annulus walls the flow direction could be markedly different to the freestream — for example for the tip of a rotor the relative flow direction at the casing wall is circumferential — and such very large excursions in blade direction cannot be allowed even were they desirable. Consequently the variation in blade angle is limited at a value appropriate to some way out from the wall.

The strategy at blade inlet is fairly straightforward, to reduce the local incidence, but at outlet it is less obvious. One approach, which has physical plausibility for blades of moderately large span (i.e. aspect ratios which are not less than about unity), is to recognize that the secondary flow is set up by the pressure field (i.e. the deflection) of the main flow and will therefore occur regardless of the way in which the blades are modified in the end regions. If the blades are not aligned with the local flow some relaxation must take place, which is likely to be a region of separated flow. It is therefore expedient to set the blades in the direction of the flow, taking account of the secondary flow component.

In carrying out the analysis to obtain the flow directions there are frequently several additional empirical or semi-empirical inputs, the most important being to include the tip clearance flow. Adkins and Smith (1982) describe one such method as part of their radial mixing scheme and the final result for the

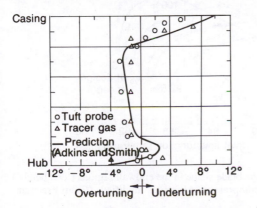

Fig. 8.24 Measured and predicted pitchwise—averaged outlet flow directions downstream of third stage stator. (From Wisler *et al.*, 1987)

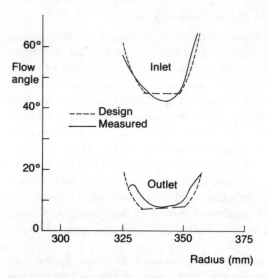

Fig. 8.25 Flow angles into and out of the 5th stage stator of a six-stage compressor where stator shape is prescribed to be compatible with flowfield in endwall regions. (From Freeman, 1985)

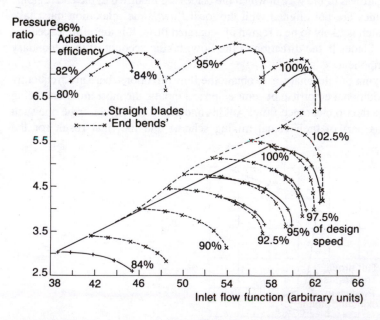

Fig. 8.26 The effect of modifying the blade shape in the endwall regions ('end bends') on the performance of a six-stage compressor. All stages modified. (From Freeman, 1985)

prediction of the radial distribution of meridional velocity and circumferential flow angle can be very good indeed, as is shown by Fig. 8.24 from Wisler *et al.* (1987).

Freeman (1985) discussed the use of 'end bends' and Fig. 8.25 shows the flow angles specified and measured into and out of one stator row in a six-stage compressor designed with 'end-bends'. The corresponding improvement in performance for the use of 'end-bends' in all rows is evident from Fig. 8.26.

An account of the Pratt and Whitney approach and experience with the end treatment of blading, so-called second-generation controlled diffusion airfoils is given by Behlke (1986). In application in a high-speed compressor these gave a 1.5 per cent improvement in efficiency and a 30 per cent increase in the operating range.

With the more widespread use of methods to solve the Navier–Stokes equations and the proper realization that most of the loss comes from the endwall region, it is certain that there will be considerable work in the future devoted to finding better shapes for the blades in the endwall regions.

An overview for endwall flows in axial compressors

The most important viscous region for all compressors is the endwall region, including the corner flows and tip-clearance flow. The endwall region has a major effect on the performance, giving most of the losses, determining most of the blockage and being involved in the initiation of stall. Present design is based on empirical information for displacement thickness and force deficit thickness but this is generally recognized to be unsatisfactory.

One of the facts which emerges very clearly from the correlations is the very great importance of tip clearance. Tip clearance alters the blade pressure distribution, removing the corner separation and increasing the loading near the tip; at the same time tip clearance introduces loss and reduces the pressure rise capability.

Much effort and many papers have been directed to endwall flows in cascades. For several reasons outlined above, these flows are not representative of the flow in compressor blade rows and ideas or data derived from them should be used with the greatest caution.

Endwall flows are not predicted at all well by the methods based on boundary layer theory, which is not surprising in view of the unsuitability of many of the boundary layer assumptions. (Many methods are able to give agreement which is satisfactory in some cases but not in general.) Endwall flows are an ideal for the application of methods of calculation based on the Navier–Stokes equations and rapid progress can be expected now that these are being applied. Although these methods are at present quite good at predicting gross or overall features of the flow, such as regions of separation and the effect of tip clearance, they are still not good for the prediction of the loss. They are likely to remain methods for analysing one blade row at a time and other approaches of a more approximate kind will be needed for the simultaneous calculation of several stages together.

At the current time there are ways for assessing the condition of a boundary layer, particularly a two-dimensional boundary layer: if, for example, the form parameter $H = \delta*/\theta$ of a turbulent boundary layer exceeds about 2 one knows that the flow is near to separation. One of the problems to be faced with the use of three-dimensional Navier–Stokes methods is that the same simple assessment cannot be used. There will need to be a period while aerodynamicists learn to appreciate the significance of features calculated and observed so that suitable design decisions can be taken on the basis of the predictions. Already it is clear that many flows which are normally considered satisfactory have extensive regions of separation flow.

8.5 Viscous effects in centrifugal compressors

The nature of the centrifugal compressor is such that there is a long narrow flow path, both in the impeller and in the downstream diffuser. The viscous effects are therefore very important as has been emphasized in Chapters 6 and 7.

The impeller

There are two particular problems for the centrifugal impeller. Very often the hub and shroud are so close together near the outlet that the flow is more nearly a fully-developed turbulent flow than a free-stream with hub and shroud boundary layers. At the same time the long flow length means that the corner flows become very important. In many cases there is also the complication of the tip clearance flow and in many centrifugal compressors the tip clearance makes up a large fraction of the total blade height, particularly towards the impeller outlet.

Moore (1973) applied conventional three-dimensional boundary layer calculation methods to a rotating diffuser with quite good results. This configuration was without a tip clearance and so modelled the shrouded impeller most nearly. The geometry also avoided the strong curvatures usually found in radial compressor impellers, both in the sense of the blade camber and incidence and in the sense of the shroud curvature. In Moore's configuration the turn from axial to radial took place at very small radius and in this respect it was like a very low specific speed compressor. In each of the passage corners the crossflow on one surface was allowed to flow around to the next. Thus, for example, the crossflow on the hub and shroud flowing towards the suction surface would be allowed to continue around to become a crossflow parallel to the suction surface. The method successfully predicted the separation on the suction surface but the geometry was better suited to the boundary layer methods than a more typical impeller. From Chapter 6 it may be recalled that separation in impeller flows normally occurs first on the shroud and the high-loss fluid is then transported around by the secondary flow, often finishing

on the suction surface. The shroud curvature is therefore of paramount importance in determining the boundary layer development. The problem of
calculating the flow in impellers is sufficiently important that there are still
attempts to calculate it with boundary layer methods. Kenny (1984) reported
the use of a boundary layer method applied to all four surfaces of an impeller.
The method was said to work well but, as with boundary layer methods applied
to axial compressors, this may not be generally valid.

Moore has continued calculating flows in centrifugal compressors but has
dropped the boundary layer approach in favour of the full three-dimensional
solution of the Navier—Stokes equations, for example Moore *et al.* (1984).
Others, including Dawes (1987), are also calculating the flows in impellers
with the Navier—Stokes equations and reproducing many of the major flow
features which are measured, Fig. 6.15. Accurate prediction of loss remains
difficult and probably requires more accurate modelling of the turbulence.

The diffuser

The flow in the diffuser of a centrifugal compressors, be it vaned or vaneless,
is highly dependent on the viscous flow on the walls, principally the hub and
shroud wall. For the vaned diffuser one of the most critical quantities required
is the blockage into the throat of the diffuser, since this determines the pressure
recovery in the diffuser and, for high pressure ratio machines, the maximum
mass flow which the compressor will pass. Most of the blockage at the throat
will be produced on the hub and shroud walls, with relatively little on the vane
surfaces themselves. The flow into the diffuser is complicated by the jump
from a rotating hub to stationary hub outside the impeller (and also on the
shroud side for a shrouded impeller) as well as by the unsteadiness and non-
uniformity of the flow out of the impeller, as has been referred to in Chapter 7.
However Baghdadi (1977) demonstrated that unsteadiness does not seem to
be important in determining the performance of vaned diffusers and it seems
that the flow in the diffuser may be treated as steady. The calculation of the
flow into the diffuser is difficult because the flow is highly three-dimensional,
the free-stream being made to curve by the presence of the diffuser vanes.
In high-speed compressors the vanes will have shock waves standing ahead
of the leading edges. Given such complexity, one approach has been to examine
a much simpler case by a two-dimensional boundary layer method, as described,
for example, by Kenny (1972). It is not clear whether these methods have had
much effect on design. If the comparison of predicted and measured blockage
does not look very good such a method can be adapted or 'calibrated' to match
the measurements. The result of this is similar to the annulus boundary layer
calculation methods sometimes used in axial compressors — the methods are
not soundly based in the physical sense but they provide a way of interpolating
(and sometimes extrapolating) measurements already obtained.

A better method for the prediction of the viscous flow into and inside the
diffuser of a centrifugal compressor must start from the Navier—Stokes equation, since almost all the assumptions of boundary layer theory are invalidated
in the region. At the time of writing, virtually all the published work based

on these methods has been for impellers, notwithstanding the fact that the loss produced in the inlet region to the vaned diffuser or in a vaneless diffuser is much higher than in the impeller.

Rotation and curvature

The nature of turbulence, both its inception and characteristics, is affected by curvature of the flow and by rotation of the frame of reference. In axial machines the effect of rotation on turbulence is normally negligible, because it is the cross-product of angular velocity and radial velocity which matters and the radial velocity is usually small. In axial compressors the effect of curvature is also usually negligible because the boundary layers on blades are thin near the leading edge and elsewhere the flow has a large radius of curvature. In the centrifugal impeller both effects can be important: the curvature will be particularly important on the shroud, which often has a thick inlet boundary layer and a small radius of curvature.

The effect of surface curvature was considered in some detail by Bradshaw (1973) and separately the rotational effects have been examined by Johnston (1973) and co-workers. For the type of flow in centrifugal impellers the gradient of relative velocity across the passage inside the boundary layer or shear layer is denoted by $\partial w/\partial y$ and this is much larger than the impeller angular velocity ω or the gradient of the mean flow about the curved surface w/R_c. For flows in impellers the scaling parameters, see Johnson and Eide (1976), are Richardson numbers given approximately by

$$Ri_{\text{curv}} = 2\frac{w/R}{\partial w/\partial y} \quad \text{and} \quad Ri_{\text{rot}} = -2\frac{\omega}{\partial w/\partial y} \tag{8.9}$$

Positive values of Richardson number correspond to stabilization and occur on the trailing (i.e. suction) side of the centrifugal impeller passage due to rotation and on the convex side of surfaces, such as the shroud, due to curvature. If some global number is required then the velocity gradient can be approximated by $\partial w/\partial y = W/d$, where d is the boundary layer or shear layer thickness or even W/h, where h is the channel width. In this form the Richardson number can be related to a Rossby number Ro by

$$Ri_{\text{rot}} = -2\frac{\omega}{W/\delta} = \frac{-2}{Ro}$$

In a series of experiments Johnston *et al.* explored the effect of Richardson number on transition and on the turbulent structure of the boundary layer. Some allowance for the effects of rotation and curvature on the turbulent shear stresses will need to be included if greater precision in the prediction of loss in centrifugal impellers is to be obtained by the Navier–Stokes methods.

9 *Stall and surge*

9.1 Introduction

Stall and surge have been a problem for as long as turbomachinery compressors (as opposed to positive displacement types) have been built. Because either phenomenon can be catastrophic for performance the topic has attracted considerable interest over the years, with an extensive literature. Unfortunately there was, and to some extent still is, considerable confusion about terminology. However, this has been much less of a problem since the middle of the 1970s when the topic was put on a much surer footing both in terms of theory and measurement. A description of the work done and the understanding at the time is given by Fabri (1967). The whole topic has been reviewed by Greitzer (1980, 1981) after the large advances in understanding had been made.

Types of stall and surge behaviour

It is normal with compressors that as the mass flow is reduced the pressure rise increases. Generally a point is reached at which the pressure rise is a maximum and further reduction in mass flow leads to an abrupt and definite change in the flow pattern in the compressor. Beyond this point the compressor enters into either a stall or a surge; regardless of the type of change occurring, the conventional terminology for the point of instability of the nominally axisymmetric flow is the 'surge point' and the line marking the locus of these points

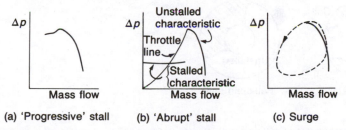

Fig. 9.1 Pressure rise—mass flow characteristics for three different types of behaviour at flow rates below that for stable operation

for different rotational speeds is known as the surge line. Three different ways in which this change in the flow can occur are shown schematically in Fig. 9.1 in terms of the pressure rise—mass flow rate characteristic. This figure is more representative of behaviour of axial compressors but the relevance to centrifugals will be discussed later. Figures 1a and b show rotating stall, which is discussed first, whilst Fig. 1c shows surge.

Rotating stall

In the first case shown, Fig. 9.1a, the drop in overall performance is quite small, and often the presence of the stall is indicated only by a change in noise or by high-frequency instrumentation inside the machine. This behaviour is sometimes called 'progressive stall'. The behaviour depicted in Fig. 9.1b leads to a very large drop in pressure rise and flow rate; when the compressor stalls, the mean operating point moves rapidly down along the throttle line to settle at the new point with a pressure rise that might be only a fraction of that before stall for a multi-stage compressor. This is sometimes referred to as 'abrupt stall'. In both case a and b the flow is no longer axisymmetric but has a circumferentially non-uniform pattern rotating around the annulus, giving rise to the term 'rotating stall'. The annulus then contains regions of stalled flow, usually referred to as 'cells', and regions of unstalled flow, see Fig. 9.2. Rotating stall is a mechanism which allows the compressor to adapt to a mass flow which is too small: instead of trying to share the limited flow over the whole annulus the flow is shared unequally, so that some blade passages (or parts of passages) have a quite large flow and some very little. In rotating stall the overall or time-mean mass flow remains constant but the local mass flow varies as the rotating cell passes the point of observation. Inside the stall cells the axial or throughflow velocity is very much less than in the unstalled flow but the circumferential component of the velocity can be large. There may be several stall cells or only one and the cells may extend right across the annulus (full-span stall) or only over part (part-span stall). When the cell is full span there normally seems to be only one cell, whereas when the stall is part span there can be multiple cells. The abrupt stall in Fig. 9.1b is characteristic of a compressor going directly into full-span stall whereas the progressive stall in Fig. 9.1a

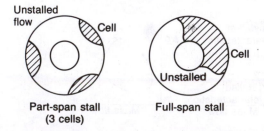

Part-span stall
(3 cells)

Full-span stall

Fig. 9.2 Different types of rotating stall. Regions of low axial velocity (cells) shown shaded

is more likely if part-span stall were to be established. If the compressor in part-span stall is throttled further it may then go into full-span stall with an abrupt drop in performance. The stall cells always rotate in the direction of the rotor. Part-span cells very often rotate at close to 50 per cent of the rotor speed, full-span cells usually rotate more slowly in the range 20–40 per cent, the speed increasing with the number of stages.

Full-span stall cells extend axially through the whole length of an axial compressor whereas part-span stall can exist in only part of the compressor, even in a single blade row. This difference arises because the blade rows are normally too close together for there to be significant circumferential redistribution of flow between the stalled and unstalled regions. The net mass flow into or out of a full span cell is very limited and the width of the cell must therefore be comparable at entry and exit. In the part-span cell, by contrast, there is the possibility of redistribution of the flow in the radial direction. Furthermore there are usually several part-span cells and with the larger number some substantial circumferential redistribution is also possible.

Surge

The type of behaviour depicted in Fig. 9.1c is quite different. Here the overall annulus averaged mass flow varies with time, so that the entire compressor changes more or less in phase from being unstalled to stalled and back again. This process is known as surge. The process may be so violent that the mass flow is reversed during the left-hand leg and previously compressed gas then emerges out of the inlet; this is sometimes called deep surge. On the other hand it may be very mild so that the operating point orbits around the top of the pressure rise–mass flow characteristic and the main evidence for the surge is an audible burble; such behaviour is quite common with small turbocharger compressors prior to a 'harder' surge when the full reversal of the flow takes place. During some parts of the surge cycle, while the compressor is forced to operate at a low mass flow, the flow may be transiently in rotating stall. The time scale for surge is much longer than for rotating stall; for a modern aircraft gas turbine the frequency of rotating stall detected by a stationary instrument is typically in the range 50–100 Hz but the corresponding surge frequency is typically about 3–10 Hz. The frequency of the rotating stall depends on the speed with which it rotates and on the number of cells and this is determined by the flow pattern in and around the compressor. The frequency of the surge cycle is set by the times for the storage volume to fill and empty.

Although the surge process may be effectively axisymmetric when fully developed, it is certainly not so in the initial transient. In fact one of the most damaging effects of surge in high pressure ratio axial compressors is the very large transverse load placed on the rotor and casing because of the non-axisymmetric nature of surge, leading to severe blade rubbing and then a range of further damage. The origin of the non-axisymmetry is that in the initial phase

of surge the reverse flow has to grow from some initiation point and the axial
length of the compressor is likely to be considerably less than the circumference.

The rotation of stall cells

A basic explanation for the rotating nature of the stall, first given by Emmons
et al. (1955), is indicated in Fig. 9.3, first for the initial stall assumed to affect
only one blade passage and then when a large stall cell is established covering
many blade passages. (Recent work by McDougall (1988) shows that not just
one blade passage is involved at the initiation but several, extending over about
15 per cent of the circumference). The blockage caused by the cell leads to
a reduction in the blade incidence on one side and an increase on the other.
The blade with increased incidence will tend to stall. The stall, therefore tends
to run in the direction in which the incidence is increased: for a rotor row
it means that the cell moves backwards relative to the rotor, for a stator it

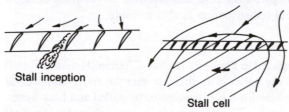

Stall inception

Stall cell

Fig. 9.3 The conventional explanation for the mechanism of rotation of stall at inception
and for a large stall cell.

moves in the direction of the rotor. Viewed by a stationary observer, the rotation of the stall cells is therefore always in the same sense as the rotor but
at a speed which must be less than the rotor speed. The blockage of a large
or fully developed stall cell also gives a deflection to the incoming flow so
as to increase incidence to one side and reduce it on the other, leading to rotation
of the disturbance.

Flow instability

The operating point at which the flow in the compressor becomes unstable
and progresses to rotating stall or surge depends not only on the compressor
but on the system in which it is operating: the duct, volume and throttle in
a simple test configuration. The simple argument in this section looks at only
one type of instability, a one-dimensional breakdown of the flow. Surge is
a one-dimensional pattern but in some cases the one-dimensional breakdown
can ultimately result in rotating stall, a multi-dimensional instability. Whether
one obtains rotating stall or surge and, if the former, part-span or full-span
rotating stall is also affected by the system.

Figure 9.4 shows three idealized systems of increasing complexity with Fig.
9.4a being the simplest possible system, the compressor followed by a throttle. The throttle is assumed to be near enough to the compressor outlet that

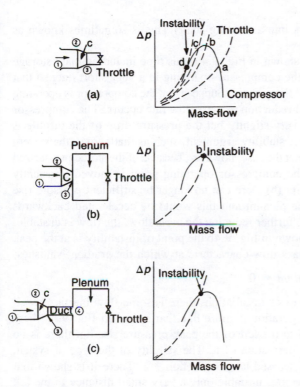

Fig. 9.4 Three different models for compressor instability. Compressor characteristic shown with solid lines, throttle with broken lines

there is effectively no storage of compressed gas and no inertia of the flow, but far enough away that any non-axisymmetric flow field of the compressor and throttle do not interact or couple. All inertia in the inlet duct will also be assumed negligible. The compressor pressure rise is shown by the solid line, the throttle pressure drop for a range of throttle sizes by broken lines. Normal operation might take place at a point such as (a) on the diagram which is unconditionally stable. This can be seen by considering a small reduction in mass flow; this leads to an increase in the compressor pressure rise and a fall in the pressure drop across the throttle so that the flow is accelerated and therefore increases until the original equilibrium is restored. For a small change at point (b) a reduction in mass flow does not alter the compressor pressure rise substantially but the throttle pressure drop is reduced considerably, so the flow is again stable. Even at point (c) the flow is stable for, as drawn, a small reduction in mass flow leads to a greater decrease in throttle pressure drop than the decrease in the pressure rise of the compressor so that the flow is again accelerated back to the former equilibrium. The limiting point of stability is point (d), here the slope of the compressor and throttle are the same and even a small reduction in mass flow will initiate an instability leading to rotating stall. (With no plenum there is no mass storage and the compressor cannot surge in the sense of causing oscillations of mass flow, but the flow 'collapses' and readjusts into a rotating stall.) This is sometimes known as 'static instability'.

A different system is shown in Fig. 9.4b, this time including mass storage in the plenum but with the compressor operating in a very short duct so that there is vanishingly small flow inertia. Suppose that the compressor is operating at point (b) when a small reduction in mass flow rate occurs. The compressor pressure rise decreases very slightly but the pressure drop of the throttle is substantial and on a static stability argument, such as that above, the system should be stable. This is not the case, however, because of the stored pressurized gas in the plenum. If the compressor operating point is moved transiently slightly to the left of point (b) there can no longer be sufficient pressure rise to confine the gas in the plenum and this would be accelerated backwards through the compressor, further reducing the mass flow: the flow is unstable. For the configuration shown in Fig. 9.4b the point of instability is at the peak on the pressure rise−mass flow characteric at which the gradient vanishes,

$$\partial(p_3 - p_{02})/\partial m = 0$$

In fact in most compressor installations there is some flow inertia as well as storage and the configuration is more like that shown in Fig. 9.4c; here the instability point lies to the left of the peak pressure rise but there is no correspondingly simple derivation of it. The stability of the type of system shown in Fig. 9.4c is discussed later in Section 9.2. There it is shown that the system normally becomes unstable only a very small distance to the left of the peak of the characterstic, so the peak of the characteristic provides a convenient working approximation for the stall or surge point.

In a test configuration it is possible to identify separately the components of the system, the most important of these, apart from the compressor and throttle, being the inertia of the flow in the ducts into and out of the compressor and the volume of the plenum containing the compressed gas, as shown schematically in Fig. 9.4c. In highly compact configurations, like a gas turbine, the simple separation of the components is no longer possible: the ducts are now so short that the inertia of the flow in the compressor itself may have a comparable importance; similarly the region in which the compressed air is stored may need to include volume of the duct and the compressor. For a gas turbine the throttle is replaced by the pressure drop in the combustion system and the turbine. Despite these problems, the modelling of the actual system by the discrete components has proved a very helpful way of considering the flow.

Centrifugal and axial behaviour

Rotating stall was first reported in a centrifugal compressor, Cheshire (1945), but by far the greatest interest and activity has been with the stall of axial compressors. In centrifugal compressors it is possible to operate fairly satisfactorily with rotating stall present. Centrifugal compressors also sometimes

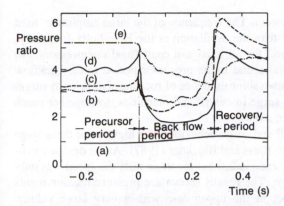

Fig. 9.5 Pressure variations during the surge of centrifugal compressor with a vaned diffuser: pressure in (a) inlet plenum; (b) vaneless space; (c) semi-vaneless space; (d) diffuser throat; (e) collector. (From Toyama *et al.*, 1977)

operate with axisymmetric stall near the inducer tips or with stationary non-axisymmetric stall produced by downstream asymmetry, notably the volute. This tolerance to stalled regions is largely because so much of the pressure rise is produced by centrifugal effects which will occur even in the presence of rotating stall cells or other forms of separated flow. Surge is usually the mode of low-flow behaviour which is of most concern in the centrifugal compressor, when the considerations are more similar to those in the axial.

There has been relatively little experimental study of surge in centrifugal compressors but much of what there is has been reported in Toyama *et al* (1977) and Dean and Young (1977). Figure 9.5 shows pressures measured in various parts of the compressor during a surge cycle. Most immediately apparent is the large low-frequency backflow. What can also be seen are the small oscillations prior to the backflow at a frequency close to the Helmholtz frequency

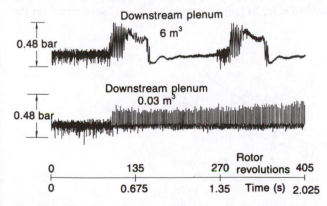

Fig. 9.6 Pressures recorded ahead of a three-stage axial compressor running at 12 000 rev/min (design speed 17 000 rev/min) for two different volumes on downstream side. (From Riess and Blöcker, 1987)

of the lumped parameter system. The frequency of the large amplitude 'hard surge' is very much lower than the oscillation at the Helmholtz frequency; this is a common observation for all axial and centrifugal compressors. No evidence of rotating stall was found in the precursor period to the backflow although other work has shown some evidence of rotating stall prior to surge. The behaviour of stall and surge in centrifugal machines is altogether much less well understood that in axials.

Figure 9.6 shows the stall and surge behaviour of a high-speed three-stage axial compressor taken from Riess and Blöckner (1987). At the design condition the stagnation pressure ratio is 2.0 but for these tests the speed was only 70 per cent of that at design. The results shown are pressure measurements upstream of the first rotor, in the upper case with a very large volume downstream of the compressor, in the other with only a very small volume. The case with a large volume shows a very clear surge whereas with the small volume only rotating stall is produced. It can also be seen that part of the surge process is a period of rotating stall. The very different frequencies of rotating stall and surge are clearly shown in Fig. 9.6.

Surge margin

There are many uncertainties concerned with stall and surge: uncertainties of prediction for a given design and uncertainties associated with the conditions of operation. The uncertainties of prediction will be discussed below. Uncertainties of operation include inlet distortion, which may be transient (for example during aircraft manoeuvres); transient throttle changes, such as occur when a gas turbine is accelerated; transient geometry changes such as tip and axial clearance changes following speed changes; and compressor mechanical damage including blade erosion and the effects of large foreign body ingestion.

To cope with these it is normal to specify a quantity known as the 'surge margin' (even though the compressor may not go into surge but into rotating stall it is conventional to refer to 'surge margin' and to a 'surge line' on the

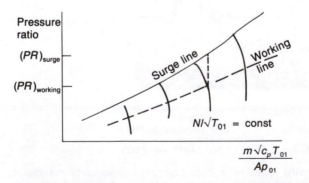

Fig. 9.7 A compressor pressure ratio—mass flow characteristic used in a method for defining surge margin

compressor operating map). There are many different ways of defining surge margin, but one of the most simple is illustrated in Fig. 9.7, with the surge margin defined by

$$SM = (PR_s - PR_w)/PR_w$$

where PR_w is the pressure ratio on the working line for a given corrected rotational speed and PR_s is the pressure ratio on the surge line for the same mass flow rate as the condition on the working line. In a multistage compressor for use with a turbojet engine it would be normal to insist on a surge margin of about 25 per cent. According to this definition the corrected speed will be higher for the points on the surge line than the working line. If operation is at a single corrected speed it is more appropriate to define a surge margin in terms of the inlet mass flow on the working line and on the surge line for that one corrected speed.

An altogether more logical definition considers the change in *outlet* flow function $F_{out} = m\sqrt{(c_p T_{02})}/A p_{02}$ between the working line (or design point) and the surge line for the same corrected rotational speed. (The flow function is equivalent to the corrected mass flow.) A suitable definition of surge margin would be $\{F_{out:working\ line} - F_{out:surge}\}/F_{out:working\ line}$. The significance of the outlet flow function is that it gives a measure of the throttle area changes, or equivalent throttling process, necessary to take the compressor into stall. Compressor performance is usually quoted in terms of $m\sqrt{(c_p T_{01})}/p_{01}$ using *inlet* variables. This parameter, denoted here by F_{in}, is of the same form as the outlet flow function but retaining the pressure and temperature ratios. In fact the temperature ratio is much smaller than the pressure ratio

$$T_{02}/T_{01} = \{p_{02}/p_{01}\}^{\gamma - 1/\eta\gamma} \approx \{p_{02}/p_{01}\}^{0.3}$$

for air and with reasonable values of polytropic efficiency η. The temperature ratio is further reduced to the half power in evaluating the flow function so the changes in temperature between the design and stall points can be neglected. A definition of surge margin which retains the physical significance of the outlet flow function but uses the more easily measured inlet flow function is

$$\text{Surge margin} = 1 - \left\{ \frac{(p_{02}/p_{01})_{working}}{(p_{02}/p_{01})_{surge}} \times \frac{F_{in:surge}}{F_{in:working}} \right\}.$$

For a given compressor the surge margin may be arbitrarily increased by lowering the working line, i.e. reducing the pressure rise in steady undistorted flow. In practice this is unattractive because it is reducing the quantity most wanted from the compressor, namely pressure rise. The necessary surge margin can be reduced by allowing less distortion into the compressor and by requiring any rise of the working line during transients, such as engine acceleration, to be small. Aerodynamic development, in the sense of improved blading, can lead to the surge line being raised, so too can mechanical improvements such as reducing the tip clearance. Special design features like casing treatment, discussed below, can also be used to raise the surge line.

Hysteresis

Most compressors require a much larger throttle opening to unstall them than was in use at the time when stall first occurred. This effect is usually referred to as hysteresis and can be a serious problem in the operation of any compressor when occasional stalls are liable to occur. The degree of hysteresis is primarily a function of the design of the compressor; for similar levels of blade loading the strongest variable for the axial compressor is the design value of the flow coefficient $\phi^* = V_x/U$. In geometric terms, fixing a large value of ϕ^* is equivalent to choosing small stagger (measured from the axial direction) for the blading. The effect of ϕ^* on the hysteresis is illustrated by Figure

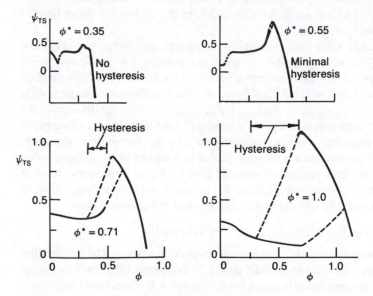

Fig. 9.8 The effect of ϕ^* (design value of V_x/U) on stall/unstall hysteresis for a three-stage compressor. (From Day and Cumpsty, 1978)

9.8, taken from Day *et al.* (1978), for four different three-stage compressors. As the design flow coefficient ϕ^* is raised, the unstalled pressure rise increases but so too does the extent of the hysteresis.

The presence of hysteresis causes a major complication in multistage compressors because it means that the pressure ratio−mass flow characteristic is no longer unique. If, for example, a rotating stall is set up in some stages near the front when the compressor is being operated at low speed, it is possible for this to persist right up to design speed if the compressor is accelerated on a high working line, i.e. with a small throttle opening. On the other hand, if the acceleration is allowed to take place with a large throttle opening the rotating stall may disappear.

Chapter layout

The most important aspect of compressor stall and surge is the prediction of the condition at which this will occur. This is considered below in the next section of this chapter, 'Instability and stall inception'. Here the discussion is complicated somewhat by the need to consider distorted (i.e. non-uniform) inlet flow since this can have a big effect on the instability point. The effect of distortion is of particular concern to the designers of jet engines, but it also has relevance to other applications as well. A quite separate concern, addressed in the section on 'Post instability behavour' is whether rotating stall or surge will be established (and if rotating stall, whether full-span or part-span) once instability occurs. The performance of the compressor in rotating stall and the flow pattern which is established can be important because there is a need to predict the pressure rise and flow rate in stall and the point at which the compressor will revert to unstalled operation. The nature of the flow in stall cells is discussed in the section 'Flow in the cell'. As already noted there are methods for raising the surge line, i.e. raising the surge margin, which are discussed in the final section 'Stability enhancement'.

9.2   Instability and the inception of stall

Predicting the condition at which instability will occur in a compressor is an essential part of the design process. Here the outcome of the instability will normally be referred to as stall but it is to be understood that stall or surge could result. The topic is confused by the terminology, with stall being used with a wide range of meanings. It is, for example, used for fully developed rotating stall; it can also refer to the fluid dynamic transient which initiates a surge; or it can be used to indicate that the flow in a blade row is heavily separated, with a large pressure loss and reduced flow turning (i.e. a large deviation in the case of an axial compressor).

It is common experience that compressors can often work quite satisfactorily with extensive regions of separated flow and that the appearance of separation (often called stall) on one blade row is not necessarily an indicator of imminent compressor instability and stall. The confusion is further increased because the behaviour of one row or stage may be quite different on its own from that as part of a complete multistage compressor. Many axial compressors operate at some conditions with the stages mismatched so that some are at a non-dimensional flow rate below and some above that for which they were designed. The stages at the low flow rate will often have large regions of separated flow and would be unable to operate on their own without entering rotating stall. The coupling of stages together can maintain the whole machine unstalled until the entire compressor crosses the criterion for instability.

To illustrate this Fig. 9.9 shows measurements made in a low-speed three-stage compressor by Longley (1988). The compressor on the left, Fig. 9.9a, was mismatched so that stages 2 and 3 were more highly staggered and therefore

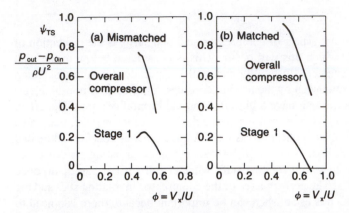

Fig. 9.9 Pressure rise–flow rate characteristics for two compressors, each of three stages. Stage 1 is geometrically identical in (a) and (b); stages 2 and 3 have 10° extra stagger in case (a). (From Longley, 1988)

designed for a lower mass flow than stage 1. Stage 1 would therefore be expected to experience loss in performance at a mass flow rate when the other two stages were operating satisfactorily. The results for this one mismatched stage are shown together with the pressure rise for the whole compressor. Because of the stabilizing influence of stages 2 and 3, stage 1 was able to operate in an axisymmetric, manner to the left of the characteristic peak in a way that would be impossible as an isolated stage. In Fig. 9.9b all the stages are staggered at the same angle as stage 1 on the left-hand diagram and all the stages become unstable together with the characteristic ending abruptly near the peak. It might be added that the performance of the stage 1 is identical in the matched and mismatched builds up to the point of stall.

The coupling of the components of a compression system in this way is important in many applications, and the combined effects can be added if the physical separation of the components is very small. The primary disturbance is one cycle around the circumference, so the coupling decays in proportion to $e^{-x/r}$, where r is the mean radius. This decay rate is much smaller than that which occurs with the blade-to-blade flow which decays as $e^{-x/l}$ where the length scale l is the blade pitch or span. The coupling also applies between separate compressors, such as the high pressure and low pressure compressors of a gas turbine. In the latter case the coupling is less than complete, depending on how closely spaced the two compressors are, but is nevertheless significant. Coupling takes place in centrifugal compressors between closely spaced components, but the theoretical relationship in this case is less thoroughly worked out. An example of the fluid dynamic coupling in a centrifugal compressor is the much wider operating range obtained by Eckardt (1975) when operating with a uniform ring throttle in the vaneless diffuser compared to a configuration with some asymmetry well downstream.

Correlation and empirical relations for stall inception

It is first useful to consider stall onset from the point of view of correlation, including the effect of distortion, before going on to consider the theoretical treatment. At present the most reliable method of estimating the stall or surge point is to use information obtained from tests of similar machines. Only when a design is radically different from a predecessor will this not provide the most accurate prediction. This does not, however, provide much guidance as to which features are most potent in contributing to the delay of stall to low flow rates.

Axial compressors

One of the approaches to assessing the likelihood of stall in an axial compressor at a given flow rate is the use of the diffusion factor, described in Chapter 4 on susbsonic blading. It is often assumed that if the diffusion factor exceeds a particular value, such as 0.6, the blade will be stalled or close to stall. In one sense this may be true, in that the boundary layer on the suction surface may be thick and there may be a region of separated flow near the suction surface trailing edge. This does not, however, indicate that the compressor is near to stall and it turns out that blade rows may be extensively stalled without the compressor itself becoming unstable. Nevertheless a compressor with most blade rows having very large values of diffusion factor at all spanwise positions would probably be unsatisfactory; each row would tend to have thick wakes and deviation rising rapidly with incidence so the compressor would therefore not be able to give a substantial increase in pressure rise if the flow rate were decreased. However the diffusion factor has not provided a satisfactory correlating parameter for compressor instability.

The ineffectiveness of diffusion factor as a predictor of stall is demonstrated by Fig. 9.10, taken from Smith (1969), which shows the pressure rise—mass flow characteristic for multistage compressors with blading of two different aspect ratios (the ratio of blade span to chord) but the same diffusion factor. The stages of each compressor were identical and the flow essentially incompressible. The compressor with long chord blades, i.e. low aspect ratio, stalls

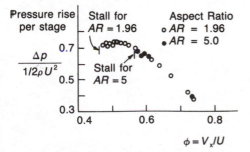

Fig. 9.10 Stage pressure rise versus flow rate for multistage axial compressors showing effect of aspect ratio (AR = blade span ÷ blade chord). (From Smith, 1969)

at a much lower flow coefficient. The explanation for this is that the limiting feature of the flow is associated with the boundary layer on the hub and casing and changes in the blade chord have a big effect on this. The importance of the blade chord can be understood when it is realized that there is a length scale associated with the boundary layers on the endwalls, for example the displacement thickness, and the tolerance of the boundary layer to separation will depend on the severity of the pressure gradient when measured in terms of the length scale relevant to the boundary layer. Increasing the blade chord therefore has the effect of reducing the severity of the pressure gradient on the endwall. The importance of having blades of sufficiently long chord was neglected for a long time because attention was focussed on measures of the blade loading, such as the diffusion factor, see Wennerstrom (1986).

The importance of the endwall flow is not a new idea and it underlies the work by de Haller (1953) who proposed, on the basis of cascade tests, that to avoid stall the pressure coefficient for each blade row should be limited so that $c_p < 0.44$, or in terms of velocity, V_2/V_1 or $W_2/W_1 > 0.75$. Fundamentally this was too simple a relation because it ignores all blade parameters. Although it has not found very wide acceptance or application it is nevertheless nearer to a proper description of the limit on loading than, say, the diffusion factor.

Much more recently Koch (1981) published a method for estimating the pressure rise at stall, making use of the very large General Electric data base of measurements in compressors. The essential idea was to relate the performance of blade passages to straight diffusers using the correlation of Sovran and Klomp (1965). For diffusers it is normal to use *inlet* width g_1 divided by diffuser length as the prime geometric variable; for the blade passage it was blade camber line length L divided by the staggered gap at *outlet* g_2 which was used. The geometrical definition is shown in Fig. 9.11 together with the velocity triangle into the stator. In devising the approach it was clearly stated

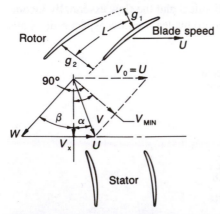

Fig. 9.11 Schematic showing stage geometry for stalling pressure loss correlation of Koch. (From Koch, 1981)

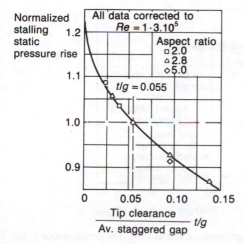

Normalized stalling static pressure rise

All data corrected to $Re = 1 \cdot 3.10^5$

Aspect ratio
□ 2.0
△ 2.8
◇ 5.0

$t/g = 0.055$

Tip clearance / Av. staggered gap = t/g

Fig. 9.12 Koch's correction for the effect of tip clearance on stalling pressure rise. (From Koch, 1981)

that this is a simple method to complement the preliminary calculations made at blade mid-height (so-called meanline or pitchline calculations). The blading was therefore averaged over a stage with no attempt made to separate rotor and stator performance, an approximation which was justified *post facto*. Quite surprisingly this approach works well even if the degree of reaction is well away from 50 per cent (so that the rotor and stator rows are very different) and gives better predictions than would have followed the consideration of rotor and stator separately.

When the stalling static pressure rise coefficient was plotted against the ratio L/g_2 the agreement with the trend for two-dimensional diffusers with different levels of inlet blockage was only fair. By using the data base it was possible to isolate the various effects which were leading to the spread in the correlation. The effect of Reynolds number on stalling pressure rise has been shown as Fig. 1.8; using this, all results were corrected to a Reynolds number based on blade chord of $1.3 \cdot 10^5$. Figure 9.12 shows the very important effect of tip clearance, t; with this all the data was corrected to a tip clearance equal to 5.5 per cent of the average staggered gap g. There is, in addition, a similar graph for the effect of axial gap between rotor and stator, but for gaps greater than about 25 per cent of pitch this effect was quite small.

Incorporating the corrections derived for Reynolds number, tip clearance and axial gap still left a pronounced trend which showed stages with low stagger blades having a markedly lower pressure coefficient at stall than high stagger blades, Fig. 9.13. (Low stagger corresponds to high values for the design flow coefficient $\phi^* = V_x/U$.) The benefits of high stagger were discussed in Chapter 2 and this appears to be the cause of the trend found by Koch. For stages of low ϕ^* the velocity triangles show that flow leaving a blade row with a low axial velocity, such as occurs in the endwall boundary layers, has

Adjusted stage static pressure rise coefficient at stall

$$\frac{\Delta p}{1/2\rho V^2}$$

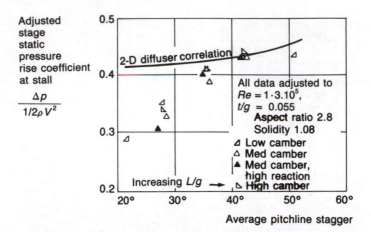

Fig. 9.13 The effect of stagger on stage pressure rise at stall. The denominator in the ordinate is the mean of the dynamic pressure into rotor and stator at mid-span. For high speed stages a compressible form is used for the ordinate. (2D diffuser correlation to Sovran and Klomp, 1967). (From Koch, 1981)

higher relative dynamic pressure into the blade row downstream as a result of the change of frame of reference. In effect the endwall boundary layer fluid is re-energized by the change in frame of reference for stages with high stagger. Koch defined an empirical correlation factor F_{ef} to be applied to the inlet dynamic pressure for each stage:

$$F_{ef} = V_{ef}^2/V^2 = \{V^2 + 2.5V_{min}^2 + 0.5U^2\}/(4.0V^2)$$

where $\quad V_{min}^2/V^2 = \sin^2(\alpha + \beta)$, if $(\alpha + \beta) \leq 90°$ and $\beta \geq 0°$

(low stagger)

but $\quad V_{min}^2/V^2 = 1.0 \quad\quad\quad$ if $(\alpha + \beta) > 90°$ (high stagger)

and
likewise $\quad V_{min}^2/V^2 = U^2/V^2, \quad\quad$ if $\beta < 0°$ (outlet flow turned beyond axial, as might occur in a low-stagger build.)

Here α and β are the mean absolute and relative flow direction out of the rotor, defined, together with other quantities, in Fig. 9.11. With only a small amount of scatter the stalling pressure rise coefficient from a very large number of stages showed a direct proportionality to F_{ef} which was found to lie in the range 0.78 to 1.11, although some compressors designed for very high flow coefficient were found to have values of F_{ef} as low as 0.39.

The final form of the correlation for stalling pressure rise coefficient is shown in Fig. 9.14. This figure shows high-speed data from a range of compressors and fans together with regions containing numerous low-speed results. The one data point which greatly exceeds the correlation was obtained with a tandem rotor which may explain the difference. The two points to the right of the diagram with very large values of L/g_2 are for non-typical designs and may either be outside the scope of the correlating procedure or may have been

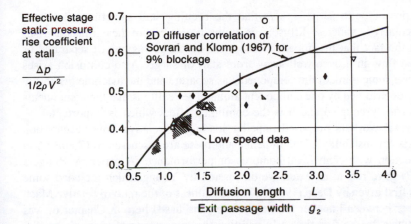

Fig. 9.14 The stage static pressure rise coefficient at stall versus diffusion length: points high speed data; shaded areas show low speed data. The denominator in the ordinate is the mean of the dynamic pressure into rotor and stator at mid-span. For high speed stages a compressible form is used for the ordinate. (From Koch, 1981)

aerodynamically inadequate in other ways. The solid line shown in the figure as a reference is that given by Sovran and Klomp (1967) for a two-dimensional diffuser with an inlet blockage of 9 per cent. With this quite high level of blockage reasonable agreement with the compressor data is obtained; the relatively low pressure rise is, however, a result of the choice of quite large tip clearance, axial gap (38 per cent of stator chord) and low Reynolds number $(1.3 \cdot 10^5)$ as the reference condition to which all compressor data was adjusted by quite large corrections. If very small tip clearance and very small axial gap had been selected as the reference condition a diffuser with perhaps 1 per cent of inlet blockage might have been more appropriate.

The line drawn in Fig. 9.14 gives a good estimate for the maximum possible pressure rise at stall for a stage and as such is a useful tool in the design of compressors. Many compressors have been observed to have maximum pressure coefficients less than this correlation but very few with more. As a basis for design a realistic assumption would be that 80 per cent of the value corresponding to the line in Fig. 9.14 confidently achieved. Koch (1981) used the correlation to examine multistage high-speed compressors with the stages mismatched and correctly identified those stages known from other evidence to be near their limiting pressure rise. In summary, the Koch correlation method provides a reasonable way of estimating the maximum stage pressure ratio. It also provides a rationale for the greater stability found for compressors designed for high stagger (low ϕ^*); such compressors operate efficiently and are stable with diffusion factors of 0.6 or more at design, far higher than previously assumed feasible.

Centrifugal compressors

In the case of the centrifugal compressor there is great controversy surrounding the cause and condition for stall or surge. Some maintain that rotating stall is a precursor to surge; others that the rotating stall sometimes detected, particularly in vaneless diffusers, is unimportant. Part of the confusion arises from the fact that the impeller most often has large regions of separated or stalled flow in it even well away from stall or surge. Also confusion arises because some think of the components as separate and the machine instability being determined by one component alone. The reality is that the components are strongly coupled and it is the combined effect which is important.

Several workers proposed impeller loading or relative flow deceleration criteria for instability of the impeller and these are discussed in Chapter 6 in connection with centrifugal compressor performance. A paper by Rodgers (1977) gave rise to an extensive discussion by Young which included some ideas first given by Dean (1974). (The usefulness of the ratio of relative Mach number between impeller inlet and outlet, discussed here in Chapter 6, was described for the first time.) In addition the importance of recognizing that the combined effect of all the components of the compressor must be considered in deciding if the compression system is stable was stressed and a method for looking at the contributions of all the parts to the overall stability was given.

The simplest compression system of practical interest from the point of view of stability of centrifugal compressors is that shown in Fig. 9.4b with a plenum and throttle downstream of the compressor (no inertia) and this was the case treated by Dean (1974). For this case the flow is stable to one-dimensional disturbances if

$$\partial(PR)/\partial m \leq 0$$

where PR is the overall compressor total-to-static pressure ratio and m is the mass flow rate. The overall pressure ratio is the product of that for each of the component parts

i.e. $$PR = PR_1 PR_2 \ldots PR_n$$

The differential then becomes

$$\frac{1}{PR} \frac{\partial PR}{\partial m} = \sum_i \frac{1}{PR_i} \frac{\partial PR_i}{\partial m}$$

and it is convenient to define a stability parameter for each component

$$SP_i = \frac{1}{PR_i} \frac{\partial PR_i}{\partial m}$$

The overall condition for stability is therefore

$$\Sigma SP_i \leq 0$$

where components for which SP_i is positive are destabilizing and those for which it is negative are stabilizing.

Figure 9.15 from Dean (1974) shows the values of SP derived from

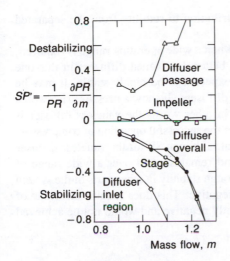

Fig. 9.15 The variation in the stability parameter ($SP = 1/PR \ \partial PR/\partial m$) with flow rate for the components of a centrifugal compressor. The impeller was without backsweep. (From Dean, 1974)

measurements plotted against the mass flow for the components of a high-speed centrifugal compressor with an impeller having no backsweep. Of immediate note is that in this example the impeller itself is close to neutral stability and its value of SP varies little with flow rate. This is surprising since, although the flow in the impeller contains large regions of separated flow, the inducer pressure rise might be expected to increase with incidence, giving some contribution to stability. The vaned diffuser passage is highly unstable, reflecting the tendency of the pressure rise to decrease if the inlet blockage rises and the blockage to rise as the mass flow is decreased. The inlet to the diffuser is very highly stable and it is this which keeps the whole system stable.

   Figure 9.15 is a single test case and must be generalized with some caution, in particular the approximately neutral stability of the impeller. Evidence that impellers can be stable comes from the success of Jansen *et al.* (1980) in using casing treatment over the impeller inducer, as described below in Section 9.5; if the casing treatment extends the operating range it implies that stability has been restored to the impeller which had disappeared near the stall point. Good designs of centrifugal impeller, though often separated in the downstream parts of the impeller, have attached flow in the inlet, allowing the pressure rise to increase with incidence, $SP \ll 0$, and giving a substantial contribution to stability. In some configurations, particularly those for multistage industrial compressors, the meridional curvature in the inlet makes attached flow there unlikely; consequently there is little chance of pressure rise increasing as the flow rate is decreased and little or no contribution to stability. Backswept impellers are naturally more stable, since the pressure rise inevitably increases with decreasing flow rate, and an important reason for using them in industrial compressors is the absence of a contribution to stability from the separated flow in the inlet region.

Vaneless diffusers tend to be used when a wide operating range is required, because they have been found to be better than vaned diffusers in this one respect, and consequently might be expected to be highly stable. It may be recalled from Fig. 7.10 that vaneless diffusers all show a pressure rise which increases with increasing flow rate; in other words the vaneless diffuser is inherently unstable. The explanation for the successful operation of compressors with vaneless diffusers is probably that, although SP for the vaneless diffuser is positive, the magnitude is small and remains small over a wide range of flow rates. There must therefore be enough stability in other parts of the system to maintain the overall value of *SP* negative. This does require the value of *SP* for the impeller to be at least slightly negative and would not be achieved in the example shown in Fig. 9.15.

Inlet distortion

The performance of compressors, particularly axial compressors, can be seriously affected by inlet distortion, that is by non-uniform inlet flow. A full summary of this complicated and important field, with an extensive reference list, is given by Williams (1986). The distortion can be in static pressure or stagnation temperature, but the most common distortion is stagnation pressure and it is this which will be addressed here. Such distortions often occur naturally because of the unsatisfactory nature of the inlet or because of operational effects,

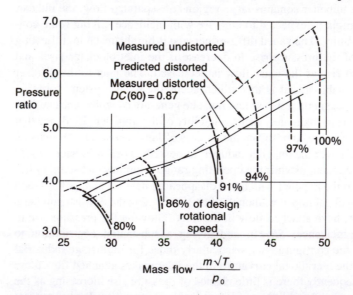

Fig. 9.16 Compressor performance with undistorted and distorted inlet flow. Prediction for distorted flow by 'compressors-in-parallel' using measured compressor performance in undistorted flow. (From Reid, 1969)

such as cross-wind into aircraft engine intakes. Very often the distortion is transient, for example when a gust of wind passes an intake; it is generally recognized that the distortion will have little or no effect unless it persists for at least one revolution of the rotor. If it persists for longer than this it is normally treated as quasi-steady, i.e. the process is long in relation to the time between passage of rotor blades. The distortion pattern is normally non-uniform in the circumferential and the radial sense but experimental tests and theoretical treatments are usually performed with idealized patterns, either circumferential or radial. Circumferential distortion seems to be the most serious (radial distortion can in some circumstances bring an improvement in performance, including an increase in surge margin.) Because of this only circumferential distortion will be considered here. Consideration will also be restricted to the performance of axial compressors.

Figure 9.16, taken from Reid (1969), shows the effect of circumferential distortion on the performance of a nine-stage axial compressor. The most significant change is the loss in surge margin, with surge occurring at very much lower pressure ratios for all speeds. Distortions with a large circumferential extent have a more pronounced effect on compressor performance. This is very clearly indicated by the results in Fig. 9.17 for the multistage compressor obtained by Reid (1969) (and reproduced by Bowditch and Coltrin, 1983) for tests carried out with upstream screens of uniform and constant porosity. Figure 9.17a shows the effect of varying the circumferential extent of a single lobe, the extent forming the abscissa. The ordinate is the outlet pressure at the surge point, expressed as a percentage of the pressure in undistorted flow. The effect of the distortion of the same level increases as the circumferential extent increases until an angle of nearly 90° is reached, after which there is no further worsening. In Fig. 9.17b the effect of taking the 90° distortion and subdividing it is shown. This confirms the trend that distortions of small circumferential extent have a much reduced effect on the stability of compressors, in other words high harmonics of distortion are less important. A common way of allowing for this in the assessment of distortion

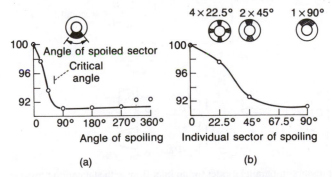

Fig. 9.17 The effect of distortion size on outlet static pressure at the surge point, expressed as percentage of that in undistorted flow. (From Bowditch and Coltrin, 1983)

is to define a coefficient which averages the low stagnation pressure over the worst arc of the circumference, with the extent of the arc being the angle at which it is judged that the plateau in Fig. 9.17a is reached. This gives rise to a distortion correlation parameter such as $DC(\theta_{crit})$ where θ_{crit} is the circumferential angle at which the full effect of the distortion is felt. Very often θ_{crit} has been taken to be 60° and this gave rise to the so-called DC(60) parameter defined by

$$DC(60) = \frac{\begin{array}{c}\text{average inlet stagnation pressure} - \\ \text{average inlet stagnation pressure over lowest 60°}\end{array}}{\text{average inlet dynamic pressure}}$$

The subject of distortion assessment is complicated, particularly when radial distortion is included as well; it is beyond the scope of this chapter but is well covered by Williams (1984) and by Seddon and Goldsmith (1985). For contract guarantee purposes a $DC(60)$ of 0.5 would be typical but many engine compressors can tolerate a value of 1.0 or more.

Compressors in parallel

The most common method of predicting the effect of inlet distortion on the performance of the compressor is an approximation known as the method of compressors in parallel and this is shown schematically in Fig. 9.18. As well as being a method in practical use it is a vehicle for explaining some of the simple ideas of distortion.

At its most simple the method considers a distorted inlet flow in two parts,

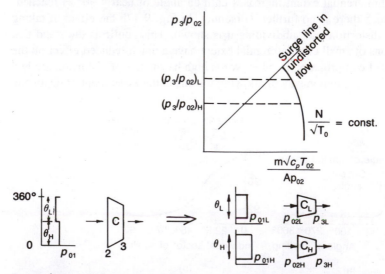

Fig. 9.18 The compressors-in-parallel model for an inlet flow with stagnation pressure distortion. Station 1 is well upstream where the *static* pressure is effectively uniform.
Note: $p_{01L} = p_{02L}$, $p_{01H} = p_{02H}$, $p_{02L} \neq p_{02H}$, $p_{3L} = p_{3H}$

one with high inlet stagnation pressure, the other low. (Compressors in parallel could likewise deal with temperature distortion.) The compressor is similarly divided into two separate parts, each assumed to act as a separate machine, that is to say, in parallel. The pressure rise—flow rate relation for each machine is assumed to be identical to that for the whole machine with uniform flow.

At some distance upstream of the compressor inlet face, designated station 1, the static pressure is assumed uniform, though the stagnation pressure, and therefore the axial velocity, are not uniform. At the outlet of the compressor the static pressure is assumed uniform, a reasonable assumption if the compressor discharges into a region where there is little pressure recovery, such as a plenum, and the outlet static pressure is therefore equal for each of the compressors in parallel. Because the outlet static pressure is uniform it is convenient to specify the compressor pressure ratio in terms of the outlet static and inlet stagnation pressures, p_3/p_{02}. For a given non-dimensional rotational speed of the compressor the pressure rise can be written in terms of the inlet flow function

$$p_3/p_{02} = f\{m(c_p T_{02})^{1/2}/A p_{02}\}$$

This relation for uniform inlet flow is assumed to be valid locally in the distorted flow too, effectively ignoring unsteady effects and upstream swirl.

Inside the compressor there can be no flow circumferentially from one 'machine' to the other (in other words from the part of the compressor accepting high pressure flow to that accepting low pressure flow) because the blade rows are usually separated by sufficiently small axial gaps that this is precluded. Upstream of the compressor, between stations 1 and 2, there are circumferential velocities and the area of the high and low stagnation pressure flow can change, although the mass flow in each must stay constant.

At the compressor inlet face, station 2, the static pressure will not be uniform. In the region where the inlet velocity to the compressor is low the pressure rise will be greater, as dictated by the compressor pressure rise—mass flow characteristic. Since the outlet static pressure is assumed to be uniform the inlet static pressure will thus be low in this region. Upstream of the compressor, but downstream of the distortion creating element, viscous losses and unsteady effects are small so that the stagnation pressure can be taken to convect along streamlines. Therefore where there is a local low stagnation pressure, implying a low axial velocity well upstream, there must also be an increase in velocity towards the compressor inlet face brought about by the action of the compressor itself. Although the relative area of high and low stagnation pressure well upstream are normally taken as initial given conditions, and with it the corresponding mass flows, the relative areas at compressor inlet are not known but depend on the pressure rise—mass flow characteristic of the compressor itself.

The compressor therefore acts to reduce the velocity difference in the distortion by locally lowering the static pressure where the stagnation pressure is low and raising it where stagnation pressure is high. In fact if the pressure rise—mass flow characteristic is vertical (i.e. the pressure ratio can vary but for a given rotational speed the corrected mass flow rate is constant, as often occurs in high-speed compressors) the compressor will entirely eliminate the velocity distortion and the velocity (though not the static or stagnation pressure) will be uniform at the compressor inlet face.

The calculations of the compressor-in-parallel method are easily performed iteratively to find the operating points in the low and high inlet stagnation pressure regions. In the simple compressor-in-parallel method stall or surge is *assumed* to occur when the operating point for the low inlet total pressure fluid crosses the surge line for undistorted flow. Surprisingly good agreement can sometimes be obtained, as the results predicted by this method in Fig. 9.16 show. In general, however, as shown in Fig. 9.19, the measured surge pressure ratio change is qualitatively indicated but the quantitative prediction is not good. The discrepancy is larger for distortions of small circumferential extent and the principal cause for this is the neglect of unsteady effects in assuming that the compressor performance can be adequately expressed by the relation for uniform inlet flow. It is possible to make some allowance for the unsteady

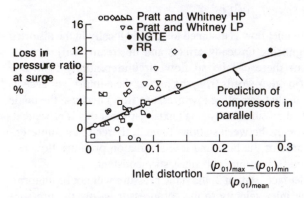

Fig. 9.19 The sensitivity of various compressors to an inlet distortion occupying 180° of circumference. (From Mazzawy, 1977)

effects by a one-dimensional treatment of the flow in the blade passages and this has been incorporated in a multi-segment compressor-in-parallel method by Mazzawy (1977).

Theoretical investigation of instability

The conditions for the breakdown of the unstalled flow into a rotating stall or surge have been investigated theoretically by many people, extending back to the early 1950s. There are essentially two different approaches, one which looks at the flow field in a two-dimensional manner and the other which analyses the system in a one-dimensional way (as yet there are no very general three-dimensional methods). The two approaches predict a very similar condition for instability in many cases, although the manner of flow breakdown is very different, and only recently has some unification been achieved. The early work is summarized and thoroughly referenced by Greitzer (1980, 1981). Most of this work has been directed to the axial compressor but some, particularly the one-dimensional treatment, can be applied to the centrifugal as well. A particularly economic treatment by Stenning (1980) forms the basis of the treatment here.

One-dimensional system stability

The system to be analysed is shown in Fig. 9.4c. There are four components: compressor; duct (to give flow inertia); plenum (to give transient mass storage); and throttle. The flow is treated as one-dimensional so that at each station there is a single instantaneous static pressure and the flow can be characterized by a single instantaneous mass flow or velocity. The variables such as pressure and mass flow are taken to vary about the time-mean values so that at the compressor outlet, for example,

$$m_3 = \bar{m} + m_3' \text{ and } p_3 = \bar{p}_3 + p_3'$$

where $\bar{m}$ and $\bar{p}_3$ are the time-mean values and m_3' and p_3' are the unsteady perturbations, assumed to be small. The time mean mass flow rate $\bar{m}$ is constant throughout the system. An equation is derived for the perturbations in each of the components of the system.

The compressor. There is assumed to be no inertia or mass storage in the compressor itself, that is

$$m_2 = m_3.$$

The compressor is assumed to be axisymmetric and to behave in unsteady flow as it does in steady flow so that for a given rotational speed and inlet temperature the pressure rise—mass flow characteristic is specified by

$$p_3 - p_{02} = C(m_2) \tag{9.1}$$

It is convenient to write the gradient in pressure rise as

$$\mathrm{d}(p_3 - p_{02})/\mathrm{d}m_2 = \mathrm{d}C/\mathrm{d}m_2 = c$$

and since the inlet stagnation pressure is assumed to be constant one can therefore write

$$p_3' = cm_2' \tag{9.2}$$

For normal operation of a compressor the gradient c is, of course, negative. The linearization is equivalent to replacing the curved pressure rise—mass flow characteristic by a straight line at the local slope; it is a good approximation provided the amplitude of the changes is sufficiently small that the local value does not differ significantly from the linear value.

The duct. All the inertia is taken to be in the duct between stations 3 and 4.

In steady flow the static and stagnation pressures will be equal at 3 and 4; in unsteady flow the static pressure difference $p_3 - p_4$ will be proportional to the length of the duct L, the density of the gas $\rho_3 \approx \rho_4$ and the rate at which the mass flow along it changes. The pressure difference along the duct can then be written with sufficient accuracy for low Mach number flow in the duct as

$$p_3' - p_4' = L/A \; \mathrm{d}m_3'/\mathrm{d}t \qquad (9.3)$$

The plenum. The compressed gas is stored in the plenum and it is normally assumed that the variation in its mass may be related to the variation in plenum pressure p_4 by treating the process as isentropic. Thus

$$m_3' - m_4' = V \, \mathrm{d}\rho_4'/\mathrm{d}t$$
$$= V(\rho_4/\gamma p_4)\mathrm{d}p_4'/\mathrm{d}t = V/(a_4)^2 \mathrm{d}p_4'/\mathrm{d}t \qquad (9.4)$$

where a_4 is the speed of sound inside the plenum.

The throttle. It is assumed in this simple analysis that the pressure difference across the throttle is a function only of the instantaneous mass flow through it, or

$$p_4 - p_5 = G(m_4) \qquad (9.5)$$

where G depends on the throttle area and the properties of the gas. If the pressure downstream of the throttle is constant one obtains

$$p_4' = g \, m_4' \qquad (9.6)$$

It is important to note the throttle parameters G and $g = \mathrm{d}G/\mathrm{d}m_4$ are always positive.

The four equations, 9.2, 9.3, 9.4 and 9.6 together with $m_2' = m_3'$ define the perturbations for the compressor, the duct, the plenum and the throttle. There are effectively four variables (m_3', m_4', p_3' and p_4') and a solution may be obtained in terms of any one of these. The four equations can be rearranged to give a single second-order ordinary differential equation, written here in terms of z, which could be any one of the perturbation quantities.

$$\frac{LV}{Aa_4^2} g \frac{\mathrm{d}^2 z}{\mathrm{d}t^2} + \left(\frac{L}{A} - \frac{cgV}{a_4^2} \right) \frac{\mathrm{d}z}{\mathrm{d}t} + (g-c)z = 0 \qquad (9.7)$$

The analogy of the compressor, duct, plenum and throttle with a mass-spring-damper system is clear and this is brought out in the schematic shown in Fig. 9.20. Figure 9.20 also shows inertia in the throttle but normally this is small enough to be ignored. The frequency of the undamped perturbation is

$$\omega^2 = \frac{Aa_4^2}{LV} \left(\frac{g-c}{g} \right) = \omega_{\text{Helm}}^2 \left(\frac{g-c}{g} \right)$$

where ω_{Helm} is the Helmholtz frequency of the duct and plenum in the absence of the compressor and throttle. When the compressor is operating normally

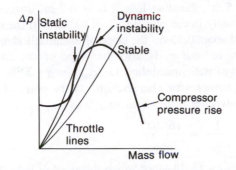

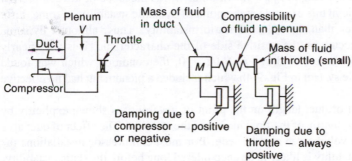

Fig. 9.20 The mechanical analogue of the compression system

the slope of the characteristic $c = \mathrm{d}(p_3 - p_{02})/\mathrm{d}m$ is negative and the effect of this is to 'stiffen' the system and raise the natural frequency.

The stability of the system may be examined by looking at the coefficients of z and the first derivatives of z in equation 9.7. If the coefficient of z becomes negative, that is

$$c > g \tag{9.8}$$

it is equivalent to the mass-spring system in which the spring has a negative rate, in other words the spring would act to increase any displacement. Physically the instability occurs when the slopes of the compressor and throttle lines are equal. This was found earlier for a system consisting of only a compressor and throttle in Fig. 9.4a and was denoted there as a static instability.

If the coefficient multiplying the first derivative of z in equation 9.7 is negative it is equivalent to negative damping in the mass-spring analogy. Small disturbances will grow in time in an exponential manner. This is referred to as dynamic instability and the conditions occur at

$$\frac{L}{A} < \frac{cgV}{a_4^2} \quad \text{or} \quad c > \frac{La_4^2}{AgV} \tag{9.9}$$

The criteria for static and dynamic stability, equations 9.8 and 9.9, reveal more when some numerical values are inserted. Suppose, for example, that the plenum volume $V = 0.5$ m^3, the duct length $L = 0.2$ m, duct area $A = 0.4$ m^2 and the sonic velocity in the plenum $a_4 = 400$ m/s. If the throttle is choked and has an area of about 0.35 m^2 the slope of the throttle characteristic $g \approx 2200$. This gradient, of course, is also the slope of the compressor characteristic at the point of static instability, i.e. $c_{static} = g \approx 2200$. The corresponding slope of the compressor characteristic at the point of dynamic instability C_{dyn} can be rewritten from equation 9.9 as

$$c_{dyn} = \frac{La_4^2}{AV} \frac{1}{c_{static}} = \frac{16 \cdot 10^4}{c_{static}}$$

so that for this example $c_{dyn} \approx 73$. In other words the slope of the compressor characteristic in this example at the point of dynamic instability is some thirty times smaller than the value of static instability. Thus, although dynamic instability occurs on the positive side of the characteristic, it is very nearly at the point where $c = d(p_3 - p_{02})/dm = 0$, the condition which was found for the simple system in Fig. 9.4b which included a plenum but had no effective inertia.

The effect of duct length on the point of instability is shown explicitly by equation 9.9; inertia of the flow increases the stability. The effect of duct area and plenum volume can also be seen. For most compressor installations the dynamic instability is likely to be encountered long before the static instability. Testing with low Mach number machines, with very small plenum volumes or with long ducts are ways of inhibiting the dynamic instability and extending operating range to the point of static instability.

Two-dimensional stability theory

A quite different approach to the stability of compressor flow is to consider the two-dimensional flow field in the compressor. This has more relevance

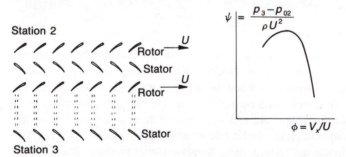

Fig. 9.21 A developed view of a multistage axial compressor. Station 1 is well upstream, 2 at the compressor inlet face and 3 at the outlet face

to the axial compressor than the centrifugal. It gives no consideration to other elements of the system such as the duct inertia, the plenum or the throttle. A compressor of high hub—casing radius ratio is considered (so that variation in the radial direction may be ignored) and a developed view in the axial—

circumferential plane is given in Fig. 9.21. Stations 2 and 3 again are at the compressor inlet face and exit face respectively.

It is assumed that the static pressure at the outlet plane is uniform so that perturbations in this, p_3' are equal to zero. This is a convenient assumption but not a correct one: more detailed analysis shows that it does not affect the prediction of the point of instability although it does lead to an overestimate of the rotational speed of the disturbances by a factor of up to two, depending on the precise geometry.

The compressor is taken to behave locally as it would in uniform flow, so that at any particular position the non-dimensional inlet-stagnation to outlet-static pressure rise, $\psi_{TS} = (p_3 - p_{02})/\rho U^2$, depends only on the local flow coefficient $\phi(y) = V_{x2}/U$, as indicated in Fig. 9.21. It is also assumed that there is no crossflow inside the compressor, in other words, the circumferential distribution of mass flow is the same at inlet and outlet to the compressor, even though displaced in the y-direction. Small perturbations to the flow in the region upstream of the compressor are considered and it is found that these grow, so the flow becomes unstable, if

$$\frac{d\psi_{TS}}{d\phi} = \frac{1}{\rho U} \frac{d}{dV_x} (p_3 - p_{02}) > 0 \qquad (9.10)$$

This important result was first obtained by Dunham (1965) although similar results, expressed rather differently, had been obtained earlier by others. (Dunham also extended the analysis to three dimensions for low hub—casing radius ratio compressors with free-vortex velocity distributions but there was no longer a very simple stability limit comparable to equation 9.10.) The point of instability therefore occurs at the peak of the stagnation-to-static pressure rise characteristic, at the point where the gradient vanishes. This seems to give a good indication of the point at which rotating stall is first noticed in a number of compressors but by no means all. It is also very close to the condition for dynamic instability in the one-dimensional model and it is not always clear which mechanism is leading to the instability. Certainly few compressors stall at positions on the characteristic where $d\psi_{TS}/d\phi > 0$, although this would be the case for low-speed compressors if the one-dimensional static or dynamic instability were responsible.

Some compressors stall at a condition where the gradient $d\psi_{TS}/d\phi < 0$, to the right of where the gradient would vanish. A possible reason for this is that the pressures plotted as the ψ_{TS}–ϕ characteristic are the time-mean values averaged over the annulus. Close to stall, compressors often exhibit pronounced unsteadiness, similar to ordinary two-dimensional or conical diffusers which achieve peak pressure recovery in the intermittent stall regime. If the pressure rise is fluctuating with respect to time, it implies that for part of the time the instantaneous ψ_{TS} will be lower than the time-mean and the magnitude of the gradient $d\psi_{TS}/d\phi$ therefore smaller. It is possible that for a time long enough for the stall to develop the condition for instability of the flow, $d\psi_{TS}/d\phi > 0$,

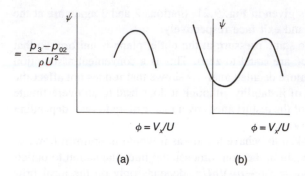

Fig. 9.22 Alternative axisymmetric compressor characteristics used in theoretical calculations of distorted and stalling performance

is satisfied over a substantial part of the annulus. This remains unproven at present.

Recent theoretical developments

It is clearly most unsatisfactory to have two different models for compressor instability, described above as the one-dimensional and the two-dimensional models, in which it often cannot be said unequivocally which is responsible for the stall or surge.

The reconciliation of the two models was made possible after a fresh approach by Moore (1984), later extended by Moore and Greitzer (1986) to look at compressor systems with uniform inlet flow. Hynes and Greitzer (1987) and Hynes *et al* (1986) went on to look at the effect of flow distortion. The physical system analysed is that shown schematically in Fig. 9.4c, that is a compressor (with an inlet duct), a long outlet duct, a plenum and a throttle.

The modelling of the compressor is crucially important. It is assumed that every compressor has a pressure rise−flow rate characteristic

$$\psi_{TS} = \psi_{TS}(\phi)$$

for circumferentially uniform steady flow with the form in either Fig. 9.22a or b. The more extensive form in Fig. 9.22b is only used when post-stall behaviour is being studied. These characteristics appear unfamiliar because the region to the left of the peak is normally unobtainable because the flow becomes unstable and breaks down into rotating stall. Experimental evidence for the form shown in Fig. 9.22a was given for a stage in Fig. 9.9, where the instability was suppressed by other stages in close proximity, and the transient existence of the form shown in Fig. 9.22b during surge cycles has also been confirmed.

The forms in Fig. 9.22 define the pressure rise capability of the compressor in steady axisymmetric flow; once the flow becomes perturbed ϕ will be a function of circumferential position θ. As soon as transients or non-uniform

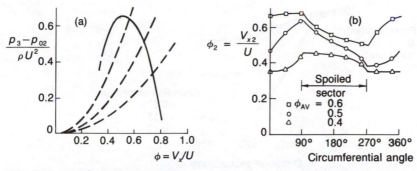

Fig. 9.23 (a) Assumed compressor characteristic for uniform flow. (b) Local flow coefficient at compressor inlet face for three different mean flow rates; inlet distortion $(p_{01})_{MAX} - (p_{01})_{MIN} = 0.2\rho U^2$. (From Hynes and Greitzer, 1987)

flow are considered, the flow in the blade passages becomes unsteady, so in modelling the compressor the inertia of the fluid in every blade passage must be included. The inertia gives rise to an unsteady pressure difference across each blade row which can be approximated by a simple one-dimensional flow inside each passage. The unsteady pressure differences are,

across a stator $\Delta p_{unsteady} = -\rho c \partial V/\partial t$ (9.11)
and across a rotor $\Delta p_{unsteady} = -\rho c \{\partial W/\partial t + U/r \partial W/\partial \theta\}$ (9.12)

where here c denotes the blade chord, for simplicity taken to be equal for rotor and stator. The second term on the right-hand side of equation 9.12 for the rotor arises because even a stationary non-uniformity will give rise to unsteady fluctuations in rotor blade passages.

Using this approach the analysis of circumferential distortion was in two parts, first the computation of the steady circumferentially non-uniform flow and then second the calculation of stability of this non-uniform flow to small disturbances. The compressor could be operating locally at different points on the compressor characteristic of the type shown in Fig. 9.22a or b; for example in some regions the compressor might be operating to the left of the peak and at others to the right of it. Figure 9.23, from Hynes and Greitzer, shows on the left the assumed compressor characteristic and to the right the result of the first part of the calculation, the predicted circumferential variation in the coefficient of the steady flow at the inlet face when a 180° distortion of large amplitude is imposed on the compressor. Three cases are shown, corresponding to different overall mean flows; the top trace is for the compressor operating with the mean flow to the right of the characteristic peak and the bottom trace with it to the left of the peak. The stagnation pressure distortion imposed is a square wave and the circumferential variation in the inlet flow rate is a result of the time lags in the blade passages. It should be noted how the distortion produces a lowering of flow rate when to the right of the peak ($\phi_{AV}=0.6$) and a raising of the flow rate to the left ($\phi_{AV}=0.4$).

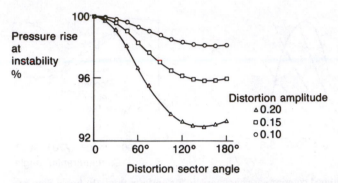

Fig. 9.24 Predicted loss in pressure rise at the instability point (normalized by pressure rise for uniform flow) with different levels and extent of inlet distortion. (From Hynes and Greitzer, 1987)

In the second part of the calculation the stability of this non-uniform type of flow was analysed.

Figure 9.24 shows loss in stability margin predicted by Hynes and Greitzer (1987) versus the size of the spoiled sector for the compressor whose characteristic is shown in Fig. 9.23. The trend is very similar to that shown from measurements in Fig. 9.17, and could be made to agree even better by the adjustment of parameters. As it is, quite good correlation has been found between the predictions and the well-known distortion coefficient, $DC(\theta_{crit})$, where θ_{crit} is the circumferential angle at which the full effect of a distortion is felt, often taken as 60°. A useful rule has emerged from the calculations which is that for distorted inlet flow the point of instability will occur when

$$\int_0^{2\pi} \frac{d\psi_{TS}}{d\phi} \, d\theta = 0 \tag{9.13}$$

which reverts to the condition $d\psi_{TS}/d\phi = 0$ (equation 9.10) for instability in undistorted flow. Hynes *et al* (1986) found that this is a good predictor of instability except for conditions when the rotating stall type of disturbance modes have a frequency close to that of the surge type of disturbance. The rotating stall frequency is fixed mainly by the compressor itself whereas the surge frequency depends more on the system volume and inertia. When the two frequencies are similar the instability occurs when the integral is less than zero, in practical terms stability is lost at a higher flow coefficient.

The same approach has been used to examine a number of other aspects of stalling and surging behaviour. Moore and Greitzer (1986) used it to study the compressor behaviour after instability has occurred and have computed flow behaviour of the rotating stall or surge type very similar to that observed experimentally.

9.3 Post stall behaviour

There are essentially two questions concerning the behaviour of a compressor subsequent to the onset of instability. First, does the compressor go into rotating stall or surge and second, if into rotating stall, is it full-span or part-span stall? During surge cycles the compressor often operates transiently in rotating stall but this is distinct from the steady or continuous operation of the compressor in rotating stall.

Stall or surge

There was extensive confusion surrounding the different nature of stall and surge and this legacy remains with the mixture of names still used. It was Greitzer (1976) who definitively established the circumstances for their occurrence. The different behaviour can only be understood when the entire system is considered. The minimum system for a proper consideration is that shown in Fig. 9.4c, consisting of compressor, a duct to give flow inertia, a plenum to give storage of the compressed gas and a throttling device. As noted earlier, in compact configurations such as gas turbines it is not possible to separate out the components in this way, the inertia and storage occurring in the same physical region and inside the compressor itself. Nevertheless the separation implied by Fig. 9.4c has allowed a new understanding to be brought to the topic.

The system shown in Fig. 9.4c was used to examine the stability of the system in a one-dimensional manner. The same equations 9.1 9.3, 9.4 and 9.5, which were used earlier to look at stability, can be used to look at the post-stall behaviour. What is no longer admissible is the use of linearization, because the disturbances are now large, but the four equations can readily be solved numerically. An input is the measured compressor pressure rise−mass flow characteristic, including a section with the compressor operating in rotating stall. Of the different non-dimensional parameters one emerged as determining, and this is

$$B = \frac{U}{2\omega_{\text{Helm}}L} \tag{9.14}$$

where the Helmholtz frequency

$$\omega_{\text{Helm}} = a\sqrt{\frac{A}{LV}} \tag{9.15}$$

is the natural frequency of the duct−plenum system in the absence of the compressor and throttle. A and L are the duct cross-sectional area and length respectively, V the plenum volume and a the velocity of sound. The parameter B can therefore also be rewritten as

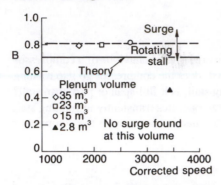

Fig. 9.25 Comparison of measured and predicted boundary between rotating stall and surge for a 3-stage low-speed compressor showing overriding dependence on the B-parameter. (From Greitzer, 1977)

$$B = \frac{U}{2a}\sqrt{\frac{V}{AL}} \qquad (9.16)$$

The choice of the blade speed in the parameter B was a result of the way Greitzer nondimensionalized his equations and it could be replaced by, for example, axial velocity.

For the compressors examined by Greitzer the calculations showed that if B was greater than about 0.7 the system would go into surge, with oscillations in the overall mass flow, whereas at lower values of B the compressor would sink to the point on the stalled part of the compressor characteristic where it crossed the throttle line through the stall inception point. Figure 9.25 shows the predicted and measured values of B at the boundary between surge and rotating stall. The measurements were made using a low-speed three-stage compressor with a large plenum to make B large enough for surge to be possible. The measured stall/surge boundary lies at about $B = 0.8$, which is in reasonable agreement with the theoretical value. Equation 9.16 shows that increasing the blade speed leads to an increase in B for the same system parameters so that a given configuration is more likely to surge as the rotational speed is increased.

The parameter B can be given a physical significance by rewriting it as

$$B = \frac{(\rho U^2/2)A}{\rho A L U \omega_{\text{Helm}}} \qquad (9.17)$$

The numerator gives a measure of the pressure rise capability of the compressor multiplied by the duct area. The denominator contains the product ρAL, which is the mass of the gas in the duct, the velocity U and the frequency of natural oscillation ω_{Helm}. The axial velocity V_x for a given compressor will be approximately proportional to the blade speed U. The denominator is therefore approximately equal to $\rho A L V_x \omega_{\text{Helm}}$, the force required to produce small oscillations of the flow in the duct at the natural frequency. The parameter B can therefore be viewed as the ratio of the compressor pressure rise capability to the pressure rise required to induce mass flow oscillations. If B is low it is equivalent to saying that the compressor does not have much pressure rise capability compared to what is required to begin a surge.

A more complete description of the post-stall behaviour would include some description of the compressor pressure rise−flow rate characteristic. The value of B found to be critical by Greitzer is dependent on the performance of the compressor and different values would be obtained for different designs or different numbers of stages. A crude way to account for the number of axial compressor stages is to use NB as the critical parameter, where N is the number of stages. Since Greitzer used a three-stage machine this would suggest surge if NB is greater than about 2.5. The critical value of B has indeed been found to drop with the number of stages, but a bit more slowly than in proportion to N. For engines the value is frequently sufficiently close to the critical value that the inaccuracy or arbitrariness in assessing volume, area and length become important. (Proprietary expressions with similar meanings to B have been devised using variables more relevant to high speed operation.) Although aircraft engines can either surge or move into rotating stall, the clear trend is, as predicted by the B-parameter, that the engine is more likely to surge as the speed is increased. However the importance of the B-parameter is not so much that it allows one to predict whether an engine will surge or stall but that it has altered the way that surge and stall are thought about.

Greitzer found the B parameter not only to determine whether the compressor

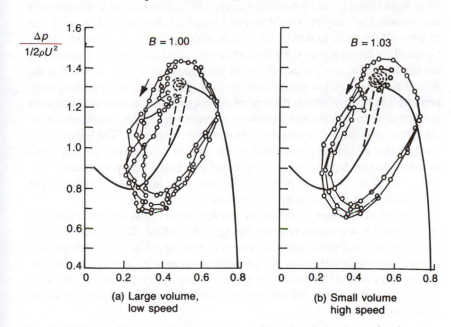

Fig. 9.26 Measured locus of pressure versus flow rate during surge for the same compressor but different plenum sizes and speeds, showing dependence on the B-parameter. (From Greitzer, 1976)

would surge but to have an overwhelming effect in determining the path of the compressor with respect to time on the pressure−flow rate traces. This is demonstrated in Fig. 9.26 in which the value of B is essentially equal ($B \approx 1.0$) — although the speed and plenum size are different the locus of the surge is virtually identical. As B is increased further above the critical value there is a tendency for the surge cycle to encompass larger excursions of mass flow rate. An additional effect is associated with the throttle setting; as the throttle is closed and the surge is forced to take place about a smaller average mass flow rate there is a greater tendency for reversed flow to occur during the surge cycle, producing what is called 'deep surge'. Deep surge is characterized by the following cyclic process: stall of the compressor leading to (a) a rapid reduction in mass flow rate at almost constant pressure rise; (b) a reduction of the pressure in the plenum (during which time the mass flow rate may be negative) until the pressure of the stalled part of the characteristic is reached; (c) a rapid change in mass flow at almost constant pressure rise to the unstalled part of the characteristic; and finally (d) a slow pumping up of the system along the unstalled characteristic until the stall point is reached.

The treatment by Greitzer was essentially incompressible, except for storage in the plenum, but it has been applied successfully to high-speed, high-pressure ratio situations. A complication still existed, however; some compressors in bypass; jet engines encountered rotating stall when they were expected to experience surge. Full-span rotating stall is generally far more serious than surge in jet engines, mainly because it is very difficult to recover from; recovery may require the complete shut-down and restart of the engine. This gave rise to names like 'lock-in-stall', 'stagnation stall' and 'non-recoverable stall'. Gradually it has begun to seem that what was different with the geometry of the bypass engine was the possibility of the imposition of low pressure at the rear of the low-pressure compressor and the inlet of the high-pressure compressor. The low pressure at the rear of the low-pressure compressor stabilizes it whilst at the inlet to the high-pressure compressor it imposes a quite unsustainable pressure difference across the high-pressure compressor. The highly stable low-pressure compressor prevents the high-pressure compressor surging, with the result that the high-pressure compressor goes into a full-span rotating stall. The principal remedy is to isolate the inlet face of the high-pressure compressor from the bypass flow as much as possible.

The work of Greitzer was directed at axial compressors and the magnitude of the critical B was appropriate for this type of machine. The underlying idea, that surge can only occur when the pressure rise capability of the compressor is large enough in relation to the pressure difference needed to oscillate the flow in the ducts at the natural frequency of the system, is quite general to all types of compressor. An appropriate value of B could be found for radial compressors.

Part-span or full-span stall

Full-span rotating stall is probably the most serious condition for an axial compressor to be in. The pressure rise and flow rate are generally low but the

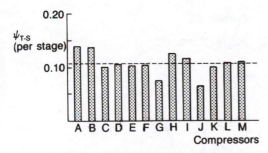

Fig. 9.27 Stage pressure rise coefficient $\psi_{T-S} = (p_2 - p_{01})/\rho U^2$ for thirteen different compressors at shut-off (zero net mass flow rate). (From Day *et al.*, 1978)

work input is high so that very high temperatures can develop very quickly. The existence of hysteresis means that a large opening of the throttle, or equivalent process in an engine, may be needed to unstall the machine. By contrast part-span stall is usually much less serious, with a relatively small drop in performance and little if any hysteresis. Part-span stall occurs quite frequently in the front stages of multistage axial compressors as a result of mismatching at low corrected rotational speeds and, provided no serious vibrations are excited, it can normally be tolerated. There is therefore a need to be able to predict whether full-span or part-span rotating stall will be produced and to have some guidance on the extent of the opening of the throttle needed to unstall the machine.

The only model which gives guidance on this topic is an approximate one due to Day *et al* (1978) which primarily relied on two experimental observations. The first followed an observation by A B McKenzie, that in full-span rotating stall the non-dimensional inlet stagnation to outlet static pressure rise is independent of the compressor blading and the unstalled pressure rise. Expressed symbolically the stalled pressure rise is

$$\psi_{TS} = \frac{p_{exit} - p_{0inlet}}{\rho U^2} \approx 0.11N \qquad (9.18)$$

where N is the number of stages. Figure 9.27 shows measured values of ψ_{TS}: despite the large differences between the various compressors this range is smaller than that for the unstalled performance. (In all treatments of stalled

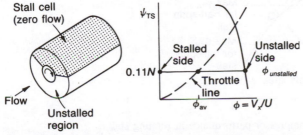

Fig. 9.28 A compressor with N stages operating in rotation stall

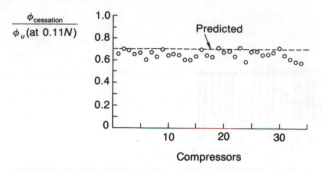

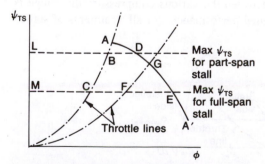

Fig. 9.29 Comparison of predicted and measured flow rate at stall cessation, normalized by flow rate in unstalled region at stage total-to-static pressure rise of 0.11. General Electric measured data. (From discussion by Harman of Day *et al.*, 1978)

flow it has to be accepted that there can be less precision than in the treatment of unstalled flow; the flow is more complicated in every way and has been studied far less.) The values of pressure rise shown are for shut-off (zero net mass flow through the compressor) though the variation in stalled pressure rise with flow rate is usually small.

The second observation was that there is a critical size of stall cell, expressed as a fraction of the annulus area, λ, below which full-span stall cannot exist. The size of the stall cell was deduced from the blockage by assuming that the stalled compressor operates as shown in Fig. 9.28. The blockage was then given from the flow coefficient ($\phi = V_x/U$) by

$$\phi_{average} = (1-\lambda)\phi_{unstalled} \tag{9.19}$$

Experimental evidence suggested that λ = 30 per cent is the *minimum* for full-span stall. Figure 9.29 shows that this is a quite good approximation in the case of cessation (that is just before the compressor unstalls) and similar agreement exists at inception.

Additional information was collected about part-span stall; it was deduced that there was a *maximum* blockage of 30 per cent for part-span stall and that the inlet stagnation to outlet static pressure rise was given by

Fig. 9.30 Model for compressor performance in rotating stall

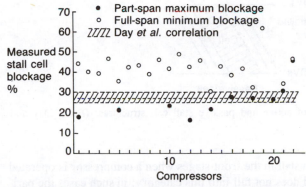

Fig. 9.31 Stall cell blockage, maximum for part-span and minimum for full-span stall. General Electric measured data. (From discussion by Herman of Day *et al.*, 1978)

$$\psi_{TS} = 0.17N$$

though the amount of data to support this was fairly small.

The method for using the model is indicated in Fig. 9.30. Stall occurs at point A and the pressure rise and flow rate are constrained to follow the throttle line through it. Point B is the intersection of the throttle line with the pressure rise for part-span stall, $\psi_{TS} = 0.17N$. If the blockage at this condition, λ_B = BD/LD, is less than the critical value, taken to be 30 per cent, then the compressor can operate in part-span rotating stall. If λ_B would exceed 30 per cent at this pressure rise for part-span stall then the compressor collapses to point C at the full-span stall pressure rise, $\psi_{TS} = 0.11N$. Figure 9.31, showing data quoted by Harman of General Electric in the discussion of the paper by Day *et al.*, indicates that the method does indeed provide a good prediction of the occurrence of full-span or part-span stall.

With the compressor in full-span stall the effect of opening the throttle can be visualized by moving from point C to point F. The compressor will come out of full-span stall when λ = FE/ME drops below 30 per cent; if point F is a long way to the right of point C then there is large hysteresis.

The possibility of part-span stall and the extent of the hysteresis both depend on the slopes of the compressor *and* the throttle line. With steep lines the compressor will, when stalled, tend to go into part-span stall; if it is further throttled so that it goes into full-span stall there will only be small hysteresis between stall and unstall. The steepness of the pressure rise−mass flow characteristic increases for compressors designed for low V_x/U (small ϕ^*) and Fig. 9.8 confirms how much smaller the hysteresis is for the low ϕ^* compressors. For reasons also explainable in terms of the blockage part-span stall extending the whole length of the compressor becomes much less probable when the number of stages is increased.

The Day *et al.* (1978) model considered multistage compressors which were correctly matched, as they might be at the design speed and mass flow. The

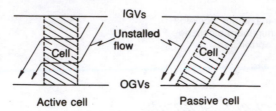

Fig. 9.32 Comparison of active and passive stall cell structures. (From Day and Cumpsty, 1978)

formation of part-span stall in the front stages when a compressor is operated below its design speed does not fall into this category; in such cases the part-span stall cells provide a redistribution within the compressor.

9.4   The flow in the rotating stall cell

Although rotating stall in axial compressors has been a problem for many years, until fairly recently it was difficult to make measurements inside the cell itself. Before the measurements were available it was believed that the stall cell was rather like a wake with very little axial velocity through it and very little flow crossing the boundaries on its sides: this may be described as the passive stall cell. To satisfy the absence of flow across the boundaries it is necessary that the cell twist around the annulus, as indicated in Fig. 9.32. Measurements by Day and Cumpsty (1978) in multistage compressors of a wide range of design flow coefficient showed that the stall cell is essentially axial and of constant width through the length of the machine; this is similar to the observation that the disturbance from inlet distortion tends to be in an axial strip. It is a necessary corollary of this that fluid must enter one side of the stall cell and leave on the other, giving rise to the model of the active stall cell shown in Fig. 9.32. Much of the flow entering and leaving the cell is carried in the rotor blade passages rather than in the axial gaps between the blade rows. In the case of a compressor of 50 per cent reaction the average circumferential velocity in the unstalled region is equal to one half blade speed; full-span stall cells generally rotate at less than 50 per cent of blade speed so there must be a net transfer *into* the cell on its trailing edge and transfer *out* on its leading edge. Quantitative estimates by Day and Cumpsty showed that for the 50 per cent reaction compressors used the mean flow circumferentially though the cell was indeed very close to the value of $0.5U$ expected.

Day and Cumpsty reported measurements in three compressors of different design values of $\phi^* = V_x/U$ (essentially different blade stagger angles), each with three stages and of 50 per cent reaction. The stall cells remained axial through the length of the machine and the velocity pattern in the first stage was very similar to that in the last. Figure 9.33 shows the velocity magnitude and direction measured at mid-span upstream and downstream of the first-stage rotor for the build of $\phi^* = 0.55$ (for different values of ϕ^* the pattern was

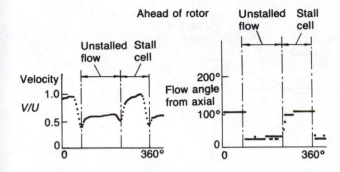

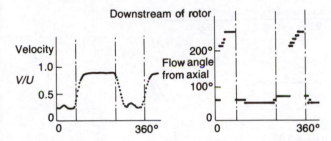

Fig. 9.33 Measured velocity magnitude and direction upstream and downstream of rotor in stalled flow. Measurements at mid-span. $\phi_{test} = (V_x/U)_{mean} = 0.33$; $\phi^* = (V_x/U)_{design} = 0.55$. (From Day and Cumpsty, 1978)

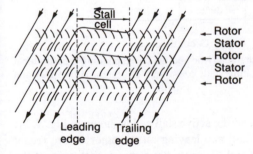

Fig. 9.34 Sketch in the *absolute* frame of reference of the active stall cell structure showing unstalled flow entering cell on its trailing edge and leaving on its leading edge. (From Day and Cumpsty, 1978)

found to be broadly similar). The velocity is seen to be very high ahead of the rotor inside the stall cell, equal to the rotor speed at its highest. Upstream of the rotor the flow is at approximately 90° to the axial, that is swirling in the same direction as the rotor. Downstream of the rotor the velocity is generally low inside the stall cell, the direction being near 90° over part of the cell and 270° over the remainder, 270° being in the opposite sense to the rotor direction. (The flow directions are not exactly 90° or 270° and there is a small axial velocity inside the cell, but much less than that outside it.)

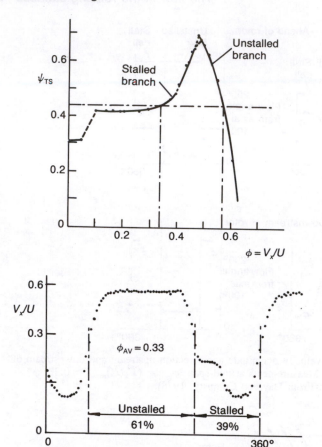

Fig. 9.35 Three-stage compressor characteristic and the corresponding measurement of instantaneous axial velocity in rotating stall; $\phi^* = 0.55$ at design. (From Day and Cumpsty, 1978)

The general pattern of flow around the active stall cell is shown in Fig. 9.34; flow entering on one side of the cell and leaving on the other. More recant measurements than those of Day and Cumpsty have shown that most of the circumferential flow across the stall cell occurs carried in the blade passages and not in the axial gaps.

The mid-span axial velocity is shown in Fig. 9.35, which also shows the relation of the operating point to the overall characteristic. If the pressure rise in the stalled region is fixed, as assumed in the model of Day *et al.* (1978), it can be seen that the axial velocity in the unstalled region is determined by the requirement for the unstalled section of the compressor to give the same pressure rise. Variation in the net flow rate through the stalled compressor is produced in the main by varying the relative size of the stalled and unstalled regions.

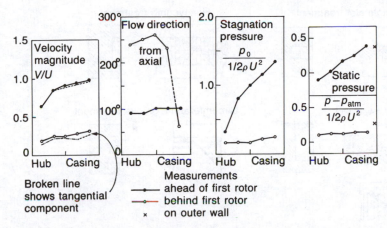

Fig. 9.36 Measured flow properties across the span at the circumferential centre of a full-span stall cell; $\phi^* = 0.55$, $\phi = 0.33$. (From Day and Cumpsty, 1978)

Some measurements were made at different radii inside the cell, all at the circumferential mid-point of the cell. These are shown for the $\phi^* = 0.55$ compressor in Fig. 9.36. It is clear, particularly from the measurements of flow direction, that the pattern is not simple. There were strong radial gradients of stagnation and static pressure ahead of the rotor, consequent on the high circumferential velocity there. The static pressure variation is well predicted by

$$p_{\text{casing}} - p_{\text{hub}} = \int_{\text{hub}}^{\text{casing}} \rho \, \frac{V_\theta^2}{r} \, dr$$

Measured through the machine the static pressure at the casing appeared to rise over each stator and stay more or less constant across each rotor. The cause of this is the high radial pressure gradient ahead of each rotor attributable to the high whirl velocity there. Some explanation is needed for the observation that the non-dimensional total-to-static pressure rise across each stage in full-span rotating stall is approximately equal to 0.11, but so far none has been forthcoming.

The measurements reported by Day and Cumpsty were for one combination of hub–casing ratio and axial gap between the blade rows. Das and Jiang (1984) also reported measurements inside stall cells at two axial gaps, the differences in flow pattern for the two cases being large. The variation in the pattern radially and circumferentially was greater than apparent from the earlier measurements of Day and Cumpsty, though the newer measurements support many of the earlier conclusions.

9.5 Stability enhancement: casing treatment

There is every incentive to obtain the widest possible operating range for a compressor, provided that it does not compromise other aspects of the

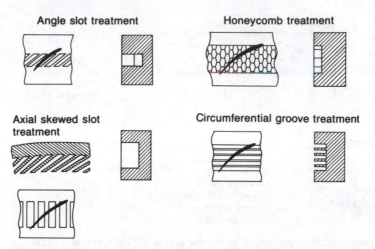

Fig. 9.37 A selection of different casing treatments for the delay of stall

performance too seriously. One of the approaches which has been adopted with some success is known as casing treatment. It has mainly been applied to axial compressors and is usually installed in the casing over the rotor tips. It can also be installed in the rotating hub under the stators if these are cantilevered in from the casing. It can be put in the casing over the inducer of a centrifugal impeller and has even been mounted in the extreme outboard region of an impeller so that it passes the inlet region to vaned diffusers.

Axial compressors

The benefits of casing treatment seem to have been discovered more or less by accident around 1970. The discovery occurred on a transonic fan and this type of machine was used for many of the subsequent studies mainly at NASA Lewis. A fairly complete list of the early references is given by Greitzer *et al.* (1979) and a few of the types of treatment used are shown in Fig. 9.37.

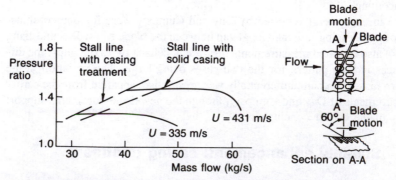

Fig. 9.38 The effect of casing treatment on the performance of a transonic fan. (From Mikolajczak and Pfeffer, 1974)

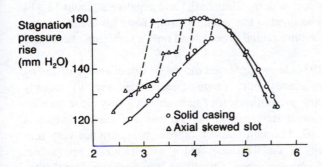

Fig. 9.39 The effect of casing treatment on the performance of a low-speed compressor. Also shows stall-unstall hysteresis. (From Takata and Tsukuda, 1977)

Perhaps surprisingly, a treatment which is effective at high inlet relative Mach numbers, like 1.5, is also likely to be effective at a low Mach number, such as 0.15. Figure 9.38, taken from Mikolajczak and Pfeffer (1974), shows a high-speed example with a diagram of the slightly modified axial skewed slot shown to the side; Fig. 9.39, taken from Takata and Tsukuda (1977), shows the effect of a conventional axial skewed slot for a low-speed rotor. The modification shown in Fig. 9.38 is essentially the introduction of a baffle half-way along the slot which was found to have the effect of reducing the efficiency penalty whilst retaining most of the stall margin improvement. Casing treatment also acts as a diagnostic tool: when it is effective over the rotor tips it may be inferred that the cause of stall is in the tip region of rotors. Furthermore, since casing treatment is effective at high and low Mach numbers, the mechanism of stalling may also be very similar at these different speeds.

Not all treatments are effective and some can in fact worsen the flow. It will be noticed that the axial skewed slot is inclined so that the slots are inclined to meet the flow. Takata and Tsukuda (1977) tested axial skewed slots not only inclined as drawn in Fig. 9.36, but also with the slots radial and with them inclined in the opposite sense. Whereas the stall margin improvement with small rotor tip clearance was about 20 per cent with the correctly skewed

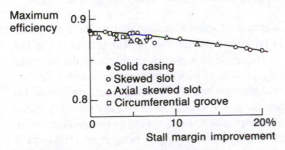

Fig. 9.40 The correlation between efficiency and stall margin improvement for three types of casing treatment. (From Fujita and Takata, 1984)

slot it was only 13 per cent with the slots radial and negative at about 11−14 per cent with the slots inclined in the wrong sense. They also found that the blade angle slots and circumferential grooves had only a very small beneficial effect.

Fujita and Takata (1984) have addressed the major disadvantage of casing treatment, the loss in efficiency. For a range of geometries, varying not only the type but the geometric proportions, they have shown a very close correlation between the increase in stall margin and the loss inefficiency, which is summarized in Fig. 9.40. The stall margin improvements are not very sensitive to the dimensions of the treatment and a wide range is acceptable. Although quite a large number of types of treatment have been considered these were all geometrically simple, made up of straight lines and sharp corners with no clear aim of matching the geometry to the flow. It may be that with the proper specification of slot shape points well above the line in Fig. 9.40 could be obtained, but at present logical design is hampered by ignorance of what initiates stall in the first place and how casing treatment delays it in the second.

Even the most effective type of treatment does not always bring about an improvement in surge margin. Casing treatment is only effective when it is

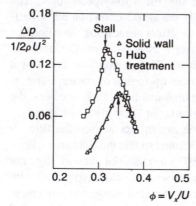

Fig. 9.41 The static pressure rise across a stator versus flow rate with and without hub treatment (rotating hub with axial skewed slots). (From Cheng *et al.*, 1984)

installed where the rate of increase of local blockage and perhaps deviation with decrease in flow rate is rapid; very often this is near the rotor tips but it need not always be. For example, if in a particular machine the cause of the instability is predominantly growth of blockage around the hub then casing treatment over the rotor will offer no benefit. If casing treatment is installed in only one place in a compressor and if it is to have a really decisive effect on compressor stalling it is necessary that this one region be the overwhelming cause of the blockage growth leading to instability. Cheng *et al.* (1984) have tested a rotating hub treatment underneath a cantilevered stator, partly to test if this was useful and partly as a vehicle for carrying out research into the

mechanism of stall delay with an axial skewed slot but without the complication of working in the rotor frame of reference. Figure 9.41 shows the static
pressure recovery for the stator; it is very clear that the stator flow is radically
altered by the hub treatment.

One of the problems in deciding on the usefulness of casing treatment, and
in speculating how best to design it, is the ignorance which exists regarding
the flow in the region which is most often critical, the endwall region near
the blade tips. Sometimes casing treatment has been ineffective even when
the rotor tips were thought critical. Greitzer *et al.* (1979) divided conditions
into wall stall and blade stall. In wall stall it was envisaged that the annulus
boundary layer would grow approximately axisymmetrically to form a thick
and possibly separated layer. In blade stall it was envisaged that the blades
stall more or less as they would in a two-dimensional cascade. The measure
of the likelihood of wall stall was the non-dimensional static pressure rise across
the blade row, in the manner of de Haller (1953): the measure of blade loading
was the familiar diffusion factor. For the same static pressure rise (i.e. the
same de Haller number) the diffusion factor can be reduced by increasing the

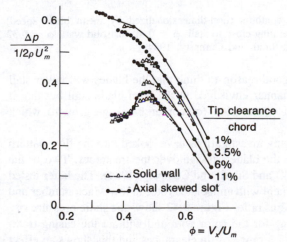

Fig. 9.42 The effect of tip clearance on the static pressure rise of a rotor with and
without casing treatment. (From Smith and Cumpsty, 1984)

blade solidity. Only the case of wall stall was imagined to be amenable to
improvement by casing treatment.

Greitzer *et al.* tested this by using a rotor with two different solidities (1.0
and 2.0) but the same camber and stagger. The casing treatment was highly
effective with a solidity of 2.0, for which the diffusion factor was not high,
but ineffective with a solidity of 1.0 when diffusion factor was much higher.
Although the wall stall/blade stall hypothesis offers a convenient method for
assessing the likely performance of casing treatment it is not the only explanation for the different behaviour. Furthermore, the division into blade and wall
stall, though convenient, does not properly describe the flow behaviour, even

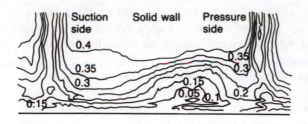

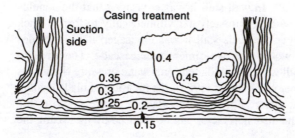

Fig. 9.43 Axial velocity contours (non-dimensionalized by mean blade speed) downstream of rotor tip operating close to stall: $\phi = 0.40$ for solid wall, $\phi = 0.32$ for casing treatment. (From Smith and Cumpsty, 1984)

approximately. There is good reason to think that the blades will never stall in the two-dimensional manner envisaged in the term blade stall but that it will always be a type of corner stall (or corner blockage growth) which precipitates stall.

Relatively few of the many investigations have looked into the flow pattern in the treatment or inside the blade passage over the treatment. Two of the few are Prince *et al.* (1974) and Smith and Cumpsty (1984). The latter tested an axial skewed slot treatment with an isolated rotor at low Mach number and with a low hub−casing radius ratio. Figure 9.42 shows the static pressure rise characteristic on the casing for the rotor with and without the casing treatment. Results are given for a range of tip clearances and this shows an effect

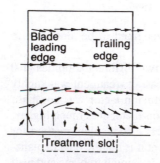

Fig. 9.44 Spanwise flows with axial skewed slot, measured in plane 20 per cent of pitch from blade pressure surface. Very close to stall. (From Smith and Cumpsty, 1984)

noticed in other investigations — the stall margin improvement increases as the tip clearance is raised, at least until the clearance becomes very large. The ability to mitigate the effect of larger than optimum tip clearance may be one of the most significant advantages of casing treatment. Axial velocity contours measured downstream of the rotor tip are shown in Fig. 9.43. The rotor tip was highly staggered and the high loss flow with low axial velocity tends to pile up in the casing—pressure surface corner, as is evident with the solid wall build. With casing treatment the region of low axial velocity is replaced by a region of high velocity and also high relative stagnation pressure. It seems that the casing treatment works by withdrawing fluid near the pressure surface trailing edge (where the pressure is high) and reinjecting it near the leading edge. The radial velocities measured in the tip region are shown in Fig. 9.44 for a section 20 per cent of pitch away from the pressure surface of the blade. The radial flow out near the front of the slot and flow in near the rear is very clear. With the skewed axial slots, the slots are angled so that it is easy for the flow to enter, pushed in by the rotor, and the ejected flow comes out so that it has high stagnation pressure relative to the rotor.

Smith and Cumpsty were able to conclude rather tentatively that the unsteadiness of the blades passing the slots was unimportant but could not resolve whether the withdrawal of low axial velocity fluid at the rear or the ejection out of the front of the slot was crucial. Recent measurements by Lee (1988) have been carried out with the rotating hub treatment used earlier by Cheng *et al.* (1984). Suction into the rear of the skewed axial slot and blowing from the front were tried separately, the suction and blowing produced by an external machine. The conclusion from this is that both sucking away the blockage at the rear and blowing from the front (inclined at a large angle to the radial, 60° here) are able to bring about a significant improvement in blade row performance and both are potentially important for increasing the stall margin.

It does seem probable that a slot geometry could be devised more suitable than the simple and rather crude types used hitherto. Curved passages which direct the flow out of the slots so that its direction is more similar to that of the free-stream might be an improvement. Otherwise it seems impossible to break from the constraint indicated by Fig. 9.40: the loss in efficiency as the price for surge margin improvement has been the main reason that designers have been reluctant to use casing treatment. When casing treatment has been used in axial compressors it has nearly always been the axial skewed slot type because this has been shown to be highly effective. If the mechanism by which the axial skewed slot operates is only relatively well understood, there is no satisfactory explanation for the operation of treatments such as the circumferential groove, the honeycomb or the blade angle slots.

Centrifugal compressors

There has been little application of endwall treatment to the centrifugal compressor, but a programme of tests on two compressors has been reported by

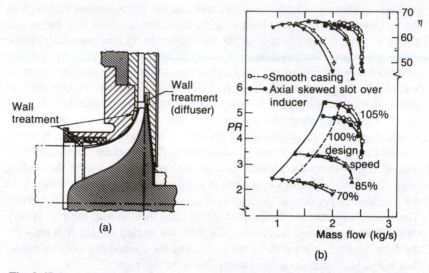

Fig. 9.45 Drawing shows location of treatment in centrifugal compressor. Performance map shows stagnation-to-static pressure ratio for the stage with vaneless diffuser versus mass flow. Results for smooth casing and for skewed axial slot over inducer. (From Jansen *et al.*, 1980)

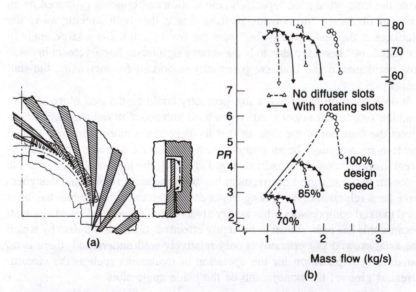

Fig. 9.46 Drawing shows arrangement of rotating slots at diffuser inlet. Performance map shows stagnation to static pressure ratio versus mass flow. Results for no diffuser slots and for rotating slots at vaned diffuser inlet. (From Jansen *et al.*, 1980)

Jansen *et al*. (1980). A total of thirty different ideas were tried to delay stall and, as the paper makes explicit, most were unsuccessful. The paper describes those which were successful and interesting. The layout is indicated in Fig. 9.45a. Two impellers were used, one which gave a pressure ratio of about 3 and had backswept blades, the other with radial blades giving a pressure ratio of about 6; both the examples shown here were obtained for the high pressure compressor. When tested with a vaneless diffuser the stall was initiated in the impeller and in this case the treatment just over the inducer was highly effective in delaying the onset of stall, Fig. 9.45b, with very little loss in efficiency. This is very convincing evidence that the stability of the stage with the vaneless diffuser is primarily brought about by the inducer. Put another way, the vaneless diffuser is known to be unstable, the main part of the impeller is more or less neutrally stable so it is only the inducer which offers a real contribution to the overall stability. Treatment near the impeller outlet gave no improvement in stall margin but lowered the efficiency by about 15 per cent.

The treatment over the inlet to the vaned diffuser, shown in Fig. 9.46a and b, had an unexpected effect; the stall point was not moved to smaller flow rates but the mass flow at choking was markedly increased, Fig. 9.46b. The higher mass flow at choking may be understood when it is remembered that the axial skewed slot is effective by reducing the blockage.

Two other investigations in the use of wall treatment with centrifugal compressors should be mentioned. Amman *et al*. (1975) were able to extend the range of a centrifugal compressor whose range was limited by the vaned diffuser by putting a circumferential groove in the shroud near the impeller tip. Wiggins and Walz (1977) obtained an improvement by placing circumferential grooves near the inlet of the impeller. There is no obvious explanation for the improvement in either case. With the evidence from the tests by Jansen *et al*., however, it may be concluded that casing treatment has considerable scope for application to centrifugal compressors, particularly in the inducer region.

10 <u>*Vibration and noise*</u>

10.1 Introduction

Blade vibration and noise are conveniently considered together for two reasons; both are unsteady processes and each can be a serious nuisance. Of the two, the problems of blade vibration are likely to be far more worrying to the designer, for high levels of vibration can lead to part or all of the blade breaking off. Enormous amounts of downstream damage are then likely to occur, possibly causing total destruction of the compressor. Noise is a less serious threat to the compressor itself and only in exceptional cases is the level of pressure fluctuation high enough to cause mechanical damage. The attenuation of noise by acoustic treatment on the inlet and outlet duct walls can be very great and the aerodynamic penalty which acoustic treatment introduces is remarkably small. The modern jet engine demonstrates just how effective the quite short ducting ahead and behind the fan can be at attenuating noise without seriously compromising the aerodynamic performance.

If the problem of vibration is likely to be more serious than noise to most compressor aerodynamicists, it is appropriate to comment on the greater length of this chapter devoted to noise. This is justified by the recent publication of AGARDograph No. 298 (1987) in which world authorities give accounts of many different aspects of the aerodynamics associated with aeroelasticity in axial turbomachines. A thorough background is also given by Carta (1986). The section in the present chapter devoted to vibration is a basic introduction to the topic intended to serve a reader who needs some knowledge but does not want to engage the details.

10.2 Vibration

The vibrations of blades in compressors and turbines have many features in common and in axial turbines vibration was a problem before the first axial compressor was built. Campbell's (1924) work on axial turbines carries over to today's compressors and is remembered in the Campbell diagram, to be referred to below. The vulnerability of turbomachines to vibration, particularly axial machines, is not surprising in view of the use of long slender blades (particularly when high aspect-ratio blades are used), the large gas loads, the proximity of moving and stationary components and the small amount of mechanical damping.

The mechanical damping is often crucial and configurations which are potentially unstable are saved by having quite small amounts of damping. An important example of this is the stators in axial compressors. Vibration problems are nearly always restricted to rotors and the reason for this is believed to be the high load at the rotor root, arising from the centrifugal loading. This provides a very tight contact with little rubbing and therefore little damping. The stator root, on the other hand, does not normally tighten up to anything like the same extent and the small amount of movement is able to damp vibration. In a case when the stators of an axial compressor were welded into the casing there were unusually severe vibration problems in the stator. Recognizing that vibration of stator blades in axial compressors or stationary diffuser vanes in centrifugals is possible but less likely than vibration of the rotor, in what follows the discussion will proceed assuming that it is the rotor which is the vibrating element.

There are two quite different types of vibration to be considered, forced vibration and flutter. The former, which is easier to understand and to obviate, arises from the movement of the rotor through disturbances which are usually stationary, for example the wakes or potential field of an upstream stator, the wake of some upstream element such as a strut or the inlet distortion. It can also arise from the excitation produced by passing through rotating stall cells. Forced vibration becomes a problem when the excitation frequency coincides with a natural frequency of the blades. Flutter, on the other hand, is a self-excited oscillation or instability at or close to the natural frequency of the mechanical system which does not require any disturbance of finite amplitude to excite it.

For forced vibration almost all the sources, other than those from rotating stall, must be at harmonics of the rotational frequency of the rotor itself. This gives rise to the presentation of the information in the form of the Campbell diagram, Fig. 10.1, with rotor speed shown along the abscissa and frequency as the ordinate. This example was taken from Sisto (1987). Each of the lines springing from the origin is a harmonic of the shaft frequency and represents a possible excitation frequency; they are often referred to as engine orders. The lines which are nearly horizontal in Fig. 10.1 are the natural frequencies of the blades, which in this range are actually the lowest natural frequencies. Where the natural frequency coincides with one of the engine orders there is a potential problem of resonance; the circles on the figure show where high vibrational stresses were encountered as a result of this.

Figure 10.1 also shows regions of high stress which occur at frequencies not coincident with orders of the rotor frequency, shown by crosses. These could be a result of forced vibration excited by rotating stall or of flutter. Whereas rotating stall in multistage compressors is common at low speeds,

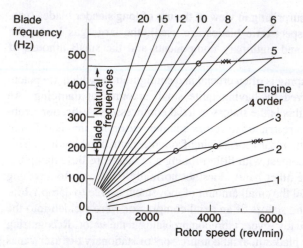

Fig. 10.1 Campbell diagram for a rotor blade. Circles indicate forced resonances (coincidence of natural frequency and engine rotational frequency), crosses show *either* rotational stall or flutter. (From Sisto, 1987a)

at higher speeds it is far more likely to be flutter. The vibration problems of rotating stall can be removed by improving the aerodynamics to get rid of the stall, for instance by fitting variable guide vanes to improve the matching at low speeds. In many cases the blades are able to tolerate excitation by rotating stall cells, which typically occurs at low rotational speeds and for fairly short periods. By far the most difficult problem is flutter. It is trying to understand flutter which has most occupied workers in this field, and the largest part of this section will therefore be devoted to it.

Mechanical vibration modes

Whether in forced vibration or flutter the blades vibrate in their natural modes at their natural frequencies, a consequence of the small damping and small unsteady aerodynamic forces involved. There are very many modes possible and therefore many natural frequencies. The natural frequencies can be changed by changing the design of the blade (for example reducing the chord from hub to tip raises the natural frequencies) or by changing the material from which it is made. So far as material is concerned it is the ratio of Young's modulus to density, E/ρ which determines the frequency and this is more or less equal for aluminium alloy, steel and titanium so that the effect of material on natural frequency is relatively small. The rotation of the rotor brings about changes in the natural frequencies, discussed below, but once the blade is made the natural frequencies are essentially fixed parameters for the aerodynamic investigation, with the system free to 'choose' the mode and natural frequency most favourable to vibration.

The damping of most blades, either by mechanical or aerodynamic effects, is usually sufficiently small that it is possible to consider the undamped modes

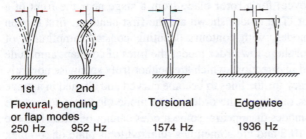

1st   2nd

Flexural, bending   Torsional   Edgewise
or flap modes
250 Hz   952 Hz   1574 Hz   1936 Hz

Fig. 10.2 Schematic of simplest modes for axial blades. (From Armstrong and Stevenson, 1960)

of the blades as an adequate description of the motion. Related to this is the high mass density of the blades, so that the influence of the aerodynamic effective mass is negligible. At this stage each blade will be assumed to be an independent entity mounted on a rigid disc and mechanical coupling between blades will be considered later.

The earliest studies of blade vibration considered axial blades in terms of beam theory and this has given rise to the method of classifying the simpler modes. Figure 10.2, taken from Armstrong and Stevenson (1960) shows the four lowest modes sketched schematically with the corresponding frequencies shown underneath; the absolute levels of the frequencies do not matter here but merely the relative levels. The frequencies may be distributed differently and, for example, the frequency of the 2F mode may be higher than the 1T in some designs. The flexural modes are sometimes known as flap or as bending modes. Actual blades are twisted, vary in camber from root to tip and have thickness distributions which are not symmetric about mid-chord. This means that the actual displacements of the blade are more complicated than the sketch in Fig. 10.2 would imply. Figure 10.3 shows computed mode shapes

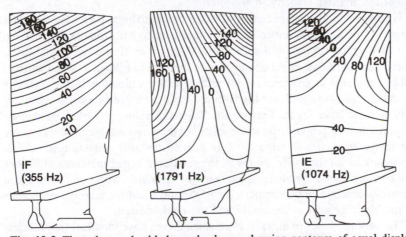

Fig. 10.3 Three low order blade mode shapes showing contours of equal displacement amplitude in first flap (IF), first torsional (IT) and first edgewise (IE) modes. (Reproduced with permission of Rolls-Royce)

(supplied by Rolls-Royce) for a rotor blade from a stage near the front of a multistage compressor. The modes shown are the first bending, first torsion and first edgewise modes, with contours denoting constant amplitudes of displacement. Even for these low order modes the lines of constant amplitude are curved and aligned in directions which are neither truly spanwise nor truly chordwise. The tendency for the lines to become curved and aligned in a range of directions increases rapidly as the order of the mode increases so that the simple picture of the modes disappears; some modes cannot be fitted into the classification of Fig. 10.2 and are sometimes referred to as plate or complex modes.

The prediction of mode shape and natural frequency using finite element packages is now very accurate for the low order modes but becomes relatively inaccurate at the high orders. At the high orders the inaccuracy or uncertainty may be greater than the difference between the natural frequencies of different modes. The measurement of natural frequencies can be carried out as a bench-top experiment but the accuracy of this is compromised by the difficulty of obtaining the representative root attachment and the neglect of centrifugal effects. The most accurate measurement of frequency is obtained from measurements made with the blades rotating at the correct speed so that the root fixing is tightened by the centrifugal load; measurement and excitation is then more difficult to arrange.

Rotation introduces extra problems, as mentioned above. The centrifugal loading has the effect of stiffening the blade to bending so that the flexural or bending natural frequency tends to increase with rotational speed, approximately as

$$\omega^2 = \omega_0^2 + k\Omega^2$$

where ω is the natural frequency, ω_0 the natural frequency with no rotation and Ω the angular velocity of the shaft.

The torsional modes are hardly affected by centrifugal stiffening and their frequency does not therefore increase significantly with rotational speed, so that they would appear on the Campbell diagram as almost horizontal lines. There is an additional effect observed with increase in rotational speed attributable to the rise in temperature bringing about a reduction in the Young's modulus; this is insignificant for fans and the front stage of a compressor but may be quite large for the rear stages. The temperature rise of a compressor is approximately proportional to the square of the rotational speed so that where the temperature rise is large there is a trend towards reduction in natural frequency of *all* the blade modes as the rotational speed increases. Whether the bending mode frequencies increase or decrease with rotational speed therefore depends on the temperature of the stage and the material properties of the blade as well as the mode and the blade design.

The various blades of an axial rotor stage do not have identical frequencies, an effect normally referred to as mistuning. In cases when the problem is forced vibration, mistuning is deleterious because all the energy can be concentrated into a few blades of the coincident frequency and catastrophe requires only that one blade breaks off. When the vibration problem is from flutter, then mistuning is desirable and may be as important as damping in keeping the amplitudes down. Groups of blades tend to flutter together with very large variations in amplitude from blade to blade. The most desirable configuration is to have alternate blade mistuning, see Crawley (1985). One effect of having part-span shrouds, often called snubbers, is to couple the blades so that the variation of frequency is greatly reduced.

If axial blades cause the disk to vibrate, more complicated coupled modes can be excited. When coupling occurs the natural frequencies are different from those without coupling. With centrifugal impellers coupling is common and the mode shape may be very complicated, particularly for shrouded impellers, and involve a coupling between blades. The descriptions of bending and torsion do not fit here at all well. It is normal to excite the statonary impeller to verify that the natural frequencies do not coincide with obvious orders of high amplitude excitation. In some cases material will be removed to alter the natural frequency to move it away from the frequency of a known strong source of excitation.

Forced vibration

The Campbell diagram, Fig. 10.1 is a convenient way to view the possible coincidence of blade natural frequency with excitation of the rotor blades by non-uniformities of the flow. Some of the engine orders are more important than others: the lowest engine orders tend to produce quite high excitation because the flow around the annulus is never truly uniform. There are several reasons for this. The inlet flow has non-uniform stagnation pressure because of ingested distortion or intake boundary layer effects or has static pressure distortion due to such effects as upstream bends. Inlet distortion tends to produce excitation at the low engine orders, the precise excitation being obtained from a Fourier analysis of the circumferential pattern, Danforth (1975). Struts upstream or downstream excite the blades at engine orders corresponding to their number and harmonics thereof. Upstream obstructions cause excitation by their potential flow effect and by their wakes; downstream bodies can only affect the rotor by the potential flow effect. The potential flow effect decreases approximately exponentially with distance in the flow direction (i.e. $e^{\pm nx/r}$ where n is the circumferential harmonic of the disturbance). This means that large single struts or changes to the duct shape upstream or downstream of the compressor can give an excitation which can be felt a long way away upstream. The effect of the wake decreases more slowly than the potential effect and downstream of an obstruction the wake is usually more important. The strength of excitation from adjacent blade rows falls off very rapidly with axial spacing, because the number of blades is high, and increasing the spacing is one way of reducing the forcing. Increasing the gap between moving and stationary components also has the effect of reducing the strength of the high harmonics more rapidly than the fundamental: if there are for example 57 stator vanes there will also be excitation of the rotor at the 114th, 171st etc. orders, but these higher harmonics will decrease much more rapidly with distance.

In the radial compressor the vanes in a vaned diffuser are a powerful source of excitation of the impeller and one of the principal reasons for mounting the vanes some way out from the impeller is to reduce the strength of the excitation of the impeller by the vanes, see Chapter 7. The excitation is from the static pressure field of the vanes and if the flow into these is subsonic there is a relatively smooth distribution, whereas if it is supersonic there will be shock waves propagating into the impeller giving a strong impulsive excitation, rich in harmonics. The presence of a volute tongue is also a source of excitation for the centrifugal impeller; the fundamental component of excitation is at first engine order but the most serious excitation may be obtained at the higher harmonics. The amplitudes of the higher orders decrease more rapidly than the fundamental as the radius at which the tongue is placed increases in relation to the impeller outlet radius.

The design problem is to avoid the resonances which occur when there is coincidence of the engine order excitations with natural frequencies of the blades. Although it is impossible to avoid all coincidences, some are more serious than others, such as the coincidence of the low order natural frequencies with the very low engine order excitations. Not only is the level of excitation likely to be high for the low engine orders but the low order modes are easier to excite to high amplitude because with low order modes large parts of the blade are moving in phase. As Fig. 10.1 shows it is not in fact possible to avoid some coincidence for even the low order modes over the whole speed range; an acceptable design might have the coincidence at a speed which is encountered only briefly during acceleration or at a speed so low that the level of excitation will be enormously reduced. Similarly if it is known that there will be a high level of excitation because of some feature of the design, for example six large struts holding the front bearing, the blade natural frequencies can be chosen so as not to coincide with the corresponding engine order in the normal operating range.

Excitation of rotor blades will occur at high engine orders corresponding to the number of stator blades in adjacent rows. The harmonics involved with the interactions with stators will normally be much higher than the 20th engine order shown in Fig. 10.1 and very high order modes may be excited. The natural frequencies of these modes, it will be recalled, are not normally predictable with great accuracy and it is not normally possible to design to avoid resonance. Moreover in a multistage compressor there are so many possible excitations that it becomes impossible to avoid all resonances over a range of operating speeds. The selection of stator numbers in relation to the natural frequencies of the rotors has the same difficulties, particularly for multistage axial compressors required to operate over a range of speeds. There remain many possible resonances which it is impossible to design away and whether or not these are a serious problem sometimes can still only be determined by testing.

The force and moment imparted to a blade as it experiences a non-uniform flow can be calculated in the case of inviscid two-dimensional flow, for example Mani (1970) or Smith (1973). To be useful, such calculations need there to be an accurate input of the flow disturbance and only rarely is this available. To know whether the excitation will be a problem in turn requires an estimate of the mechanical damping. For these reasons the normal assessment of whether a blade or impeller is satisfactory is done entirely on the basis of the possible coincidence of natural frequency and major excitation frequency. Part of the costly process of development is determining if those which have been allowed to remain are damaging.

Flutter

Flutter, which is more of a problem for axial compressors than radials, is much more worrying than forced vibration because its occurrence is generally so much less predictable. Flutter in axial compressors is often comparatively ill-defined, occurring at frequencies which are not multiples of engine order and at different places in the operating map of the compressor. As speed is increased towards conditions for flutter the vibration character alters gradually from forced random vibration to self-excited vibration with the amplitudes controlled by nonlinear effects. There is not even unanimity in the terminology of the different types of flutter, some referring to it by its mechanical characteristic (bending or torsion) and some by its supposed aerodynamic origin (stall or choke). If both are quoted the flutter is fairly well described. Flutter sometimes occurs on only a few blades in a row, or with widely different amplitudes on the individual blades, but it is always true that as the amplitude rises the flutter tends to be more coherent and to involve all the blades at a common frequency with a fixed phase angle between the motion of adjacent blades.

Flutter occurs as a rule at the natural frequencies of the blades although even this is sometimes complicated by the possibility of torsion and bending modes at similar frequencies coalescing to a flutter at a single frequency. It is common to use a reduced frequency in discussing the flutter given by $\omega c/V$ where ω is the natural frequency of the blade in radians per second, c is the chord and V the incident velocity. (In American literature it is common to use a reduced frequency based on the half-chord. In some work the reduced velocity, $V/\omega c$, being the inverse of the reduced frequency, is used.) The reduced frequency can be given a physical significance by thinking of the circulation around a particular blade. As the lift changes there has to be a circulation shed downstream and to a first approximation this will convect away at the exit velocity V_2. The distance this will travel in one period of the vibration is $L = 2\pi V_2/\omega$ so the reduced frequency can be written as $2\pi c/L$. A reduced frequency of unity implies that the shed vortices move away 2π chord lengths downstream in one period. As the reduced frequency rises the vortices move a shorter distance downstream so that their effect on the blade is more pronounced and the blade behaves in a manner less like a quasi-steady flow.

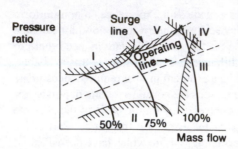

Fig. 10.4 The various flutter regions on the operating map of a transonic compressor. (After Mikolajczak *et al.*, 1975)

The complication of flutter in an axial compressor stage or fan with a high relative inlet Mach number is demonstrated by Fig. 10.4, derived from the diagram originally given by Mikolajczak *et al.* (1975). It shows most types of flutter which have been identified but it is not suggested that they all occur for the same stage. Nor is it as complete as it might be, for subdivisions have been proposed and there is no unanimity in the definitions. With the exception of region III and possibly region IV all the regions for flutter are off the design working line for the stage. For a multistage compressor off-design operation may cause some stages to operate in one of the other regions even while the compressor as a whole is on the normal working line; the ideal remedy in such cases is to improve the off-design matching to move the stage out of trouble.

Region I is often called subsonic stall flutter and is encountered at similar conditions to rotating stall so that it is sometimes difficult to decide whether the blades are vibrating because of flutter or forced vibration from the rotating stall. When flutter does occur in this region it seems that flow separation from the blades is an essential part of the mechanism. Flutter in this region was one of the first types to be encountered and is the type of behaviour referred to by Armstrong and Stevenson (1960). In their experience it was always the lowest order torsional and bending modes which fluttered. Because the flutter involves the blades operating with undesirable regions of separated flow it was possible to remove the flutter by small amounts of restaggering of the rotor tips, that is by twisting the tips so that locally the stagger was greater. From measurements of vibration in compressors they concluded that there were magnitudes of frequency parameter above which flutter was unlikely to be a problem if the compressor was operated near to the surge line. These are

$$\omega c/V > 1.6 \text{ for fundamental torsional mode}$$
and $\qquad \omega c/V > 0.3$ for fundamental bending mode.

Nowadays it is normal to use a figure like 0.33 or 0.35 as the limit for bending flutter. Working with these entirely empirical values it has been possible in many cases to avoid flutter for subsonic inlet flow near the surge line and they have remained a widely accepted correlation for this type of instability. When high Mach number stages with low hub—casing ratio and high aspect ratio blades were designed it was found that the frequency parameters were unacceptably low unless part-span shrouds were fitted to stiffen them up. It is these part-span shrouds which give rise to the instability shown here in region III.

Region II occurs at negative incidence at speeds below design speed and is sometimes referred to as choke flutter. At relatively high rotational speeds, choking is possible with the supersonic region in the passages terminated with a quite strong shock. The shock can cause boundary layer separation and set up an excitation of the blades. At still lower rotational speeds a different type of flutter, unrelated to choking, is possible. This has been fairly exhaustively studied theoretically and will be discussed in more detail below.

Region III sometimes called supersonic unstalled flutter or low backpressure supersonic flutter, is where the most worrying type of flutter occurs. This is because a realistic working line is likely to encounter this region at high speed: flutter of this type represents an effective operational limit with no discernible change in aerodynamic performance. Contours of stress run more or less parallel to the boundary sketched for the region so that increasing the rotational speed or decreasing the pressure rise both lead to an increase in stress, Halliwell (1975). Supersonic flutter in region III is usually associated with high aspect ratio blades with part-span shrouds, for it requires the mechanical coupling between the blades so that the whole rotor (blades, part-span shrouds and disc) vibrates as an integral assembly. Because of its importance and the success there has been in predicting this type of instability it is discussed in more detail below.

Region IV is best called high back-pressure supersonic flutter and is thought to be driven by an in-passage shock. It appears that high blade loading is a necessary part of this type of instability but without there necessarily being a massive amount of separation of the flow from the blades.

Region V is termed supersonic stall flutter and is described by Adamczyk *et al.* (1982). The blade deformation which has been observed is primarily bending at the highest speeds and torsional at lower speeds, where it may well overlap with the subsonic stall flutter of region I. Adamczyk *et al.* (1982) were able to use an actuator disc method to get reasonably good predictions of the flutter boundary. Halliwell (1975) reports that this type of flutter, which usually disappears before full design speed is reached, is most common amongst fan or front-stage compressor blades which are without part-span shrouds, suggesting that it is a vibration of the blades rather than of the entire rotor assembly. Halliwell does not report a flutter of the type described here as region IV.

In all types of flutter it is necessary that there be an input of energy from the gas stream to the blade large enough to overcome the mechanical damping, including that due to the material properties. The magnitude of the mechanical damping is small and generally unknown, so that when calculations are performed for the stability of possible designs it is usual to make the conservative assumption that the mechanical damping is zero. If the net work input is positive the aerodynamic damping is taken to be negative and the blade is assumed unstable in that mode, whilst negative net work is assumed to give positive damping and a blade free from flutter for that particular mode.

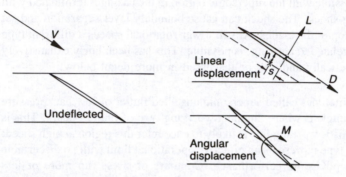

Fig. 10.5 The notation for coupled flutter

In assessing the work input it is usual to consider the blades in the two-dimensional blade-to-blade surface and then to integrate along the whole span. As in Fig. 10.5 suppose that L is the lift force normal to the blade chord with h the displacement of the torsional axis in the same direction; D the drag force with s the displacement parallel to the chord; M the pitching moment with α the angular movement of the blade. All of L, D and M are per unit spanwise height. Then the work input from just the lift force at a particular spanwise position is the integral of L with respect to dh over a complete cycle, taking the component of force in phase with the vibration velocity,

i.e. work $= \oint L . \mathrm{d}h$

The total work input for all the components, integrating from hub to casing, is then

$$W = \int_{r_{\text{hub}}}^{r_{\text{casing}}} [\oint L \cdot \mathrm{d}h + \oint D \cdot \mathrm{d}s + \oint M \cdot \mathrm{d}\alpha] \mathrm{d}r \qquad (10.1)$$

The difficulty of implementing equation 10.1 is that the lift, drag and moment each depends on h, s and α but the form of the dependence is not in general known. For most of the flutter analyses the contribution due to drag is neglected. For blades with unseparated flow the variation in lift and moment can often be calculated, but for stalled blades there is still no satisfactory method for calculating or accurately estimating these and therefore no satisfactory method to find the work input for the so-called subsonic stall flutter of region I in

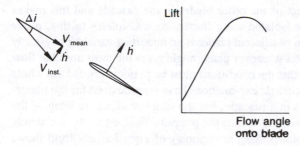

Fig. 10.6 Schematic diagram for isolated aerofoil showing change in instantaneous incidence as a result of heaving motion with velocity $\dot{h}$

Fig. 10.4, see Sisto (1987b). A very simple quasi-steady argument for excitation of a stalled blade can be given for bending motion, sometimes called heaving motion. Suppose the blade in Fig. 10.6 has an instantaneous vibration velocity $\dot{h}$ normal to the chord in the same direction as the lift. As the sketch shows this implies that there is a negative increment in flow incidence. If the blade is being operated to the right of the peak lift in the diagram in Fig. 10.6, in the region generally associated with blade stall when there will be large amounts of separated flow on the blades, a negative increment in incidence will give an increase in the lift. With the lift force and the velocity in phase there will be a net energy input from the gas to the blade, i.e. an excitation. (For torsional motion a more refined argument is necessary because a lag between incidence and lift is necessary to get a work input in torsion and quasi-steady theory gives no help.) The same argument shows that unseparated blades, operating to the left of the peak in Fig. 10.6, are not inherently unstable to the bending type of motion. This simple argument is intended only to show how radically the direction of the work transfer can change for stalled blades and help explain why flutter is so common near to the surge line.

Most of the analytical work concentrated on unstalled subsonic flutter, which may be relevant to the flutter in region II of Fig. 10.4 and the methods have been reviewed in some detail in AGARDograph (1987). The early methods were for incompressible flow over cascades of flat-plate aerofoils using singularity methods. More recently field methods have been developed to calculate the steady flow and unsteady perturbations are then considered to be superimposed on this. Using such field methods the flow can be subsonic or supersonic (provided the Mach number is low enough for the shock to be weak) and there is no restriction to small turning of the steady flow. Some methods include strong shocks by patching these in. All the methods are inviscid, which is a serious gap since many types of flutter, as already mentioned, occur in the region close to the surge line where significant boundary layer behaviour is very important. In particular, the nature of transonic flow can be radically altered by the blockage of the boundary layer and this too could have a major effect on the prediction of flutter.

In considering flutter of blades in a compressor it is important to take account of the aerodynamic effect of the other blades in the cascade and this makes a major difference from isolated wing flutter. As a corollary to this, phase angle between the motion of adjacent blades is an important variable; physically the system is able to select whatever phase angle gives the most unstable flow with only the constraint that the conditions must be periodic around the whole annulus. A number of heuristic explanations have been devised for the importance of the other blades in a cascade, but the simplest idea is to imagine the vibration of one blade moving it into the pressure field set up by the steady loading of the adjacent blades. Simple reasoning of a quasi-steady form shows that with some interblade phase angles the system is unstable and with others stable. Some important findings from the classical theory of flutter, obtained using a method of distributed vortices, have been given by Whitehead (1966, 1987). For cascades of blades in subsonic flow able to vibrate in bending (i.e. the only motion is h normal to the chord in Fig. 10.5) the flow is always stable if there is no steady lift on the blades. If, however, there is steady lift, then there may be flutter with bending motion only. Flutter can occur below critical values of reduced frequency $\omega c/V$, the value of this increasing as the mean flow deflection increases, the blade row solidity increases or the stagger decreases. For example in incompressible flow, for a blade row with solidity of unity, stagger 30° and flow deflection 30°, the critical reduced frequency is calculated to be about 0.1. For cascades free to exercise torsional motion there *can* be flutter with no steady loading on the blades. The reduced frequency below which flutter is possible tends to be rather higher than in the bending case and to increase as the stagger angle increases. With this torsional flutter there is a large dependence on the position of the torsional axis; having the torsional axis around 20 per cent of chord is most favourable for the avoidance of flutter, around 50−60 per cent least favourable. The effect of compressibility was highly favourable and torsional flutter of unstalled blades in subsonic flow was most unlikely for flow Mach numbers in excess of 0.5. The type of flutter that was being considered was therefore that occurring in region II of Fig. 10.4, towards the low-speed end of the region but not so far down the constant corrected speed line that choking occurs.

Supersonic unstalled flutter

Although this type of flutter originally looked most perplexing, partly because it did not fit in with the correlation parameters or conventional ideas of the time, it is now probably the best understood and most carefully documented. Since this type of flutter could occur at the design working point it was particularly important that it should be possible to predict it and design to avoid it.

The mechanism was first unravelled by Carta (1967) who examined the consequence of coupling between the blades, the disc and the part-span shroud. The part-span shroud is actually the dominant coupling element. In some designs there may be more than one ring of part-span shrouds. The motion of the blade and the forces involved are shown in Fig. 10.5 with a plunging or heaving motion perpendicular to the chordline, a pitching motion about the torsional axis and a motion parallel to the chordline in the same direction as the drag force. What is crucial to this type of flutter is that the plunging motion and the pitching motion are coupled, and it is easy to show that the two are out of phase by 90° at the part-span shroud radius (but not at the tip). If the vibration mode is rotating forward, that is in the same direction as the rotor rotates, the plunging motion lags the pitching motion by 90°, whilst if it is rotating backward it leads by the same amount. It is the forward rotating mode which appears to be unstable, Halliwell (1975).

The assembly vibrates with nodes, that is lines of zero plunging displacement, which are both circumferential and diametral. (The pitching motion does

 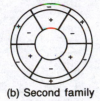

(a) First family   (b) Second family

Fig. 10.7 The families of coupled modes

not exhibit a circumferential node but increases in magnitude monotonically from root to tip.) In the simplest case, sometimes referred to as the first family of modes, there is no circumferential displacement node; this is sketched diagrammatically in Fig. 10.7a for a mode with three diametrical nodes. If there is a single node in the circumferential direction it is described as a second family mode and this appears to be the dominant set for fans with a single part-span shroud, the node occurring at about 80–90 per cent of span. A second family mode is sketched in Fig. 10.7b. Typically the second family modes have natural frequencies which are two or three times greater than the first family, with the frequency also rising as the number of diametral modes increases. The phase angle between the motion of the adjacent blades also rises as the number of diametral modes increases. A hologram interference picture revealing the amplitude of plunging motion for a non-rotating fan assembly, taken from Parker and Jones (1987), is shown in Fig. 10.8. This is for a fan vibrating in the second-family mode with three diametral nodes. (Great care is needed with such stationary tests to simulate the effect of the centrifugal acceleration in twisting the blades and tightening the root fixing and shrouds.) The mode shapes, giving the variation in the plunging and pitching amplitude with radius and the phase variation between blades can be predicted with reasonable accuracy. The prediction of the mode natural frequency is also good. Figure 10.9, taken from Halliwell (1975), shows a Campbell diagram on which are drawn the frequencies of the coupled modes for the first and second families

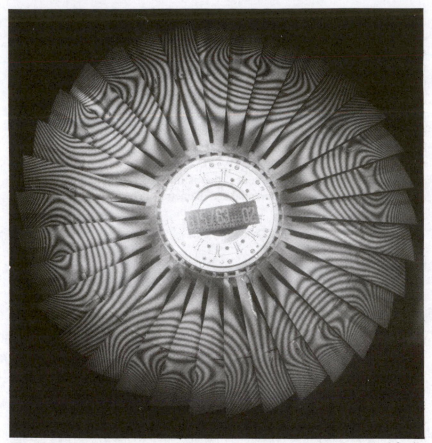

Fig. 10.8 A stationary fan rotor with part-span shrouds excited in the second family mode with three diametral nodes (i.e. second family 3D Mode). Obtained by double-pulse holography. (From Parker and Jones, 1987)

with the numbers of diametral modes shown as 1D, 2D, etc. Flutter was found in this case in the 3rd and 4th diametral modes of the second family and the agreement with the predicted frequencies is clearly excellent.

In reality the coupled vibration is more complicated than first appears because of the slight mistuning of the blades: Ford and Foord (1980) have shown that it really consists of two spatially orthogonal modes of slightly different natural frequency which combine to form a single pattern of rotating disturbance. Because the aerodynamic behaviour can be understood without this subtlety it will not be pursued further here.

To evaluate the unsteady aerodynamic work input to the blade vibration it is necessary to integrate equation 10.1 along the span. For this it is necessary to know the blade forces and moments, both the amplitude and phase, in terms of the pitching and plunging amplitudes at each spanwise position. Mikola-jczak *et al.* used the method of Verdon (1973) to get the force and moment,

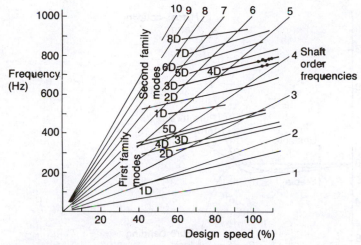

Fig. 10.9 Campbell diagram for a transonic fan with part-span shrouds. Flutter shown in 3rd and 4th diametral modes of second family. (From Halliwell, 1975)

whilst Halliwell used one by Nagashima and Whitehead (1974): both are linear methods for supersonic flow and give identical results. Some years earlier Carta (1967) had obtained a good prediction of the boundary between stable and unstable for the coupled mode of high-speed rotors. To get the force and moment he used a method for isolated aerofoils in incompressible flow — evidently the method of calculating the force and moment is not the most sensitive part of the prediction procedure! The prediction of mode shape is needed to obtain the relative magnitude and phase of the plunging and pitching of the blade; the drag force D and the corresponding motion s are often omitted as small quantities. The non-dimensional lift force L and pitching moment M may be written symbolically in terms of the pitching motion $\alpha = \bar{\alpha}e^{i\omega t}$ and plunging motion $h = \bar{h}e^{i(\omega t + \phi)}$ as

$$L = L_\alpha \alpha + L_h h$$

and
$$M = M_\alpha \alpha + M_h h.$$

(Here the denominator making L_h non-dimensional is $\rho V_1 \omega c \bar{h}$, which can be rewritten in terms of the reduced frequency λ as $\rho V_1^2 \lambda \bar{h}$. The corresponding denominator in the case of M_α is $\rho V_1^2 \lambda c^2 \bar{\alpha}$.) Each of the coefficients has a real and imaginary part, for example

$$L_\alpha = L_{\alpha R} + iL_{\alpha I}.$$

The work equation can then be written in terms of amplitudes $\bar{\alpha}$ and $\bar{h}$ of the pitching and plunging motion, taken to be at 90° to one another, in the form

$$W = \pi\rho\omega V_1 c \int_{r_{\text{hub}}}^{r_{\text{casing}}} [L_{hI}\bar{h}^2 + (M_{hR} + L_{\alpha R})c\bar{\alpha}\bar{h} + M_{\alpha I}(c\bar{\alpha})^2]dr \qquad (10.2)$$

The variation in the size of the three terms in the equation are compared in

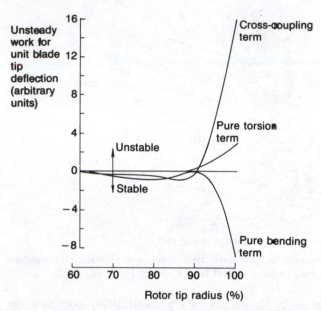

Fig. 10.10 The unsteady work predicted along the span of a transonic rotor with part-span shrouds. (From Halliwell, 1975)

the case of a fan running at design speed in Fig. 10.10, taken from Halliwell (1975). The term which gives the large energy input is the middle, cross-coupling term. The inter-blade phase angle, fixed by the number of diametral nodes, also has a large effect on the work input; Mikolajczak *et al.* showed for a particular fan rotor with an inlet relative Mach number of 1.42 that there is positive work into the vibration only in the range of phase angles from about 18° to 100°; outside this range the work is negative and vibration therefore suppressed.

To decide which are the most likely modes to vibrate it is common to take the expression given by Carta (1967) for the aerodynamic damping, the logarithmic decrement, which assumes that the rotor assembly is a simple linear system,

$$\delta_{\text{aero}} = \frac{-nW}{4KE} \tag{10.3}$$

Here n is the number of rotor blades, W is the work input per blade and KE is the average vibrational kinetic energy of the rotor assembly. If δ_{aero} is positive, i.e. aerodynamic damping is positive, the blades do unsteady work on the gas. There is in addition mechanical damping, δ_{mech}, which, though small and difficult to measure, is important; Mikolajczak *et al.* estimate it to be about 0.03 whereas Halliwell estimates between 0.002 and 0.005. Because of its smallness and uncertainty, designers usually ignore the mechanical

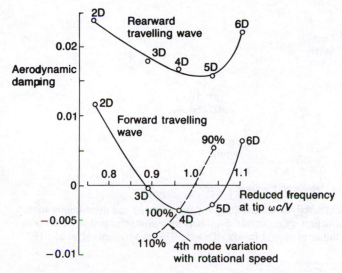

Fig. 10.11 The variation of aerodynammic damping for second family modes in a transonic rotor having part-span shrouds with reduced frequency. The number of diametral nodes is shown in number preceding D. (From Halliwell, 1975)

damping as a conservative stance, but rotors operating with small negative aerodynamic damping may be rendered stable by virtue of it. The mechanical damping is unaffected by gas density, whereas the possible work input to the vibration is proportional to density. This means that in the case of aircraft engines the reduction in density with altitude makes testing at sea level essentially conservative; an engine which is flutter free at maximum corrected speed at sea-level conditions will not flutter at the same corrected speed at high altitude. (This is not necessarily true for high Mach number flight, for which the inlet stagnation pressure can exceed that at sea level.)

Figure 10.11 from Halliwell shows the predicted aerodynamic damping versus tip reduced frequency for a fan which was found under test to flutter at between 100 and 110 per cent of design speed in the 4D forward travelling mode. The predictions are good. Figures 10.12 and 10.13 taken from Mikolajczak *et al.* show predicted aerodynamic damping for a rotor with a high aspect ratio of 4 and a high tip-speed of 550 m/s. The rotor was found to flutter at 89 per cent of design speed in the second family, four-diameter (4D) mode. The prediction clearly indicates the second family as the most unstable although the 5D or 6D modes are shown as less stable than 4D: it is likely that prediction has increased in accuracy since then. Figure 10.13 shows the predicted damping for the second family modes at three different rotational speeds; the fan was stable at both the lower speeds shown, despite the small negative aerodynamic damping predicted at 77 per cent speed, which is an indication of the importance of the mechanical damping. In the same paper Mikolajczak *et al.* describe experience with another rotor of the same

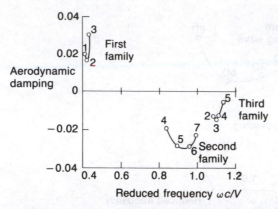

Fig. 10.12 Predicted aerodynamic damping for first, second and third family modes of transonic rotor with part-span shrouds. Numbers by points refer to diametral nodes. Rotor observed to flutter in second family 4Dmode. (From Mikolajczak *et al.*, 1975)

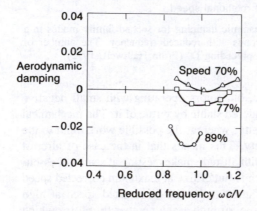

Fig. 10.13 The variation in predicted damping of second family modes for a transonic rotor with part-span shrouds as a function of reduced frequency. The rotor is that shown in Fig. 10.12 and flutter occurred in the second family. The speeds are percentage of design tip speed, 549 m/s. (From Mikolajczak *et al.*, 1975)

design tip speed but an aspect ratio of only 2.87. No flutter was observed experimentally up to the design speed whilst the predictions showed substantial amounts of positive aerodynamic damping in all modes. Good predictions of the onset of coupled flutter were found for a range of different machines. It can be concluded that the prediction of coupled flutter is reasonably satisfactory.

10.3 Noise

During the 1960s and 1970s there was frantic activity in the area of turbomachinery noise research, mainly directed towards reducing aircraft engine noise. Although the fundamental understanding of turbomachinery noise generation is still generally weak the goal of reducing aircraft noise has been largely achieved and research in the field has settled to a more reasonable pace. The aeronautical aspect of the majority of the research and thinking on compressor noise will be obvious in this chapter; specifically non-aeronautical noise is considered in a short final section.

Consideration of turbomachinery noise rests on an understanding of acoustics, a large and distinguished field of physical science to which a good introduction is provided by Dowling and Ffowcs Williams (1983). In this chapter it is convenient to derive some elements of the mathematical formulation of acoustics in a way which leads itself to the needs of turbomachinery, in particular to examine propagation of sound along simple ducts. Before doing this, however, it is fruitful to consider how noise is measured or assessed, for only then can one consider more specifically the noise of compressors.

Scales and rating of noise

Because acoustic signals vary by such large amounts it is the almost invariable practice to use logarithmic scales, expressing the level in terms of decibels (dB). Acoustic signal levels may either refer to the local level of pressure fluctuation or to the acoustic power generated. In the former it is known as the sound pressure level (SPL) defined by

$$\text{SPL} = 20 \, \log_{10}(p/p_{\text{ref}}) \, \text{dB}$$

where p is the RMS pressure fluctuation and p_{ref} is the reference RMS pressure fluctuation equal to $2 \cdot 10^5 \, \text{N/m}^2$.

The acoustic power, or power watt level (PWL) is defined by

$$\text{PWL} = 20 \, \log_{10}(W/W_{\text{ref}}) \, \text{dB}$$

where W is the acoustic power in watt and W_{ref} is the reference acoustic power equal to 10^{-12} watt (occasionally 10^{-13} watt is used as the reference).

The signals of pressure or power will have a frequency spectrum, and the level referred to may either be the total signal (i.e. the spectrum integrated over frequency) or else components of the spectrum at particular frequencies. If the overall signal is used to obtain the RMS pressure or power the levels are generally known as the overall sound pressure level (OASPL) or overall power watt level (OAPWL) respectively. Very frequently the signal is broken down into its spectral components and in this case it is necessary to specify the bandwidth over which the signal is summed. Mathematically the process is a Fourier transform. Fourier transforms produce power-spectral density, for which the bandwidth is 1 Hz, but electrical analogue filters or computer algorithms generally have a different bandwidth. The bandwidth may be constant, regardless of the frequency being examined, or it may vary with frequency, usually so that the bandwidth is proportional to frequency. The most common bandwidths of this type are the octave and third-octave bandwidths; an octave corresponds to a doubling of frequency, a third-octave

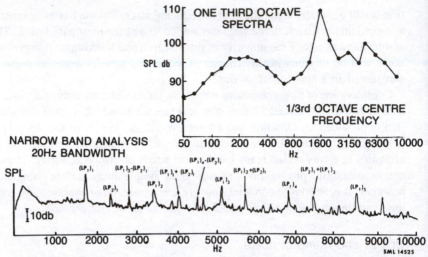

Fig. 10.14 Compressor noise spectra on 1/3rd octave and constant bandwidth basis. Rolls-Royce Conway engine at part power, noise measured at 30° to intake axis

bandwidth is therefore one where the upper frequency is equal to $2^{1/3}$ times the lower frequency. Depending on the choice of bandwidth very large differences in the displayed spectrum can be obtained, which is illustrated by Fig. 10.14 when the same signal was analysed with a constant bandwidth and with a third-octave filter. Evidently quite different information is emphasized in the two traces. The narrow, constant bandwidth filter is useful for study and diagnosis, whilst the third-octave filter is more similar in its discernment to the human ear.

The spectra in Fig. 10.14 show two characteristics, broad band noise and tones. A tone has the character of a whistle at a discrete frequency or over a very narrow range of frequency. Broad band noise is characteristic of the noise of a jet where the energy is spread fairly uniformly over a wide range of frequencies. By analogy with light it is common to refer to 'white' noise when the broad band noise is of virtually constant level over an extensive range of frequency.

So far the discussion has referred to objective measures, but noise is inherently subjective and to assess noise one requires an assessment of the annoyance produced on a person. The different ratings used throughout the world have been collected and described by Pearson and Bennett (1974). The simplest ratings weight the overall sound pressure in a manner related to the sensitivity of the human ear. The most common of these, the A-weighted sound pressure level, finds wide use in non-aeronautical as well as aeronautical applications throughout the world and the levels are seen expressed as dBa, dB(A) or, more recently, LA. Though less common there are also, B, C and D scales intended for different levels and conditions.

A more complicated noise rating, which is described in detail by Pearson and Bennett, is the perceived noise level (PNL) for which the units are PNdB. The numerical value of PNL was intended to represent the sound pressure level of the octave band of 'white noise' centred at 1 kHz which would be judged equally noisy as the sound to be rated. A sound is judged equally noisy to another when one would just as soon have or not have it in one's home during the night or day. The perceived noise level was found to be less reliable as an indicator of annoyance when the spectrum was no longer broad-band but contained a strong tone. A tone-corrected perceived noise level (PNLT) was therefore produced. A third-octave spectrum is first examined to determine if tones are present, evident by the protrusion of one band above its neighbours by a prescribed ratio. If a tone is present a correction is added to the PNL calculated in the normal way, the amount of the correction depending on frequency and on the extent of the protrusion.

Elementary acoustics

The wave equation

Acoustics is the study of pressure waves through fluids and the assumption most generally applied is that the amplitude of the waves is very much less than the atmospheric pressure. Only in cases where shock waves are present is this restriction normally invalidated. The importance of this restriction cannot be overemphasized, in particular it leads to a simple interrelation of density, pressure and velocity perturbations. It will be recalled that the speed of sound c is given by $c^2 = (\partial p/\partial \rho)_s$ and it therefore follows that the small perturbations p' and ρ' are related by $p' = c^2 \rho'$. For conciseness perturbed variables will be denoted by p' and ρ' only when necessary: often this is unnecessary, as in $\partial p/\partial \rho$.

It is helpful to begin by considering the case of sound in a stationary medium. The continuity equation is

$$\frac{\partial \rho}{\partial t} + \nabla \cdot (\rho u) = 0 \qquad (10.4)$$

In this case of a stationary medium the velocities u are produced only by the sound field and are therefore small so that the term $u \cdot \nabla \rho$ is of second order and may be neglected.

The momentum equation may be written as

$$\frac{\partial u}{\partial t} + u \cdot \nabla u = -\frac{\nabla p}{\rho} \qquad (10.5)$$

and in a *stationary fluid* $u \cdot \nabla u$ is second order and negligible. The two relevant equations for small amplitude disturbances in a stationary fluid can therefore be combined as

$$\frac{\partial^2 p}{\partial t^2} - c^2 \nabla^2 p = 0 \tag{10.6}$$

which is the wave equation in three-dimensions. In a cartesian system

$$\nabla^2 = \frac{\partial^2}{\partial x^2} + \frac{\partial^2}{\partial y^2} + \frac{\partial^2}{\partial z^2} \tag{10.7}$$

but of generally more use in turbomachines are the spherical form or, in particular, the cylindrical form,

$$\nabla^2 = \frac{\partial^2}{\partial r^2} + \frac{1}{r}\frac{\partial}{\partial r} + \frac{1}{r^2}\frac{\partial^2}{\partial \theta^2} + \frac{\partial^2}{\partial x^2} \tag{10.8}$$

In the special case of a one-dimensional wave, which also approximates the spherical wave at very large radii from the source, the wave equation simplifies to

$$\frac{\partial^2 p}{\partial t^2} - c^2 \frac{\partial^2 p}{\partial x^2} = 0 \tag{10.9}$$

for which the solutions are of the form $p = p(x+ct)$ for waves travelling in the negative x-directions at speed c and $p = p(x-ct)$ in the positive x-direction. The velocity perturbation will also be of the form $u = u(x \pm ct)$ and substituting into the linearized momentum equation gives, for one-dimensional cases,

$$\frac{du}{c} = \frac{1}{c^2}\frac{dp}{\rho} = \frac{dp}{\gamma p} \tag{10.10}$$

which relates the velocity and pressure perturbation amplitudes. This inter-relation is also valid for waves in the more complex geometries when du is taken to be the velocity in a direction normal to the wave points.

Acoustic sources

The homogeneous wave equation 10.6 that was obtained by using the mass continuity and momentum conservation equations describes propagation in a stationary medium but does not give any information about sources of sound. Suppose that there is a source of material Q (units kg m^{-3} s^{-1}) in the region; the continuity equation is therefore

$$\frac{\partial \rho}{\partial t} + \rho \nabla u = Q \tag{10.11}$$

Likewise suppose that there is a force vector F per unit volume (units N m^{-3}) in the region; the momentum equation is therefore

$$\rho \frac{\partial u}{\partial t} - \nabla p = F \tag{10.12}$$

The corresponding wave equation is

$$\frac{1}{c^2}\frac{\partial^2 p}{\partial t^2} - \nabla^2 p = \frac{\partial Q}{\partial t} - \nabla F \tag{10.13}$$

The terms on the right-hand side of equation 10.13 represents the acoustic sources. The term $\partial Q/\partial t$ is usually referred to as a simple or monopole source and represents the local rate of change of mass addition; the mass addition can be provided by puffs of air, a pulsating sphere or by a piston, such as a loudspeaker diaphragm, where during each cycle as much is removed as is added. The simple source does not have any directionality and in an unbounded spherical geometry gives rise to spherical wave-fronts; to conserve the acoustic power it is necessary that the intensity of the pressure waves decreases in inverse proportion to the surface area of the wave fronts.

In general the solution of an equation as simple as equation 10.13 is non-trivial even with only the simple source $\partial Q/\partial t$ present, the amount of complexity depending on the boundary conditions. In a compressor the boundary conditions are formidably complicated and not much progress is likely from integrating an equation as simple as 10.13. It may also be added that source terms such as $\partial Q/\partial t$ are unlikely in compressors since they imply a fluctuating mass flow rate. Fluctuating forces, F in equation 10.13, are very common in compressors because of the interactions between blade rows and give what are often called dipole sources. Some workers have evaluated the pressure amplitude from a fluctuating point force by treating it as an isolated element in an unbounded environment; with this very special idealized boundary condition the integration of equation 10.13 is trivial. The simple result for pressure amplitude is

$$p = \frac{\cos\alpha}{4\pi r}\frac{\omega}{c}f \tag{10.14}$$

where f is the amplitude of the oscillating force with frequency ω at large distance r from the point of observation. The angle α is that between the direction of the force and the line between the point of observation and the force. Although the dipole representation of a fluctuating force has been used for estimating noise from turbomachine blade rows the true situation is so complex that it is possible to regard this merely as a method able to produce the appropriate scaling and a *very* approximate estimate of magnitude. It does, however, lead to the reasonably accurate prediction that the acoustic power generated by the predominantly dipole sources rises in proportion to the sixth power of the blade speed.

Transmission along ducts

The analysis so far has been concerned with the simplest possible cases, sources in an unbounded stationary medium, which are very different from anything found in a turbomachine. However, the analysis which has had an enormous influence on turbomachinery noise quite remarkably ignores the sources altogether but considers only propagation effects. Mathematically it is concerned with the homogeneous wave equation, equation 10.6, or with the appropriate equation in the case of sound in ducts with mean flow. Non-uniform flow introduces great difficulties, both mathematical and conceptual, but with a uniform flow it is possible to analyse the sound field as if the medium were stationary and introduce the flow later.

The most common geometry for turbomachinery inlets or outlets is obviously a duct of circular cross-section and the most natural case to analyse is therefore the propagation along a circular cylinder or circular annulus. This gives a solution in terms of Bessel functions which, being less familiar than sines and cosines, tend to obscure the principles. The strategy here will be to use the approach adopted by pioneers in this field, Tyler and Sofrin (1962), and analyse a two-dimensional cartesian system and merely quote the results for a cylindrical geometry.

1. *Two-dimensional duct*

Consider a duct in the $x-y$ plane where the height normal to x and y is small. Suppose x is the distance along the duct and y the distance across it, the total width in the y-direction being L. For convenience the pressure is assumed to vary harmonically with frequency at ω rad/s, and is denoted by $p(t) = pe^{i\omega t}$ where the real component is implied. The method of solutions is the standard separation of variables so that the formal dependence is expressed by

$$p(x,y,t) = X(x)Y(y)T(t)$$

where X, Y, and T are functions only of x, y and t respectively. By the original choice of harmonic time dependence $T(t) = e^{i\omega t}$.

The wave equation for a stationary homogeneous medium in two-dimensional coordinates is

$$\frac{\partial^2 p}{\partial t^2} - c^2 \left(\frac{\partial^2 p}{\partial x^2} + \frac{\partial^2 p}{\partial y^2} \right) = 0 \tag{10.15}$$

and $p = XYT$ may be substituted in to give,

$$\frac{1}{T}\frac{d^2 T}{dt^2} - c^2 \left(\frac{1}{X}\frac{d^2 X}{dx^2} + \frac{1}{Y}\frac{d^2 Y}{dy^2} \right) = 0 \tag{10.16}$$

Although $1/T d^2 T/dt^2 = -\omega^2$ the other expressions are unknown. With y denoting the coordinate across the width of the duct the function Y must 'fit' the duct, i.e. $p(y=0) = \pm p(y=L)$, so that Y can be expressed as a Fourier series for a range of wavelengths

$$\lambda_1 = L, \lambda_2 = L/2. \ldots \lambda_n = L/n.$$

Sometimes it is more convenient to use the wavenumber

$$k_{yn} = 2\pi/\lambda_n = 2\pi n/L.$$

The form for Y is therefore

$$Y = \sum_n a_n e^{ik_{yn}y} \tag{10.17}$$

Putting expressions for T and Y into equation 10.16 and rearranging yields

$$\frac{1}{X_n} \frac{d^2 X_n}{dx^2} = k_{yn}^2 - \frac{\omega^2}{c^2} \tag{10.18}$$

Now the solution of this ordinary (not partial) differential equation, is

$$X_n = A_n \exp\left\{ \pm \left(k_{yn}^2 - \frac{\omega^2}{c^2} \right)^{1/2} x \right\} \tag{10.19}$$

where A_n is a constant.

A crucial difference arises depending on the sign of the expression in brackets. If

$$\omega > k_{yn}c \tag{10.20}$$

the exponent in equation 10.19 is imaginary and the function X varies harmonically in the x-direction without attenuation.

If, however

$$\omega < k_{yn}c \tag{10.21}$$

the form of X is a decaying real exponential with no variation in phase. The condition at which the change takes places, $\omega = k_{yn}c = 2\pi nc/L$, is usually referred to as the 'cut-off' condition; waves which propage without attenuation are above cut-off, whereas disturbances which decay exponentially are referred to as cut-off.

The condition $\omega = k_{yn}c$ can be given a physical significance by multiplying both sides by the period of one cycle τ; thus $\omega\tau = 2\pi c\tau/L$. During one period $\omega\tau$ changes by 2π so that

$$2\pi = 2\pi nc\tau$$

or $\qquad c\tau = L/n$

In other words the cut-off frequency is such that during one period the perturbation can travel in the transverse, y, direction a distance of exactly one wavelength, L/n. This is a sort of resonance condition and is also important in the consideration of blade vibration. If there is a source of excitation at a similar frequency to that for resonance, such as vortex shedding, intense amplitudes can be generated, Parker (1967).

A second physical interpretation can be given to the cut-off condition, which will be very convenient for later considerations. The speed of propagation of a wave is equal to frequency times wavelength, and in the *y*-direction this may be written as

$$V_s = \omega/2\pi \times L/n = \omega L/2\pi n.$$

But since at cut-off $\omega = 2\pi nc/L$ the condition is therefore

$$V_s = c,$$

i.e. the phase velocity in the *y*-direction is equal to the velocity of sound. Phase velocities greater than this lead to unattenuated propagation, below this to exponential decay. Defining the Mach number of a disturbance in the *y*-direction by $M_s = V_s/c$, the waves from the disturbance propagate axially without attenuation if $M_s > 1$. Because the *x*-dependence below cut-off is the decaying exponent of a real number there is no phase variation in the *x*-direction, but for waves above cut-off the phase does vary in the *x*- as well as the *y*-direction and the wave fronts form a set of oblique lines. These effects are illustrated in Fig. 10.15 for cases below and above cut-off. As the phase speed in the *y*-direction becomes very large the wave fronts swing round to be nearly parallel to the *y*-axis and the waves tend to travel in the *x*-direction. In the case of no mean flow the wave fronts are parallel to the *x*-axis at the point of cut-off and move in a direction parallel to the *y*-axis.

2. *Cylindrical duct*

The above two-dimensional treatment avoided all possible complications including variations in the direction mutually perpendicular to *x* and *y*. In the concentric cylindrical duct of low hub−casing ratio the variation in the radial

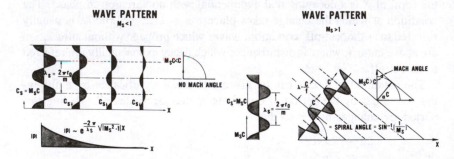

Fig. 10.15 Wave patterns in a two-dimensional duct below cut-off ($M_s < 1$) and above cut-off ($M_s > 1$). (From Sofrin, 1973)

and circumferential directions must be considered; from these it is possible to arrive at the *x*-dependence.

The appropriate wave equation for the cylindrical geometry is obtained by putting the form for ∇^2 from equation 10.8 into equation 10.6 and the same approach to that used above is adopted, namely putting

$$p = R(r)\,\Theta(\theta)\,X(x)\,T(t) \qquad (10.22)$$

As before $T(t) = e^{i\omega t}$ and $\Theta(\theta) = \sum_m a_m \cos m\theta$, although it is convenient to take a single value of m, i.e. a single circumferential mode, analogous to the single value of ω. This time a form for $R(r)$ must be chosen and the appropriate characteristic or eigen solutions in the radial sense are Bessel functions so that

$$R(r) = J_{m\mu}(k_{m\mu}r) + Q_{m\mu} Y_{m\mu}(k_{m\mu}r) \qquad (10.23)$$

$J_{m\mu}$ and $Y_{m\mu}$ are Bessel functions of the first and second kind of circumferential order m and radial order μ. $k_{m\mu}$ and $Q_{m\mu}$ are eigenvalues depending on m, μ and the inner and outer radii of the duct; for a cylindrical duct with no inner radius the solution reduces to $R(r) = J_\mu(k_\mu r)$.

The eigenvalues are found by solving the simultaneous equations for the boundary conditions at the inner and outer walls, $dR/dr = 0$. Figure 10.16 shows the radial distribution in a duct with hub—casing radius ratio of 0.5 for three values of m and μ. Whilst m denotes the number of cycles around the circumference, the radial order, μ, denotes the number of zero crossings in the radial direction. (A combination of Bessel functions may be used to represent an arbitrary radial variation just as in the more familiar two-dimensional geometry sines and cosines may be used to fit an arbitrary distribution by Fourier analysis.)

The speed with which the pattern sweeps the outer wall is Ωr_0 and expressed as the outer wall Mach number of the pressure field this is $M_s = \Omega r_0 / c$. By methods analogous to those used in two dimensions the x-dependence of pressure can be shown to be of the form $\exp\{k_x x\}$ where

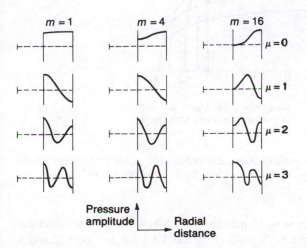

Fig. 10.16 Radial variation in pressure amplitude in a cylindrical duct with hub—casing ratio 0.5. Circumferential mode order denoted by m, radial mode order by μ. (From Tyler and Sofrin, 1962)

$$k_x = \pm \frac{i}{r_0} (m^2 M_s^2 - k_{m\mu}^2 r_0^2)^{1/2} \tag{10.24}$$

The wall Mach number at which cut-off occurs may be conveniently expressed as

$$M_s = k_{m\mu} r_0 / m$$

where $k_{m\mu}$ is also a function of hub–casing ratio. Figure 10.17 shows the predicted axial decay rates in the case of $\mu = 0$ for a hub–casing ratio of 0.5. What is particularly noteworthy is the enormous rate of decay for large values of m. If m is related to blade number then values of $m = 16$ or more are usual and the pressure field of subsonic rotor blades decays by more than 100 dB (i.e. a pressure amplitude ratio of 10^{-5}) for each radius length in the axial direction.

Figure 10.17 also shows how the cut-off Mach number varies with the circumferential order m, tending to unity as m becomes large. Likewise as the hub–casing ratio becomes large the cut-off Mach number tends to unity. Higher radial modes, that is $\mu > 1$, lead to even more rapid decay rate and higher cut-

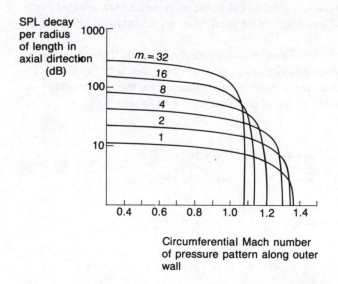

Circumferential Mach number
of pressure pattern along outer
wall

Fig. 10.17 Predicted decay rates for zero radial order mode ($\mu = 0$) for various circumferential orders. Hub–casing ratio 0.5. (From Tyler and Sofrin, 1962)

off Mach number: for an $m = 16$ pattern in a 0.5 hub–casing ratio duct the cut-off Mach numbers are 1.13, 1.46, 1.71 and 1.95 for $\mu = 0, 1, 2$ and 3 respectively. Moore (1972) compared the decay ahead of a subsonic fan with the prediction of equation 10.24 and demonstrated that the agreement is indeed very good.

3. *Effect of mean flow on duct propagation*

So far the effect of mean flow has been neglected, even though most duct systems upstream and downstream of turbomachines have flows with quite significant Mach numbers. In general the problem is highly intractable, but in the case of a uniform axial velocity the mode shapes (i.e. the variation of pressure with r and θ) remain the same and the axial variation can be written $\exp\{k_x x\}$ although k_x is modified. The presence of flow always reduces the outer wall Mach number of the pressure pattern at cut off. In the simple two-dimensional geometry the cut-off condition reduces to

$$V_x^2 + \frac{\omega L}{2\pi n} = c^2,$$

physically that the resultant of the flow axial velocity and the disturbance phase velocity should be equal to the speed of sound.

The axial mean flow is a convenient idealization for the upstream and downstream situations in turbomachines. Between the blade rows the flow is usually swirling, which in general makes it impossible to separate the variables. For certain special distributions of swirl, such as forced or free vortex distributions, it is possible to handle the pressure wave propagation after some restriction to small swirl. With swirl in the flow inhomogeneities such as entropy or vorticity also give rise to pressure fluctuations, making the problem most complicated, see Kerrebrock (1974).

The idealization adopted so far has been the constant-diameter circular or annular duct. In practically all cases this is fulfilled for only a short distance and the diameter or shape of the duct is changed. Several methods have been suggested for considering propagation in ducts of varying shape, see for example Nayfeh *et al.* (1975). In many cases the principal effects can be uncovered without considering the change of shape because, if cut-off occurs, the attenuation is generally so rapid that the amplitudes are very small before the change of shape is significant.

Radiation

The modal structure in a duct is often of less interest than the radiated field of it. The most common approach to this is that adopted by Tyler and Sofrin (1962) in which the mean flow is ignored and the duct exit is supposed to be surrounded by a large baffle. This baffle is merely a mathematical convenience, but it appears that it does not seriously distort the radiation field or total acoustic power. Lansing *et al.* (1970) calculated the radiation field from an intake represented rather more realistically by an unflanged pipe and have shown very close agreement with the Tyler—Sofrin method except in the region near 90° to the intake axis, where the discrepancy resulting from the baffle is evident. It is only for modes in the duct which are close to cut-off that the calculation using flanged exits is significantly in error.

The analyses of radiation have omitted refraction effects which are frequently strong in intakes because of the high Mach number flows involved. On the exhaust side the radiated field shape is radically altered by the refraction produced by the velocity and temperature gradients in the exhaust, with the sound field turned away from the exhaust axis, as calculated by Savkar (1975). In fact, the turbulence in the jet frequently disturbs the field shape, so that the distinct lobed pattern seen from intakes is missing from the exhaust.

Compressor and fan noise

Having laid the basis of duct acoustics, it is possible to consider the noise from compressors and fans. Because the interest in reducing noise from aircraft engines funded most of the research it is axial machinery for this application which will provide most of the examples here. The earliest work was concerned with noise from multistage compressors used with engines having low (or zero) bypass ratios. The compressors for these engines were designed with no regard for noise whatsoever and almost invariably used inlet guide vanes. Even after it was recognized that the compressor noise could be as disturbing as the jet noise, the mechanical constraints of these existing engines prevented very much being done to ameliorate matters. The advent of the high bypass-ratio fan allowed, for the first time, noise to be considered from design inception, but introduced a new type of noise from the supersonic blade tips. With this design, the engine is dominated by a large fan at the front, the inlet guide vanes (IGV) have disappeared, the fan tip speeds are normally supersonic at high-power conditions and the fan flow is split into core and bypass streams, each with very different geometries. Thus, the new fans introduced a marked change in emphasis and the largest amount of work was expended on them. It took some time for it to be realized how important the rearward-radiated noise is from fans and for it to receive equal attention.

The field of axial turbomachinery noise has been thoroughly reviewed, with an excellent and readable account by Morfey (1973) and one from a different point of view by Cumpsty (1977). It should be noted that Morfey referenced no fewer than 180 papers in 1973 and since then the number has increased.

Figure 10.18 shows typical spectra of the noise radiated from the inlet (i.e., forward arc) of a fan. The subsonic tip speed case shows tones at the blade passing frequency and its harmonics and broadband noise of a lower level elsewhere. The supersonic spectrum shows tones at multiples of shaft passing frequency with rather ill-defined broadband noise between them; these tones at shaft frequency are known as multiple pure tones, combination tones, or 'buzz-saw' noise after the subjective impression gained by listening to this supersonic phenomenon. In the rear arc the spectrum at both subsonic and supersonic tip speeds is generally similar to the forward-arc subsonic spectrum.

The spectra illustrated in Fig. 10.18 correspond to a particular angle to the axis of the machine and, since the field shape of each element of the spectrum is different, a different spectrum could be found at another angle. A 'complete' picture requries spectra at a wide range of angles, but the amount of data very

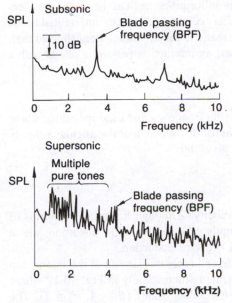

Fig. 10.18 Typical forward arc noice spectra from a small transonic fan

rapidly becomes too large to assimilate. The normal approach is to consider the spectrum at one angle and to look at the field shape of the particular spectral components, such as the blade passing frequency tone.

The discussion of compressor and fan noise will be under three headings: interaction tone noise (producing the forward-radiated tones from subsonic machines and the rearward-radiated tones from all machines), multiple pure tone noise (a supersonic phenomenon), and broadband noise.

Interaction tone noise

Interaction tone noise has been the most widely studied aspect of compressor noise and relates to the forward and rearward arcs of subsonic machines, as well as to the rear arc of supersonic machines. The kinematics of the process are first examined, then the experiments relating to rotor—stator interaction are considered, which leads naturally into the analytic studies. The theory has attracted many researchers and there is a perhaps disproportionate amount of information to describe here. Finally the effect of inlet flow distortion is considered.

1. *Kinematics of interaction noise*

From the discussion of duct acoustics on page 438 it will be recalled that the pressure field rotating with a wall Mach number less than unity decays rapidly in the axial direction — i.e., is cut off. At first sight, therefore, it is remarkable that tones are produced by machines with subsonic tip speeds. The essential explanation, which was presented by Tyler and Sofrin (1962), is that the rotors interact with stationary non-uniformities, such as the wakes or potential field of stators, to set up an interference pattern. The circumferential Mach number of the pattern can be greater than the cut-off value. They showed that, for B rotor blades and V stator vanes, an interaction pattern is set up with a circumferential order m given by

$$m = nB \pm kV$$

where n is the harmonic of blade passing frequency and k the spatial harmonic of the distortion produced by the stators. It follows that the angular velocity of the interaction pattern Ω_{pattern} is given by

$$\Omega_{\text{pattern}} = \Omega \, \frac{nB}{nB \pm kV}$$

where Ω is the rotor shaft angular velocity. Although formulated in terms of interacting rotors and stators, it is equally valid for the inflow distortion, where V would denote the circumferential order of distortion.

This expression is purely kinematic and takes no account of the interaction process involved. The effect is illustrated schematically in Fig. 10.19 where a four-blade rotor interacts with a three-blade stator ($B = 4$, $V = 3$). The fundamental blade-passing frequency interaction, $n = 1$, would therefore be expected to rotate $4/(4 \pm k3)$ times the rotor speed. The largest value is achieved taking $k = 1$ and the negative sign to give a ratio of 4. Figure 10.19 shows that, after the rotor moves 30° from the top interaction, the next interaction moves round by 120° or four times as far. This is continued so that, after the rotor moves 90°, the pattern has rotated a full 360°. If there had been five stators ($V = 5$) the smallest modulus of the denominator would be achieved again when $k = 1$ but in this case the value of m is equal to -1 and the pattern would again rotate at four times the rotor speed, but in the opposite sense to the rotation of the rotor.

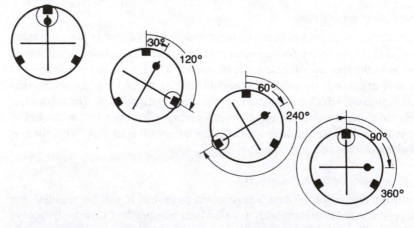

Fig. 10.19 Schematic representation of rotor–stator interaction; $B=4$, $V=3$. (From Lowrie, 1976)

A schematic representation by Sofrin (1973) of the way in which the waves from interactions above the cut-off add to produce wave fronts, whereas those below cut-off overlap and cancel, is shown in Fig. 10.20. The analogy can be drawn between supersonic and subsonic projectiles, the former producing disturbances that coalesce and the latter waves that interact and cancel.

2. *Rotor–stator interaction experiments*

Some insight into aspects of the interaction processes can be obtained from Fig. 10.21, taken from Cumpsty (1972), which shows the forward arc field shapes of the second harmonic of the blade-passing frequency from a high-speed fan. These field shapes were obtained by slowly traversing a microphone along the arc of a circle drawn about a point in the inlet plane. In this case,

PROPAGATING INTERACTION FIELD

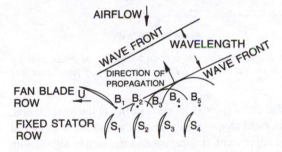

DECAYING INTERACTION FIELD

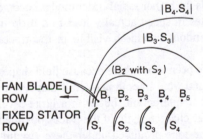

Fig. 10.20 Schematic representation of wave fronts for rotor–stator interaction above and below cut-off. (From Sofrin, 1973)

the fan had no inlet guide vanes, 25 rotor blades, and 39 outlet guide vanes placed very close behind the rotor. From the blade numbers it will be noted that at the blade-passing frequency a mode of circumferential order $m = 14$ will be set up rotating in the *opposite* direction to the rotor at 25/14 times the rotor speed, while at the second harmonic an $m = 11$ mode is produced rotating 50/11 times faster than the rotor in the *same* direction. Even at the lower speed shown in Fig. 10.21, the second harmonic was well above

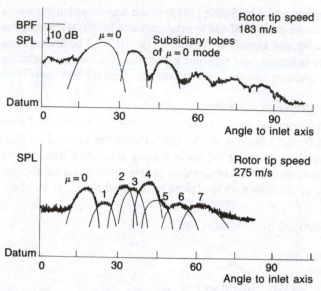

Fig. 10.21 Forward arc field shapes of tones at second harmonic of blade-passing frequency (25 rotor blades, 39 stator vanes; $m = 2 \times 25 - 39 = 11$). (From Cumpsty, 1972)

cut-off. The strong zero-radial order ($\mu = 0$) mode is shown at the lower speed, identified by overlaying the field shape predicted by the radiation model of Tyler and Sofrin. The good agreement is apparent and the steady signal with characteristic shape is evidence of rotor—stator interaction. At the lower speeds the field is dominated by the $\mu = 0$ mode, even though higher radial orders are above cut-off, but at the higher speeds the higher radial-order modes become much more important. Further increases in speed actually lead to a drop in the level and to an increase in the random fuzziness visible in the traces presented.

The second harmonic field shapes have been shown here because field shapes at the blade passing frequency fundamental were smooth in overall shape, with no evidence of lobes due to rotor—stator interaction, but with a large level of fuzziness, or unsteadiness of the tone, characteristic of the traces obtained when a rotor is run without stators. The tone produced under these circumstances is sometimes called by the misleading name of 'rotor-alone' tone. In fact, it is the result of interaction of the rotor with random, turbulent distortions drawn into the machine and will be discussed below. The absence of the rotor—stator interaction at the blade passing frequency fundamental (even when this is above the cutoff and the rotor and stator are close together) and the reduction in the level noted for the second harmonic as the speed is increased are a result of the rotor blocking the sound generated on the outlet guide vanes. The blocking is common for modes rotating in the opposite sense to the rotor when the rotor blades are highly staggered.

There have been many tests run to determine the level of the tones produced by the interaction of rotors and stators. In the early tests using multistage compressors, the principal interaction was produced between the inlet guide vanes and the rotor. For such interactions Smith and House (1967), for example, recommended a variation in tone level of the form $10\log_{10}(\Delta x/c)^2$, where Δx is the axial gap between the blade rows and c is the blade chord. Considerable variation in the trend with axial gap–chord ratio occurs for different configurations. The interaction at very small gaps can be by the potential fields around the blades whereas at larger spacings it is mainly the wake that interacts with the downstream row since the viscous decay is much slower than that of the potential field.

The generation of the interaction tone by high bypass-ratio fans is rather different from most multistage compressors because of the absence of inlet guide vanes and the low hub–tip ratio, but also because the large spacing is usually included between the rotor and the bypass section outlet guide vanes to reduce the interaction. Figure 10.22 shows forward-radiated acoustic power from tests of a high bypass-ratio fan reported by Burdsall and Urban (1971). The reference case shown is for a build with a sufficient number of outlet guide vanes to ensure that the rotor–stator interaction is cut off; these results show the blade passing frequency tone from an aerodynamically equivalent fan with the rotor–stator interaction effectively eliminated from the acoustic point of view. Any tone produced in this reference case must be related to other sources of distortion, almost certainly the atmosphere. At 4-chord spacing, the tone

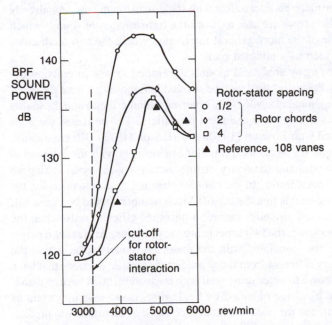

Fig. 10.22 Radiated acoustic power at blade passing frequency in forward arc. Blade tip relative flow sonic at approx. 5000 rev/min. (46 rotor blades, 76 stator vanes). (From Burdsall and Urban, 1971)

can therefore be seen to be due to the interaction with atmospheric distortion, with the rotor—stator interaction playing no part, in the forward arc (and also in the rear arc). At higher speeds, this is also true for the forward-arc noise at 2-chord spacings, because the rotor tends to block the forward propagation of the sound generated on the outlet guide vanes. The same blocking is not evident in the rear arc, since the sound is being generated on the furthest downstream blade row. The rise in tone level for the close-spaced rows as the speed increases through the cut-off value, more than 20 dB, is quite spectacular. However although the tone levels do increase rapidly with speed, particularly close to cut-off, it may be seen from these results that the trend is not so simple at higher speeds.

3. *Analysis for interaction tone noise*

The rotor—stator interaction in a high-speed, low hub—tip ratio machine is highly complex. The blade rows are normally highly loaded, so that, in addition to significant pressure changes across the blades, large deflections are normally encountered, particularly near the root. The wakes from rotors are distinctly different from those behind isolated airfoils and the wakes in turbomachines twist because of the radial variaton in the swirl velocity. The interaction on a downstream blade will therefore not take place simultaneously along the whole span, but the wake will sweep along it. The wake also tends to collect at the hub or tip and becomes confused with the secondary flows, which will themselves produce an interaction with the downstream row. As already mentioned, the blade rows are able to block the transmission of sound, which is another example of the more general interaction where the two blade rows should really be seen as a coupled pair.

There have been many analytical models developed for the interaction between blade rows. Basically, they may be divided into those that treat the problem as a two-dimensional cascade (i.e., ignoring radial effects) and those that treat blade rows as if they were unducted propellers. The unducted methods are often associated with Lowson (1970) and Hanson (1973). Kinematically, both the cascade and the unducted propeller are similar, with the interaction between moving blades and stationary disturbances (or vice versa) setting up a pattern rotating much faster. In the two-dimensional case, the criterion for unattenuated propagation is that the cut-off Mach number should be exceeded, whilst for the unducted propeller radiation becomes efficient only when the disturbance is supersonic: the two criteria are very similar. At least two major disadvantages may be associated with the unducted models. One is that the field shapes are very different from those predicted by other methods that have themselves been shown to agree quite well with measurement. A second disadvantage is that the blockage of sound by blade rows may be as important as the generation itself, and the unducted models do not lend themselves to predicting this.

Directing attention to two-dimensional cascade methods, two basic approaches to calculating the interaction of blade rows with inlet velocity distortions or sound waves can be found. Conveniently, these can be designated the compact and distributed approaches. If the blade chords are assumed to be sufficiently short in relation to the wavelength, they may be regarded as acoustically compact sources. By treating the sources as compact, it is a simple application of classical acoustics to relate the dipole source strength to the fluctuation in the lift (or drag) of the blades and then estimate the acoustic pressure from equation 10.14. The lift fluctuation is normally much larger than the drag and is generally obtained by a method such as Kemp and Sears (1955) for incompressible flow about uncambered blades. A compact source method by Mani (1970) was used to predict the acoustic power produced by a small rotor running in the wake of a number of rods in a Freon tunnel, Lipstein and Mani (1972). When the wave fronts generated are nearly parallel to the blade chord the theory predicts that the acoustic power would be much higher than when the wave fronts are normal to the chord. With 82 rods the wave fronts were very nearly normal to the blade chords, whereas with 41 rods the wave fronts were nearly parallel. The measured and predicted levels were very similar in each case with substantially higher levels observed for 41 rods.

Despite this success, one should nevertheless be concerned about many of the steps in compact source models; the calculation of the lift fluctuation makes many assumptions that are not justified, perhaps more seriously, the blade chord and the sound wavelength turn out to be very similar for many high-speed machines.

The methods for two-dimensional cascades use the distributed source approach and do not make restrictions about chord length, but instead match the incident upwash with the source strength along the chord. They are, nevertheless, restricted to cascades of uncambered blades at zero incidence (flat plates) and become less accurate at Mach numbers approaching unity. The method due to Kaji and Okazaki (1970) was presented only for the calculation of the effects of incident pressure waves (although it may be generalized), while the methods by Whitehead (1972) and Smith (1973) admit both this and vorticity distortion as inputs. Kaji (1975) compared the predictions obtained using a distributed (i.e., noncompact) source model with that obtained from a compact source and found that large errors may be produced by the compact source methods, particularly for the wave propagating upstream.

The discrepancy found between the compact and distributed source methods for predicting the interaction of blade rows with flow distortion is not found for the interaction of blade rows with sound waves, for which the ratio of blade chord to wavelength had only a small effect. This was demonstrated by Kaji and Okazaki (1970) for the transmission and reflection of a sound wave approaching a cascade of uncambered blades from downstream. It is the inclination of the wave propagation direction to the chord line which is crucial (the 'venetian blind effect'). This makes for convenient modelling and using it Cumpsty (1972) was able to explain some aspects of the blocking of sound already referred to.

In the preceding paragraphs it has been assumed that the principal mechanism for the generation of sound is the pressure fluctuation on the blades when they pass through a distorted flow. This is frequently referred to as dipole noise and in suitably idealized conditions gives acoustic power proportional to the sixth power of the incident Mach number. Quite a different mechanism was pointed out by Ffowcs Williams and Hawkings (1969) in which the velocity perturbation v_i interacts with the steady velocity field V_j surrounding the blade to produce noise proportional to $(\rho v_i V_j)$. This is generally known as quadrupole noise and the acoustic power is proportional to the eighth power of the incident Mach number. Therefore, quadrupole noise may be expected to become more important at higher speeds than dipole noise, but it is not yet known where the crossover will occur. Experimentally the two are virtually indistinguishable because in practice it is impossible to obtain the idealized situation in which the noise varies as the sixth or eighth powers. Theoretical investigations have been attempted, but they are so limited by the assumptions required that they really indicate only that the quadrupole noise probably dominates at very high subsonic speeds.

4. *Rotor/inflow-distortion interaction*

At the beginning of this section, attention was drawn to the unsteady tone levels, which were sometimes of higher amplitude than the steady tones attributable to rotor−stator interaction. It was mentioned there that this tone is due to the interaction of the rotor with unsteady, turbulent distortions in the inlet flow. It may be added that steady inlet distortions can be present, particularly in bad installations, and that these can produce *steady* tones of high level similar to those from rotor−stator interaction.

At one time it seemed that the noise of compressors without inlet guide vanes operating at subsonic tip speeds would be dominated by the blade passing frequency tone attributable to this interaction between the rotors and the inlet distortion. It was not, however, understood why the turbulence in the calm conditions normally used for noise testing could give rise to the high levels of distortion needed to produce the high tone levels. Mani (1971) applied a compact source method to the interaction of a rotor with homogeneous, isotropic turbulence, but even with large-scale turbulence the spectrum did not contain the peaks of comparable sharpness to the tones (or, more correctly, narrow-band noise) normally measured.

The explanation appeared indirectly by comparing the tone levels inside the intake of an engine mounted on an aircraft. With the aircraft stationary on the ground, a strong blade passing tone was produced, exhibiting the familiar unsteadiness. In flight, or with appreciable forward speed, the level of the tone dropped dramatically. This is shown by Fig. 10.23 presented by Cumpsty and Lowrie (1974), where the large reduction persists until the tips are supersonic. The change with forward speed demonstrated the essential feature of the turbulence drawn into a stationary intake, that it is *not* isotropic, but is strongly elongated in the axial direction. Hanson (1974) later measured the length scales

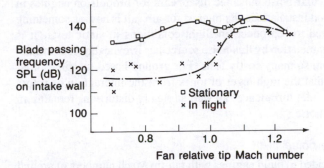

Fig. 10.23 Variation of blade-passing frequency sound pressure level (SPL) measured on intake wall with fan tip relative Mach number. Measurements stationary and in flight. (From Cumpsty and Lowrie, 1974)

and found that in the axial direction they are typically 100−200 times that in transverse directions. The long, axial eddies are necessary for the production of narrow-blade noise, which is recognizable as a tone. The contraction of the eddies in the lateral sense so as to enter the intake leads to an enormous increase in the lateral velocities (explaining why turbulent distortion remains important in even very calm conditions). When the intake possesses a significant forward speed, this effect is decreased. The two conditions are illustrated in Fig. 10.24. Large bulbous screens placed in front of the contraction of the intake have been found to reduce the tone level due to interaction of the rotor with the flow distortion and the use of such screening devices on the inlet is now quite normal for noise testing.

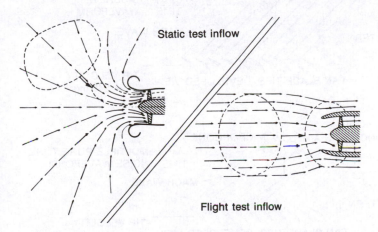

Fig. 10.24 Schematic representation of streamline patterns into an intake, static and in flight. (From Lowrie, 1975)

Because this particular noise nuisance disappears for propulsion engines in flight, at least for good installations, its interest for aircraft is now as something that must be removed to reproduce in-flight conditions in static tests. It is unfortunate that contamination by this noise source has irrevocably compromised data obtained from so many costly tests. For ground-based machines it is, of course, the case that the high level of tones from the interaction of rotors with the anisotropic inlet turbulence, as well as steady distortion, remains an important cause of noise.

Multiple pure-tone noise

When it was first decided to build engines with fan-tip Mach numbers of around 1.4, it was widely expected that the steady pressure field around the rotor, being above cut-off, would propagate unattenuated along the intake to produce an unparalleled noise nuisance. In the event, the situation was considerably less serious than this, primarily because the steady pattern toward the tip consists of shock and expansion waves that interact as they propagate, as discussed in Chapter 5. The essential processes are demonstrated schematically in Fig. 10.25, taken from Sofrin and Pickett (1970); the upper sketch showing the situation for perfectly uniform shocks ahead of the rotor and the lower the more realistic case of slightly differing shocks. The noise produced by supersonic rotors is now perhaps better understood than that from any other source, probably because it is inherently simpler, being steady with respect to the moving blades. Sofrin and Pickett clearly demonstrated the process of decay and evolution ahead of several rotors. If x denotes the axial distance ahead

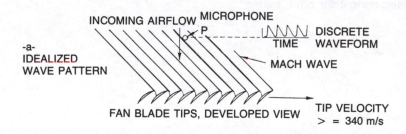

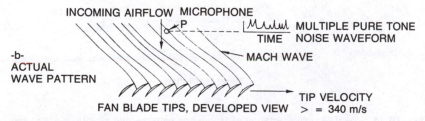

Fig. 10.25 Schematic representation of shock waves ahead of a supersonic rotor. (From Sofrin and Pickett, c 70)

of the rotor leading edge the decay of the average shock wave amplitude is proportional to $1/\sqrt{x}$ for x less than about two blade pitches and then as $1/x$ further upsteam. These two rates of decay can be predicted for uniform shocks by one-dimensional shock theory.

The prediction of the average rate of decay says nothing, however, about the spectral evolution. The changes in the pressure—time traces and in the spectrum measured by Sofrin and Pickett at small distances ahead of the rotor are clearly demonstrated in Fig. 10.26. The evolution is essentially non-linear and occurs because the stronger shocks travel faster than the weaker, eventually overtaking and absorbing them. As Fig. 10.26 shows, just ahead of the rotor the shock strengths are nearly equal, giving a spectrum with a dominant tone at the blade passing frequency. Only a few blade pitches upstream, the number of shocks is decreased and the spectrum altered to be dominated by tones at low harmonics of shaft rotating frequency.

Kurosaka (1971) considered analytically a rather idealized system of shocks and expansions around the leading edge of a cascade of uncambered blades, restricting his attention to shocks attached to the leading edges. Random variations in both blade spacing and stagger were considered. With quite small variations in stagger the spectrum only a short distance ahead of the rotor closely resembled those measured: similar variations in blade pitch produced nowhere near the correct spectrum. While this points very clearly to variations in the inclination of the forward part of the blade as being the principal cause of the spectral evolution, it must be borne in mind that supersonic rotors usually operate with the shocks somewhat detached and not in the choked condition assumed by Kurosaka.

A large number of tests have been run to try to reduce the multiple pure tone noise by altering the blade profile. It will be appreciated that subjectively a change in spectrum may be as important as a change in level. In any case, the success of acoustic treatment in the intakes at attenuating multiple pure

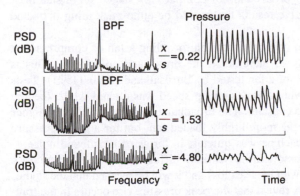

Fig. 10.26 Measured evolution of power spectral density (PSD) and pressure—time trace ahead of supersonic fan. x/s is axial distance divided by blade pitch at tip. (From Burdsall and Urban, 1971)

tone noise has reduced its importance and with it the interest in eliminating it at source.

Broadband noise

Whereas the multiple-pure-tone noise is fairly well understood, the situation for broadband noise is most unsatisfactory. One of the problems is the difficulty of pinpointing this source experimentally. Unlike the blade passing tones, which are easily identified in the spectrum and from which something can be learned of the source by the time history and field shape, almost nothing can be said unequivocally about the broadband noise. The broadband noise forms the spectral background between the tones, as in Fig. 10.18, but frequently what was believed to be broadband noise in the past has been found to contain a large number of tones when analysed with a narrower bandwidth filter. What understanding there is comes from observation of changes in the noise with alterations such as removing a stator row or altering the pressure ratio by throttling. Unfortunately, removing a downstream stator row has aerodynamic effects on the rotor that are frequently overlooked, but which are quite likely to alter the noise from it. Moreover, changing the pressure ratio by throttling also affects transmission of sound through blade rows.

The two main hypotheses for broadband noise generation are the interaction of the blades with turbulence and the self-generated unsteadiness of the flow in the blade passages themselves. Separated flow is well known for the high level of broadband noise generated. Sharland (1964) noted the higher noise from a rotor when it stalled. Gordon (1969) looked in detail at the noise from fully stalled bluff bodies in pipes and found that the broadband noise power was approximately proportional to the cube of the steady pressure drop across the obstruction, that is the sixth power of the peak velocities. One of the difficulties in turbomachines is that only one part of the machine needs to be stalled for it to dominate the broadband noise: a partially stalled set of stators or a bad rotor hub could give results that would be quite misleading if treated as general.

It has frequently been observed that, on throttling a fan or compressor at constant speed, there is an increase in the broadband noise level. This is illustrated in Fig. 10.27 for a fan tested by Burdsall and Urban (1971). Tests were also carried out when the design tip speed was altered while keeping the design pressure ratio constant. The broadband noise was markedly higher at a given tip speed for the more highly loaded fan, but for a given pressure rise the lowest-speed machine was quietest in terms of broadband noise.

The blades of compressors are essentially diffusers optimized to produce the largest pressure rise with the least loss and it will be recalled that the studies of conventional diffusers reveal that the peak pressure rise occurs in the transitory stall region. Therefore it is not surprising if the blade rows also exhibit unsteady behaviour leading to increased noise radiation as the pressure rise is raised, and that heavily loaded rows tend to produce generally higher levels

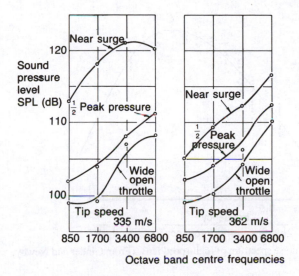

Fig. 10.27 Broadband noise spectra at 60° to inlet axis of a large fan for two transonic tip speeds and three throttle settings. (From Burdsall and Urban, 1971)

of unsteadiness. Burdsall and Urban concluded from their tests that the rotor diffusion factor was the best correlating parameter and they chose to use the value at 75 per cent of the span. Despite the rationale behind this approach, it has not been widely adopted.

Ginder and Newby (1977) used nine single-stage fans with design tip speeds of 300–475 m/s to devise a correlation. They found rotor loading at design to be a poor indicator of broadband noise and one of the quieter fans for a given tip speed was also the most heavily loaded. Their correlation used rotor incidence and tip relative Mach number as the primary correlating parameters. The sum of the forward and rearward radiated acoustic power was assumed to be proportional to the sixth power of the relative Mach number, which is equivalent to assuming a simple dipole source for the broadband noise. The broadband noise was found to vary by up to 20 dB at a given tip speed for the whole range of machines considered. If the forward arc at transonic and supersonic tip speeds is ignored (and at this condition multiple pure tone noise is dominant), the standard deviation of measured levels from the correlation curve is reduced to within 2 dB. Figure 10.28 gives the measured broadband spectra, scaled by blade passing frequency, together with the correlation curve deduced. The variation in broadband level with incidence found by Ginder and Newby was approximately 1.7 dB per degree. The correlation of Ginder and Newby does not postulate a mechanism, although it implies that incidence onto the tip is a special concern. The tip region of the rotor can be identified as a likely region of broadband noise generation, and the annulus boundary layer interacting with the tip is known to be very significant to the tone generation.

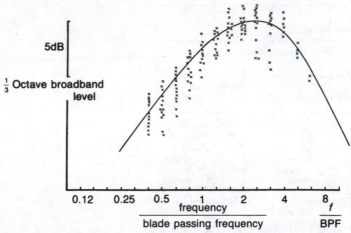

Fig. 10.28 The broadband noise spectrum of high-speed fans. (From Ginder and Newby, 1977)

For some time there has been an interest in the noise generated by unsteadiness in the flow on each blade, rather than in the diffusing combination of a cascade or the interaction with the wall at the tip. Sharland (1964) considered noise due to the interacton with the incident turbulence carried by the flow (discussed below) and separately what he called vortex noise. The vortex noise produced by the airfoil in laminar flow is believed to be due to the random shedding of vortices (analogous to the non-random von Karman vortex street) produced by the imbalance of the boundary-layer thickness toward the airfoil trailing edge. Using plausible assumptions, Sharland estimated the fluctuating lift due to vortex shedding and hence the compact dipole strength; the prediction agreed quite well with measurements made with an isolated airfoil held in the potential region of a jet. Inherent in such a prediction are estimates for the pressures associated with the vortex shedding and the correlation area over which it extends. This led to the pressure amplitude and correlation measurements by Mugridge (1971) on an isolated airfoil showing that the surface pressure fluctuations were quite different from those measured on wind tunnel walls, with much larger amplitides at low frequencies, just as one would expect if vortex shedding were important.

Sharland (1964) also investigated the noise generated by an airfoil in a turbulent jet in experiments linked to his tests in non-turbulent flow. By application of unsteady airfoil theory, with some plausible assumptions about scale, and by treating the airfoil as a compact acoustic dipole, Sharland was able to predict the sound pressure level. The agreement he obtained with measurements over a range of conditions was remarkably good. (More extensive tests were later carried out on an isolated airfoil in a jet, several results of which are described by Burdsall and Urban, 1971.) By putting a ring upstream of a fan rotor Sharland also increased the ingested turbulence to a rotor and again had some success at predicting the rise in noise. Smith and House (1967) also report a rise in broadband noise when a ring was placed upstream of a rotor.

Thus far the discussion of broadband noise has been restricted to the generation of the unsteadiness produced by the blades themselves. The evidence from this suggests that under some conditions (such as when the blades are at high incidence) the 'self-generated' noise dominates. When the blades are not in this condition, it is not clear whether 'self-generation' or interaction with the ingested turbulence is what matters.

Theoretical considerations of broadband noise by blades interacting with turbulence has frequently been an adjunct to the calculation of tones. In considering the effect of homogeneous isotropic turbulence interacting with a rotor, Mani (1971) calculated broadband noise with peaks at the blade passing frequency. When considering anisotropic turbulence, Pickett (1974) found the same type of spectrum, although with sharper tones. Later, Hanson (1974) calculated tones and broadband noise by representing the eddies as a modulated train of pulses, and obtaining rather good agreement by representing the sources as compact dipoles and deriving the fluctuating lift from an incompressible unsteady method. The objections raised above to this procedure for tones apply for the broadband noise as well, and it is usually found that the broadband noise peaks at a frequency where the wavelength is comparable to the chord. In discussing tones the possible importance of the quadrupole component was considered. The same considerations apply to the broadband noise.

Until now there has been no explicit distinction between broadband noise from subsonic and supersonic tip speed compressors. In the forward arc of a supersonic compressor it seems highly probable that the broadband noise will contain a different mechanism due to the random time variations in the bow shocks, but it must also be remembered that the flow toward the hub is subsonic and the noise from this may in some cases prevail. In fact, in the forward-arc noise spectra from many supersonic fans and compressors in which the multiple-pure-tones are dominant, it is often difficult to decide what is broadband noise.

Overall, broadband noise from compressors and fans can be summed up as badly understood. It must be recognized that the dominant source can change with the design, with the overall type of machine, and with the mode of operation. A bad or a heavily loaded design may have a region of separation as the dominant source of broadband noise, while the interaction with the turbulence ingested may be more important in a good design. The effects of tip speed are not fully appreciated; although higher speeds for the same pressure rise seem to lead to higher noise levels, higher pressure rises frequently lead to higher noise levels at the same tip speed. Finally, the mode of operation can radically alter the broadband noise, with a rapid rise in the broadband level being normal as the machine is throttled.

Non-aeronautical aspects of compressor noise

The title of this section is perhaps a misnomer and aeronautical should be replaced by high speed. Since high-speed machinery and aviation are so linked, and the thrust of so much research has been toward reducing aircraft noise, the title probably sums up the topic to best advantage.

While the availability of funding has directed research towards reducing aircraft noise, certain factors make the reduction of aircraft compressor noise particularly difficult. In the first place theoretical study is made difficult by the high Mach numbers and frequencies. In addition, both theory and experimental source location is made much more difficult because the acoustic wavelength and blade chord are typically nearly equal. The theoretical treatment of turbomachines not intended for aviation is generally very much easier. The Mach numbers are low and so realistic estimates of lift fluctuations can be found. Because the frequencies are relatively low the wavelengths are much longer than the chord and a point or line dipole description of the noise source is suitable. Mugridge (1975), for example, using simple representations of the response to distortion, showed promising agreement with measurements from low-speed fans.

Another reason for turning attention away from aeronautical machinery is the small scope there remains in this for modification. In the first place, extensive aerodynamic development has probably left relatively little separated flow which, on removal, might provide lower noise. In the second, any change giving lower noise with reduced aerodynamic or mechanical performance attracts little support. For non-aeronautical applications there is more chance of spectacular reductions in noise with improvements in aerodynamics and there is more opportunity for compromising weight and even performance to reduce noise.

Undoubtedly most of the non-aeronautical compressors are of the radial type and it is therefore surprising how little research has been carried out on these. Indeed the literature seems to indicate that there has been virtually no research on high-speed radial compressor noise. Yeow (1974a) referenced only 10 papers associated with centrifugal fan noise, most being collections of measured data. The work of one or two stands out. Embleton (1963) investigated the effect of the volute tongue on the blade passing frequency tone. As with rotor–stator interaction in axial machines, moving the tongue away reduced the noise. Hübner (1963) has carried out experiments to establish the broadband noise sources. The problem is inherently a complex one because the flow in most radial machines, even the best, is heavily separated. Yeow (1974a) carried out careful measurements on a ducted centrifugal rotor. His results were generally in line with earlier work, in particular there were high levels of broadband noise at low frequencies.

Soon after this Krishnappa (1976) carried out a survey of centrifugal fan noise covering much the same ground but describing measurements made by the National Research Council of Canada. One is very struck by the gross variations in geometry that are current practice, far wider than the range for high-speed axial machines. The lack of sophistication in measurement, analysis and prediction is also in striking contrast to the aircraft axial machine.

About the same time Neise (1976) summarized research towards reducing noise from centrifugal fans. A number of the references are identical to those cited by Yeow (1974a) but this paper considers the field only from the point of view of reducing the noise. It is made very clear how simple modifications can produce very significant reductions in blade passing frequency noise; Lyons and Platter (1963) obtained a 10 dB reduction in blade passing frequency tone without loss in efficiency by making the cut-off tongue slope and cover two impeller passages. Reducing broadband noise appears to be much less certain, reflecting the need to improve the aerodynamics, which is itself very imperfectly understood. Nevertheless Krishnappa (1977) has reported significant reductions in tone and broadband noise levels when the volute around a fan was modified to make the conditions surrounding the impeller more uniform. This programme deserves comment on two other counts. First, the modification did not significantly worsen the aerodynamic performance. Second, the noise signals for the unmodified and modified fans were subjected to narrow band analysis; just as was the case with high-speed axials a decade or more before, the apparent broadband noise was shown to contain many quite sharp tones. Progress in understanding the noise behaviour of radial machines is likely to be much faster if this signal analysis can be used more generally.

The low frequencies make the calculation of propagation away from radial fans relatively tractable, since propagation is predominantly one-dimensional along ducts. On the other hand, the duct termination impedance has a very large effect on the source and changes to either inlet or outlet can significantly change sound from the other. In tests by Yeow (1974a) the acoustic power from the outlet was 30 dB higher than from the inlet over most of the frequency range, with an acoustic termination at outlet and a bellmouth at inlet. Yeow (1974b) had some success at representing the acoustic behaviour of centrifugal rotors by a lumped impedance model and it appears that the internal impedances of the impeller can be ignored.

Acoustic treatment

The application of acoustic treatment to the inlet and outlet ducts, sometimes also over the rotor tips, has proved extremely effective in reducing the level of noise propagating from high-speed compressors without introducing large aerodynamic losses. The effectiveness of acoustic treatment is most evident with modern jet engines, where quite short lengths of treatment in the fan intake and bypass ducts bring a very large attenuation. What is normally used is a perforated metal sheet on top of a honeycomb structure which keeps it away from a solid plate. The depth of the honeycomb determines the frequency most attenuated and by varying the porosity of the perforated plate the spread of frequencies over which attenuation takes place can be altered. With acoustic treatment there is the real possibility of design in the sense of choosing a combination to give an optimum attenuation of the most serious part of the noise.

Acoustic treatment works best with high-frequency noise, that is noise for which the distance of propagation over the acoustic lining is several wavelengths. The noise from high-speed compressors tends to peak around where the wavelength of the sound is similar to the chord of the blades. This means that the wavelength is very much less than the duct circumference and it is easy to have a significant amount of acoustic treatment in even a short duct. It is likewise one of the problems of noise from low-speed machines that the wavelengths of the high amplitude noise tend to be long and long ducts are necessary to effect much reduction in level.

For some years now it has been possible to make good estimates for the attenuation for a given noise field, one of the biggest problems being to define the mode pattern of the noise field entering. Figure 10.29 taken from Mariano (1971) shows measured and predicted noise attenuation spectra for a rectangular duct with acoustic treatment on two opposite walls 0.375 m apart. The length of the acoustic treatment in the propagation or flow direction was 0.41 m. Three cases are shown: with no flow ($M = 0$); with flow in the same direction as

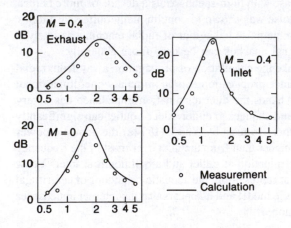

Fig. 10.29 The spectra of attenuation produced by lining two sides of a rectangular duct with a perforated honeycomb. (From Mariano, 1971)

propagation, as in the outlet duct ($M = 0.4$); with flow against the propagation, as in the inlet duct ($M = -0.4$). In each case with flow the effect of the boundary layer is included in the prediction. Propagation with the flow reduces the amount of attenuation and moves the peak to higher frequency; propagation against the flow increases the peak attenuation and moves it to lower frequency. In the design of an acoustic treatment it would be possible to allow for the flow and choose the lining depth and porosity to give peak attenuation where it is most needed.

11 Design, measurement and computation

11.1 Introduction

An appreciable amount of time has elapsed between the appearance of the major books on the subject of compressors and the present one. The report of the staff of the staff of the NACA Lewis Flight Propulsion Laboratory was published as a 'Confidential' document in 1956 (later to be declassified in 1958 and republished as NASA SP36 in 1965). Horlock's book *Axial Flow Compressors* was first published in 1958 and, though it was updated in 1973, the flavour is still that of the earlier period. The volume in the Princeton Series edited by Hawthorne, *Aerodynamics of Turbines and Compressors* was published in 1964 but several of the articles had been written years before. Ferguson's book *The Centrifugal Compressor Stage* came out in 1963, not long after Traupel's *Die Theorie der Strömung durch Radialmaschinen* in 1961. The very comprehensive book by Eckert and Schnell *Axial- und Radialkompressoren* (1980) is based on a book originally by Eckert alone which appeared in 1953. The second edition of Wislecenus's book was published in 1965 and Vavra's book on all aspects of turbomachinery was published in 1960.

It seems appropriate to examine briefly what has happened over the last 30 years or so in the understanding of compressor flows and their design and to go on to consider the tools used to examine the flow experimentally and mathematically. It should be borne in mind that there have been changes of a non-technical nature which affect the subject as well. One of these is the *relative* decline of establishments like the NACA (now NASA) and NGTE compared to gas turbine manufacturers. The companies do release information but more selectively, chosen with a view to enhancing the companies' standing, whereas the research establishments were more committed to releasing all aspects of their research. As a result a very much smaller proportion of the knowledge about the subject is available in the open literature than was the case 20 or 30 years ago.

11.2 Understanding and design

In earlier chapters one of the words most used is 'probably' and this reflects the continued uncertainty about the nature of the flow in compressors. The knowledge which exists in the data bases of companies is largely in the form of correlations; there are few people who really understand the flow and even these people might be forced to admit ignorance of many of its fundamental aspects. Unfortunately the topics least understood are sometime those which have the greatest effect on the flow: what is usually called the endwall boundary layer, for example, affects the performance of a multistage axial compressor more than any other region.

In some cases it is most instructive to go back some years to see what was known and appreciated. It was recognized by some, Constant (1939) for example, that the endwall region of axial compressors was the critical region in determining the losses, and this was picked up by Howell (1945). de Haller (1953) had correctly identified the endwall region as putting the limit on compressor pressure rise. None of this prevented the debacle in which very high aspect ratio blades were used, an era well documented by Wennerstrom (1986). For some reason much of the wisdom seemed to get lost, perhaps submerged in other material, and it is easy to forget just how much some of the early workers did appreciate.

The change in the capability of design was addressed in Chapter 2. For axial compressors the pressure rise per stage has been increased for several reasons but amongst the most important are the use of higher tip speeds and hub speeds, as well as the choice of lower aspect ratio blading. The higher speeds are partly acceptable now because of the better materials and better mechanical design, with the more precise prediction of stress. It is also partly because of the improved aerodynamic design of blades for supersonic relative inlet flows. The use of lower aspect ratio followed from the experience that performance was better if the blade chord was long, rather than the recognition that what matters is the control of the endwall boundary layer and that this is helped if the blade chord is large in relation to the boundary layer thickness.

It has been found to be possible to have higher pressure ratios from axial compressors on a single shaft (on a single spool) because of mechanical reasons and the use of higher pressure rises per stage. It is also because off-design matching is better understood and more rows of variable stators are used. This has been accomplished without the narrowing of the operating range between choke and stall or surge and without sacrifice of high efficiency.

The trend with efficiency is not so much to a large absolute increase as to a small increase with a much greater pressure rise. The high efficiency can only be obtained if the matching of the compressor is good and if proper allowance can be made for all those regions where there is no proper theory or understanding, such as the endwall boundary layers. An overwhelming impression is that the improvement in axial compressor performance owes a great deal to the willingness to spend a lot of money and commit many people to the development of the machine, a trend which emerges in the case of new aircraft engines as a contribution to the rapidly increasing cost.

With radial compressors the trend has been similar but with a larger element of empiricism owing to the complexity of the flow. The available pressure rise is very much related to the mechanical design and the materials available. This, as much as aerodynamic design, is behind the improvements. One of the improvements is in efficiency, Kenny (1984) quoting 2/3 per cent rise per year, whilst at the same time the pressure ratios move upwards. The use of backsweep and blade wrap (also known as lean) at outlet has also contributed to the improved efficiency, but at the expense of pressure rise for a given impeller tip speed; to maintain the same pressure ratio means an increase in tip speed, which may depend on the availability of improved materials. With radial compressors it seems that there is no substitute for building and testing as a means of improving the performance and it is those organizations which have been able to do this which have had most success in improving their designs.

The thrust of the above comments is that effects other than aerodynamic knowledge and skill have contributed considerably to the improvements noted in both axial and radial compressors. In the next two sections the improvements in experimental and theoretical aerodynamic techniques are looked at a little more closely and some ideas for their future development are examined.

11.3 Experimental techniques

Two developments have had the greatest effect on the measurement of flows in compressors: the digital computer and optical techniques based on lasers. The computer has had an effect which would probably not have been anticipated in its early days — to provide a method of rapidly recording huge quantities of data and then processing it. Along with the development of the computer as a data-logging device has been the development of pressure transducers and scanning valves to replace banks of manometers. The scanning valve allows many pressures to be measured with one transducer, the scanner rapidly switching from one to the other and including a calibration at every few measurements. In the modern testing of multistage compressors there may well be hundreds of pressures being recorded continuously by a data logger and computer to produce a body of data so vast that only a computer can handle it. This will be supplemented by a large number of temperature measurements, also monitored by the computer. The extent to which this has altered the nature of compressor testing should not be minimized.

The pressures and temperatures which can be measured in a high-speed compressor are nearly always from fixed tappings in the blades, the endwall or from fixed probes. It is very common to attach probes to measure stagnation pressure and temperature to the leading edge of stator blades. To get detailed information about such quantities as the boundary layer it is usually necessary to carry out some sort of probe traverse and this is generally very difficult in a high-speed compressor. It is now widely recognized that much useful information and many correct ideas can come from running tests at low speeds with geometries similar or related to those of the high-speed machine. This topic is discussed at some length by Wisler (1985). Fundamentally, it is because the processes which limit compressor performance are related to what are usually classed as boundary layer effects and for most compressors, leaving aside the type of stage with supersonic relative inlet flow, this is only weakly affected by compressibility. Where compressibility plays an overwhelming part is in the matching of the stages, but this is not directly related to the fluid mechanics of the processes involved.

The availability of more pressure measurements does not in itself ensure a better understanding of the flow; in this respect the position is not so different to how things were 30 years ago. In most compressors the blade rows are close together, with small axial gaps, and it is difficult to make measurements which are not in some way compromised by the proximity of other blade rows. Furthermore, the measurements which are possible inside the compressor are usually of three types: stagnation pressure and temperature at a number of spanwise positions and static pressures at the casing wall. In addition the mass flow and rotational speed are known along with the geometry. There is no conventional way of measuring velocity inside the high-speed compressor and to obtain velocities it is necessary to use measured stagnation pressure and temperature with a computational scheme very similar in its physical basis and assumptions to that used in design. Most often this is a streamline curvature type of method, but configured with different input and output variables to the type of method used in design.

A recent paper by Weingold and Behlke (1987) helps to escape from these limitations by utilizing the well-developed numerical techniques for the computation of static pressure or Mach number along the surface of blades, as discussed in Chapter 4, to provide a very sensitive diagnostic tool. In their method Mach numbers derived from measured static pressures are compared with computed values. In measuring static pressures on blades in a high-speed compressor one is hampered by the blade thinness so that very few tappings can be put on any one blade. The skill is in choosing the position of the tappings to yield most information. Weingold and Behlke carried out a parametric study for a particular blade and this is reproduced here as Fig. 11.1. For this case the inlet Mach number is 0.711, inlet and outlet flow angles $\alpha_1 = 59.7°$ and $\alpha_2 = 46.1°$ respectively and axial velocity—density ratio ($AVDR$) 1.122. (The specification of the outlet flow angle as a variable is an alternative to using some form of Kutta—Joukowsky condition in the calculation method.) In Fig. 11.1a the Mach number is varied keeping the other parameters unchanged. An increase in inlet Mach number leads to a swelling of the whole diagram. Inlet angle changes, Fig. 11.1b, lead to changes mainly in the forward part of the diagram. Changes in outlet angle, Fig. 11.1c, lead to a variation in the rear part with the suction and pressure surface distributions failing to meet properly at the trailing edge at values other than the correct one. Changes in $AVDR$, Fig. 11.1d, show up as a movement of the trailing edge Mach number up or down as appropriate. With this background Weingold and Behlke were able to obtain good estimates of the flow conditions across a stator blade with

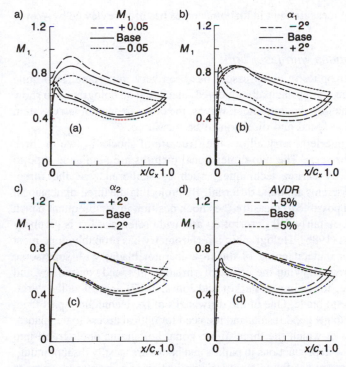

Fig. 11.1 Results of a parametric study in which Mach number was calculated about a blade section in transonic flow. Variations in inlet Mach number M_1, inlet flow angle α_1 outlet flow angle α_2 and axial velocity–density ratio *AVDR*. (From Weingold and Behlke, 1987)

only five static tappings on the suction surface and four on the pressure surface.

Hot wire techniques have been used occasionally in turbomachinery for a very long time. The more robust hot films attached to blades have found some application, though these tend to be most often used as a qualitative indicator of laminar–turbulent transition. Hot wires are vulnerable to mechanical damage in high-speed machines and have a short working life. The problem is compounded by the sensitivity of the hot wire to temperature as well as velocity changes and in high-speed compressors this has meant that hot wires too are used in a mainly qualitative manner. They do, however, find extensive use in low-speed research compressors. It is now common to take very large amounts of unsteady data and the ensemble average process is a well-established method of obtaining the steady flow field rotating with the rotor using anemometers (or transducers) which are held stationary. It may also be mentioned that more care now goes into the measurement of mean pressure when the pressure at the probe or tapping is highly unsteady, since the normal damping of a long pressure line will not produce the exactly correct average. In most compressor applications there are so many other causes of inaccuracy and uncertainty in the measurement that this is probably not paramount. The whole subject of measurement in turbomachines has been reviewed by Weyer (1984).

Optical techniques with laser light

The use of laser optics in compressors has been very important. Two quite separate techniques are principally involved, one based on holography to show up features of the flow, mainly shock waves, the other an anemometry system to allow the flow speed and direction to be measured.

Holographic interferometry allows the structure of shocks in two or three dimensions to be seen. The two-dimensional pictures are similar in type to those obtained by earlier techniques, such as schlieren, but the three-dimensional holograms are quite different. By projecting the three-dimensional image it is even possible to measure the shock position in a subsequent bench top experiment. A fairly full description and wide reference list is given by Bryanston Cross (1986). Holographic interferometry has probably done most to give a proper understanding of the flow in rotor blades with supersonic relative inlet flow, showing the effect of variation in speed, mass flow and blade shape (see, for example, Parker and Jones, 1987) as well as the variations from blade to blade. One of the drawbacks of the technique, apart from the skill needed to get good results and the need for optical access to the blades, is the poor way in which the three-dimensional holograms photograph into two dimensions — reproductions in papers and books are usually disappointing.

Laser anemometry has found a natural application in compressors because it is often possible to project light into blade rows where a solid probe could not be used. A recent summary is given by Schodl (1984). There are really two separate methods, known by various names, but both rely on having solid or liquid particles in the flow to scatter the light. This is usually deliberately introduced as seeding. For high-speed flows it is necessary to have very fine seeding, typically less that one micron in diameter.

One method, referred to variously as laser-two-focus (L2F), two-spot, time-of-flight or laser-transit anemometry, essentially measures the time between particles passing through two bright regions. It is very much associated with Schodl (1974) who developed it into a tool for use in turbomachines. Two fine laser beams are brought to separate foci a small distance apart, about 0.3 mm. The scattered light from seeding passing through each bright spot (i.e. focus) is taken to a photomultiplier and a digital system cross-correlates the photomultiplier outputs. The time lag to produce a correlation gives a direct measure of the component of flow velocity in the direction of the line separating the spots. In general it is necessary to rotate the two beams to obtain correlations at a range of angles. The magnitude of velocity and the direction of the flow is obtained from the orientation of the two spots for which the cross-correlation occurs at the shortest time difference. In practice the output from the cross-correlation is not a unique time but a spread corresponding to the range of velocities produced by flow unsteadiness, of which turbulence is one part.

The alternative system, known as laser doppler anemometry (LDA) or fringe anemometry, has two parallel beams intersecting in a very small region where the measurements will be obtained. The scattered light from particles passing through each beam is doppler shifted in frequency, with the magnitude of the shift being proportional to the component of velocity in the direction of the beam. The doppler shift will typically be of opposite sign for the two beams. The frequency of the light beams is too high to be handled electronically but by allowing both beams to interact in a photomultiplier it is possible to detect the difference in frequency. The difference frequency is proportional to the component of velocity in the plane of the beams and perpendicular to their bisector. It is also proportional to the angle between the beams. The same expression for the frequency is obtained if a system of interference fringes is assumed to exist in the intersection region of the beams and it is for this reason that it is sometimes called the fringe method.

Both the laser doppler and the two-spot systems have found application in compressors. In almost every case it is necessary to observe the scattering of light from the particles from a position close to where the light enters. This is known as back-scattering and is an inefficient process, since only very little of the light directed into the flow returns. Great care in optical design is needed when operating in the back-scattering mode to separate the large light input from the small light output. The problem is exacerbated by the need to use very small seeding particles in high-speed flows to reproduce the fine-scale motion, such as shock waves or the flow around leading edges. The use of particles which are too large leads to a smearing of these features. Since the light scattering of large particles is so much more efficient than small ones the presence of even a small proportion of large particles in the seeding can give misleading results.

The problems of seeding and the low level of light returned, in comparison with that projected into the flow, give a low signal-to-noise ratio. This becomes even worse near to solid walls if they scatter even a very small amount of the light incident onto them back into the detector. Signal-to-noise ratio is one of the most critical factors in determining the practicality of laser measurements and differences occur for the two-spot and laser doppler systems. A useful study of the relative merits of different systems is given by Kiock (1984) with a fairly full reference list for what is now quite an extensive field.

Although the use of laser anemometry has made possible classes of measurements otherwise impossible it is still a major task to carry out. The rigs need to be modified to allow transparent windows to be installed and if the window has significant curvature this affects the focus of the beams in a way which alters as the beams are rotated. Developing an adequate system of seeding may be a major task, particularly near to solid surfaces. Whichever method is used there is normally a very considerable amount of running time required to permit accurate enough data to be acquired. The time required depends on the amount of seeding, on the laser power, on the quality of the collecting optics and photomultiplier and on the amount of flow unsteadiness. In any application of laser measurement it is normal to collect huge quantities of data and the method would be quite impractical without digital computers.

A third use of lasers in compressors has been to cause a vapour present as a tracer in the gas (2,3, butanedione) to fluoresce by irradiating it with a pulse of laser light. By measuring the intensity of the light radiated by the fluorescence it is possible to get a measure of the local gas density. This technique has been successfully applied by Epstein (1977) in a blow-down transonic axial compressor.

11.4 Mathematical techniques

The analytical study of turbomachinery flows, as opposed to the numerical study, is quite old and continues to this day, but with ever fewer practitioners. The main reason for this is that not much has changed in the way of analytical ability to match the changes which have taken place in the field of numerical analysis as a result of the developments in digital computers. The drawback to analytical methods is that they frequently contain assumptions which are questionable and diminish the usefulness of the results or the confidence in them. Amongst the most common assumption is a linearization of some perturbation about a mean value. Linear techniques are still commonly adopted in the study of stall and flutter, to name two areas, but in both the growth of small disturbances is important and linearization is a natural assumption. Little interest now remains in linear techniques for the prediction of mean flow or of secondary flow, both areas once pursued actively.

Numerical analysis has transformed the consideration of turbomachinery flows and in the discussion of topics in earlier chapters of this book frequent use has been made of solutions obtained numerically. It has also been observed that some topics, for example the treatment of cascade flows by approximate methods or by conformal transformation, have been rendered virtually obsolete. For this reason it seems relevant to give some discussion of the methods and the likely direction of progress.

Computational fluid dynamics in general is a fast-moving field with large resources devoted to it. Taken as a whole, computational fluid dynamics is quite possibly a bigger field than turbomachinery in terms of the number of high-calibre people working in it, so that reviewing it as part of a chapter in a book devoted to compressors is fraught with problems.

Advances in technique are taking place which go hand-in-hand with the hardware developments; each generation of computer has greater memory and faster operation which is soon absorbed by more refined meshes and/or a less approximate handling of the governing equations. With no simplification but those relating to turbulence the flow can be described by the *unsteady* Navier–Stokes equations, so a solution is required to the three momentum equations, the mass continuity equation and an equation for the conservation of energy:

$$\frac{\partial V}{\partial t} + V \cdot \nabla V = -\frac{1}{\rho} \nabla p + \nabla \cdot (\mu \nabla V) + \nabla \left(\frac{\mu}{3} \nabla \cdot V \right) \quad \text{Momentum}$$

$$\nabla \cdot (\rho V) = -\frac{\partial \rho}{\partial t} \quad \text{Mass continuity}$$

$$\nabla q + \nabla (\rho V h_0) + \frac{\text{shear work}}{\text{terms}} = -\frac{\partial (\rho e_0)}{\partial t} \quad \text{Energy}$$

where h_0 is the stagnation enthalpy and $e_0 = c_v T + V^2/2$ is the stagnation internal energy. They are always handled in the Reynolds averaged form so that the turbulent velocities appear only in the stress terms.

The same equations apply, of course, very generally to fluid flows but there are special features of turbomachinery flows which mean that their solution needs specialized techniques. One of the important special constraints in any internal flow is the mass flow or, more precisely, the corrected mass flow or flow function $F = m\sqrt{(c_p T_0)}/A p_0$. Changes in this can bring very large changes in the flow field, particularly when the Mach numbers are near to one. It is therefore important that all the variables in F are computed accurately and the successful methods are set up in such a way that as many as possible of the variables in it are handled in a way that will not introduce serious errors.

In general any numerical scheme will introduce errors as it approximates the gradients of properties in the flow. It is possible to arrange the numerical scheme so that, for example, *exactly* equal mass flows enter and leave a mesh element. This is known as conservation form. The mass flow, momentum flux $p + \rho V^2$ and stagnation temperature can be required to be in conservation form in all the levels of simplification discussed below. In F, the flow function, m and T_0 are then exactly conserved but p_0 will not be exactly conserved unless it is specified that the entropy is constant. Because methods which assume the entropy is constant introduce no errors in m, T_0 and p_0 the flow function is preserved at its correct value and the solutions can be very accurate as well as quick. Note that accurate here means only in terms of solving the problem as posed with the simplifications adopted for the flow. The simplification that entropy is constant may introduce errors in modelling, or limitations on the method whose importance depends on the nature of the flow. However, inaccuracies arising from the lack of conservation can be very serious, predicting different mass flows in and out and changes in stagnation pressure where there is no physical mechanism to bring this change about.

Conservation of variables can be achieved in several ways. For some methods, such as the streamline curvature methods, constant stagnation enthalpy and entropy can be imposed. For the Navier−Stokes or the Euler equation solvers it is automatically achieved for several variables if a finite volume scheme is adopted: because of this the finite volume method of solution has become more or less universal in field methods for turbomachinery applications. In this the flow region is divided into little boxes or cells with common boundaries and for each equation the variable concerned (mass flow, momentum flux and stagnation enthalpy flux) entering one cell is, in the steady state, all passed to the surrounding cells. In this way no flux is lost and global conservation is ensured.

The mathematical difference between the finite volume and the finite difference schemes can be seen by considering the momentum equation for a level of simplification which involves omitting the viscous terms in the Navier–Stokes equation to get the Euler equation. In conventional differential form this is

$$\nabla p + \rho(V.\nabla)V = -\rho\frac{\partial V}{\partial t}$$

This can be written in finite difference form on a conventional grid. The same physical information can be given for a small control volume, i.e. a finite volume. The left-hand terms, which are treated as momentum fluxes, are integrated over the faces of the volume and the right-hand term is integrated over the volume, the volume integration variable being denoted here by dv. The result is

$$\iint \{p\mathrm{d}A + \rho V(V.\mathrm{d}A)\} = -\iiint \frac{\partial(\rho V)}{\partial t}\,\mathrm{d}v$$

In the numerical realization of this integrated form of equation a mean density and velocity would be used for the volume on the right-hand side; similarly, in the numerical realization of the area integral, mean values of velocity, density and pressure on each face of the volume would be taken. In many methods intended to solve for steady flow unsteady equations such as this are used to solve for the evolution of the flow in time, starting from an initial guess. The steady solution corresponds to the condition when the net flux is equal to zero and the right-hand side vanishes.

Any numerical operation involves some degree of approximation. The amount of inaccuracy or error rises rapidly as the changes in the magnitude of variables across a mesh increases; near to the leading edge of a blade the velocity changes very rapidly indeed and even with a refined mesh the magnitudes on opposite sides of a finite volume or adjacent points in a mesh may be very different. The approximation which comes with differentiation always introduces some higher order terms. These have an effect on a calculation equivalent to viscosity and are usually referred to as artificial or numerical viscosity. The error appears as a spurious prediction of stagnation pressure loss near the leading edge in inviscid methods supposedly unable to generate loss. A thorough account of numerical methods in turbomachinery, with an extensive set of relevant references, has been given by Denton (1986).

With such rapid advances being made, whatever is written risks being rapidly superseded and with such a large field it is hard to know how to provide a useful summary. The expedient adopted here is to utilize the approach adopted by Dr J D Denton in an informal lecture; any numerical treatment is some sort of approximation and the method Denton adopted is to progress through the various levels of simplification commonly adopted.

Levels of simplification

Simplification 1 is to average out the high-frequency fluctuations so that the turbulent velocities retain a significance only in the stresses and energy fluxes that they contribute: the so-called Reynolds average. This assumption is almost universal in turbomachinery. Low-frequency variations, in terms of the velocity and length scale of the problem, can be retained. The evaluation of the stresses immediately requires some sort of turbulence modelling and this is a special and very active field in its own right; in the simplest treatment the laminar viscosity is replaced by a turbulent eddy viscosity, with the magnitude of this obtained from a simple algebraic relation such as the mixing length model. In more complicated treatments of turbulent shear stress, such as the $k-\epsilon$ method, it is necessary simultaneously to integrate transport equations for the turbulence quantities themselves. The very simple models for turbulence seem to be adequate for flows which can be characterized by thin boundary layers, although the prediction of loss may need a more precise description of turbulence. When there are large regions of separated flow there seems to be no generally valid description of the turbulence.

Simplification 2 introduces the assumption of constant stagnation enthalpy along streamlines for adiabatic flow. This would be replaced by constant rothalpy in a rotating frame of reference. What this means is that the work done by one streamline on another by shear work is exactly balanced by heat conduction. This assumption is only precise if the Prandtl number is unity but, since for most gases it is close to one, it is a generally good approximation. With this assumption the need to solve the energy equation is removed. The entropy is not constant and can rise because of shear work and shocks.

The methods in the categories of simplification 1 and 2 are described conventionally as Navier–Stokes solvers. A method in category 2 written by Dawes (1987a) was used for some of the comparisons with measurement in earlier chapters. Dawes' method, like other Navier–Stokes methods developed for turbomachinery, uses only a simple mixing length approximation for the shear stress distribution but this appears adequate for the prediction of the gross features of the flow, such as the position of two- or three-dimensional separation lines. This is because turbomachinery flows are determined in the main by the pressure differences and blockage, the shear and mixing providing a relatively small contribution until the regions of separated flow become large. Such methods are not generally regarded as adequate for the quantitative prediction of loss although the trend in actual loss variation may be well predicted.

Simplification 3 involves the neglect of all viscous and heat flow terms to yield what are known as the Euler equations. They are useful when the boundary layer or viscous regions are thin and can be calculated separately. If the boundary layer thickness becomes large enough to have a significant effect on the entire flow the approach to calculating the flow in this way must be coupled interactively with a separate boundary layer calculation. Sometimes it is possible to calculate the inviscid flow, then the boundary layer flow, modify the effective shape of the surface to allow for the boundary layer displacement thickness and then recalculate the inviscid flow. Close to separation this tends to be unstable and in these cases an inverse type of calculation must be employed, for example Calvert (1983). The calculation methods almost invariably incorporate simplification 4, described below.

Simplification 4 follows on with the Euler equations (i.e. all shear stresses and heat conduction are neglected) but assumes that the stagnation enthalpy (in rotating systems the rothalpy) is constant along streamlines so that the energy equation can be omitted. Methods with this level of simplification are widely used for solving compressor flows in two and three dimensions. The assumption of constant enthalpy is only valid in steady flow; constant rothalpy is the conserved quantity for steady flow in the rotating frame of reference. (It will be recalled from Chapter 1 that the substantive rate of change of stagnation enthalpy is related to the rate of change of pressure with respect to time, i.e. $Dh_0/Dt = 1/\rho \partial p/\partial t$.) Nevertheless it is almost universal in all methods, at all levels of simplification, to neglect the unsteadiness observed on one blade row due to the adjacent rows, and to treat each as steady in its own frame of reference. This means that the circumferential non-uniformities in stagnation pressure are effectively assumed to mix out and the non-uniformities in static pressure to decay in the short distance between the blade rows.

Euler solvers, as the methods in category 3 and 4 are usually known, have found very wide application in turbomachinery flows. They are able to cope with many of the features found, including non-uniform inlet flow (i.e. inlet vorticity and entropy), shocks and complicated three-dimensional geometries. They are generally much faster than Navier–Stokes solvers; not only is the calculation simpler but with the Euler solver it is not normally necessary to have a fine mesh near to solid surfaces to model the steep velocity gradients there.

The Euler methods can be modified to include viscous (i.e. turbulent) effects by the inclusion of shear stress at the wall and a body force inside each finite volume. Each body force is the net shear force exerted on the six sides of the small volume. Using this and a very simple model for eddy viscosity, Denton (1986) has shown agreement with measurements comparable in accuracy to that obtained by a full Navier–Stokes solution for flows without extensive regions of separated flow using only a very simple model for the eddy viscosity and the results of this method have been used in Chapter 5. In that chapter the point was made that it is the blockage produced by the boundary layer displacement thickness which is most important in getting accurate predictions, the losses being relatively unimportant for the prediction of the velocities and static pressure.

Simplification 5 assumes that both the entropy and stagnation enthalpy are constant along streamlines. Since $Tds = dh - 1/\rho dp$ it follows that along a streamline

$$Tds/dm = dh/dm - 1/\rho dp/dm = 0$$

or $\qquad d(h_0 - V^2/2)/dm = 1/\rho dp/dm$

and since $dh_0/dm = 0$

$$dp/dm = -\rho V dV/dm$$

This is a momentum equation along the streamline and is automatically satisfied by the assumptions adopted.

  This lends itself to a convenient method of solution of two-dimensional flows, since one momentum equation is automatically satisfied and the energy equation has degenerated into h_0 = constant. In two-dimensional flow the original four equations have therefore reduced to two: a momentum equation in a direction other than the m-direction and the mass continuity equation. The remaining momentum equation is conveniently solved in a direction approximately normal to the streamline direction so that a dominant term is the pressure gradient which is balanced by the centripetal acceleration of the flow following the streamline curvature. This gives rise to the name streamline curvature methods. The solution follows from an integration of the flow across the stream from one solid boundary to the other. The constant of integration must be adjusted iteratively to get the correct mass flow across the passage. Another iterative process is required to adjust the streamline shape.

  Streamline curvature methods are the most common method of solving the meridional throughflow. They also find wide application with blade-to-blade flows, although they are not satisfactory in the leading edge region, often requiring some form of leading edge cusp to be applied. They can be made to cope with regions of both subsonic and supersonic flow, as well as weak shock waves, but have difficulty with choking.

Simplification 6 uses, in addition to the assumption of constant stagnation enthalpy and entropy along streamlines, a system of two-dimensional flows with a stream function defined by

$$\partial\psi/\partial r = r\rho V_x \text{ and } \partial\psi/\partial x = -r\rho V_r$$

in axisymmetric flow, or

$$\partial\psi/\partial y = \rho V_x \text{ and } \partial\psi/\partial x = -\rho V_y$$

in two-dimensional cartesian flow. The advantage of the use of a streamfunction is that it automatically satisfies the mass continuity equation. The momentum equation in the flow direction may be taken to be satisfied automatically as a result of the assumptions of constant entropy and stagnation enthalpy in that direction. For each two-dimensional flow there is then only one equation to be solved in the direction approximately normal to the streamlines. Such methods can be used for the meridional throughflow or for the blade-to-blade flow and are very fast and accurate within the limits of applicability of the Laplace equation. Nevertheless they have not found much favour and are not widely used in turbomachinery. This is probably because they can only be used where every region of the flow is either subsonic or supersonic: no mixed flow can be allowed without incorporating special techniques because differentials of streamfunction are double valued with respect to Mach number. The widely adopted streamline curvature methods, category 5, are less restrictive and are therefore more popular.

Three-dimensional flows can be handled by combining two two-dimensional flows, i.e. blade-to-blade and hub—casing flows. For this to be mathematically exact the two-dimensional flows would not be on planes or surfaces of revolution but on distorted surfaces of constant stream function determined by the flow itself. This requires an iterative solution in which the geometry on which the solution is performed changes from one iteration to another. It was the method originally proposed by Wu (1952) but it is complicated to implement on digital computers and it has been superseded by the more straightforward three-dimensional methods.

Simplification 7 assumes that both the stagnation enthalpy and entropy are uniform throughout the flow. From Crocco's theorem for steady flow

$$V \times \nabla \times V = \nabla h_0 - T\nabla s$$

so that in homentropic and uniform stagnation enthalpy flow the vorticity is zero, in other words the flow is irrotational. With the vorticity zero it is possible to define a velocity potential such that

$$V = \nabla \phi$$

The condition of irrotationality imposes constraints on the velocity gradients (e.g. $\partial V_x/\partial y = \partial V_y/\partial x$) which replaces the need to satisfy a momentum equation. Only the mass continuity equation remains to be solved, the energy equation having gone with the assumption of constant stagnation enthalpy and a momentum equation with the restriction to homentropic flow. The continuity equation is

$$\nabla.(\rho V) = 0$$

which can be expanded as

$$\nabla.(\rho \nabla \phi) = 0$$

or $\qquad \nabla^2 \phi = (\nabla \phi . \nabla \rho)/\rho$

Because the flow is homentropic the static density and temperature are related to the constant stagnation quantities by

$$\rho/\rho_0 = (T/T_0)^{1/(\gamma-1)}$$

which can be put in terms of the velocity potential as

$$\rho/\rho_0 = \{1-(\nabla\phi)^2/(2c_pT_0)\}^{1/(\gamma-1)}$$

The potential methods can be used in two or three dimensions and are very fast. The methods can adapt to supersonic as well as subsonic flow and can cope with shock waves provided they are weak enough to cause an insignificant entropy increase. They are frequently solved by the finite element method which lends itself well to the complicated shapes of turbomachines and allows the concentration of mesh points where changes are rapid, for example around leading edges of blades. A finite element method described by Whitehead and Newton (1985) was used to obtain most of the predictions of subsonic and transonic cascade flow shown in Chapter 4.

The speed of potential methods and the accuracy with which they solve the problem as posed makes it attractive to find a way of keeping some of their benefits whilst allowing the flow to be rotational. Many idealized flows are irrotational over most regions but contain vorticity concentrated in some parts. For example, a rotor in inviscid flow with non-uniform loading, such that the circulation is not inversely proportional to radius, sheds vorticity in a trailing vortex sheet. One way of handling this is to use the Clebsch transformation with the vorticity vector Ω expressed in the form

$$\Omega = \nabla\lambda \times \nabla\alpha.$$

where λ and α are both scalars. The velocity can then be expressed in the form

$$V = \nabla\phi + \lambda\nabla\alpha$$

This expression for velocity contains three scalars, just as the velocity contains three components. The surfaces of constant λ and α contain the vortex lines so their intersection defines a vortex line. Hawthorne and Tan (1987), for example, use α to represent the blade surface and trailing vortex sheet and λ is then related to the circulation on the blade. The estimates for α and λ provide an input for the first iteration of a solution of a Poisson equation for ϕ of the type

$$\nabla^2\phi = -V\cdot\nabla\ln\rho + \nabla\cdot(\lambda\nabla\alpha)$$

Simplification 8 adopts the assumptions of simplification 7 leading to potential flow, but in addition the flow is assumed incompressible so that $\nabla^2\phi = 0$. With this it is possible to use singularity methods, of which the best known in turbomachinery is that due to Martensen (1959) for two-dimensional cascades. For this method discrete vortices are distributed around the solid surfaces to represent the boundaries. Since the physical reality is often a thin viscous boundary layer, in fact a thin region of distributed vorticity, a method which uses a finite number of discrete vortices is intuitively appealing.

Singularity methods are extremely fast and can handle complicated geometries. They can also provide accurate solutions for geometries where there are rapid changes, such as the leading edge region of blades, and occasionally still find an application for this, as was the case in Chapter 4. Compressibility corrections have been devised but, given the alternative methods which include compressibility properly, the singularity methods are no longer much used in turbomachinery.

Methods of solution and application

The streamline curvature method, described above under simplification 5, finds very wide application in compressors and for a long time was about the only general method. It is still the dominant method for the meridional throughflow and is particularly widely used for the blade-to-blade flow in centrifugal compressors. Streamline curvature methods have special stability requirements and these were considered by Wilkinson (1970) and are discussed by Denton (1986).

The other methods in wide use are the field methods: Navier–Stokes solvers (simplifications 1 and 2), Euler solvers (simplifications 3 and 4) and the potential flow solvers (simplification 7). Common to all the field methods are the problems which occur when a flow contains regions of subsonic and supersonic flow. Every part of a subsonic flow is affected to some degree by the boundaries on all sides because the disturbance or effect of the boundary can always travel through the flow. In supersonic flow small disturbances cannot travel against the flow and any part of the boundary only effects the flow in a cone stretching downstream. This incompatability of boundary conditions was for a long time a major stumbling block. In mathematical terms the equations are described as elliptic in subsonic flow and hyperbolic in supersonic flow. A common remedy is to treat the flow as unsteady, even when only the steady solution is required, so that the equations are hyperbolic whether the flow is subsonic or supersonic — this approach has come to be known as 'time marching'. The calculation is continued until the unsteady part vanishes and the steady solution is obtained. Physically it can be reconciled with the way a real flow sets itself up with an unsteady process; just as with a real flow it is important to arrange the proper set of upstream and downstream boundary conditions.

As a way of solving the equations it is possible to update all the variables together or to update only one variable at a time. The former is known as an implicit method and usually requires the inversion of a matrix. The program by Dawes (1987) is of this type. The latter is known as explicit and, being much simpler to implement, is more widely used, for example Denton (1982). With the explicit approach there is a restriction on the maximum size of the time steps which can be used to maintain stability, namely

$$\Delta t \ < \ \Delta x/(V+c).$$

In other words the mesh size Δx must exceed the distance a disturbance can travel in one time increment if it is carried by the flow at its maximum velocity V while propagating at the local speed of sound c. As the mesh size is reduced to give the necessary accuracy in regions of rapid change the time steps also decrease. The total number of numerical operations therefore rises very rapidly as the number of mesh points increases. In finding the steady solution the unsteady part is, in itself, of no interest and it is common with methods in simplification 2 or 4 of the above to use an artificial time variable. This means that different time steps can be used for regions of small mesh and large mesh. Ingenious procedures called multigrids have been developed to combine the accuracy of the fine grid with the rapid diffusion of information that occurs with a coarse grid: the flow is solved simultaneously with coarse and fine grids and the results suitably combined.

Artificial viscosity is an automatic consequence of finite difference approximations for gradients. The artificial viscosity has the effect of smearing out the flow, particularly near regions of rapid change. With an Euler solver artificial viscosity is the way that the nominally inviscid method can represent shocks, since some dissipative mechanism is essential. Artificial viscosity also plays an important part in maintaining stability and is particularly useful in the early stages of a calculation when there are large gradients and large excursions in variables liable to be amplified and to lead to the solution failing. With Euler solvers it is normal to include extra artificial viscosity, either in some regions of the flow, such as where the velocity gradients are very large, or else in the early stages of the calculation before it settles down. A brief discussion of the use of artificial viscosity in the finite element method is given by Whitehead and Newton (1985).

The simple types of finite difference procedure cannot ever represent shock waves as the discontinuities they really are and it is inevitable that they will appear spread over several mesh points. This may have serious drawbacks for calculating shock boundary layer interactions where it is not only the pressure rise magnitude but the rate of change of pressure which matters to the boundary layer. The choice of artificial viscosity affects the calculated shock width: too much and the shock becomes very smeared; too little and undulations in the pressure and velocity may be found upstream and downstream of the shock.

The potential methods are inherently incapable on their own of predicting loss. The Euler methods can predict only shock loss and for compressors this is not usually the main source; the shock boundary layer interaction is likely to be more serious. The Navier–Stokes solvers can in principle predict loss but few protagonists claim that they are accurate enough at the present to do more than indicate trends. To be able to get truly believable loss estimates it will be necessary to have larger computers so that adequately fine meshes can be put near the solid surfaces and in the clearance gaps. It will also be necessary to include better turbulence modelling.

There are then a wide range of types of calculation methods available for two- and three-dimensional flow calculations in turbomachines. In one sense a balance needs to be drawn between accuracy of physical modelling and accuracy of numerical solution. If the flow can be approximated or specialized in some way, for example assuming constant entropy along streamlines, there are large advantages in terms of speed and in terms of accuracy of solution. Such gains are worthless if the flow is such that the entropy is not constant, because entirely erroneous flow patterns may be predicted. An example of this is the inviscid calculations performed in centrifugal impellers which have in many cases been predicting flows which are physically quite wrong.

Should the flow be known to be irrotational and to have no shocks or only weak ones, then the best method to use is undoubtedly one based on these restrictions, namely potential flow. If the flow is two-dimensional in nature and the flow along streamlines can be assumed to have constant or prescribed entropy and stagnation enthalpy, a method such as the streamline curvature method is quick and accurate. Only when such simplifications are inappropriate is it necessary to use methods such as the Euler solvers, which are inevitably slower. More seriously, because the Euler solvers relax the constraints on the variables, additional errors of numerical origin will be introduced in the solution. In terms of numerical accuracy of solution it seems impossible for the methods which solve the Euler equations to be as accurate as the more restricted potential methods for the same number of grid points. The crux is that the flow is often badly approximated by the potential flow model so that prediction of the flow using a potential flow method may be highly inaccurate. The Euler methods, though they solve the model posed less accurately, nevertheless allow a more realistic flow to be considered and can be applied to a wider range of flows.

The inviscid methods will be adequate if the boundary layers are thin, are simple enough to be accurately predicted by boundary layer methods, or where the main interest is not in the flow near the solid surfaces. If the prime interest is in the flow near the walls or in the loss-generating mechanism then it is necessary to take the extra step of solving the Navier–Stokes equations or carrying out an equivalent procedure to modify the Euler equations. This is particularly true when the flow contains major regions of separated flow. The Navier–Stokes equations can cope with separations and, as was shown in Chapter 4, produce a good estimate for the separation lines, whereas the boundary layer equations become singular and cannot proceed across a separation line.

At the present time the design procedures are still dominated by two-dimensional methods and thinking. The methods are modified to take account of variations in radius and streamsheet thickness, they are then often referred to as quasi-three-dimensional. They are applied successively in the meridional direction and on the blade-to-blade surfaces. (Some insight into the design procedures for axial machines is given by Stow, 1984.) Three-dimensional methods are costly and time-consuming to run, but this does not seem to be the real factor holding back the widespread application of these methods. The main reason seems to be the difficulty designers have in thinking in three dimensions and therefore the great difficulty there is in making design decisions. The problem of representing or presenting the information graphically is quite considerable. At present the design usually seems to be carried out using the quasi-three-dimensional methods and only where there is reason to suspect strong three-dimensional effects will a full three-dimensional calculation be carried out. At the present time the types of geometry assessed in this way are low hub—casing ratio high-speed fans, low aspect ratio blades where the flow entering and leaving is highly non-uniform, and centrifugal impellers. Perhaps with time a body of experience will build up so that it will be possible to design in three dimensions; the present procedures are no more than a first step in this direction.

APPENDIX *Blade profile families for axial compressors*

Although the modern approach is to use blade profiles designed to a prescribed velocity distribution, many blades for specified families are still in use and continue to be used. Also much of the early research was carried out using such blades and for this reason alone there is a need to know what the forms are. The most common families are the NACA 65 series, the British C-series and the double circular arc. The double circular arc, also known as the biconvex, is very simply defined in terms of the suction surface and pressure surface radii as well as the leading and trailing edge radii and needs no further description.

NACA 65-series

The section was designed for use as a low drag aircraft wing with a combination of camber and thickness chosen to give a virtually constant loading (i.e. pressure *difference*) between suction and pressure surface from leading edge to trailing edge. The '6' in 65 refers to the series. The static pressure was to be nearly constant for a large portion of the chord to give extensive regions of laminar flow and the '5' in 65 refers to the fraction of chord in tenths for which this was to occur. Although the shape has been extensively used in cascade applications the cascade was never part of the design philosophy and in cascade the original pressure distribution would be lost. In the specification of the isolated aerofoil the thickness and camber distribution are 'tailored' together to give the required overall lift and pressure distribution. This is not, however, followed in cascade applications where a particular thickness distribution and camber line shape are scaled to give the section desired.

The specification of this type of blade might be given as 65−(12)10. The number in brackets is the design lift coefficient of the *isolated* aerofoil at low Mach number expressed in tenths (i.e. $C_L = 1.2$ for this example). The final number is the approximate thickness-chord ratio expressed in hundredths (i.e. $t/c = 10$ per cent in this example). When the lift coefficient is less than unity the brackets can be omitted, e.g. 65−810 for an aerofoil with an isolated lift coefficient at design of 0.8 and a thickness−chord ratio of 10 per cent. As already noted the philosophy behind the 65 series was a constant pressure difference or loading from leading edge to trailing edge. Some variation on this was allowed and this was then specified as, for example, 65−(12)$A_8$10, where the A_8 denotes that the loading is uniform over the front 80 per cent of chord. (A_{10} denotes that the loading is uniform over the entire chord and could therefore be omitted without loss of information.)

478

For all 65-series blades the maximum thickness occurs at 40 per cent chord. There is, however, considerable confusion concerning the precise thickness distribution and many variations are in use. The original aerofoil had a cusped trailing edge and this was structurally very unsatisfactory. A so-called blower section was created by scaling down the thickness of an uncambered aerofoil of 16 per cent maximum thickness to 10 per cent and then adding 0.0015 times the percentage of axial chord to the ordinates for the upper and lower surfaces. Subsequently a thickness distribution for the uncambered NACA 65−010 aerofoil section was derived, differing only slightly from the scaled blower section: the ordinates of this are given here in Table A1. The difference between this and the earlier 'blower' section was judged to be insignificant. The trailing edge was still very thin, with no trailing edge radius, which was not mechanically sound and modifications of some sort were normal from the early days. Kovach and Sandercook (1961), describing designs of compressor from the early 1950s, used the 65-series thickness from the leading edge to 60 per cent chord and then a linear fairing from the 60 per cent thickness to a trailing edge radius of 0.8 per cent of chord. Other organizations have their own local modifications, mainly in the rear portion, but there is no reason to think that aerodynamically this is at all significant.

The camber line used in cascade is properly obtained by scaling the ordinate of that camber line which would give an isolated aerofoil lift coefficient of 1.0 and the camber line shape for the standard 65-series camber line for $C_L = 1.0$ is given in Table A1. This distribution does not lead to a simple analytical form and in fact the slopes are infinite at the leading and trailing edges. If the camber line is plotted out and compared with a circular arc to give the same maximum camber, as in Fig. A1, it can be seen that the actual differences are very small and for many years it has been normal when employing 65-series aerofoils for cascades to use a circular-arc camber line in place of the true one. Figure A2 derived from a similar plot by Lieblein (1965) relates camber angle for an equivalent circular arc to the C_L for the true 65-series aerofoil. If the true 65-series camber line is used it is not meaningful to refer to blade inlet and outlet angles since these turn sharply in the immediate vicinity of the leading and trailing edges.

The 65-series thickness distribution in Table A1 is for a 10 per cent thickness−chord ratio blade. In application this is normally scaled to give the required maximum thickness (and very probably modified to thicken it in the trailing edge region) and then laid along the camber line. The thickness is applied perpendicular to the local camber line but at distances specified along the *chord* line.

The C-series

This series of profiles was first codified by Power Jets Ltd in the 1940s but has since come to be thought of as the NGTE profiles (Howell, 1944). The most commonly used in early research is the C4 and this was shown in comparison with the 65-series and double circular arc in Fig. 4.3. The maximum thickness is at 30 per cent of chord compared to 40 per cent for the 65-series and 50 per cent for the double circular arc. The leading edge of the C4 is noticeably blunter and can then be expected to be more robust but less suited to high Mach number operation. With the C-series the camber line is specified as either a circular arc or parabolic arc, the camber angle and the position of maximum camber being specified. The chosen thickness distribution (scaled in magnitude from the 10 per cent thickness distribution always quoted) is then laid on the camber line so that the thickness is normal to the local camber line direction.

Table A1 Thickness and camber of NACA 65 series aerofoils and thickness of C4 aerofoils

% Chord	NACA 65 Half thickness (% chord) (see note 1)	Camber line (% chord) $C_{LO}=1.0$ (see note 2)	C4 half thickness for profile (% chord)
0	0	0	0
0.5	0.772	0.250	
0.75	0.932	0.350	
1.25	1.169	0.535	1.65
2.5	1.574	0.930	2.27
5.0	2.177	1.580	3.08
7.5	2.647	2.120	3.62
10	3.040	2.585	4.02
15	3.666	3.365	4.55
20	4.143	3.980	4.83
25	4.503	4.475	
30	4.760	4.860	5.00
35	4.924	5.150	
40	4.996	5.355	4.89
45	4.963	5.475	
50	4.812	5.515	4.57
55	4.530	5.475	
60	4.146	5.355	4.05
65	3.682	5.150	
70	3.156	4.860	3.37
75	2.584	4.475	
80	1.987	3.980	2.54
85	1.385	3.365	
90	0.810	2.585	1.60
95	0.306	1.580	1.06
100	0	0	0
L.E. Radius	0.687	—	1.2
T.E. Radius	0	—	0.6

1. There are several NACA-65 series thickness distributions quoted and in use, all differing slightly. This is the one quoted by Emery *et al.* (1958) intended for use in turbomachines. For structural reasons it has been common to modify this thickness distribution further by, for example, adding a trailing edge radius of 0.8% chord and connecting this with straight lines to merge with the tabulated thickness distribution at 60% chord.
2. This camber distribution has been rarely used in recent times but a circular arc one instead.

Table A2 C-series blade sections Table of coordinates

Section	x/c%	0	1.25	2.5	5	7.5	10	15	20	30	40	50	60	70	80	90	95	100
C1	±y/c%	0	1.375	1.94	2.675	3.225	3.6	4.175	4.55	4.95	4.81	4.37	3.75	2.93	2.05	1.125	0.65	0
C2	+y/c%	0	1.49	2.08	3.00	3.58	4.01	4.55	4.90	4.98	4.76	4.30	3.70	2.91	2.02	1.05	0.60	0
	−y/c%	0	1.63	2.26	3.12	3.66	4.06	4.58	4.89	5.02	4.79	4.31	3.72	3.00	2.15	1.20	0.68	0
C3	+y/c%	0	1.42	1.84	2.31	2.72	3.04	3.60	3.83	4.35	4.76	5.00	4.00	2.93	1.62	0.92	0	
	−y/c%	0	1.42	1.86	2.545	3.03	3.41	3.80	4.18	4.82	4.945	5.00	4.00	2.93	1.62	0.92	0	
C4	±y/c%	0	1.65	2.27	3.08	3.62	4.02	4.55	4.83	5.00	4.89	4.57	4.03	3.37	2.54	1.60	1.06	0
C5	±y/c%	0	1.65	2.27	3.08	3.62	4.02	4.55	4.83	5.00	4.89	4.58	4.09	3.48	2.75	1.95	1.52	0
C7	±y/c%	0	1.51	2.04	2.72	3.18	3.54	4.05	4.42	4.86	5.00	4.86	4.43	3.73	2.78	1.65	1.09	0

Futher data

Section	Position of t_{max}(%c)	$\dfrac{LE\ radius}{max\ thickness}$ %	$\dfrac{TE\ radius}{max\ thickness}$ %	Remarks
C1	33	8	2	Modified RAF 27 section — low speed
C2	30	12	2	Derived from Göttingen 436 aerofoil, note thicker LE for ease of manufacture
C3	50	12	2	t_{max} well back for high speed
C4	30	12	6	Note thicker TE for ease of manufacture
C5	30	12	12	Modified C4 with still thicker TE, more robust, slightly higher loss
C7	40	12	6	Similar to C3 but t_{max} further forward. For higher speed operation, 'obtained from NACA series of aerofoils'

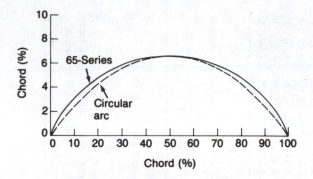

Fig. A1 A comparison of the NACA 65-series camber line and a circular arc camber line with the same maximum camber. Note the scale normal to the chord line is five times that along the chord. (From Lieblein, 1965)

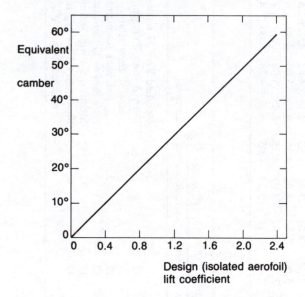

Fig. A2 The circular arc camber angle to give a good approximation to the camber line for a NACA-65 aerofoil defined in terms of lift coefficient for an isolated aerofoil. (From Lieblein, 1965)

In laying out the thickness the distance is measured along the *camber* line. (cf. the 65-series where thickness is positioned on the camber line at distances measured along the chord line.)

The C4 is one of a series, and the leading parameters for these and the corresponding thickness distributions are set out in Table A2. According to unpublished reports the sections C3, C4 and C5 were obtained from aerofoils calculated mathematically in a similar manner to the NACA sections, then modified slightly to give supposedly better performance in cascade. The C3 was intended for high-speed use, the C4 and C5 with increased trailing edge radius, 6 per cent and 12 per cent respectively, for low speeds.

The C3 had its maximum thickness at 50 per cent chord and some experiments suggested that this was too far back so the C7, also intended for high Mach number use, was designed with maximum thickness at 40 per cent, the same as the NACA 65-series upon which it was based. The C7 is nevertheless blunter in the leading edge region than the 65-series with a leading edge radius of 12 per cent of maximum thickness compared with 7 per cent for the 65-series. Although much of the research and systematic testing was performed on the C4 section the C7 found wider use in compressors because it was more suited to high-speed operation.

The system of designation for the C-series can be illustrated by two aerofoils:

10C4/30C50 The first 10 denotes the thickness–chord ratio in percent. C4 denotes the profile thickness distribution. The number after the '/' is the camber angle, 30° in this case; the following C denotes that it is a circular arc camber line and the 50 that maximum camber is at 50 per cent chord (with a circular arc camber line the maximum camber is always at 50 per cent chord and this information is redundant).

6C7/25P40 The blade is of 6 per cent maximum thickness with a C7 thickness distribution. The camber angle is 25°, the camber line is parabolic and maximum camber occurs at 40 per cent of chord.

For any camber line shape it is true that the inclination of the blade to the axial direction at inlet and outlet, χ_1 and χ_2, are given by $\chi_1 = \xi + \theta_1$ and $c_2 = \xi - \theta_2$. Here ξ is the blade stagger angle, the angle between the chord line and the axial direction, whilst θ_1 and θ_2 are the angles between the tangent to the camber line and the chord line at inlet and outlet. The total camber $\theta = \theta_1 + \theta_2$. If the camber line is a circular arc it follows at once that the maximum camber occurs at $a = c/2$, i.e. at mid-chord and that $\theta_1 = \theta_2 = \theta/2$.

For the parabolic arc camber line it is less straightforward. It can be shown (Howell, 1942), that

$$\tan \theta_1 = 4b/(4a - c)$$

and $\qquad \tan \theta_2 = 4b/(3c - 4a)$ where

$$\frac{b}{c} = \frac{1}{4 \tan \theta} \left\{ -1 + \sqrt{\left[1 + (4 \tan \theta)^2 \left\{ \frac{a}{c} - \left(\frac{a}{c} \right)^2 - \frac{3}{16} \right\} \right]} \right\}$$

where b is the maximum distance of the camber line from the chord line. As an approximation, good when the position of maximum camber is near mid-chord,

$$\theta \approx 8b/c$$

$$\theta_1 \approx \theta/2[1 + 2(1 - 2a/c)]$$

and $\qquad \theta_2 \approx \theta/2[1 - 2(1 - 2a/c)]$

Bibliography

ABDELHAMID A N 1982 Control of self excited flow oscillations in vaneless diffuser of centrifugal compressor systems. ASME: Paper 82-GT−188

ABDELHAMID A N, BERTRAND J 1979 Distinctions between two types of self excited gas oscillations in vaneless radial diffusers. ASME: Paper 79-GT−58

ABU-GHANNAM B J, SHAW R 1980 Natural transition of boundary layers — the effects of turbulence, pressure gradient and flow history. *Journal of Mechanical Engineering Science* 22(5)

ADAMCZYK J J, STEVENS W, JUTRAS R 1982 Supersonic stall flutter of high-speed fans. *Trans ASME Journal of Engineering for Power* 104: 675−82

ADKINS G G, SMITH L H 1982 Spanwise mixing in axial flow turbomachines. *Trans ASME Journal of Engineering for Power* 104: 97−110

ADLER D 1980 Status of centrifugal impeller internal aerodynamics Part 1: Inviscid flow prediction methods; Part 2: Experiments and influence of viscosity. *Trans ASME Journal of Engineering for Power* 102: 728−46

AGARD 1978 Conference on seal technology in gas turbines. Conference proceedings AGARD CP237

AGARDOGRAPH 1987 AGARD manual on aeroelasticity in axial-flow turbomachines Volume 1: Unsteady turbomachinery aerodynamics. M F Platzer, F O Carta (eds). AGARDograph 298

AMMAN C A, NORDENSEN G E, SKELLENGER G D 1975 Casing modification for increasing the surge margin of a centrifugal compressor in an automotive turbine engine. *Trans ASME Journal of Engineering for Power* 97: 329−36

ANDREWS S J 1949 Tests related to the effect of profile shape and camber-line on compressor cascade performance. Aeronautical Research Council R and M 2743

ARMSTRONG E K, STEVENSON M A 1960 Some practical aspects of compressor blade vibration. *Journal of the Royal Aeronautical Society* 64: 117−30

BAGHDADI S 1977 The effect of rotor blade wakes on centrifugal compressor diffuser performance — a comparative experiment. *Trans ASME Journal of Fluids Engineering* 99: 45−52

BAGHDADI S, McDONALD A T 1975 Performance of three vaned radial diffusers with swirling flow. *Trans ASME Journal of Fluids Engineering* 97: 155−73

BEHLKE R F 1986 The development of a second generation of controlled diffusion airfoils for multistage compressors. *Trans ASME Journal of Turbomachinery* 108: 32−41

BENVENUTI E 1978 Aerodynamic development of stages for industrial centrifugal compressors, Part 1: Testing requirements and equipment — immediate experimental evidence. ASME Paper 78-GT−4; Part 2: Test data analysis, correlation and use. ASME Paper 78-GT−5

BÖLCS A, SUTER P 1986 *Transsonische Turbomachinen*. G Braun, Karlsruhe

484

BOLDMAN D R, BUGGELE A E, SHAW L M 1983 Experimental evaluation of shockless supercritical airfoils in cascade. Paper AIAA-83—0003, AIAA 21st Aerospace Meeting, Reno, Nevada

BOWDITCH D N, COLTIN R E 1983 A survey of inlet engine distortion capability. NASA TM 83421 prepared for AIAA Joint Propulsion Specialists Conference

BRADSHAW P 1973 Effects of streamline curvature on turbulent flow. AGARDograph 169

BRILEY W R 1971 A numerical study of laminar separation bubbles using the Navier—Stokes equations, *J Fluid Mech* **47**: 713—36

BROWN W B, BRADSHAW G R 1949 Design and performance of diffusing scrolls with mixed flow impellers and vaneless diffuser. NACA Report 936

BRYANSTON CROSS P J 1986 High speed flow visualisation. *Progress in Aerospace Sciences* **23**: 85—104

BULLOCK R O 1961 Critical highlights in the development of the transonic compressor. *Trans ASME series D Journal of Engineering for Power* **83**: 243

BURDSALL E A, URBAN R H 1971 Fan-compressor noise: prediction, research and reduction studies. Final report to Dept of Transportation, Federal Aviation Agency, Report FAA-RD—71—73

BUSEMANN A 1928 Das Förderhöhenverhältniss radialer Kreiselpumpen mit logarithmisch-spiraligen Schaufeln. *Zeitschrift fur Angewandte Mathematik und Mechanik* **8**: 371—84

CALVERT W J 1983 Application of an inviscid—viscous interaction method to transonic compressor cascades. AGARD Conference *Viscous Effects in Turbomachines*, Copenhagen. AGARD CP351

CAME P M, HERBERT M V 1980 Design and experimental performance of some high pressure ratio centrifugal compressors. Paper 15, AGARD Conference *Centrifugal Compressor Flow Phenomena and Performance* Brussels. AGARD CP282

CAME P M, MARSH H 1974 Secondary flow in cascades: two simple derivations for the components of vorticity. *Journal of Mechanical Engineering Science* **16**: 391—401

CAMPBELL W 1924 Protection of steam turbine disk wheels from axial vibration. Paper 1920, Spring meeting ASME Cleveland Ohio.

CARRARD A 1923 On calculations for centrifugal wheels. *La Technique Moderne* T.XV 3 and 4 trans J Moore Cambridge Univ CUED/A-Turbo/TR73 1975

CARTA F O 1967 Coupled blade-disk-shroud flutter instabilities in turbojet engine rotors. *Trans ASME Journal of Engineering for Power* **89**: 419—26

CARTA F O 1986 Aeroelasticity and unsteady aerodynamics. Lectures in ASME Turbomachinery Institute, Ames, Iowa

CARTER A D S 1950 The low speed performance of related aerofoils in cascade. Aeronautical Research Council CP29

CARTER A D S 1955 The axial compressor. In Roxbee-Cox H (ed) *Gas Turbine Principles and Practice*. Newnes, ch 5

CARTER A D S 1961 Blade profiles for axial flow fans, pumps and compressors etc. *Proc I Mech E* **175**(15): 775—806

CARTER A D S, HUGHES H P 1946 A theoretical investigation into the effect of profile shape on the performances of aerofoils in cascade. Aeronautical Research Council R and M 2384

CASEY M V 1983 A computational geometry for the blades and internal flow channels of centrifugal compressors. *Trans ASME Journal of Engineering for Power* **105**: 288—95

CASEY M V 1987 A mean line prediction method for estimating the performance characteristic of an axial compressor stage. IMechE International Conference *Turbomachinery — Efficiency Prediction and Improvement*, Cambridge, Paper C264/97

CASEY M V, MARTY F 1986 Centrifugal compressors — performance at design and off design. *Proc. Institute of Refrigeration*

CASEY M V, ROTH P 1984 A streamline curvature throughflow method for radial turbocompressors. In *Computational Methods in Turbomachinery*. IMechE Publications, London

CETIN M, UCER A S, HIRSCH Ch, SEROVY G K 1987 Application of modified loss and deviation correlations to transonic axial compressors. AGARD Report No. 745

CHENG P, PRELL M E, GREITZER E M, TAN C S 1984 Effects of compressor hub treatment on stator stall and pressure rise. *AIAA Journal of Aircraft* **21**: 469–75

CHESHIRE L J 1945 The design and development of centrifugal compressors for aircraft gas turbines. *Proc IMechE* **153**

CHIMA R V, STRAZISAR A J 1983 Comparison of two and three-dimensional flow computations with laser anemometer measurements in a transonic compressor rotor. *Trans ASME Journal of Engineering for Power* **105**: 596–605

CONSTANT H 1939 Note on performance of cascades of aerofoils. Royal Aircraft Establishment, Note E 3696 (unpublished)

CRAWLEY E F 1985 Optimisation and mechanism of mistuning in cascades. *Trans ASME Journal of Engineering for Gas Turbines and Power* **107**: 418–26

CUMPSTY N A 1972 Tone noise from rotor/stator interaction in high speed fans. *Journal of Sound and Vibration* **24**: 393–409

CUMPSTY N A 1977 Critical review of turbomachinery noise *Trans ASME Journal of Fluids Engineering* **99**: 278–93

CUMPSTY N A, LOWRIE B W 1974 The cause of tone generation by aero-engine fans at high subsonic speeds and the effect of forward speed. *Trans ASME Journal of Engineering for Power* **96**: 228–34

DANFORTH C E 1975 Distortion-induced vibration in fan and compressor blading *Journal of Aircraft* **12**: 216–25

DAS D K, JIANG H K 1984 An experimental study of rotating stall in a multistage axial flow compressor. *Trans ASME Journal of Engineering for Gas Turbines and Power* **106**: 542–51

DAWES W N 1987a A numerical analysis of the three-dimensional viscous flow in a transonic compressor rotor and comparison with experiment. *Trans ASME Journal of Turbomachinery* **109**: 83–90

DAWES W N 1987b Application of a three-dimensional viscous compressible flow solver to a high-speed centrifugal compressor rotor — secondary flow and loss generation. IMechE International Conference *Turbomachinery — Efficiency Prediction and Improvement*, Cambridge, Paper C261/87

DAY I J, CUMPSTY N A 1978 The measurement and interpretation of flow within rotating stall cells in axial compressors. *Journal of Mechanical Engineering Science* **20**: 107–14

DAY I J, GREITZER E M, CUMPSTY N A 1978 Predictions of compressor performance in rotating stall. *Trans ASME Journal of Engineering for Power* **100**: 1–14

DEAN R C 1954 The influence of tip clearance on boundary layer flow in a rectilinear cascade. *MIT Gas Turbine Laboratory Report* 27–3

DEAN R C 1959 On the necessity of unsteady flow in fluid machines. *Trans ASME, Journal of Basic Engineering* **81**: 24–8

DEAN R C 1971 On the unresolved fluid dynamics of the centrifugal compressor. In Dean R C (ed) *Advanced Centrifugal Compressors*. ASME, ch 1

DEAN R C 1974 The fluid dynamic design of advanced centrifugal compressors. Creare TN-185, Creare Inc, Hanover, New Hampshire, presented as lectures at von Karman Institute, Brussels, 1974

DEAN R C, SENOO Y 1960 Rotating wakes in vaneless diffusers. *Trans ASME Journal of Basic Engineering* **82**: 563–74

DEAN R C, YOUNG L R 1977 The time domain of centrifugal compressor and pump stability and surge. *Trans ASME Journal of Fluids Engineering* **99**: 53–63

DENTON J D 1978 Throughflow calculations for transonic axial flow turbines. *Trans ASME Journal of Engineering for Power* **100**: 212–18

DENTON J.D. 1982 An improved time-marching method for turbomachinery flow calculation. ASME International Gas Turbine Conference and Exhibit, Wembley, England. Paper 82-GT–239

DENTON J D 1986 The use of a distributed body force to simulate viscous effects in 3D flow calculations. ASME 31st International Gas Turbine Conference and Exhibit, Dusseldorf, Paper 86-GT–144.

DENTON J D 1987 Computational methods for turbomachinery flows. Lectures to ASME Turbomachinery Institute, Ames, Iowa

DHAWAN S, NARASIMHA R 1958 Some properties of boundary layer flow during the transition from laminar to turbulent. *Journal of Fluid Mechanics* **3**: 418–36

DONG Y 1988 Boundary layers on compressor blades. PhD Disseration, University of Cambridge

DONG Y, GALLIMORE S J, HODSON H P 1987 Three-dimensional flows and loss reduction in axial compressors. *Trans ASME Journal of Turbomachinery* **109**: 354–61

DONOVAN L F 1984 Calculation of transonic flow in a linear cascade. NASA Tech Memo 83697, presented at 20th Joint Propulsion Conference AIAA/SAE/ASME

DOWLING A P, FFOWCS WILLIAMS J E 1983 *Sound and sources of sound*. Ellis Horwood

DRING R P, JOSLYN H D, WAGNER J H 1983 Compressor rotor aerodynamics. AGARD Conference *Viscous Effects in Turbomachines,* Copenhagen, CP351

DUNCAN W J, THOM A S, YOUNG A D 1970 *The Mechanics of Fluids* 2nd edn. Arnold

DUNHAM J 1965 Non-axisymmetric flows in axial compressors. *Mechanical Engineering Science Monograph* No 3, IMechE

DUNHAM J 1972 Predictions of boundary layer transition on turbomachinery blades. AGARD AG164

DUNHAM J 1974 The effect of stream surface convergence on turbomachine blade boundary layers. *Aeronautical Journal*: 90–92

ECKARDT D 1975 Instantaneous measurements in the jet wake discharge flow of a centrifugal compressor impeller. *Trans ASME Journal of Engineering for Power* **97**: 337–46

ECKARDT D 1976 Detailed flow investigations within a high speed centrifugal compressor impeller. *Trans ASME Journal of Fluids Engineering* **98**: 390–402

ECKARDT D 1977 Vergleichende Stromungsuntersuchungen an drei Radialverdichter-Laufradern mit konventionellen Messverfahren. Forschungsbericht Verbrennungskraftmaschinen, Vorhaben 182, Heft 237.

ECKARDT D 1978 Radialverdichter Untersuchung der Laufradströmung in hochbelaster Radialverdichterstufen Rad B. Forschungsbericht Verbrennungskraftmaschinen, Vorhaben 182, Heft 243

ECKARDT D 1979 Jet-wake mixing in the diffuser entry region of a high speed centrifugal compressor. *Joint Symposium on Design and Operation of Fluid Machinery* **1**: 301–20 IAHR/ASME/ASCE

ECKARDT D 1980 Flow field analysis of radial and backswept centrifugal compressor impellers, Part 1: Flow measurement using a laser velocimeter. 25th ASME Gas Turbine Conference and 22nd Annual Fluids Engineering Conference, New Orleans. Symposium *Performance Prediction of Centrifugal Pumps and Compressors* Proceedings published by ASME

ECKERT B, SCHNELL E 1980 Axial- und Radialkompressoren 2nd ed. Springer-Verlag

EMBLETON T W F 1963 Experimental study of noise reduction in centrifugal blowers. *Journal of the Acoustical Society of America* **35**: 700–5

EMERY J C, HERRIG L J, ERWIN J R, FELIX A R 1958 Systematic two-dimensional cascade tests of NACA 65-series compressor blades at low speeds. NACA Report 1368

EMMONS H W, PEARSON C E, GRANT H P 1955 Compressor surge and stall propagation. *Trans ASME* **77**: 455–69

EPSTEIN A H 1977 Quantitative density visualisation in a transonic compressor rotor. *Trans ASME Journal of Engineering for Power* **99**

EVANS B J 1971 Free-stream turbulence effects in a compressor cascade. Ph.D. Dissertation, University of Cambridge

FABRI J 1967 Rotating stall in axial flow compressors. *Internal Aerodynamics (Turbomachinery)*. IMechE

FALCHETTI F, BROCHET J 1987 Calcul des écoulements secondaires dans un compresseur axial multiétage. *AGARD PEP Meeting*, Paris June 1987

FELIX A, EMERY J C 1953 A comparison of typical Gas Turbine Establishment and NACA axial-flow compressor blade sections in cascade of low speed. NACA TN **3937**

FERGUSON T B 1963 *The centrifugal compressor stage*. Butterworth, London

FERNHOLZ H 1964 Halbempirische Gesetze zur Berechnung turbulenter Grenzschichten nach der Methode der Integralbedingungen. *Ingenieur Archiv* **33**: 384–95

FINK D A 1988 Surge dynamics and unsteady flow phenomena in centrifugal compressors. PhD Thesis, Massachusetts Institute of Technology. Also Gas Turbine Lab Report 193

FISCHER K, THOMA D 1932 Investigation of the flow conditions in a centrifugal pump. *Trans ASME* HYD-54–8

FISHER E H, INOUE M 1981 A study of diffuser rotor interaction in a centrifugal compressor. *Journal of Mechanical Engineering Science* **23**: 144–56

FOWCS WILLIAMS J E, HAWKINGS D L 1969 Sound generation by turbulence and sources in arbitrary motion. *Phil. Trans, Royal Society of London*, Series A **264**: 331–42

FORD R A J, FOORD C A 1980 An analysis of aeroengine fan flutter using twin orthogonal vibration modes. *Trans ASME Journal of Engineering for Power* **102**: 367–81

FOWLER H S 1968 The distribution and stability of flow in a rotating channel. *Trans. ASME Journal of Engineering for Power* **90**: 229–36

FOZI A A 1985 Discussion of paper by R A Huntingdon. *Trans ASME Journal for Gas Turbines and Power* **107**: 878

FREEMAN C 1985 Effect of tip clearance flow on compressor stability and engine performance. von Karman Institute for Fluid Dynamics, Lecture series 1985–05

FREEMAN C and CUMPSTY N A 1989 A method for the prediction of supersonic compressor blade performance. Paper to the 34th ASME Gas Turbine Conference and Exposition, Toronto

FREEMAN C, DAWSON R E 1983 Core compressor developments for large civil jet engines. Paper 83−Tokyo−IGTC−46. *International Gas Turbine Congress* Tokyo October

FRIGNE P, VAN DEN BRAEMBUSSCHE R 1983 Distinction between different types of impeller and diffuser rotating stall in a centrifugal compressor with vaneless diffuser. ASME 28th International Gas Turbine Conference and Exhibit, Phoenix Arizona. Paper 83-GT-61

FUJITA H, TAKATA H 1984 A study of configurations of casing treatment for axial flow compressors. *Bulletin of Japan Society Mechanical Engineers* **27**: (230) 1675−81

GALLIMORE S J, CUMPSTY N A 1986 Spanwise mixing in multistage axial flow compressors Part I: Experimental investigation. *Trans ASME Journal of Turbomachinery* **108**: 2−9

GALLIMORE S J 1986 Spanwise mixing in multistage axial flow compressors Part II: Throughflow calculatioins including mixing. *Trans ASME Journal of Turbomachinery* **108**: 10−16

GARABEDIAN P, KORN D 1976 A systematic method for computer design of supercritical airfoils in cascade. *Communications in Pure and Applied Mathematics* **XXIX**

GINDER R B, NEWBY D R An improved correlation for the broadband noise of high speed fans. *Journal of Aircraft* **14**: 844−9

GINDER R B, CALVERT W J 1987 The design of an advanced civil fan rotor. *Trans. ASME Journal of Turbomachinery* **109**: 340−5

GOLDSTEIN A W, MAGER A 1950 Attainable circulation about airfoils in cascade. NACA Report 953

GORDON C G Spoiler-generated flow noise II: Results. *Journal of the Acoustical Society of America* **45**: 214−23

GOSTELOW J P 1971 Design and performance evaluation of four transonic compressor rotors. *Trans ASME Journal of Engineering for Power* **93**: 33−41

GOSTELOW J P 1984 *Cascade Aerodynamics*. Pergamon

GREITZER E M 1976 Surge and rotating stall in axial flow compressors Parts I & II. *Trans ASME Journal of Engineering for Power* **98**: 190−217

GREITZER E M 1980 Review−axial compressor stall phenomena. *Trans ASME Journal of Fluids Engineering* **102**: 134−51

GREITZER E M 1981 The stability of pumping systems. *Trans ASME Journal of Fluids Engineering* **103**: 193−242

GREITZER E M, NIKKANEN J P, HADDAD D E., MAZZAWAY R S, JOSLYN H D 1979 A fundamental criterion for the application of rotor casing treatment. *Trans ASME Journal of Engineering for Power* **101**: 237−43

HAH C A 1986 Numerical modeling of endwall and tip-clearance flow of an isolated compressor rotor. *Trans ASME Journal for Gas Turbines and Power* **108**: 15−22

DE HALLER P 1953 Das Verhalten von Tragflugelgittern in Axialverdichtern und in Windkanal. *Brennstoff-Warme-Kraft* **5**, Heft 10

HALLIWELL D G 1975 Fan supersonic flutter: prediction and test analysis. Aeronautical Research Council R and M 3789

HAMRICK J T, GINSBURG A, OSBORN W M 1951 Method of analysis for compressible flow through mixed flow centrifugal impellers of arbitrary design. NACA Report 1082

HANSON B 1973 Unified analysis of fan stator noise. *Journal of the Acoustical Society of America* **54**: 1571–91

HANSON B 1974 Measurements of static inlet turbulence. Unpublished report of Hamilton Standard, Windsor Locks Connecticut, Report 6558

HARADA H 1985 Performance characteristics of shrouded and unshrouded impellers of a centrifugal compressor. *Trans ASME Journal of Engineering for Gas Turbines and Power* **108**: 32–7

HAWTHORNE W R 1951 Secondary circulation in fluid flow. *Proceedings of the Royal Society*. London Series A **206**: 374–87

HAWTHORNE W R (ed) 1964 *Aerodynamics of Turbines and Compressors*. Volume X of the Princeton Series on High Speed Aerodynamics and Jet Propulsion, Princeton University Press

HAWTHORNE W R 1965 The applicability of secondary flow analyses to the solution of internal flow problems. *Proc. of the Symposium on the Fluid Mechanics of Internal Flow* General Motors Research Labs, Michigan 1965. Elsevier Publishing Company 1967

HAWTHORNE W R 1974 Secondary vorticity in stratified compressible fluids in rotating systems. Cambridge University Engineering Department, CUED/A-Turbo/TR 63

HAWTHORNE W R, TAN C S 1987 Design of turbomachinery blading in three-dimensional flow by the circulation method: a progress report. *Second International Conference on Inverse Design Concepts and Optimisation in Engineering Sciences* (ICIDES-II), State College, Pennsylvania

HAYWOOD R W 1980 *Analysis of Engineering Cycles* 3rd edn. Pergamon Press

HEAD M R 1958 Entrainment in the turbulent boundary layer. *Aeronautical Research Council* R and M 3152

HERBERT M V, CALVERT W J 1982 Description of an integral method for boundary layer calculation in use at NGTE with special reference to compressor blades. *NGTE Memorandum* 82019

HERZIG H G, HANSEN A G, COSTELLO G R 1954 A visualisation study of secondary flows in cascades. NACA Report 1163

HETHERINGTON R Compressor noise generated by fluctuating lift resulting from rotor-stator interaction *AIAA Journal* **1**: 473–4

HIRSCH C, DENTON J D 1981 Throughflow calculations in axial turbomachinery. *AGARD* AR 175

HIRSCH C. WARZEE G 1979 An integrated quasi-3D finite element calculation program for turbomachinery flows. *Trans ASME Journal of Engineering for Power* **101**: 141–8

HOBBS D E, WEINGOLD H D 1984 Development of controlled diffusion aerofoils for multistage compressor applications. *Trans ASME Journal of Engineering for Gas Turbines & Power* **106**: 271–8

HOHEISEL H and SEYB N J 1986 The boundary layer behaviour of highly loaded compressor cascade at transonic flow conditions. AGARD Conference *Transonic and Supersonic Phenomena in Turbomachines* Munich, AGARD CP401

HORLOCK J H 1958 *Axial Compressors*. Butterworth, reprinted with supplementary material, Krieger Publishing Co Inc 1973

HORLOCK J H 1969 Boundary layer problems in turbomachinery. *Proceedings of the Symposium on Flow Research on Blading*. Brown Boveri Co Ltd, Baden, Switzerland 1969. In Dzung L S (ed) *Flow Research on Blading*. Elsevier Publishing Company 1970

HORLOCK J H 1978 *Actuator Disc Theory*. McGraw-Hill

HORLOCK J H 1971 On entropy production in adiabatic flow in turbomachines. *Trans ASME Journal of Basic Engineering* **93**: 587–93

HORLOCK J H, LAKSHMINARAYANA B 1973 Secondary flows: theory, experiment and application in turbomachinery aerodynamics. *Annual Review of Fluid Mechanics* **5**: 247–80

HORLOCK J H, MARSH H 1971 Flow models for turbomachinery. *Journal of Mechanical Engineering Science* **13**: 358–68

HORLOCK J H, PERKINS H J 1974 Annulus wall boundary layers in turbomachines. AGARDograph 185

HORTON H P 1969 A semi-empirical theory for the growth and bursting of laminar separation bubbles. Aeronautical Research Council CP 1073

HOWELL A R 1942 The present basis of axial flow compressor design, Part I: cascade theory and performance. *Aeronautical Research Council* R and M 2095

HOWELL A R 1944 A note on compressor base aerofoils C1 C2 C3 C4 C5 and aerofoils made up of circular arcs. Power Jet Memorandum NOM 1011

HOWELL A R 1945 Fluid dynamics of axial compressors. *Proc. I. Mech. E.* **153**: 441–82

HOWELL A R 1948 The aerodynamics of the gas turbine. *Journal of the Royal Aeronautical Society* **52**: 329–56

HOWELL A R, CALVERT W J 1978 A new stage stacking technique for axial-flow compressor performance prediction. *Trans ASME Journal of Engineering for Power* **100**: 698–703

HOWELL A R, CARTER A D S 1946 Fluid flow through cascade of aerofoil. 6th International Congress of Applied Mechanics, Paris 1946

HÜBER G H 1963 Noise of centrifugal fans and rotating cylinders. *Journal of the American Society of Heating Refrigeration and Air-conditioning Engineers* **5**: 87–94

HUBERT G 1963 Untersuchungen über die Sekundärveluste in axial Turbomachinen. *VDI Forschungshelt* 496, VDI Verlag, Dusseldorf

HUNTER I H, CUMPSTY N A 1982 Casing wall boundary layer development through an isolated compressor rotor. *Trans ASME Journal of Engineering for Power* **104**: 805–17

HUNTINGDON R A 1985 Evaluation of polytropic calculation methods for turbomachinery performance. *Trans ASME Journal of Engineering for Gas Turbines and Power* **107**: 872–9

HYNES T P, GREITZER E M 1987 A method for assessing effects of circumferential flow distortion on compressor stability. *Trans ASME Journal of Turbomachinery* **109**: 371–9

HYNES T P, CHUE R, GREITZER E M, TAN C S 1986 Calculation of inlet distortion induced compressor flow field instability. AGARD Conference *Engine Response to Distorted Inflow Conditions*, Munich CP400

INOUE M 1983 Radial vaneless diffusers: a re-examination of the theories of Dean and Senoo and Johnston and Dean. *Trans ASME Journal of Fluid Engineering* **105**: 21–7

INOUE M, CUMPSTY N A 1984 Experimental study of centrifugal impeller discharge flow in vaneless and vaned diffusers. *Trans ASME Journal for Gas Turbines and Power* **106**: 455–67

INOUE M, KUROUMARU M, FUKUHARA M 1986 Behaviour of tip clearance flow behind an axial compressor rotor. *Trans ASME Journal of Engineering for Gas Turbines and Power* **108**: 7–14

ISHIDA M, SENOO Y 1981 On the pressure losses due to the tip clearance of centrifugal blowers. *Trans ASME Journal of Engineering for Power* **103**: 271

JANSEN W 1964a Steady fluid flow in a radial vaneless diffuser. *Trans ASME Journal of Basic Engineering* **86**: 607–19

JANSEN W 1964b Rotating stall in a radial vaneless diffuser. *Trans ASME Journal of Basic Engineering* **86**: 750–8

JANSEN W, CARTER A F, SWORDEN M C 1980 Improvement in surge margin for centrifugal compressors. Presented at AGARD 55th specialists' meeting *Centrifugal Compressors, Flow Phenomena and Performance*. Brussels

JANSEN W, MOFFATT W C 1967 The off-design analysis of axial flow compressors. *Trans ASME Journal of Engineering for Power* **89**: 453–62

JAPIKSE D 1980 The influence of diffuser inlet pressure fields on the range and durability of centrifugal compressor stages. AGARD Conference *Centrifugal Compressors, Flow Phenomena and Performance*, Brussels

JAPIKSE D 1984 *Turbomachinery Diffuser Design Technology*. The Design Technology Series DTS-1, Concepts ETI Inc., Norwich, Vermont, USA

JAPIKSE D 1987 A critical evaluation of three centrifugal compressors with pedigree data sets, Part 5—Studies in component performance. *Trans ASME Journal of Turbomachinery* **104**: 1–9

JAPIKSE D, OSBORNE C 1986 Optimisation of industrial centrifugal compressors, Part 6A and 6B. ASME Gas Turbine Conference, Dusseldorf, Papers 86-GT-221 & 222

JENNIONS I K, STOW P 1985 The quasi-three-dimensional turbomachinery blade design system, Part I: Throughflow analysis, Part II: Computerised system. *Trans ASME Journal of Engineering for Gas Turbines and Power* **107**: 308–16

JENNIONS I K, STOW P 1986 The importance of circumferential non-uniformities in a passage-averaged quasi-three-dimensional turbomachinery design system. *Trans ASME Journal of Engineering for Gas Turbines and Power* **108**: 240–5

JOHNSEN I A, GINSBURG A 1953 Some NACA research on centrifugal compressors. *Trans ASME* **75**: 805–17

JOHNSON M C, GREITZER E M 1987 Effects of slotted hub and casing treatments on compressor endwall flow fields. *Trans ASME Journal of Turbomachinery* **109**: 380–7

JOHNSON M W 1978 Secondary flows in rotating bends. *Trans ASME Journal of Engineering for Power* **100**: 553–60

JOHNSON M W, MOORE J 1983a Secondary flow mixing losses in a centrifugal impeller. *Trans ASME Journal of Engineering for Power* **105**: 24–32

JOHNSON M W, MOORE J 1983b The influence of flow rate on the wake in a centrifugal impeller. *Trans ASME Journal of Engineering for Power* **105**: 33–9

JOHNSON J P 1973 The suppression of shear layer turbulence in rotating systems. *Trans ASME Journal of Fluid Engineering* **95**: 229–36

JOHNSTON J P 1986 Radial flow turbomachinery. Lecture in series *Fluid Dynamics of Turbomachinery*. ASME Turbomachinery Institute, Ames, Iowa

JOHNSTON J P, DEAN R C 1966 Losses in vaneless diffusers of centrifugal compressors and pumps. *Trans ASME, Journal of Engineering for Power* **88**: 49–62

JOHNSTON J P, EIDE S A 1976 Turbulent boundary layers on centrifugal compressor blades: prediction of the effect of surface curvature and rotation. *Trans ASME Journal of Fluid Engineering* **98**: 374–81

KAJI S 1975 Noncompact source effect on the prediction of tone noise from a fan rotor. AIAA Paper No. 75–446

KAJI S, OKAZAKI T 1970 Propagation of sound waves through a blade row. *Journal of Sound and Vibration* **11**: 355–75

KATSANIS T 1965 Use of quasi-orthogonals for calculating flow distributions on a blade-to-blade surface in a turbomachine. NASA TN D2809

KEENAN J H, KAYE J 1948 *Gas Tables*. John Wiley, New York.

KEMP N H, SEARS W R 1955 The unsteady forces due to viscous wakes in turbo-machines. *Journal of Aeronautical Science* **22**: 478–83

KENNY D P 1972 A comparison of the predicted and measured performance of high pressure ratio centrifugal compressor diffusers. Lecture notes for *Advanced radial compressors* course held at von Karman Institute, Brussels

KENNY D P 1972 A comparison of the predicted and measured performance of high pressure ratio centrifugal compressor diffusers. ASME Paper 72-GT–54

KENNY D P 1984 The history and future of the centrifugal compressor in aviation gas turbines. *1st Cliff Garrett Turbomachinery Award Lecture*. Society of Automotive Engineers paper SAE/SP-804/602

KERREBROCK J L 1974 Waves and Wakes in turbomachine annuli with swirl. AIAA Paper 74–87

KERREBROCK J L 1981 Flow in transonic compressors. 1980 Dryden Lecture, *AIAA Journal* **19**(1): 4–19

KERREBROCK J L and MIKOLAJCZAK A A 1970 Intra-stator transport of rotor wakes and its effect on compressor performance. *Trans ASME Journal of Engineering for Power* **92**: 359–69

KIOCK R 1984 Comparative review of laser-doppler and laser-two-focus anemometry in view of turbomachinery applications. Technical Memorandum 34 of the von Karman Institute for Fluid Dynamics, Rhode Saint Genese, Belgium

KLINE S J, COCKRELL D J, MURKOVIN M V 1968 Proceedings of the AFOSR-IFP-Stanford 1968 Conference on Turbulent Boundary Layer Prediction

KOCH C C 1964 Charts of adiabatic compression efficiency versus polytropic efficiency using properties of real air. General Electric unpublished report R64FPD229

KOCH C C 1981 Stalling pressure rise capability of axial flow compressor stages. *Trans ASME Journal of Engineering for Power*, **103**: 645–56

KOCH C C, SMITH L H 1976 Loss sources and magnitudes in axial-flow compressors. *Trans ASME Journal of Engineering for Power* **98**: 411–24

KOVACH K, SANDERCOCK D M 1961 Aerodynamic design and performance of five stage transonic axial flow compressor. *Trans ASME Journal of Engineering for Power* **83**: 303–21

KRAIN H 1984 A CAD method for centrifugal compressor impellers. *Trans ASME Journal of Engineering for Gas Turbines and Power* **106**: 482–88

KRAIN H 1985 Interdependence of centrifugal compressor blade geometry and relative flow field. ASME 30th International Gas Turbine Conference and Exhibit, Houston Texas, Paper 85-GT–85

KRAIN H 1987 Swirling impeller flow. 32nd ASME Gas Turbine Conference and Exhibit, Anaheim, California. Paper 87-GT–19

KRIMERMAN Y, ADLER D 1978 The complete three-dimensional calculation of the compressible flow field in turbo impellers. *Journal of Mechanical Engineering Science* **20**: 149–58

KRISHNAPPA G 1976 Centrifugal blower noise studies, literature survey and noise measurements. National Research Council of Canada, Mechanical Engineering

Report ME-244 NRC15679

KRISHNAPPA G 1977 Some experimental results on centrifugal blower noise. International Conference on Noise Control Engineering, Zurich.

KUROSAKA M 1971 A note on multiple pure tone noise. *Journal of Sound and Vibration* **19**: 453–62

LAKSHMINARAYANA B 1970 Methods of predicting the tip clearance effect in axial flow turbomachinery. *Trans ASME Journal of Basic Engineering* **92**: 467–82

LAKSHMINARAYANA B, HORLOCK J H 1967 Leakage and secondary flow in compressor cascades. Aeronautical Research Council R and M 3483

LAKSHMINARAYANA B, HORLOCK J H 1973 Generalised expressions for secondary vorticity using intrinsic coordinates. *Journal of Fluid Mechanics* **59**: 97–115

LAKSHMINARAYANA B, SITARAM N, ZHANG J 1986 Endwall and profile losses in a low speed axial flow compressor rotor. *Trans ASME Journal of Engineering for Gas Turbines and Power* **108**: 23–31

LANSING D L, DRISCHLER J A, PUSEY C G 1970 Radiation of sound from an unflanged circular duct with flow. 79th Meeting of the Acoustical Society of America

LAWSON T V 1953 An investigation into the effect of Reynolds Number on a cascade of blades with parabolic arc camber line. National Gas Turbine Est. Memo M195

LEE N K W 1988 Effects of compressor endwall suction and blowing on stability enhancement. Massachusetts Institute of Technology, Gas Turbine Laboratory Report 192

LIEBLEIN S 1956 Experimental flow in 2D cascades. Chapter VI of *The Aerodynamic Design of Axial Flow Compressor*. Reprinted NASA SP 36 in 1965. (Originally NACA RME 56B03)

LIEBLEIN S 1960 Incidence and deviation-angle correlations for compressor cascades. *Trans ASME Journal of Basic Engineering* **82**: 575–87

LIEBLEIN S, JOHNSEN I A 1961 Resumé of transonic compressor research at NACA Lewis Laboratory. *Trans ASME Journal of Engineering for Power* **83**(3): 219–34

LIEBLEIN S, ROUDEBUSH W H 1956 Theoretical loss relations for low speed two-dimensional cascade flow. NACA TN 3662

LIEBLEIN S, SCHWENK F C, BRODERICK R L 1953 Diffusion factor for estimating losses and limiting blade loadings in axial flow compressor blade elements. NACA RM E53D01

LIEPMANN H W, ROSHKO A 1957 *Elements of Gas Dynamics*. Wiley

LINDSAY W L, CARRICK H B, HORLOCK J H 1987 Three-dimensional calculation of wall boundary layer flows in turbomachines. ASME Paper 87-GT–133, 32nd ASME Gas Turbine Conference and Exhibit, Anaheim, California

LIPSTEIN N J, MANI R 1972 Experimental investigation of discrete frequency noise generated by unsteady blade forces. *Trans ASME Journal of Basic Engineering* **92**: 155–64

LONGLEY J 1988 Inlet distortion and compressor stability, PhD Dissertation, University of Cambridge

LOWRIE B W 1975 Simulation of flight effects on aero engine fan noise. AIAA Paper 75–463

LOWSON M V 1970 Theoretical analysis of compressor noise. *Journal of the Acoustical Society of America* **47**: 371–85

LOWSON M V, WHATMORE A R, WHITFIELD C E 1973 Source mechanisms for rotor noise radiation. NASA CR-2077

LYONS L A, PLATTER S 1963 Effect of cut-off configuration on pure tones generated by small centrifugal blowers. *Journal of Acoustical Society of America* **35**: 1455–56

MCDONALD H 1985 Computation of turbomachinery boundary layers. In Oates G C (ed) *Aerothermodynamics of Aircraft Engine Components*. AIAA Education series,

MCDONALD H, FISH R W 1973 Practical calculations of transitional boundary layers. *Journal of Heat & Mass Transfer* **16**: 1729—44

MCDOUGALL N M 1988 Stall inception in axial flow compressors. PhD dissertation, University of Cambridge

MACDOUGALL N M, DAWES W N 1987 Numerical simulation of the strong interaction between a compressor blade clearance jet and stalled passage flow. Seventh GAMM conference on Numerical Methods in Fluid Mechanics, Louvain, Belgium

MALLEN M, SAVILLE G 1977 Polytropic processes in the performance prediction of centrifugal compressors. Institution of Mechanical Engineers, Paper C183/77

MANI R 1970 Discrete frequency noise generation from an axial flow fan blade row. *Trans ASME Journal of Basic Engineering* **92**: 37—43

MANI R 1970 Compressibility effects in the Kemp-Sears problems. *Proceedings of the Penn State Conference on Fluid Mechanics and Design of Turbomachinery*. NASA SP304

MANI R 1971 Noise due to interaction of inlet turbulence with isolated stators and rotors. *Journal of Sound and Vibration* **17**: 251—60

MARBLE F E 1964 Three-dimensional flow in turbomachines. Section C. In Hawthorne W R (ed) *Aerodynamics of Turbines and Compressors*. Volume X of the Princeton Series on High Speed Aerodynamics and Jet Propulsion, Princeton University Press

MARIANO S 1971 Effect of wall shear layers on the sound attenuation in acoustically lined rectangular ducts *Journal of Sound and Vibration* **22**: 261—75

MARSH H 1968 A digital computer program for the through-flow fluid mechanics in an arbitrary turbomachine using a matrix method. Aeronautical Research Council R and M 3509

MARTENSEN E 1959 Berechnung der Druckverteilung an Gitterprofilen in ebener Potentialströmung mit einer Fredholmschen Integralgleichung. *Arch Rat Mech Anal* **3**: 235—70

MAZZAWY R S 1977 Multiple segment parallel compressor model. *Trans ASME Journal of Engineering for Power* **99**: 288—97

MAZZAWY R S 1980 Surge-induced structural loads in gas turbines. *Trans ASME Journal of Engineering for Power* **102**: 162—8

MELLOR G 1956 (Unpublished) The 65-series cascade data. Gas Turbine Laboratory, Massachusetts Institute of Technology

MELLOR G L, WOOD G M 1971 An axial compressor end-wall boundary layer theory. *Trans ASME Journal of Basic Engineering* **93**: 300—16

MIKOLAJCZAK A A, MORRIS A L, JOHNSON B V 1971 Comparison of performance of supersonic blading in cascade and in compressor blades. *Trans ASME Journal of Engineering for Power* **93**: 42—8

MIKOLAJCZAK A A, PFEFFER A M 1974 Methods to increase engine stability and tolerance to distortions In *Distortion Induced Engine Instability*. AGARD lecture series 72

MIKOLAJCZAK A A, ARNOLDI R A, SNYDER L E, STARGARDTER H 1975 Advances in fan and compressor blade flutter analysis and prediction. *Journal of Aircraft* **12**: 325—32

MILLER G R, LEWIS G W, HARTMANN M J 1961 Shock losses in transonic rotor rows. *Trans ASME Journal of Engineering for Power* **83**: 235—42

MOECKEL W E 1949 Approximate method for predicting form and location of detached shock waves ahead of plane or axially symmetric bodies. NACA TN 1921

MOORE C J 1972 In duct investigation of subsonic 'rotor alone' noise. *Journal of Acoustical Society of America* **51**: 1471–81

MOORE C J 1975 Reduction of fan noise by annulus boundary layer removal. *Journal of Sound and Vibration* **43**: 671–81

MOORE F K 1984 A theory of rotating stall of multistage compressors Parts I–III. *Trans ASME Journal of Gas Turbines & Power* **106**: 313–36

MOORE F K, GREITZER E M, also GREITZER E M, MOORE F K 1986. A theory of post-stall transients in axial compression systems Parts I & II. *Trans ASME Journal of Engineering for Gas Turbines and Power* **108**: 231–9

MOORE J 1973 A wake and an eddy in a rotating, radial-flow passage Part 1: Experimental observations; Part 2: flow model. *Trans ASME Journal of Engineering for Power* **95**: 205–19

MOORE J 1976 Eckardt's impeller — a ghost from ages past. Cambridge University Engineering Department, Report CUED/A — Turbo/TR83

MOORE J, MOORE J G 1980 Calculation of three-dimensional viscous flow and wake development in a centrifugal impeller. 25th Annual International Gas Turbine Conference & 22nd Annual Fluids Engineering Conference, New Orleans 1980, reproduced in *Performance Prediction of Centrifugal Pumps and Compressors*. ASME.

MOORE J, MOORE J G, TIMMIS P H 1984 Performance evaluation of centrifugal compressor impellers using three-dimensional viscous flow calculations. *Trans ASME Journal for Gas Turbines and Power* **106**: 475–81

MORFEY C L 1970 Broadband sound radiated from subsonic rotors. *International Symposium on the Fluid Mechanics and Design of Turbomachines*. NASA SP-304 Part II

MORFEY C L 1973 Rotating blades and aerodynamic sound. *Journal of Sound and Vibration* **28**: 587–617

MORISHITA E 1982 Centrifugal compressor diffusers. A dissertation submitted for the MSc, University of Cambridge

MORRIS R E, KENNY D P 1971 High pressure ratio centrifugal compressors for small gas turbine engines. In Dean R C (ed) *Advanced Centrifugal Compressors*. ASME

MUGRIDGE B D 1975 Axial flow fan noise caused by inlet flow distortion. *Journal of Sound and Vibration* **40**: 497–512

MUGRIDGE B D 1971 Acoustic radiation from aerofoils with turbulent boundary layers. *Journal of Sound and Vibration* **16**(4)

MUGRIDGE B D, MORFEY C L 1975 Sources of noise in axial flow fans. *Journal of the Acoustical Society of America* **5**(5)

NAGASHIMA T, WHITEHEAD D S 1974 Aerodynamic forces and moments for vibrating supersonic cascades of blades. Cambridge University Report CUED/A-Turbo/TR59

NAMBA M, ASANUMA T 1967 Theory of lifting lines for cascades of blades in subsonic shear flow. *Bulletin Japan Society of Mechanical Engineering* **10**: 920–38

NARASIMHA R 1985 The laminar-turbulent transition zone in the boundary layer. *Progress in Aeronautical Science* **22**: 29–80

NASH J F 1965 Turbulent boundary layer behaviour and the auxiliary equations. AGARDograph **97**: 245–79

NATHOO N S, GOTTENBERG W G 1983 A new look at performance analysis of centrifugal compressors operating with mixed hydrocarbon gases. *Trans ASME Journal of Engineering for Power* **105** (for discussion see *Trans ASME Journal of Engineering for Gas Turbines and Power* **108**: 225)

NAYFEH A H, KAISER J E, TELIONIS D P 1975 The acoustics of aircraft engine duct systems. *AIAA Journal* **13**: 130–53

NEISE W 1976 Noise reduction in centrifugal fans: a literature survey. *Journal of Sound and Vibration* **45**: 373−403

NOVAK R A 1967 Streamline curvature computing procedures for fluid flow problems. *Trans ASME Journal of Engineering for Power* **89**: 478−90

NOVAK R A 1976 Flow field and performance map computation for axial-flow compressors and turbines. AGARD Lecture Series 83, ch 5

NOVAK R A, HEARSEY R M 1977 A nearly three-dimensional intrablade computing system for turbomachinery. *Trans ASME Journal of Fluid Engineering* **99**: 154−66

PAMPREEN R C 1972 The use of cascade technology in centrifugal compressor vaned diffuser design. *Trans ASME Journal of Engineering for Power* **94**: 187−92

PAMPREEN R C 1973 Small turbomachinery compressor and fan aerodynamics. *Trans ASME Journal of Engineering for Power* **95**: 251

PAPAILIOU K 1971 Boundary layer optimisation for the design of high turning axial flow compressor blades. *Trans ASME Journal of Engineering for Power* **103**: 147−55

PARKER R 1967 Resonance effects in wake shedding from compressor blading. *Journal of Sound and Vibration* **6**: 302−9

PARKER R J, JONES D G 1988 The use of holographic interferometry for turbomachinery fan evaluation during rotating tests. *Trans ASME Journal of Turbomachinery* **110**: 393−400

PEACOCK R E 1982 A review of turbomachine tip gap effects: Part 1 Cascades. *International Journal of Heat and Fluid Flow* **3**: 185−93

PEACOCK R E 1982 A review of turbomachine tip gap effects, Part 2: Rotating machines. *International Journal of Heat and Fluid Flow* **4** 3−16

PEARSONS K S, BENNETT R 1974 *Handbook of Noise Ratings*. NASA CR-2376

PFLEIDERER C 1961 *Die Kreisel Pumpen*. 5th edn Springer Verlag

PICKETT G F 1974 Effects of non-uniform inflow on fan noise. Spring meeting of Acoustical Society of America

PIERZGA M J, WOOD J R 1985 Investigation of three-dimensional flow field within transonic fan rotor. Experiment and analysis. *Trans ASME Journal of Engineering for Gas Turbines and Power* **107**: 436−49

POLLARD D, GOSTELOW J P 1967 Some experiments at low speed on compressor cascades. *Trans ASME Journal of Engineering for Power* **89**: 427−36

POUGARE M, GALMES J, LAKSHMINARAYANA B 1984 An experimental study of compressor rotor blade boundary layers. *Trans ASME Journal of Engineering for Gas Turbines and Power* **107**: 364−73

PRESTON J H 1954 A simple approach to the theory of secondary flows. *Aeronautical Quarterly* **V**: 218

PRIAN V D, MICHEL D J 1951 An analysis of flow in rotating passage of large radial inlet centrifugal compressor at tip speed of 700 ft/s. NACA TN 2584

PRINCE D C, WISLER D C, HILVERS D E 1974 Study of casing treatment stall margin improvement phenomenon. NASA CR 134552

PRINCE D C 1980 Three-dimensional shock structures for transonic/supersonic compressor rotors. *Trans AIAA Journal of Aircraft* **17**: 28−37

RAINS D A 1954 Tip clearance flows in axial flow compressors and pumps. California Institute of Technology, Hydrodynamics Laboratory Report 5

RECHTER H, STEINERT W, LEHMANN K 1984 Comparison of controlled diffusion airfoils with conventional NACA 65 airfoils developed for stator application in a multistage axial compressor. *Trans ASME Journal of Engineering for Gas Turbines and Power* **107**: 494−8

REID C 1969 The response of axial flow compressors to intake flow distortion. *ASME Paper* 69-GT−29

REYNOLDS W C 1979 *Thermodynamic Properties in SI*. Department of Mechanical Engineering, Stanford University, California

RHODEN H G 1952 Effects of Reynolds Number on the flow of air through a cascade of compressor blades. Aeronautical Research Council R and M 2919

RIBAUD Y 1987 Experimental aerodynamic analysis of three-dimensional high pressure ratio centrifugal compressors. ASME Paper No. 87-GT−153, 32nd Gas Turbine Conference and Exhibit, Anheim

RIESS W, BLÖCKER U 1987 Possibilities for on-line surge suppression by fast guide vane adjustment in axial compressors. AGARD 69th Propulsion and Energetics Symposium, Paris. AGARD-CP-421

ROBBINS W H, JACKSON R J, LEIBLEIN S 1956 Blade element flow in annular cascades. Chapter VII, *The Aerodynamic Design of Axial Flow Compressors*. Reprinted NASA SP 36 in 1965. (Originally NACA RME 56B03)

ROBERTS W B 1980 Calculation of laminar separation bubbles and their effect on airfoil performance. *AIAA Journal* 18: 25−31

RODGERS C 1977 Impeller stalling as influenced by diffusion limitations. *Trans ASME Journal of Fluids Engineering* 99: 84−97

RODGERS C 1978 A diffusion factor correlation for centrifugal impeller stalling. *Trans ASME Journal of Engineering for Power* 100: 592−603

RODGERS C 1980 Efficiency of centrifugal compressor impellers. Paper 22 of AGARD Conference Proceedings No 282 *Centrifugal Compressors, Flow Phenomena and Performance*. Conference in Brussels May 1980

RODGERS C 1982a The performance of centrifugal compressor channel diffusers. *ASME Paper* 82-GT−10

RODGERS C 1982b Static pressure recovery characteristics of some radial vaneless diffusers. Canadian Aeronautics and Space Institute International Symposium on Centrifugal Compressor Design, Toronto, Canada May 1982

ROTHE P H, JOHNSTON J P 1976 Effects of system rotation on the performance of two-dimensional diffusers. *Trans ASME Journal of Fluids Engineering* 98: 422−30

ROTTA J C 1967 Development of turbulent boundary layers. *Proceedings of the Symposium on the Fluid Mechanics of Internal Flow*. General Motors Research Labs, Michigan 1965, Elsevier Publishing Company

ROUDEBUSH W H 1956 Potential flow in two-dimensional cascades. Chapter IV of *The Aerodynamic Design of Axial Flow Compressors*. Reprinted NASA SP 36 in 1965. (Originally NACA RME 56B03)

RUDEN P 1937 Untersuchungen uber eintufige Axialgelblase. *Luftfahrtforschung* 14: 325−46, 458−73 (trans NACA TM 1062 1944)

RUNDSTADLER P W, DOLAN F X, DEAN R C 1975 *Diffuser Data Book*. Creare TN-186. Hanover, New Hampshire

DE RUYCK J, HIRSCH C 1981 Investigations of an axial compressor endwall boundary layer prediction method. *Trans ASME Journal of Engineering for Power* 103: 20−33

SAKAI T, WATANABE I, FUJIE E, TAKAYANAGAI I 1967 On the slip factor of centrifugal and mixed flow impellers. *ASME Paper* 67−WA/GT-10

SAVKAR S D 1975 Radiation of cylindrical duct acoustic modes with flow mismatch. *Journal of Sound and Vibration* 42: 363−86

SCHÄFFLER A 1980 Experimental and analytical investigation of the effects of Reynolds number and blade surface roughness on multistage axial flow compressors. *Trans ASME Journal of Engineering for Power* **102**: 5−13

SCHLAMAN U, TEIPEL I, RIESS W 1985 Interdependence of rotating stall and surge in axial compressors. VDI-Fortschritt-Berichte, Reihe 7, Nr 91

SCHLICHTING H 1979 *Boundary Layers,* 7th edn. McGraw-Hill

SCHLICHTING H 1954 Problems and results of investigations on cascade flow. *Journal of Aeronautical Science* **21**: 163

SCHLICHTING H, DAS A 1969 On the influence of turbulence level on the aerodynamic losses of axial turbomachinery. *Proceedings of the Symposium on Flow Research on Blading.* Brown Boveri and Co Ltd, Baden, Switzerland 1969. In Dzung L S (ed) *Flow Research on Blading.* Elsevier Publishing Company, 1970

SCHMIDT J F, GELDER T F, DONOVAN L F 1984 Redesign and cascade tests of a super-critical controlled diffusion stator blade section. Paper AIAA-84-1207, AIAA/SAE/ASME 20th Joint Propulsion Conference

SCHODL R 1974 A laser dual beam method for flow measurement in turbomachines. ASME Paper No 74-GT−157

SCHODL R 1984 Optical techniques for turbomachinery flow analysis. Lecture to NATO Advanced Study Institute on *Thermodynamics and Fluid Mechanics of Turbomachinery,* held in Turkey

SCHOLZ N 1977 Aerodynamics of Cascades. AGARDograph No 220. (Translated and revised by A Klein from *Aerodynamic der Schaufelgitter.* Verlag Braun 1965)

SCHREIBER H-A, STARKEN H 1981 Evaluation of blade element performance of compressor rotor blade cascades in transonic and low supersonic flow range. 5th International Symposium on Airbreathing Engines, Bangalore, India

SCHULZ J M 1962 The polytropic analysis of centrifugal compressors. *Trans ASME Journal of Engineering for Power* **84**: 69−82

SCHWEITZER J K, GARBELOGLIO J E 1984 Maximum loading capability of axial flow compressors. *AIAA Journal of Aircraft* **21**: 593−600

SEDDON J, GOLDSMITH E L 1985 *Intake Aerodynamics.* Collins, London

SENOO Y 1987 Pressure losses and flow field distortion induced by tip clearance of centrifugal and axial compressors. *International Journal of Japan Society of Mechanical Engineers* **30**: 375−85

SENOO Y, HAYAMI H, KINOSHITA Y, YAMASAKI H 1979 Experimental study on flow in a supersonic centrifugal impeller. *Trans ASME Journal of Engineering for Power* **101**: 32−41

SENOO Y, ISHIDA M 1986 Pressure loss due to the tip clearance of impeller blades in centrifugal and axial blowers. *Trans ASME Journal of Engineering for Gas Turbines and Power* **108**: 32−7

SENOO Y, ISHIDA M 1987 Deterioration of compressor performance due to tip clearance of centrifugal impellers. *Trans ASME Journal of Turbomachinery* **109**: 55−61

SENOO Y, KINOSHITA Y 1977 Influence of inlet flow conditions and geometries of centrifugal vaneless diffusers on critical flow angle for reverse flow. *Trans ASME Journal of Fluids Engineering* **99**: 98−103

SENOO Y, KINOSHITA Y, ISHIDA M 1977 Asymmetric flow in vaneless diffusers of centrifugal blowers. *Trans ASME Journal of Fluids Engineering* **99**: 104−14

SEROVY G K 1976 Compressor and turbine performance prediction system development — lessons from thirty years of history. AGARD Lecture Series 83, ch 3.

SEROVY G 1981 Deviation/turning angle correlations. Section II.2.8 of *AGARD Advisors Reports* No 175 of Propulsion and Energetics Panel Working Group 12 on Throughflow calculation in axial turbomachinery. Ed Hirsch and Denton

SEYB N J 1972 The role of boundary layers in axial flow turbomachines and the prediction of their effects. *AGARDograph* AG 164: 241−59

SHARLAND I J 1964 Sources of noise in axial flow fans. *Journal of Sound and Vibration* 1: 302−22

SIDERIS M TH, VAN DEN BRAEMBUSSCHE R A 1987 Influence of a circumferential exit pressure distribution on the flow in an impeller and diffuser. *Trans ASME Journal of Turobmachinery* 109: 48−54

SIMON H, WALLMANN T, MONK T 1987 Improvements in performance characteristics of single-stage and multistage centrifugal compressors by simultaneous adjustments of inlet guide vanes and diffuser vanes. *Trans ASME Journal of Turbomachinery* 109: 41−7

SISTO F 1987a Introduction and overview. ch I *AGARD Manual on Aeroelasticity of Axial Flow Turbomachines*. AGARDograph 298

SISTO F 1987b Stall flutter. ch VII *AGARD Manual on Aeroelasticity of Axial Flow Turbomachines*. AGARDograph 298

SMITH G D J, CUMPSTY N A 1984 Flow phenomena in compressor casing treatment. *Trans ASME Journal of Engineering for Gas Turbines and Power* 106: 532−41

SMITH L H 1954 A note on the NACA diffusion factor. *General Electric AGT Development Department Memorandum* CD-9 (9-54) (unpublished)

SMITH L H 1955 Secondary flow in axial flow turbomachinery. *Trans ASME* 77: 1065−76

SMITH L H 1958 Recovery ratio — a measure of the loss recovery potential of compressor stages. *Trans ASME* 80(3)

SMITH L H 1966 The radial-equilibrium equations of turbomachinery. *Trans ASME Journal of Engineering for Power* 88: 1−12

SMITH L H 1969 Casing boundary layers in multistage compressors. *Proceedings of the Symposium on Flow Research on Blading*. Brown Boveri & Co Ltd, Baden, Switzerland 1969. In Dzung L S (ed) *Flow Research on Blading*. Elsevier Publishing Company 1970

SMITH L H 1974 Some aerodynamic design considerations for high bypass ratio fans. 2nd International Symposium on Air Breathing Engines, Sheffield

SMITH L H, YEH H 1963 Sweep and dihedral effects in axial flow turbomachinery. *Trans ASME Journal of Basic Engineering* 85: 401−16

SMITH M J T, HOUSE M E 1967 Internally generated noise from gas turbine engines, measurements and prediction. *Trans ASME Journal of Engineering for Power* 89: 177−90

SMITH S N 1973 Discrete frequency sound generation in axial flow turbomachines. Aeronautical Research Council R and M 3709

SOFRIN T G 1973 Aircraft turbomachinery noise. Lecture to ASME Turbomachinery Institute, Ames Iowa.

SOFRIN T G, PICKETT G F 1970 Multiple pure tone noise generation by fans at supersonic tip speeds. *International Symposium of the Fluid Mechanics and Design of Turbomachinery*, NASA SP-304 Part II

SOFRIN T G, MCCANN J C 1986 Pratt and Whitney experience in compressor noise reduction. 72nd Meeting of Acoustical Society of America

SOVRAN G, KLOMP E D 1965 Experimentally determined optimum geometries for rectilinear diffusers with rectangular, conical or annular cross-section. Symposium on the fluid mechanics of internal flow, General Motors Research Laboratories, Michigan. In Sovran G (ed) *Fluid Mechanics of Internal Flow,* Elsevier, Amsterdam 1967

SPRENGER H 1959 Experementelle Untersuchungen an geraden und gerammten Diffusoren. Mitteilung Nr.27 aus dem Institut für Aerodymaik an der ETH Zürich

SPURR A 1982 A prediction of 3D transonic flow in turbomachines using a combined throughflow and blade-to-blade time marching method. *International Journal of Heat & Fluid Flow* 2(4)

SQUIRE H B, WINTER K G 1951 The secondary flow in a cascade of aerofoils. *Journal of Aero Science* 18: 271

STANITZ J D 1952 Some theoretical aerodynamic investigations of impellers in radial and mixed flow centrifugal compressors. *Trans ASME* 74: 473–97

STANITZ J D 1953 Effect of blade-thickness taper on axial velocity distributions at the leading edge of an entrance rotor blade row with axial inlet and the influence of this distribution on the alignment of the rotor row for zero angle of attack. NACA TN 2986

STANITZ J D, ELLIS G O 1950 Two-dimensional compressible flow in centrifugal compressors with straight blades. NACA Report 954

STANITZ J D, PRIAN V D 1951 A rapid approximate method for determining velocity distribution in impeller blades of centrifugal compressors. NACA TN 2421

STARKEN H 1986 Utilization of linear cascades in turbomachine research and development. Lectures 19 & 20. ASME Turbomachinery Institute, Ames, Iowa

STARKEN H 1987 Private communication

STARKEN H, LICHFUSS H J 1970 Some experimental results of two-dimensional compressor cascades at supersonic inlet velocities. *Trans ASME Journal of Engineering for Power* 92: 267–74

STENNING A H 1980 Rotating stall and surge. *Trans ASME Journal of Fluids Engineering* 102: 14–20

STEWART W L 1955 Analysis of two-dimensional compressible flow characteristics downstream of turbomachine blade rows in terms of basic boundary layer characteristics. NACA TN 3515

STIEFEL W 1972 Experiences in the development of radial compressors. Lecture notes for *Advanced Radial Compressors* course held at von Karman Institute, Brussels 1972

STODOLA A 1927 *Steam and Gas Turbines.* McGraw-Hill, New York

STORER J A 1988 Private communication

STOW P 1984 Turbomachinery blade design using advanced calculations. Lecture to NATO Advanced Study Institute on *Thermodynamics and Fluid Dynamics of Turbomachinery,* Turkey

STRATFORD B S 1959 The prediction of separation of the turbulent boundary layer. An experimental flow with zero skin friction throughout its region of pressure rise. *Journal of Fluid Mechanics* 1–16 and 17–35

STRATFORD B S 1967 The use of boundary layer techniques to calculate the blockage from the annulus boundary layer in a compressor. *ASME Paper 67-WA/GT-7,* Winter Annual Meeting 1967

STRAZISAR A 1986 Laser anemometry in compressors and turbines. Lecture 18 ASME Turbomachinery Institute, Ames, Iowa

STRUB R A, BONCIANI L, BORER C J, CASEY M V, COLE S L, COOK B B, KOTZUR J, SIMON H, STRITE M A 1987 Influence of the Reynolds number on the performance of centrifugal compressors. Final report of a working group of the Process Compressor Sub-committee of the International Compressed Air and Allied Machinery Committee (ICAAMC). *Trans ASME Journal of Turbomachinery* **109**: 541–4.

STURGE D P, CUMPSTY N A 1975 Two-dimensional method for calculating separated flow in a centrifugal impeller. *Trans ASME Journal of Fluids Engineering* **97**: 581–97

SULAM D H, KEENAN M J, FLŸNN J T 1970 Single stage evaluation of highly-loaded high-Mach-number compressor stages II Data and performance multiple-circular-arc rotor. NASA Contract report CR-72694

TAKATA H, TSUKUDA Y 1977 Study on mechanisms of stall margin improvement of casing treatment. *Trans ASME Journal of Engineering for Power* **99**: 121–33

TAYLOR G I 1936 *Proc. Royal Society,* Series A, **156**: 307

TAYLOR E S 1964 The centrifugal compressor. In Hawthorne W R (ed) *Aerodynamics of Turbines and Compressors.* Volume X of the Princeton Series on High Speed Aerodynamics and Jet Propulsion, Princeton University Press, Section J

THWAITES B 1949 Approximate calculation of the laminar boundary layer. *Aeronautical Quarterly* **I**: 245–80

TOYAMA K, RUDSTADLER P W, DEAN R C 1977 An experimental study of surge in centrifugal compressors. *Trans ASME Journal of Fluids Engineering* **99**: 115–31

TRAUPEL W 1942 *Neue Allgemeine Theorie der Mehrstuffigen Axialen Turbomaschinen.* Leeman, Zurich

TRAUPEL W 1962 *Die Theorie der Strömung durch Radialmaschinen.* Verlag Braun

TRAUPEL W 1977 *Thermische Turbomaschinen.* 3rd edn, Springer Verlag

TRUCKENBRODT E 1952 Ein Quadraturverfahren zur Berechnung der laminaren und turbulenten Reibungsschicht bei ebener und rotationssymmetrischer Strömung *Ing Archiv* **20** H4:211–22

TYLER J M, SOFRIN T G 1962 Axial compressor noise studies. *Trans Society of Automotive Engineers* **70**: 309–22

TYSL E R, SCHWENK F C, WATKINS T B 1955 Experimental investigation of a transonic compressor rotor with a 1.5 inch chord length and an aspect ratio of 3.0: Part 1. NACA RM E54L31

VAVRA M H 1960 *Aero-Thermodynamics and Flow in Turbomachines.* Wiley

VERDON J M 1973 The unsteady aerodynamics of a finite supersonic cascade with subsonic axial flow. *Trans ASME Journal of Applied Mechanics* **40**: 667–71

WALKER G J 1974 The unsteady nature of boundary layer transition on an axial-flow compressor blade. ASME Paper No 74-GT–135

WALKER G J 1987 Transitional flow on axial turbomachine blading. AIAA 25th Aerospace Sciences Meeting, Reno. Paper AIAA-87–0010

WALL R A 1976 Axial flow compressor performance prediction. AGARD Lecture Series 83, ch 4

WASSELL A B 1968 Reynolds number effects in axial compressors. *Trans ASME Journal of Enginering for Power* **90**: 149–56

WEBER A, FADEN M, STARKEN H, JAWTUSCH V 1987 Theoretical and experimental analysis of a compressor at supercritical conditions. Paper 87-GT–256 presented at the 32nd ASME Gas Turbine Conference and Exhibit, Anaheim, California

WEBER C R, KORONOWSKI M E 1986 Meanline performance prediction of volutes in centrifugal compressors. ASME 31st Gas Turbine Conference and Exhibit, Dusseldorf, W. Germany. Paper No 86-GT–216

WEINGOLD H D, BEHLKE R F 1987 The use of surface static pressure data as a diagnostic tool in multistage compressor development. *Trans ASME Journal of Turbomachinery* **109**: 123−9

WEINIG F 1935 *Die Strömung und die Schaufeln von Turbomaschinen.* J H Borth, Leipzig

WENNERSTROM A J 1986 Low aspect ratio axial flow compressors: why and what it means. Third Cliff Garrett Turbomachinery Award Lecture, Society of Automotive Engineers SP-86/683

WENNERSTROM A J 1984 Experimental study of a high-throughflow transonic axial compressor stage. *Trans ASME Journal of Engineering for Gas Turbines and Power* **106**: 552−60

WENNERSTROM A J, PUTERBAUGH S L 1984 A three-dimensional model for the prediction of shock losses in compressor blade rows. *Trans ASME Journal of Engineering for Gas Turbines and Power* **106**: 295−99

WEYER H 1984 Fundamentals of flow field measurement techniques. Lecture to NATO Advanced Study Institute on *Thermodynamics and Fluid Mechanics of Turbomachinery,* Turkey

WHITEHEAD D S 1966 Aerodynamic aspects of blade vibration. *Proceeding of the Institute of Mechanical Engineers* **180**: (31)

WHITEHEAD D S 1972 Vibration and sound generation in a cascade of flat plates in subsonic flow. Aeronautical Research Council R and M 3685

WHITEHEAD D S 1987 Classical two-dimensional methods. ch III *AGARD Manual on Aeroelasticity of Axial Flow Turbomachines.* AGARDograph 298

WHITEHEAD D S, NEWTON S G 1985 A finite element method for the solution of two-dimensional transonic flows in cascades. *International Journal for Numerical Methods in Fluids* **5**: 115−32

WIESNER F J 1967 A review of slip factors for centrifugal impellers. *Trans ASME Journal of Engineering for Power* **89**: 558−72

WIGGINS J V, WALTZ G L 1977 Centrifugal compressor vaneless space casing treatment. US Patent No 4 063 848

WILCOX W W, TYSL E R, HARTMANN M J 1959 Resumé of the supersonic compressor research at NACA Lewis Laboratory. *Trans ASME Series D Journal of Basic Engineering* **81**: 559−69

WILKINSON D H 1970 Stability, convergence and accuracy of two-dimensional streamline curvature methods using quasi-orthogonals. Proc IMechE

WILLIAMS D D 1984 Inlet engine compatibility. Short course — Workshop on engine− airframe integration, Bangalore, India

WILLIAMS D D 1986 Review of current knowledge on engine response to distorted inflow conditions. AGARD Conference, Munich, AGARD CP-400

WISLICENUS G F 1965 *Fluid Mechanics of Turbomachinery* 2nd edn. McGraw-Hill, New York

WISLER D C 1977 Shock wave and flow velocity measurements in a high-speed fan rotor using the laser velocimeter. *Trans ASME Journal of Engineering for Power* **99**: 181−88

WISLER D C 1985 Loss reduction in axial-flow compressors through low-speed model testing. *Trans ASME Journal of Engineering for Gas Turbines and Power* **107**: 354−63

WISLER D C 1988 *Advanced Compressor and Fan Systems.* GE Aircraft Engines, Cincinnati, Ohio, USA. Copyright © 1988 by General Electric Co USA, All Rights Reserved. (Also 1986 Lecture to ASME Turbomachinery Institute, Ames Iowa)

WISLER D C, KOCH C C, SMITH L H 1977 Preliminary design study of advanced multistage axial flow core compressor. Final report NASA CR 135133

WISLER D C, BAUER R C, OKIISHI T H 1987 Secondary flow, turbulent diffusion and mixing in axial-flow compressors. *Trans ASME Journal of Turbomachinery* **109**: 455–82

WOOD J R, STRAZISAR A J, SIMONYI P S 1986 Shock structure measured in a transonic fan using laser anemometry. AGARD CP401. AGARD Conference *Transonic and Supersonic Phenomena in Turbomachines*, Munich

WRIGHT L C 1970 Blade selection for a modern axial flow compressor. In *Fluid Mechanics, Acoustics and Design of Turbomachinery, Part II*. NASA SP 304

WU C H 1952 A general theory of three-dimentional flow in subsonic or supersonic turbomachines of axial-, radial- and mixed-flow type. NACA TN 2604

WU C H, WOLFENSTEIN L 1949 Application of radial equilibrium condition to axial-flow compressor and turbine design. NACA TN 1795

YEOW K W 1974a Acoustic modelling of ducted centrifugal rotors: (I) The experimental acoustic characteristics of ducted centrifugal rotors. *Journal of Sound and Vibration* **32**: 143–52

YEOW K W 1974b Acoustic modelling of ducted centrifugal rotors: (II) The lumped impedance model. *Journal of Sound and Vibration* **32**: 203–26

YOUNG L R 1977 Discussion of paper by Rodgers. *Trans ASME Journal of Fluids Engineering* **99**: 94–5

YORK R E, WOODARD H S 1976 Supersonic compressor cascades — An analysis of the entrance region flow field containing detached shock waves. *Trans ASME Journal of Engineering for Power* **98**: 247–57

Index